Nonlinear Control of Uncertain Systems

Mazen Alamir

Nonlinear Control of Uncertain Systems

Conventional and Learning-Based Alternatives with Python

Springer

Mazen Alamir
CNRS, Grenoble INP, Gipsa-Lab
University Grenoble Alpes
Grenoble, France

ISBN 978-3-031-93286-1 ISBN 978-3-031-93287-8 (eBook)
https://doi.org/10.1007/978-3-031-93287-8

Mathematics Subject Classification: 93-1, 93-2, 93-4, 93-5, 93-8, 93-1, 93B07, 93B35, 93B51, 93B52, 93B53, 93Cxx

This Springer imprint is published by the registered company Springer Nature Switzerland AG
The registered company address is: Gewerbestrasse 11, 6330 Cham, Switzerland

To my parents, Haifa and Adib,
my beloved Hayate, Rim and Kinann,
my brother and sister Najib and Sawssan

To all those who left too soon,
leaving me orphaned from their smiles,
adrift without their beautiful minds,

To all those who,
by the miracle of their being,
offered me—even for a fleeting moment—
a glimpse of happiness.

Preface

Beyond the equations and the algorithms it contains, this book relates the state of mind and the boiling thoughts of a control-systems researcher at the crossroads of several disciplines and cultures.

If it were possible for someone to hide in the corridors of a control research laboratory and to listen to the conversations of younger and older fellows, during the last five years or so, he/she would heard, repeatedly, recurrent heated discussions about the *Pros* and *Cons* of using Machine Learning or data-driven approaches and when arguments fly loudly in the air, it is probably that someone is bringing the topic of Deep Learning to the discussion.

Two *extreme* points of view might be easily identifiable: The one according to which the *learning stuff* is a true *game changer*, the sort of revolution that relegates old control solutions to the status of antiques. The other point of view states that learning is a set of buzzwords designating one subset of options among many others to solve a set of old problems. Even worse, many of these old problems were already addressed in a more *serious* manner by control-related or applied mathematics-related sub-topics such as identification, nonlinear programming and so on. Between these two extreme positions, opinions spanning all the spectrum of mixed feelings would be heard.

The first time I came to understand that something serious is potentially *threatening us*, as a control community, was when one of my industrial partners, head of Intelligence, Optimization and Analytics at SCHNEIDER ELECTRIC company wrote to me about the subject he proposed for the future upcoming Ph.D. thesis to be funded by the company[1]:

—*This time we need to investigate Reinforcement Learning-based solutions!*

making it clear that the Funding decision was conditioned by this specific choice.

---

[1] We used to co-supervise industrially funded Ph.D. works about building and district Energy Management Systems (EMS) since many years.

The reason for which this request alarmed me is that it settles the choice in terms of the solution rather than the problem to address!

I was used to work on industrial and real-life problems with different partners, and it was always the problem to solve that implied the ultimately selected solutions and the terms of the proposal to write and not the kind of solutions to be deployed. The latter was always the consequence of the peculiarities of the former.

When it is the solution that is pushed forward from the beginning, it generally reveals a *dogmatic stance* and this should be felt as a threat! and this is the feeling I immediately had.

The reader might shortly discover a more detailed description of what I personally think about reinforcement learning when it comes to address control-related engineering problems, but it is a fact that, by the time I received this request, I was not against experimenting *new* tracks.[2] More precisely, as a matter of fact I was digging into the topic of data-science and Machine Learning for totally different reasons.

Indeed, as many researchers in Model Predictive Control (MPC), my attention partially shifted to Moving-Horizon Estimation as these two paradigms are notoriously dual.[3] Similarly, as many researchers in state estimation, my research partially *bifurcated* to studying the problem of parametric estimation via state extension and from that to fault and anomaly detection.

After the derivation of some seemingly successful algorithms and features generation heuristics targeting the problem of *blind* and model-free normality characterization, we founded, with Katia Hilal, a close friend of mine, a company[4] that specializes in developing software solutions for blind[5] and *model-free* anomaly detection for industrial equipments.

The *model-free* and data-driven aspects of the solutions, proposed by the company, which were dictated by scalability concerns in the first place, made the issue of *mastering* (or at least, being accustomed to) data-driven and machine-learning tools mandatory should I remain on-board of this *interesting and refreshing* adventure, beyond the first original bunch of algorithms.

This was how I came to discover the python language as it was[6] the language of Data-science. But Python comes with more than a syntax of coding. It comes with a state of mind which is the one of a code/problem sharing community and *make-it-available* practices. It takes you away from the MATLAB® proprietary, closed and expensive world of tools.

The two foundations of this book were therefore simultaneously present and ready to fertilize, namely, the urgent need to think about where we are and where

---

[2] I am still in favor of that!

[3] Both use time intervals, one for future prediction and one for the reconciliation of past measurements as it is explained in Chap. 2.

[4] Amiral Technologies.

[5] In the sense that it addresses the design of anomaly detector without the need for a faulty measurements in the training data.

[6] By the time I wrote this Preface! one should be careful with the rapid developments in the area!

to go as control community in a scientific moment that is dominated by the ML and data-driven reflexes on one hand, and the tools that enable us to better examine the strengths of our *presumed threat* from the inside on the other hand. Mastering these tools enables to question this status of *threat* in the first place and above all, to take informed and solidly assessed moves in the right direction.

While digging into the wide ML literature, I came over the famous *no-free-lunch* conjecture according to which,

> *There is no single algorithm that wins on all problems!*

and this book is about suggesting that this conjecture extends to the context of algorithmic solutions to control problems. Some control problems require standard control approaches while others might hugely benefit from ML and Data-driven tools. Moreover, this is not a binary choice, all the spectrum of combinations of these two worlds should be considered in order to fine-tune the *appropriate mix* for each specific problem.

Now it is one thing to acknowledge the above statement that can hardly be denied; it is a totally different thing to grasp the ability to determine, from the components of a problem, the best mix of tools and paradigms to adopt in order to address the problem in the most appropriate way. This requires a specific kind of wisdom or intuition that needs some hard work to acquire. This is what this book is intended to provide.

This book is written for students, engineers, Ph.D. and researchers addressing control-related challenging problems. Almost no a priori control-related knowledge is required as the associated paradigms and concepts regarding control and Machine Learning are adequately recalled in the first chapters of the book. Nevertheless, a minimal mathematical culture is necessary that covers general concepts such as functions, derivatives, continuity, matrices, linear algebra, ordinary differential equations, optimization-related keywords and so on that pop up here and there in the book.

Regarding the Python language, a whole chapter is dedicated to summarize the basic syntax, data types and main modules that are used throughout the book. Using this material together with a careful reading of the provided scripts might even be an interesting way to concentrate on the elements of the Python language that are sufficient for a control-related use. At least, The scripts provided in the book might be used as a guideline for a personal *motivated* parallel investment in a Python language's learning process without which the task for total beginners might seem insurmountable. It is always easier to learn anything when the process is motivated by a specific objective and context of use.

Throughout the book, framed boxes are used, when appropriate, to particularly attract the reader's attention. These are contents of three different kinds that are shown below:

> Important
This refers to an important item to keep in mind. It can be a definition, a notation or a working assumption that lies behind the development to follow.

Question?

This refers to a question that needs an answer that generally immediately follows in the forthcoming text.

! Warning
This refers to a possible danger that might be induced by misunderstanding or forgetting some underlying issues.

Grenoble, France Mazen Alamir

Acknowledgements This book could not have been possible without the total freedom provided by the French National Research Center (CNRS) to which I am infinitely grateful. The freedom of investigating new tracks *whatever it takes!* I know how this research management model is rare and the extent to which it is under threat in a world focused on performance measuring via *dummy* and *blind* numbers.

It could not have been neither possible without the long-term collaborations with industrial partners on real-life problems. This is because, fighting against such problems helps illuminating the right paths to borrow in the jungle of competing alternatives. Among these partners, special thanks should go to Claude Le Pape-Gardeux, Patrick Béguery, David Gualino and Mustapha Amine Rahmani (Schneider Electric), Patrick Bonnay and François Bonne (CEA), Paolo Tona (IFPEN), Jean-Luc Thomas (Areva), Jean Claude Alacoque (Alstom Transport), N. Perrissin-Fabert and Q. Alloin (General Electric), Sebastien Dedieu (ST-Micro) and Franck Mandica (SEB Group).

I address also my grateful acknowledgement to some of my former brilliant Ph.D. students which are behind some of the results discussed in this book. In particular, special thanks go to Peter Pflaum for the work on the Electric Vehicles Charging Stations (Chap. 14), Rachid Amari who contributed to the control design for the Automated Manual Transmission system (Chap. 13), Jean Dobrowolski for his contribution to the decentralized control design of micro-grids (Chap. 18).

Beyond the above cited former students, I am naturally grateful to all my former students for their confidence and hard works that provided me with the wonderful feeling of utility and the pleasure of the encounter with young intellects. In particular, I will never forget Mohammed Yacine Lamoudi, whose tragic departure left me bereft, orphaned of his astonishingly brilliant and beautiful mind.

As I write these words, I hold tender thoughts for my former Ph.D. supervisor, Guy Bornard, to whom I feel I have always, perhaps unknowingly, aspired to resemble intellectually—only to realize how far I have fallen short of that goal. His rigor, astonishing level of curiosity, genuine kindness and open-mindedness have always been for me a *multi-variate* target that one can reach only *asymptotically*.

I also extend my gratitude to colleagues from around the world who have crossed my path, contributing either to specific technical results or to the realization of that mysterious yet profound sense of fulfillment—of being a researcher on a shared planet: David Q. Mayne, Graham C. Goodwin, James Welsh, Tielong Shen, Gabriele Pannocchia, Rolf Findeisen, Teo Alamo, Lorenz T. Biegler, Fermi Guererro Castellanos, Frédéric Boyer, Seddik Bacha, Nida Sheibat Othman, Jean-Pierre Corriou, and I certainly forget so many.

The list of acknowledgement would not be complete without mentioning my fellow researchers, in the control system's department of gipsa-lab at the University of Grenoble Alpes as well as all the technical staff that manages to simplify, to the extent of what is humanly possible, the delicious french administrative *mille-feuilles*!

Special thanks to Akila Mokhtary, Sonia Nogueira, Lucia Bouffard-Tocat, Virginie Jalliffier-Talmat and Marielle Di-Maria as well as to all former and present directors of our lab who accepted the difficult task of having their precious times and minds occupied by rather *dirty and thankless tasks* in order to free our minds for what does really matter for us: thanks to Jean Michel Dion, Luc Dugard, Jean Marc Chassery, Jean-Marc Thiriet, Jerome Mars, Nicolas Marchand and all the ones to come.

Finally, I am grateful to my editor Oliver Jackson from Springer for his insightful and active role in shaping the audience, profile, and spirit of this book, as well as for his meticulous review and thoughtful suggestions on many aspects that a researcher does not always consider. Every mistake that was avoided is thanks to him, while any that remain are solely mine.

Grenoble, France
January 2025

Mazen Alamir

Competing Interests The author has no competing interests to declare that are relevant to the content of this manuscript.

Contents

Abbreviations

AI	Artificial Intelligence
AMT	Automated Manual Transmission
BIS	Bispectral Index
CL	Closed Loop
CPU	Central Processing Unit
CV	Cross Validation
DNN	Deep Neural Network
DP	Dynamic Programming
EKF	Extended Kalman Filter
EMS	Energy Management System
EV	Electric Vehicle
EVCS	Electric Vehicle Charging Station
FP	Fixed-Point
FT	Fourier Transform
GPU	Graphical Processing Unit
iid	Independent Identically Distributes
IP	Interior Point
KF	Kalman Filter
KNN	K-Nearest Neighbors
LP	Linear Programming
LQR	Linear Quadratic Regulator
MHE	Moving-Horizon Estimator
ML	Machine Learning
MP	Maximum Principle
MPC	Model Predictive Control
NLP	Nonlinear Dynamic Programming
NN	Neural Network
OCP	Optimal Control Problem
OL	Open Loop
PID	Proportional–Integral–Derivative
PVTOL	Planar Vertical Take-Off and Landing
QP	Quadratic Programming

RF	Random Forest
RL	Reinforcement Learning
SDP	Stochastic Dynamic Programming
SGD	Stochastic Gradient Descent
SMPC	Stochastic Model Predictive Control
SS	State Space
SVM	Support Vector Machine
SVR	Support Vector Machine Regressor
TF	Transfer Function
XGBOOST	EXtreme Gradient Boosting

Part I
Definitions, Notation, Concepts, and Tools

In Part I, entitled *Definitions, Notation, Concepts, and Tools*, the main control paradigms are introduced and discussed in a single chapter (Chap. 2) in order to get the *big picture* by connecting all these paradigms together. This helps justifying the need for each paradigm in the context of nonlinear systems affected by uncertainties.

Given that these concepts are recalled or introduced in Chap. 2 for later use in the book, it is advised to come back to this chapter frequently when a seemingly unclear notation is encountered in later chapters although specific reminders and references are appropriately provided.

Moreover, given the central role played by the Python scripts in the concrete implementation of the different ideas and algorithms, Chap. 3 is dedicated to some extensively used concepts of the language (packages, classes, functions, etc.). Moreover, a brief presentation of the main Python-packages used in the book is provided in this chapter. In particular: numpy, pandas, scikit-learn, and casadi to cite but the main ones.

1 What Is This Book About?

1.1 Chapter's Content at a Glance

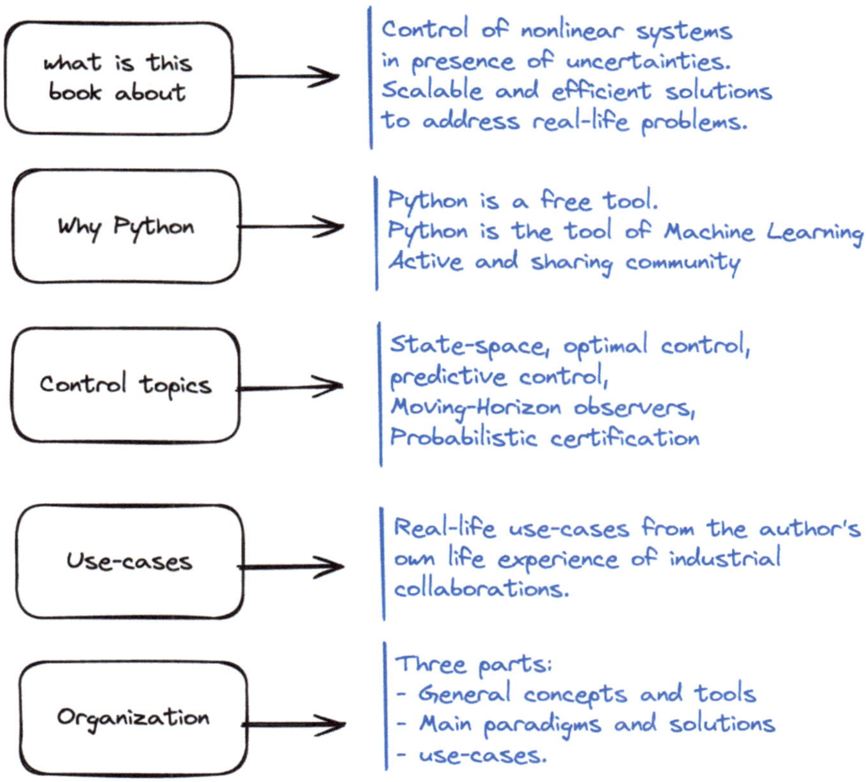

M. Alamir, *Nonlinear Control of Uncertain Systems*,
https://doi.org/10.1007/978-3-031-93287-8_1

1.2 What Is This Book About?

This book covers a part of my personal journey in the domain of nonlinear control under uncertainties. It is not intended to be an *academic survey* of existing results regarding the use of learning to tackle uncertainties in control engineering. Rather, it relates my own experience in this specific field.

The idea of making this book was suggested by some of my colleagues who kindly mentioned their own interest in finding the material I presented through talks here and there over the last two decades, gathered in a structured way within a single and coherent book.

As the title suggests, two keywords define the content of this book:

✓ **Uncertainties** in controlled systems. This arises when the control designer's knowledge of the system's dynamics is incomplete or uncertain (mainly, all the time!).

✓ **Learning** methods when used to address the above-mentioned issue. These methods are commonly referred to as **Machine Learning** (**ML**) methods.

The link between feedback control and uncertainty lies in the very definition of the concept of feedback itself. Indeed, the raison d'être of feedback is precisely the presence of uncertainties and/or disturbances in real-life systems. In the absence of uncertainty and disturbances, one can simply apply in *open loop* some scheduled sequence of actions that might be computed by some *open-loop* optimal control problems (OCPs) solver. Consequently, mentioning *control under uncertainty* is almost identical to simply mentioning feedback control.

This means that without further specific clarification, the materials of this book should cover all of the feedback control theory. This is because any feedback law enhances a natural robustness against uncertainties and/or disturbances.[1] The uncountable amounts of success stories of control methods applied to a wide variety of real-life uncertain contexts might raise the question about the need to invoke learning methods to address an issue that has been successfully addressed by standard means.

As it is shown in this book, there are two kinds of uncertainties and control problems.

1. Those problems in which the uncertainties can be handled by *standard* control techniques such as **high gain feedback** or **extended state estimators**. In the former case, a high gain feedback is applied to *compensate or kill* the influence of the uncertainties on the behavior of some key quantity one would like to control. In the later case, the state of the system is extended and an algorithm called *observer* is used to reconstruct the values of the uncertain quantities so that the problem bowls down to a deterministic or uncertainty-free problem. For this to

[1] Disturbances is viewed as uncertainties in this book, which they really are!

be possible, some conditions on the mathematical structure of the model as well as on the control objective and the underlying problem's formulation need to be satisfied. Examples of such situations are given in the book and some associated possible *standard* control methods are briefly recalled.

2. Those problems in which the impact of the uncertainties on the control objective is such that the above mentioned techniques do not apply. This might be the case when the control problem is not a simple regulation/tracking problem for which high gain techniques apply. Another situation arises when the dynamical model involves **too many parameters** compared to the excitation level which renders impossible the reconstruction of the vector of unknown parameters via extended observers.
   A typical example is the control of drug injection in cancer treatment where the model involves more than ten parameters while the excitation level is very low. Moreover there is no specific output to be tracked since the control problem amounts at minimizing the final value of the tumor at the end of the treatment period while monitoring the health condition expressed by a specific variable being above some safety level.

The goal of this book is to give some tools and heuristics that help handling the second situation mentioned above. More precisely, focus is made on these approaches that enable an explicit handling of the presence of uncertainties with known statistics. This means that one disposes of an algorithm that can randomly sample instances of realizations of the uncertainties in a relevant way. Examples of such statistics are: uniform distribution within some hypercube, Gaussian distribution around some nominal values or a non parametric statistics that might be identified (fitted) from measured real-life realizations of the uncertainties.

Obviously, the knowledge of these statistics is a strong assumption that is almost never rigorously met in real-life. This knowledge is itself generically uncertain and one should keep in mind that any statistical certification[2] that might be derived from any heuristic is conditioned by the real uncertainties being compatible with the statistics used in the sampling of realizations. Notice also that identifying some non parametric statistical laws is nowadays made possible through dedicated *learning algorithms* adding another instantiation of learning tasks that are made accessible by freely available algorithms.

This explicit handling of the uncertainties distinguishes the associated approaches from the widely used *nominal approach* in which the controller, or more generally, the decision making bloc, is designed based on the nominal values of the parameters and the robustness of the results to uncertainties is only *a posteriori checked* by simulating the closed-loop system under a variety of realizations of the uncertainties.

In the examples provided in this book, the advantages of using the explicit approach over the standard nominal design is shown when appropriate. This val-

[2] An example of certification might take the form of the probabilities of some properties being satisfied by the closed-loop behavior under the designed feedback strategy.

idation is mandatory because the explicit handling of uncertainties comes at the price of a higher design complexity and should therefore be justified by an assessed increment in the quality of the achievement of the control objective, at least on a statistical basis.

On the other hand, the book also provides examples where the standard *conventional* approaches are more appropriate than any sophisticated learning-based approach. It is the conjunction of all these examples that induces a kind of *wisdom* regarding the way uncertainties should be handled on a problem-dependent basis.

The previous discussion can be summarized as follows:

This Book
This book is about providing scalable, effective and real-life compatible methodologies and algorithms for the control of nonlinear systems in presence of uncertainties and non conventional problems settings.

It also attempts at providing a kind of wisdom regarding when to implement such or such family of solutions depending on the control objective, the nature of the uncertainties and their impact on the resulting closed-loop performance.

1.3 Python-Based Hands-on Effective Control Design Code

Beside the above mentioned objective, this book aims at providing a hands-on regarding the algorithms that implement the suggested solutions and heuristics. This is done using the `python` language. The reason for this is twofold:

1. First of all, `python` is totally free!
2. Moreover, `python` is a part of an ecosystem of free softwares supported by an active community of developers who continuously provide an increasingly efficient and *bugs-free* solutions to any problem one can think of or needs to solve.
3. More importantly regarding the topic of the present book, `python` is the language of ML thanks to the availability of powerful and freely available libraries such as `scikit-learn` [1] dedicated to ML algorithms, `Pytorch` [2] which is a modeling and computation framework mainly dedicated to Deep Neural Networks (DNNs) definition and fitting made freely available by `Meta`, `tensorflow` [3] which is another similar framework proposed by `Google`, etc.

```python
# create an instance of the PLARS class
pl = PLARS(deg=1, window=100, nModes=15, nModels=10, eps=1e-2)

# Initialize the dictionary of solutions.
solx = {f'xhat{i+1}':None for i in range(o.nx)}
solp = {f'phat{i+1}':None for i in range(o.np)}

# Create the noisy features matrices.
Xn = X * (1 + noise*np.random.randn(X.shape[0], X.shape[1]))
nTrain = int(len(Xn)/2)

# fit the models for all the components of the state vector.
# and assign the corresponding dictionary values.
for i in range(o.nx):

    print(f'Estimating x{i+1}')
    print('-----------------')
    solx[f'xhat{i+1}'], cpu = pl.fit(Xn[0:nTrain,:], yx[0:nTrain,i])
```

```
Estimating x1
-----------------
100%|██████████| 10/10 [00:00<00:00, 800.07it/s]
100%|██████████| 11/11 [00:02<00:00,  4.38it/s]
Estimating x2
-----------------
100%|██████████| 10/10 [00:00<00:00, 851.86it/s]
100%|██████████| 11/11 [00:02<00:00,  4.30it/s]
Estimating x3
-----------------
100%|██████████| 10/10 [00:00<00:00, 778.09it/s]
100%|██████████| 11/11 [00:02<00:00,  4.57it/s]
```

Fig. 1.1 Screenshot of a Jupyter notebook showing a cell of code followed by the results of its execution

In addition to the previously mentioned resources and libraries, the possibility of sharing `Jupyter-notebook`[3] (see Fig. 1.1) on the `Google-Colab` platform[4] enables everybody to execute bunches of `python` codes without even installing the python interpreter although this is obviously strongly recommended to take the full benefit of the contents and suggestions including in the forthcoming chapters.

On the other hand, the control community's researchers and developers propose a large number of `Python` libraries that are of great help to address OCP, nonlinear predictive control computation, moving-horizon state estimators as well as an almost

[3] As it is shown in the sequel, Jupyter-notebook is a dynamic `html` pages where code can be executed insides cells containing each either python code or markdown text (see Fig. 1.1).
[4] https://colab.google.

complete set of utilities compatible with MATLAB® syntax for a smoother transition of control researchers to the `Python` language. Some of these packages are mentioned throughout the book.

Now the price to pay in order to exploit all the previously mentioned advantages is to have an intermediate level of knowledge of `python` coding. This is required should one aims at extracting all the potential of the following chapters.

As a matter of fact, the topics of this book is more about getting real-life usable algorithms and heuristics than solid theory on abstract concepts. This means that it is more about practice than theory. This puts the coding experience at an important level of interest and hence whatever is the chosen support for coding, this latter should be mastered. The choice of `Python` is justified above and I do believe that making the effort to master it, at least partially, can be beneficial beyond the question of getting the maximum out of this specific book and therefore, *it is worth the effort.*

Although a concise and selective reminder of `python`'s key modules and syntax rules are provided (Chap. 3), this book cannot be the place to learn `python`. There are many resources for `python` learning that are suggested at the official website https://www.python.org/ (see the tutorial section). Moreover, many excellent tutorials can be found on YouTube (I strongly recommend the excellent `python` course by Corey Shaffer.[5])

Notice however that fully mastering everything about `python` is not possible neither necessary. Once the basics of the language are roughly mastered, it might be appropriate to jump to a `Python`-based ML book so that one can simultaneously improve the `Python` skills while learning interesting and useful topics on ML.

One excellent source for this is given by Sebastian Raschka [4]. Indeed, this book represents an excellent introduction to ML topics, key concepts and major critical issues while offering extensive bunches of `python` code that can be of great help in completing an initial introduction to the basics of the language.

Once again, although reminders concerning ML keywords, tools and best practices are provided in the following chapters (see in particular Chap. 3), this book cannot be considered as a starting point to the vast domain of ML. It is more about how to combine the ML tools and the control design paradigms and formulations to come out with efficient solutions to the problem of control design in the presence of uncertainties that cannot be handled by traditional control design frameworks.

Beyond the code snippets that are explicitly given in the book, some of the larger modules used in this book are available at the `GitHub` site (https://github.com/mazenalamir) for free download. Some of it might also be made available as a `Google-Colab` Jupyter notebooks. It goes without saying, however, that this access is provided without guarantee or responsibility. On the other hand, any comments,

[5] https://www.youtube.com/@coreyms.

suggestions or corrections would be more than welcomed in order to improve the content, the efficiency or the elegance of the codes.[6]

The previous discussion regarding the use of `Python` can be summarized as follows:

Why Python?

The motivation in using the `Python` language lies in the following facts:

(1) `Python` is free!
(2) `Python` is the tool of ML.
(3) `Python` is supported by an active and sharing-friendly community to which an increasing number of control researchers belong.

1.4 Control Background Involved

Control theory is an old topics and many results, methodologies and frameworks exist and the question that might be asked is the following:

Control Background Required?

What is the control-related background that is needed to smoothly follow the content of the present book?

The answer is: *Not that much!*

The reason for that is twofold: Firstly, the needed topics are briefly recalled anyway. Secondly, this is a book about nonlinear dynamical systems meaning that many topics and results that are specific to linear systems are not invoked since they are not needed as far as this book is concerned.[7]

To be more specific, the following concepts and topics are recalled and used in the following chapters either as a part of the design algorithms or as necessary means to understand the rationale behind such or such aspects of the discussed solutions:

- The general discrete-time **State-Space representation** of nonlinear dynamical systems. This representation takes the form of a system of nonlinear one-step-ahead difference equations. The latter generally arises from the discretization of continuous-time knowledge-based representations of some fundamental physical laws such as: Ordinary-Differential-Equations (ODEs), Algebraic-Differential-Equations (DAEs) or Partial Differential-Equations (PDEs).

[6] I must confess that I am not a native `python` programmer (I learned it from scratch 6 years earlier after too many years of being a MATLAB® user.)
[7] Such as Linear Matrix Inequalities (LMI) intensively used in Robust linear control design or H_∞ control design.

- The paradigm of **Optimal Control** is central to many of the solutions proposed in this book. This is the reason for which a clear understanding of the associated keywords such as: **cost function, constraints, decision variables, feasible solutions** is necessary and hence the needed reminders are appropriately provided.
  The reason for which optimal control is relevant to the uncertainty handling as discussed above comes from the fact that the computation of some appropriately formulated optimal control trajectories provides a long-term insight on the consequences of the actions on the system which is mandatory when the uncertainties induce long-term impact that cannot be accommodated for by using instantaneous actions via *standard methods* as explained above.
- As a matter of fact, optimal control is used as a building block of the **Model Predictive Control (MPC)** design methodology [5]. MPC is today the most widely used nonlinear control design methodology due to its ability to handle relevant optimality requirements, constraints, coupled high dimensional nonlinear models. Indeed, modern control problems can rarely be stated as output regulation or output tracking problems. Rather, long term optimization of economic objectives is needed to be performed under a set of operational and/or safety constraints.
  Within this paradigm, there is no a priori known output trajectories to be tracked. Moreover this optimization has to be done over a so called prediction horizon during which disturbances and uncertainties operate making the only possible way to compute the *best current actions* is to consider the optimization of the expected cost for a presumed probability of the uncertainties realization [6]. This gives rise to the so-called **Stochastic Model Predictive Control (SMPC)** design which comes with many variants and different formulations. This obviously put the SMPC design in the very heart of this book.
- Since focus is made on realistic solutions, only (dynamic) output feedback control are of interest. In other words, only control feedback laws that exclusively use the measured information can make their way to a concrete implementation in real-life situations. This renders crucial to understand the concept of state estimation even if the learning-based construction of feedback solutions does not explicitly rely on the construction of such estimators, it remains however necessary to have in mind the observability concept in order to understand the rationale behind the solutions.
  Moreover, as it is mentioned earlier, extended state/parameters estimation can be in the heart of standard approaches to handle the uncertainties by reconstructing, when possible, the values of the unknown parameters and hence bringing back the design problem to the deterministic realm. This is the reasons why the principle of nonlinear **Moving-Horizon Estimators** (MHEs) is recalled in this book.
- A book on the topic of control under uncertainties that does not contain a presentation of the **Stochastic Dynamic Programming (SDP)** paradigm and the associated *Q-Learning* version would be obviously incomplete. That is the reason why this paradigm is briefly stated since some of the problems of moderate size might be nicely addressed using this approach as it is shown in the part of the book dedicated to applications.

- Finally one of the frameworks that is intensively used in this book in order to provide certifiable results in spite of the presence of high level of uncertainty is the **Probabilistic certification paradigm** [7]. This paradigm found its way to the control community two decades ago to replace the too restrictive paradigm of robustness in control design.
  To say it shortly, robustness paradigm asks for a property to be satisfied *for all* possible instantiations of the uncertainties while the probabilistic certification targets a prescribed level of risk which can be expressed in probabilistic terms. This methodology is described in this book and applied to many use-cases hoping that by doing so, it becomes a kind of *design reflex*.
  The set of concepts and paradigm that are worked out and/or recalled in the present book can therefore be summarized as follows:

The Main Paradigms Involved

The main paradigms discussed in this book are

✓ state-space representation;
✓ optimal control;
✓ deterministic/SMPC;
✓ extended moving-horizon estimation;
✓ SDP/Q-learning; and
✓ probabilistic certification.

1.5 A Large Set of Use-Cases

Since the aim of this book is to provide a rich set of tools that enable to address real-life problems, the different solutions and algorithms are applied to a variety of such problems. These use-cases spread here and there along the book. Particularly involved use-cases are given in the last part of the book. In this part, I tried to summarize a set of real-life examples I had the chance to work on, together with some PhD students and/or academic colleagues and co-workers. The majority of these works corresponds to projects partially funded by industrial companies such as: Schneider-Electric, General-Electric, Alstom, IFPEN, Atomic Energy Commission and Renewable Energies (CEA) to cite but few ones.

While in the first two parts of the book, the algorithms are explained (and provided) in detail; in the last part dedicated to industrial and or real-life use-cases, the presentation is less technical in order to focus on the context, the problem statement and a brief presentation of the kind of solutions (among the ones presented in the first two parts of the book) involved and together with the kind of achievable results.

The use-cases span a wide set of domains of application. Below is a non exhaustive list:

- ✓ control of Automated Manual Transmission (AMT);
- ✓ combined chemotherapy-immunotherapy-(vaccine) of cancer;
- ✓ optimal operation of Electric Vehicle Charging Stations (EVCSs);
- ✓ energy management of hybrid vehicles;
- ✓ decentralized frequency control in micro-grids;
- ✓ propofol-based control of anesthesia; and
- ✓ some other applications that are very briefly described with references to appropriate citations showing how the problems can be addressed using the materials of different chapters.

This leads to the structure of the book depicted in the next section.

1.6 Organization of the Book

The book is divided into **three-parts**:

1. In the first part, entitled *Definitions, notation, concepts and tools*, the main control paradigms as recalled in the previous section are introduced and discussed in a single chapter (Chap. 2) in order to get the *big picture* by connecting all these paradigms together. This helps justifying the need for each paradigm in the context of nonlinear systems affected by uncertainties.
   Given that these concepts are recalled or introduced in Chap. 2 for later use in the book, it is advised to come back to this chapter frequently when ambiguity or seemingly unclear notation is encountered in later chapters. This might be helpful although specific reminders and references are appropriately provided for a smoother reading.
   Because of the central role played by the `Python` scripts in the concrete implementation of the different ideas and algorithms, Chap. 3 is dedicated to some extensively used concepts of the language (packages, classes, functions, etc.). Moreover, a brief presentation of the main `Python`-packages used in the book is provided in this chapter. In particular: `numpy`, `pandas`, `scikit-learn` and `casadi` to cite but the main ones.
2. In the second part, entitled *Methodologies and Algorithms*, the topics mentioned in Chap. 2 are successively discussed at a deeper level. More precisely:

   - **Chapter** 4 starts by recalling that control theory is a now quite established domain of knowledge which is mainly dedicated to mitigate the effect of uncertainties in dynamical systems. As such, control theory provides many concepts and tools that are worth to keep in mind should one be intended

to design a feedback control law for a nonlinear dynamical system showing high level of uncertainties.
The chapter uses some illustrative examples to assess this message and gives some specific control design methods and their associated closed-loop related results. Some of these schemes are extensively used in some of the use-cases discussed in the third part of the book.
The ambition of this short chapter is not to give an overview of all existing theories and methods that have been invented during decades of research on control system theory, but simply to recall that **for a class of regulation/ tracking problems, standard approaches might never be surpassed by so-called smart or data driven design approaches**. This makes knowing standard approaches mandatory for an optimal handling of such problems.

- Among those problems for which standard high-gain approaches such as the ones described in Chap. 4 are less efficient, one can think of problems that involve long-term consequences of uncertainties that do not boil down into a *simple* regulation or tracking problem. Rather, the control task is to achieve some economic cost under constraints for which the outcome of the closed-loop operation can only be evaluated on the long term and for which there is no obvious optimal trajectories to be tracked.
This is the reason why MPC is described in detail in Chap. 5 together with the dedicated freely-available `Python` package `MPC_solver`. The MPC setting studied in this chapter is **deterministic** as this setting needs to be mastered before any uncertainty-related extra complication is added.
The proposed implementation is based on the `Casadi Python` framework which is used to implement two options for the solver, namely: the widely used interior point (IP) algorithm `IPOPT` and a hand-made version of the `fast-gradient` algorithm [8] that uses the exact differentiation utility provided inside `Casadi`.
Here again, the intention is not to cover all possible available optimization frameworks that are developed by a very dynamic research community, but to give the interested reader a fair comprehension of the ingredients and keywords that are involved in any MPC problem formulation. This understanding which is enabled through a single meticulously introduced implementation might be helpful to the reader to make his/her way to other possible solvers and frameworks such as `Pyomo` [9] and `Geckko` [10] to cite but two alternatives that are not detailed in this book.
Beyond familiarizing the reader with the ingredients and keywords of MPC, Chap. 5 also underlines the very important topic of real-time implementation of MPC through the examination of the change in the computation time and hence in the implementability of the resulting controller when some options in the implementation are changed. Hopefully, this should smoothly prepare the reader to examine this *real-time implementability* issue in much more details in Chap. 6.
- Indeed, Chap. 6 capitalizes on the understanding of the MPC implementation hopefully gained in Chap. 5 in order to explore in depth the different details

of the implementation settings of an MPC. Moreover, the impacts of these implementation details on the success/failure of the resulting real-time closed-loop feedback are examined.
The success/failure of the resulting closed-loop depends on the certification of the satisfaction some conditions that are related to three issues, namely:

✓ the real-time implementability in terms of computation time on a specific target,
✓ the satisfaction of the constraints, and
✓ the contraction of the optimal cost which plays generally the role of the Lyapunov function in the analysis of the stability of the closed-loop under the MPC feedback.

More precisely, although a detailed definition of the implementation parameters listed below is given in Chap. 6, a quick statement of these parameters is given here to help the reader getting a rapid hint:

- the under-sampling used in the integration inside the MPC computation;
- the terminal weighting;
- the soft constraints penalty;
- the maximum number of iterations allowed for the optimizer;
- the control updating period ;
- the control horizon; and
- the prediction horizon's length.

Chapter 6 describes in detail a heuristic that enables to appropriately search the domain of definition of the above design parameters in order to determine admissible settings with regards to the three above mentioned criteria. Moreover, the freely available `Python` package `MPC_tuner` that implements the above algorithm is also presented and assessed through an illustrative example of the Planar Vertical Take-of and Landing (PVTOL) aircraft which involves six states, two regulated outputs and two control inputs.

- **Chapter** 7 gives a first example where solving **deterministic** OCP is used as a step in the design of **stochastic** optimal control feedback.
More precisely, this chapter proposes a control design framework that enables to derive *uncertainty-aware* dynamic output feedback from the solution of a cloud of optimal **deterministic** OCP. The framework assumes that the system's equations are known while the true values of the parameters being involved in the expressions inside the model are very badly known. In this context, the heuristic can be viewed as a scalable way to approximately solve stochastic MPC problems as formulated in Chap. 2.
The design methodology discussed in this chapter is very general and scalable in the sense that it can be applied to dynamical systems with high number of states, the number of uncertain parameters and the number of control inputs.

However, it is rather dedicated to economic-like control problems that show no critical stability or high precision issue.[8]
To rephrase it in a more evocative way, the solution is worth to be used in problems such that cancer treatment, maximizing production or minimizing energy rather than in the stabilization of multiple inverted pendulum on a cart.
The good news is that the latter problems generally admit more efficient standard control solutions via problem-dependent handling through the use of extended state/parameter reconstruction as it is discussed earlier in the present chapter.
The proposed framework is illustrated in Chap. 7 via the problem of maximizing the outcome of a parallel reactor while in Chap. 16, it is shown how it can be used to solve the problem of power management in a hybrid vehicle using real-life data representing a road path in the suburb of Tokyo, Japan.

- **Chapter** 8 considers the situation where there is **no available mathematical model** that governs the behavior of the nonlinear system to be controlled. Nevertheless, it is possible to gather time-series representing the evolution of the measured quantities either using open-loop control (open-loop stable systems) or by operating the system under a loosely defined sub-optimal feedback control law. The chapter propose a control design feedback that is based on the **identification of the cost function** as a map depending on the past measurements of the input and output on one hand and the future control sequence on the other hand.
  A specific dedicated solver is used to address the optimization of the possibly non continuous ML map representing the cost function. The whole framework is illustrated on a simple example and the impact of the different choices are shown.
- **Chapter** 9 addresses in detail the problem of designing a constrained extended state/parameter estimator using the **Moving-Horizon Estimation** (MHE) principle. In particular, the updating rate of the MHE is addressed that might be necessary for the real-time implementability of the MHE.
  More precisely, it is a fact that almost all textbooks on MHE consider the sampling time involved in the discretization of the dynamics and the measurement acquisition period as equal. This assumption obviously greatly simplifies the expressions and the implementation of the MHE. Nevertheless, this assumption is not realistic and might lead to some implementation issues. In particular, this might lead to an updating period for the MHE that is not compatible with a real-time implementability of the MHE.
  That is the reason why in the implementation studied in Chap. 9, this assumption is not considered leading to the MHE updating period being an implementation parameter that can be tuned depending on the dynamics of the system and the time needed for the solver to perform the underlying computation.

[8] This is the case of the majority, if not all, of data-driven solutions by the way!

In addition to the formalization of the extended state/parameters MHE, Chap. 9 introduces the freely-available `Python` package `MHE_solver` which is progressively built and commented along the chapter.

- **Chapter** 10 addresses the problem of deriving sparse representations of nonlinear relationships via the sparse solution of linear least squares problem. This corresponds, for instance to the situation when some nonlinear functions basis is used to represent the nonlinear relationship as linear combinations of the latter functions.
  The chapter discusses why the number of features might become extremely high as the number of arguments of the nonlinear functions to be identified grows. It also underlines the relevance of sparsity in the induced solutions in the industrial context where finding these relationships is a main step towards addressing many problems where uncertainties prevail.
  Based on this two arguments, a new sparse solver is proposed and its performances are compared to the state-of-the-art algorithms showing very promising results. The `scLars` developed in this chapter is used in Chap. 11 to address a specific partial state estimator in the presence of measurement noise parametric uncertainties.
- **Chapter** 11 gives an example of a framework to solve the problem of partial observation target's design via ML fitting methods. More precisely, the chapter shows a use-case where the sparse nonlinear identification algorithm `scLars` presented in Chap. 10 is used to fit mathematical structures that solve instances of the partial state estimation problem.
  Two examples are used in Chap. 11 to convey the methodology and to discuss the impact of the presence of measurement noise and parametric uncertainties on the quality of the partial estimation.
  The results that can be achieved using the `scLars` algorithm of Chap. 10 are compared to those that might be obtained using some standard ML libraries showing the relevance of the sparse identification in terms of extrapolation capability which is intimately linked to the concept of over-fitting that is fundamental to grasp in all ML-based attempt to derive data driven solutions.
- **Chapter** 12 briefly overviews some additional advanced topics, ideas and frameworks that are less invoked in the following chapters but which nevertheless deserve to be mentioned in a book on *control under uncertainty*. This includes the following topics:

  – Reinforcement Learning (RL);
  – clustering-based implementation of stochastic MPC;
  – learning-based sensitivity analysis and associated feedback design, and
  – scenario-based implementation of stochastic MPC.

  Some of these topics might be very promising and their inclusion in a *brief* mode simply results from the need to operate some choices in order to come up with a book of reasonable size.

3. In the third part of the book, entitled *Applications*, some relevant use-cases are presented that discuss the implementation of some of the methodologies, heuristics and ideas described in the previous parts of the book.
   The aim of this part is twofold:

   - First, this part aims at putting in practice the methodologies and the *way of thinking* in presence of uncertainties on real-life (or so) situations and contexts.
   - More importantly, the aim of this last part is to underline the importance of addressing each use-case for its own sake by analyzing the specificity of the problem being addressed. This leads to a systematic pattern where a problem-dependent part of the design is mandatory to come out with efficient and dedicated solutions to the problem.

   More precisely:

   - **Chapter** 13 shows how the standard control methods enable to design a robust controller for the automotive Automated Manual Transmission (AMT) system, provided that appropriate choices and techniques are mixed together. The problem in AMT control is to provide smooth and oscillation-free speed transitions while meeting the *transparency* constraint. The latter expresses the need to respect the driver's request[9] despite the high level of uncertainties. These uncertainties stem from the road characteristics and the bad knowledge of the values of the mechanical parameters involved in the model being used. Another source of uncertainties lies in the unknown settings of the other controllers which manage the motor's torque and clutch's position behavior. Although the final step of the solution involves the use of parameterized MPC (which is not detailed in the chapter), the main reason of the success comes from the simplified modeling choices and the use of state estimator in order to cope with the many unknown sources of uncertainties on the equations governing the main quantities to be controlled.
     This use-case is an example of the efficiency of standard control, learning-free, methodologies in handling the presence of high level of uncertainties. It can be conjectured that no data-driven solution can provide comparable efficiency in addressing this control problem. This is precisely why this use-case is included.
   - **Chapter** 14 addresses the *burning* problem of power management in EVCSs with application to the EVCS of the university of Berlin.
     This example shows the huge opportunities in terms of decision making and parameters tuning that is offered by the probabilistic certification paradigm presented in Sect. 2.11.

[9] As expressed through the pedal's position.

Moreover, this chapter shows another example of the relevance of the problem-dependent steps in the derivation of a successful design. In the specific case of EVCS, the problem-dependent choice concerns the definition of the degrees of freedom used in the design as well as the very problem-specific definition of a successful day in the life of the EVCS. This example also underlines the role of the freely available tools for non parametric fitting of statistical laws[10] that lies in the heart of the successful certification-based design.
The example studied in this chapter helps enforcing the message according to which, this book is not about a list of recipes that one can blindly apply to any problem. Rather, it is about formulating specific settings for each use-case in which some of the tools and concepts are invoked, mixed and used with caution in order to appropriately solve the specific problem at hand.

- **Chapter** 15 addresses the problem of controlling the level of anesthesia through Propofol injection monitoring under large parametric dispersion. Here, the control objective is to rationally deliver the Propofol so that the depth of anesthesia reaches its appropriately prescribed zone without undershoot and within a prescribed interval of time.
Here again, it is the probabilistic certification that plays the crucial role provided that an appropriately chosen parameterization of the control is adopted leading to acceptable guaranteed results in terms of the settling time and the probability of constraints violation. In this use-case, the control design methodologies in presence of uncertainties developed in Sect. 4.4 is used to design the control law on the top of which the randomized optimization is used in order to deliver certified performances.
The work in this use-case is an example of mixing traditional control design with the rather modern control objective statement which is expressed in terms of probability of success. It can be conjectured without risk that this kind of problem's formulation and frameworks of solutions will be increasingly used in the future. Note that no real experiment on real patients have been conducted, the materials of the chapter exclusively involves numerical simulations.
- **Chapter** 16 addresses the problem of power management in hybrid vehicles using the control design proposed in Chap. 7. The objective here is to minimize the energy used to travel on a specific path in the absence of a priori knowledge regarding the traffic conditions and the driver's patterns. This leads to a very rich set of possibilities in terms of speed and acceleration profiles.
In this study, real-life data instances are used which are collected on a specific path in Tokyo in order to illustrate the efficiency of the so-designed feedback law that minimizes the overall energy used while guaranteeing a terminal State of Charge (SOC) of the vehicle's battery.

[10] In this specific example, these are the laws that govern the arrival time, the duration of connection to the EVCS and the State-of-Charge upon arrival.

Here again, given the specific problem at hand which lies in the quite long path to be considered for each realization of the uncertainty, Some specific problem-dependent steps are again required in order to reduce the computation time needed to solve the underlying deterministic OCP based on which, the uncertainty-aware feedback design proposed in Chap. 7 is built.
This shows another instance of a seemingly general rule according to which, efficient design cannot avoid, at least at some stage, the use of problem-dependent treatment.[11]

- **Chapter** 17 uses the Q-learning framework presented in Sect. 2.8 to address the problem of rationalizing drug's injection in the combined (Chemotherapy /Immunotherapy) of cancer.
The objective here is to induce a decrease (contraction) in the tumor size while maintaining the level of lymphocyte's population above some health-related threshold as the size of this population of cells reflects the patient's level of resistance and health.
The results of this chapter emphasize the relevance of using a non zero weighting on the standard deviation of the cost function (augmented by the exact penalty on the constraints) in order to handle the risk management that is mandatory in this kind of applications. This variance-related term is rarely used in the applications which mainly focus on economic criteria in which only the expectation is relevant.
While the general framework follows the guidelines recalled in Chap. 2, the specific mathematical structure of the equations governing the cancer/drug dynamics is exploited in order to come out with a more tractable heuristic in terms of the evaluation of the statistical moments involved in the definition of the cost function. Note that no real experiments on real patients have been conducted, the materials of the chapter exclusively involves numerial simulations.
- **Chapter** 18 shows an example of the use of the standard approach proposed in Sect. 4.5 in order to address the problem of decentralized frequency control in multi-source micro-grids.
The chapter starts by showing that the standard so-called *swing frequency equation*[12] does not truly hold in the presence of high dynamics on the produced power. This promotes the use of the model of uncertain systems with high order destabilizing dynamics, such as the one studied in Sect. 4.5, in order to address this critical problem of frequency control in micro-grids in a somehow *Plug and Play* manner.
In this practical example, the uncertainties come from the decentralized nature of the solution on one hand and from the fact that the control law to be designed should be agnostic to the specific implementation and tuning of

[11] I know that I am repeating myself, but this is really a key message that I would like to enforce!
[12] This equation states that the derivative of the frequency is proportional to the unbalance between produced and consumed powers.

the local controllers which are used at each generator's level. Moreover, the designed controller should also be unaware of the unknown and unpredictable profile of the load's power.

1.7 Disclaimer

Writing a book such as the one lying in the reader's hands (or screen) required efforts that lasted over almost a year. During such a *long period*, many ideas, tools and alternatives might come to the mind of any author making the already written material sometimes sub-optimal.

A strong attempt is therefore unavoidable that suggests to replace the *incriminated* contents with a more up-to-date ones. But this is an endless process as the raison d'être of a researcher is to generate, learn or investigate new alternatives and within a year, so many such ideas knocked *without an appointment* threatening, at least partially, the already written chapters!

Consequently, it seemed obvious that giving in to such an attempt would have threatened the whole project of delivering a manuscript in due time.

Nonetheless, it is not too long to write this brief disclaimer asking for the reader's indulgence whenever a missed opportunity to make something in a better or more appropriate way is detected.

This being said, let's start the journey!

References

1. Pedregosa, F., Varoquaux, G., Gramfort, A., Michel, V., Thirion, B., Grisel, O., Blondel, M., Prettenhofer, P., Weiss, R., Dubourg, V., Vanderplas, J., Passos, A., Cournapeau, D., Brucher, M., Perrot, M., Duchesnay, E.: Scikit-learn: Machine learning in Python. J. Mach. Learn. Res. **12**, 2825–2830 (2011)
2. Paszke, A., Gross, S., Massa, F., Lerer, A., Bradbury, J., Chanan, G., Killeen, T., Lin, Z., Gimelshein, N., Antiga, L., Desmaison, A., Kopf, A., Yang, E., DeVito, Z., Raison, M., Tejani, A., Chilamkurthy, S., Steiner, B., Fang, L., Bai, J., Chintala, S.: Pytorch: An imperative style, high-performance deep learning library. In: Advances in Neural Information Processing Systems 32, pp. 8024–8035. Curran Associates, Inc. (2019). http://papers.neurips.cc/paper/9015-pytorch-an-imperative-style-high-performance-deep-learning-library.pdf
3. Abadi, M., Agarwal, A., Barham, P., Brevdo, E., Chen, Z., Citro, C., Corrado, G.S., Davis, A., Dean, J., Devin, M., Ghemawat, S., Goodfellow, I., Harp, A., Irving, G., Isard, M., Jia, Y., Jozefowicz, R., Kaiser, L., Kudlur, M., Levenberg, J., Mane, D., Monga, R., Moore, S., Murray, D., Olah, C., Schuster, M., Shlens, J., Steiner, B., Sutskever, I., Talwar, K., Tucker, P., Vanhoucke, V., Vasudevan, V., Viégas, F., Vinyals, O., Warden, P., Wattenberg, M., Wicke, M., Yu, Y., Zheng, X.: TensorFlow: Large-scale machine learning on heterogeneous systems (2015). https://www.tensorflow.org/
4. Raschka, S.: Python Machine Learning. PACKT, Birmingham-Mumbai (2015)

5. Mayne, D.Q., Rawlings, J., Rao, C.V., Scokaert, P.O.M.: Constrained model predictive control: Stability and optimality. Automatica **36**, 789–814 (2000)
6. Mesbah, A.: Stochastic model predictive control: An overview and perspective for future research. IEEE Control Syst. Mag. **36**(6), 30–44 (2016)
7. Alamo, T., Tempo, R., Camacho, E.F.: Randomized strategies for probabilistic solutions of uncertain feasibility and optimization problems. Autom. Control IEEE Trans. **54**(11), 2545–2559 (2009)
8. Nesterov, Y.: Universal gradient methods for convex optimization problems. Math. Program. **152**(1), 381–404 (2015)
9. Hart, W.E., Laird, C.D., Watson, J.P., Woodruff, D.L., Hackebeil, G.A., Nicholson, B.L., Siirola, J.D.: Pyomo–Optimization modeling in Python, 2 edn. Springer Optimization and Its Applications. Springer (2017). https://doi.org/10.1007/978-3-319-58821-6, http://www.pyomo.org
10. Beal, L.D., Hill, D.J., Martin, R.S., Hedengren, J.D.: Gekko optimization suite. Processes **6**(8), 106 (2018). https://doi.org/10.3390/pr6080106, https://www.mdpi.com/2227-9717/6/8/106

2 Definitions, Notation and Main Concepts

M. Alamir, *Nonlinear Control of Uncertain Systems*,
https://doi.org/10.1007/978-3-031-93287-8_2

2.1 Chapter's Content at a Glance

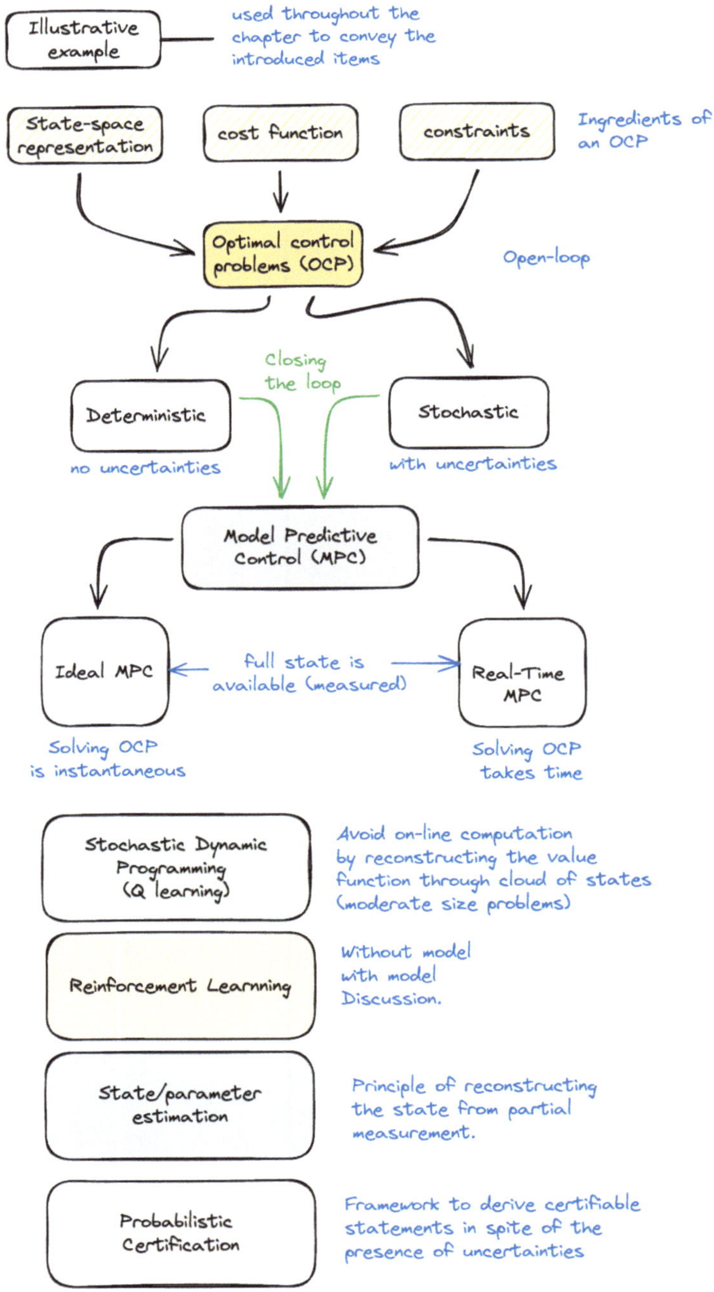

This chapter introduces the basic definitions, notation as well as the main concepts and paradigms that are discussed and used throughout the book. For the sake of clarity, an illustrative example is used to smoothly convey the introduced items. This example is discussed in the next section.

2.2 The Illustrative Example: The Planar Vertical Tack-Off and Landing

Let us consider the Planar Vertical Take-Off and Landing (PVTOL) aircraft depicted in Fig. 2.1.

We are interested in controlling the position-related planar coordinates (y, z) as well as the angular position θ by means of the two control actions u_1 and u_2 representing respectively the normal lift and the rotational torque.

Using the principles of Mechanics, it can be shown that this system is governed by the following set of *normalized* ODEs:

$$\ddot{y} = -u_1 \sin\theta + \epsilon u_2 \cos\theta, \tag{2.1a}$$
$$\ddot{z} = +u_1 \cos\theta + \epsilon u_2 \sin\theta - 1, \tag{2.1b}$$
$$\ddot{\theta} = u_2. \tag{2.1c}$$

These equations show that for any given control profile (time stamped trajectories of the actions u_1 and u_2), the future evolution, over some time interval $[t_0, t_1]$, of the aircraft can be obtained by *integrating* the ODEs (2.1) by an appropriate ODEs integrator. The latter would need to know the initial conditions, namely the values

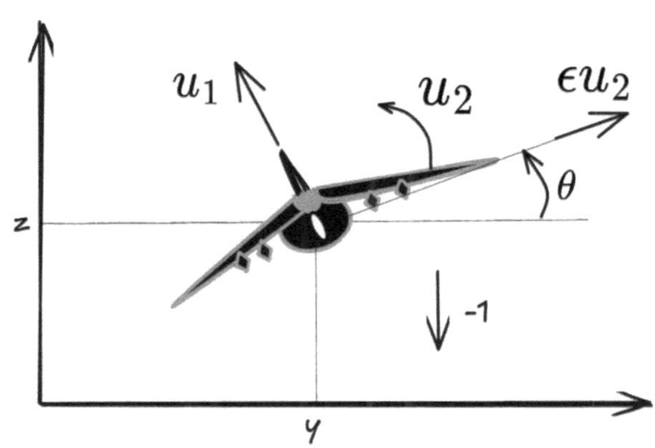

Fig. 2.1 The PVTOL and associated key variables

at instant t_0 of y, z, θ and their respective derivatives; namely, we need to know the value at instant t_0 of the following 6-dimensional vector:

$$x := \begin{bmatrix} y \\ z \\ \theta \\ \dot{y} \\ \dot{z} \\ \dot{\theta} \end{bmatrix}. \tag{2.2}$$

The vector $x(t_0) \in \mathbb{R}^6$ represents the **state of the system** at the initial instant t_0.

We say that the system is of dimension 6 since we need six scalars to describe its state. Now combining (2.1) and (2.2), one can write the ODEs governing the dynamics of the PVTOL using the state vector $x \in \mathbb{R}^6$ and the control vector $u \in \mathbb{R}^2$ as follows[1]:

$$\dot{x}_1 = x_4, \tag{2.3a}$$
$$\dot{x}_2 = x_5, \tag{2.3b}$$
$$\dot{x}_3 = x_6, \tag{2.3c}$$
$$\dot{x}_4 = -u_1 \sin x_3 + \epsilon u_2 \cos x_3, \tag{2.3d}$$
$$\dot{x}_5 = +u_1 \cos x_3 + \epsilon u_2 \sin x_3 - 1, \tag{2.3e}$$
$$\dot{x}_6 = u_2, \tag{2.3f}$$

which can obviously be put in the following more condensed form:

$$\dot{x} = f(x, u, \epsilon). \tag{2.4}$$

The ODE (2.4) is referred to as the continuous-time state-space equation representing the dynamics of the PVTOL.

Having this illustrative example in mind, the following section provides a more general set of definitions and notation.

2.3 The State-Space Representation of a Dynamical System

The laws of nature are generally expressed using the derivatives of the dynamic variables they involve. On the other hand, any modern algorithm needs to behave while taking into account *the situation at the instant this algorithm is called* and since the computation lasts a non vanishing amount of time, only *discrete-time* representation of the system can be effectively used.

These two versions regarding the representation of dynamical systems are discussed hereafter.

[1] Since $\dot{x}_1 = \dot{y} = x_4$ and so on.

2.3.1 Continuous-Time State-Space Representation

The general form of a continuous-time state-space representation of a dynamical system is given by[2]:

$$\dot{x} = f(x, u, w), \tag{2.5}$$

where

- $x \in \mathbb{R}^{n_x}$ is the state vector.
- $u \in \mathbb{R}^{n_u}$ is the control vector (referred to sometimes as the vector of manipulated variables).
- $w \in \mathbb{R}^{n_w}$ is a vector gathering all[3] the parameters or variables that modify the evolution of the state vector other than the state vector and the control input vector. In the example of the previous section, $w = \epsilon$.

2.3.2 Discrete-Time State-Space Representation

The One-Step-Ahead Prediction Map

In a computer-based digital control context, only piece-wise constant profiles of actions can be used despite the fact that the sampling time can be quite small. This is the reason why one needs to work with the discrete-time form of the state-space equation. The latter can be derived from the continuous-time form (2.5) using a fixed step (sampling period) method such as the well known Runge–Kutta method that is recalled in Fig. 2.2.

The discrete-time state-space evolution map is shortly referred to using the following form[4]:

$$x^+ = F(x, u, w), \tag{2.6}$$

where for the sake of simplicity, the sampling time `dt` is omitted as it is generally known from the context.[5]

The discrete-time evolution map F defined by (2.6) is of great importance in model-based control design since it encompasses the direct consequences of the choice of the current action u_k given the current state x_k.

In the early years of nonlinear control design, it was common to look for some so-called energy (storage) function of the state, say $V(x)$, and look for a **state feedback**

---

[2] Note that in the case where the evolution depends explicitly on time, the state can be extended using one additional component with the dynamic $\dot{x}_{n_x+1} = \dot{t} = 1$.

[3] When it will come to consider uncertainties and/or disturbances, it is w that will play this role of the unknown or uncertain variable.

[4] Or $x_{k+1} = F(x_k, u_k, w_k)$ where k refers to the k-the sampling period.

[5] See Fig. 2.2 to examine the role played by `dt` in the integration process.

4-th Order Runge–Kutta Scheme

```
input: x, u, w, dt

k1 = f(x,u,w)
k2 = f(x+0.5*k1*dt,u,w)
k3 = f(x+0.5*k2*dt,u,w)
k4 = f(x+k3*dt,u,w)
x+ = x + dt*(K1+2(k2+k3)+k4)/6

output: x+
```

Fig. 2.2 The algorithm of the 4th-order Runge–Kutta method that can be used to compute the discrete-time state evolution map F from the continuous-time state-space map f. This integration scheme leads to an integration precision that is of order 4 in the sampling period `dt`

law $K(x)$ such that the following inequality holds true for some desired target value x_d of the state[6]:

$$V\Big(\overbrace{F(x_k, K(x_k), w_k}^{x_{k+1}}\Big) - V(x_k) \le -\ell(x_k - x_d) + \gamma|w_k|, \tag{2.7}$$

for some **positive definite function** ℓ since in this case, summing over the future instants from the instant indexed by $k = 0$ to the one indexed by $k = N$ leads to the following inequality:

$$V(x_{k+N}) \le V(x_0) - \sum_{k=0}^{N} \ell(x_k - x_d) + \gamma \sum_{k=0}^{N} \|w_k\|^2, \tag{2.8}$$

and since V is positive, the last inequality implies the following one:

$$\sum_{k=0}^{N} \ell(x_k - x_d) \le V(x_0) + \gamma \sum_{k=0}^{N} \|w_k\|^2, \tag{2.9}$$

which obviously means that in the absence of disturbance, one necessarily gets that:

$$x_k \to x_d,$$

since otherwise, the infinite sum of positive quantities cannot be bounded.

[6] This kind of techniques is referred to as **Lyapunov-based approaches**. This is definitively not the kind of approaches used in the remainder of this book. The objective here is to show an example of use of the one step-ahead-prediction map F.

On the other hand, in the presence of disturbances and provided that there is $\kappa > 0$ such that $\ell(r) \geq \kappa\|r\|^2$, one can use the last inequality to write:

$$\sum_{k=0}^{N} \|x_k - x_d\|^2 \leq \frac{1}{\kappa} V(x_0) + \frac{\gamma}{\kappa} \sum_{k=0}^{N} \|w_k\|^2, \tag{2.10}$$

which expresses a sort of **finite L_2-gain between the energy of the disturbance and the L_2-distance to the desired target**.

Such an analytical method in control design was dominant in the late 80s where the book [1] by Khalil was the bible of many researchers on nonlinear systems.[7]

Notice that the computation involves only the **one-step-ahead** prediction map F since the condition (2.7) involves only the relationship between the values of the positive definite map V at two successive instants, namely k and $k+1$.

It became quickly (20 years actually!) clear however that finding such Lyapunov function V together with the associated couple of state feedback K and a stage cost function ℓ such that the inequality (2.7) is satisfied is horribly difficult if not impossible in the majority of realistic situations, let alone the fact that such an approach is incompatible with the handling of constraints or non regulation-like problems settings. In particular, the resulting function V was dictated exclusively by the analytical tricks needed for the proof rather than relevant requirement regarding a performance indicator to be maximized or a cost function to be minimized.

The Multi-step-Ahead Prediction Map

A part of the difficulty in finding a map V stems from the fact that the condition (2.7) is point-wise while its implication should hold on the long term. This puts too many requirements on the function V that make it difficult to find if only it exists.

As a matter of fact, as it is shown later on when discussing MPC, such a map V can be indirectly obtained by solving a constrained Nonlinear Programming (NLP) problem[8] which explains the difficulty to finding it by point-wise *analytical* expressions since the optimal solutions of such NLP problems are not necessarily smooth nor even continuous for general nonlinear systems and general problem's formulations. This is the reason why using the one-step-ahead prediction map F revealed insufficient to address advanced control problem statements.

In the general case, one should look over a longer prediction horizon to see the *big picture*! This brings to light the need to look for control input profiles[9] $\mathbf{u}_k$ that

---

[7] As a matter of fact, the analytical developments were performed on the continuous-time form (2.3) of the dynamics since the resulting feedback laws, denoted by $K(x)$ in the above discussion, were extremely cheap to compute (although very hard to find for many real-life systems) so that the sampling period could be considered as infinitely short.

[8] As a matter of fact, $V(x)$ would be the integral of the stage cost over the optimal solution starting at state x.

[9] Note that profiles are denoted by **boldfaced** notation $\mathbf{u}_k$ unlike the point-wise vector notation u_k.

are defined over a time interval $[k, k+N]$ rather than a single control u_k, namely:

$$\mathbf{u}_k := \begin{bmatrix} u_k \\ u_{k+1} \\ \vdots \\ u_{k+N-1} \end{bmatrix} \in [\mathbb{R}^{n_u}]^N \quad ; \quad u_{k+i} \in \mathbb{R}^{n_u}. \tag{2.11}$$

Notice that the same notation is used to denote the trajectory of the uncertainties vector $\mathbf{w}_k \in [\mathbb{R}^{n_w}]^N$ as well as the profile of the state $\mathbf{x}_k \in [\mathbb{R}^{n_x}]^N$ over the same prediction horizon.

Indeed, disposing of $\mathbf{u}_k$ and $\mathbf{w}_k$ representing the behavior of the manipulated variables and the disturbances respectively over the future that starts at the present instant k together with the initial state x_k, it is possible to iteratively apply the one-step-ahead map F in order to predict the trajectory of the controlled system over a prediction horizon of length N, namely:

$$\begin{aligned} x_{k+1} &= F(x_k, u_k, w_k) \\ x_{k+2} &= F(x_{k+1}, u_{k+1}, w_{k+1}) \\ \vdots \quad & \quad \vdots \\ x_{k+N} &= F(x_{k+N-1}, u_{k+N-1}, w_{k+N-1}) \end{aligned} \quad \Leftrightarrow \quad \mathbf{x}_{k+1} = \mathbf{F}(x_k, \mathbf{u}_k, \mathbf{w}_k), \tag{2.12a}$$

where in accordance with the notation (2.11), the state profile can be denoted by[10]:

$$\mathbf{x}_{k+1} := \begin{bmatrix} x_{k+1} \\ \vdots \\ x_{k+N} \end{bmatrix} \in [\mathbb{R}^{n_x}]^N, \tag{2.13}$$

is the resulting predicted state trajectory.

The so-defined map $\mathbf{F}$ is the multi-step-ahead prediction map. This map enables to **associate to each triplet** $(x_k, \mathbf{u}_k, \mathbf{w}_k)$ composed of:

- an initial state x_k,
- a sequence of future actions $\mathbf{u}_k$ over the time interval $[k, k+N]$, and
- a sequence of future realizations $\mathbf{w}_k$ of the uncertainty vector over the time interval $[k, k+N]$,

a state trajectory of the system over the same time interval $[k, k+N]$.

[10] Note that N is not used in the notation for the sake of simplicity. Keep in mind that in the sequel, the definition of the trajectories depends on the prediction horizon length N. When necessary, the notation $\mathbf{F}_N(x_k, \mathbf{u}_k, \mathbf{w}_k)$ might be used to refer to the prediction horizon's length being used.

To summarize:

The Multi-step-Ahead Prediction Map
It is the map that provides the future state trajectory given the current state and the future trajectories of the control and the uncertainty:

$$\mathbf{x}_{k+1} = \mathbf{F}(x_k, \mathbf{u}_k, \mathbf{w}_k),$$

where $\mathbf{F}$ is defined by (2.12).

Now depending on the control objective, the resulting sequence of states $\mathbf{x}_{k+1} := \mathbf{F}(x_k, \mathbf{u}_k, \mathbf{w}_k)$ can be used to *give a score or a performance index or a cost value* induced by the use of the sequence of possible future actions, represented in $\mathbf{u}_k$, for the specific realization $\mathbf{w}_k$ of the uncertainty profile over the prediction horizon.

Before going further, let us examine an instantiation of the above discussion, definitions and notation, applied to the illustrative example introduced in Sect. 2.2.

Example of Definition of a Cost Function

Assume that we would like to regulate the position and the orientation of the PVTOL aircraft around some desired values $y = y_d$, $z = z_d$ and $\theta = 0$.

By the very definition of the state $x := (y, z, \theta, \dot{y}, \dot{z}, \dot{\theta})$, this objective leads to a desired state vector defined by[11]:

$$x_d := \begin{bmatrix} y_d \\ z_d \\ 0 \\ 0 \\ 0 \\ 0 \end{bmatrix}.$$

On the other hand, given the dynamics (2.3d)–(2.3e), it comes clearly that the associated steady control vector is defined by:

$$u_d := \begin{bmatrix} 1 \\ 0 \end{bmatrix},$$

since this is the value of the control vector that makes stationary the targeted state x_d, namely

$$0 = f(x_d, u_d, \epsilon)$$

[11] Since we also need the variations of y, z and θ to be equal to 0.

regardless of the value of the parameter ϵ.

Consequently, we can define a *cost function* that penalizes the distance to this desired state x_d over the prediction horizon of length N, namely (Recall that $\mathbf{x}_{k+1} := \begin{bmatrix} x_{k+1} \\ \vdots \\ x_{k+N} \end{bmatrix}$):

$$J(\mathbf{u}_k|x_k, \epsilon) := \sum_{i=0}^{N-1} \|x_{k+i+1} - x_d\|_Q^2 + \|u_{k+i} - u_d\|_R^2 \tag{2.14a}$$

$$\mathbf{x}_{k+1} = \mathbf{F}(x_k, \mathbf{u}_k, \epsilon), \tag{2.14b}$$

for some non negative weighting matrices[12] $Q \in \mathbb{R}^{n_x \times n_x}$ and $R \in \mathbb{R}^{n_u \times n_u}$.

Note that Q and R enable to assign different costs on the distances to the desired values on y, z, θ or on the distance of their derivatives to 0.

Recall that the notation $\|v\|_Q^2$ stands for the scalar resulting from the matrix operation $v^T Qv$. For instance, if a penalty of the form $\|\varepsilon\|_Q^2$ is applied where $\varepsilon \in \mathbb{R}^2$ is normalized and the following values of the penalty matrix Q is used:

$$Q = \begin{bmatrix} 10 & 0 \\ 0 & 1 \end{bmatrix},$$

then it comes out that it is 10 times more expensive to have high values on ε_1^2 than on ε_2^2.

To summarize:

The Cost Function

The cost function $J(\mathbf{u}_k \mid x_k, \epsilon)$ should be understood as:
The cost associated to the control sequence $\mathbf{u}_k$ GIVEN the initial state x_k and the realization ϵ of the uncertain parameter.

Note that among the arguments of the cost function J defined at instant k where the state is given by x_k, only the sequence of future actions represents a decision variable that one might choose. The current state x_k and the real value of the parameter ϵ cannot be modified.

It is sometimes interesting to use a higher penalty on the terminal state (the state reached after N steps) since after all, one might consider that during some intermediate steps the state of the system cannot be immediately brought close enough to the desired state but over the long term, one would like the reached state to be really close to the desired one. This is the reason why the cost function can sometimes be

[12] Here $n_x = 6$ and $n_u = 2$.

augmented by a so-called **terminal penalty** such as the one used in the following expression:

$$J(\mathbf{u}_k \mid x_k, \epsilon) := \|\mathbf{x_N} - \mathbf{x_d}\|^2_{\mathbf{Q_f}} + \sum_{i=0}^{N-1} \|x_{k+i+1} - x_d\|^2_Q + \|u_{k+i} - u_d\|^2_R, \quad (2.15)$$

with some terminal weighting matrix $Q_f \in \mathbb{R}^{n_x \times n_x}$.

Note that the cost function is defined through some a priori fixed parameters such as Q, R and Q_f but also through varying parameters such as x_d and u_d that might change during the operations of the aircraft.[13] For the sake of rigor, these parameters should therefore appear in the left-hand side of the last definition.

Let us then gather the varying parameters in a single vector of parameters denoted hereafter by p_{cost}, namely:

$$p_{\text{cost}} := \begin{bmatrix} x_d \\ u_d \end{bmatrix}, \quad (2.16)$$

so that the cost function can be more rigorously denoted by:

$$J(\mathbf{u}_k \mid x_k, \epsilon, p_{\text{cost}}). \quad (2.17)$$

This is quite enough for the illustrative example, let us go back to the general setting.

2.4 The Cost Function

Following the discussion above, based on the Multi-Step-Ahead map **F** introduced in Eq. 2.12, let us assume that the following items are available:

1. the current state $x_k \in \mathbb{R}^{n_x}$ at the sampling instant k,
2. a sequence $\mathbf{u}_k \in [\mathbb{R}^{n_u}]^N$ of future actions over the time interval $[k, k+N]$,
3. a vector $\mathbf{w}_k \in [\mathbb{R}^{n_u}]^N$ representing the realizations of the uncertainty vector over the time interval $[k, k+N]$, and
4. a vector p_{cost} of cost-related possibly modifiable parameters.

These components enable to define a cost function that reflects the control objective and which takes the following form[14]:

$$J(\mathbf{u}_k \mid x_k, \mathbf{w}_k, p_{\text{cost}}), \quad (2.18)$$

[13] The aircraft is not supposed to remain indefinitely at (y_d, z_d), we hope!

[14] Recall that in order to define the cost function map J, one disposes of the resulting trajectories of the state that is given by $\mathbf{x}_{k+1} = \mathbf{F}(x_k, \mathbf{u}_k, \mathbf{w}_k)$ provided by the multi-step-ahead map **F** previously defined by (2.12).

which represents the cost of using the control sequence $\mathbf{u}_k$ over the future time interval $[k, k+N]$ when the initial state is x_k, the realization of the uncertain vector of parameters over the time interval is given by $\mathbf{w}_k$ and when the modifiable parameters of the cost function are given by the vector p_{cost}.

It is important to underline that since the trajectory of the state $\mathbf{x}_{k+1} = \mathbf{F}(x_k, \mathbf{u}_k, \mathbf{w}_k)$ is available, it is possible to imagine many formulations of the cost function that are not necessarily obtained by summing over the sampling periods some stage cost $\ell(x_{k+i+1}, u_{k+i})$. Rather, some **functional of the state trajectories** can be used as a part of the definition of the cost function. For instance, a penalty on the quantity

$$\max_{i\in\{0,\dots,N-1\}} \|x_{k+i+1}\|^2$$

might be used to penalize the maximum norm of the state. This is just an example and the number of possible variations is unlimited!

By now we defined the potentially uncertain dynamic model of the controlled system as well as the cost function that reflects the control objective. It is time to invoke the constraints that should be taken care of while designing any realistic feedback control law.

This is the object of the following section.

2.5 The Constraints

Any control problem comes with constraints of different kinds that are discussed in this section.

- Box-Constraints on the Control Inputs.
  This kind of constraints represent limitations on the actuators or, when the decision variables are set-points for instance, these constraints might represent admissible allowable values that are compatible with some comfort criteria or safety-related constraints for instance. These constraints can be denoted as follows:

$$u^{min} \le u \le u^{max} \quad \text{where} \quad u^{min} \in \mathbb{R}^{n_u} \quad \text{and} \quad u^{max} \in \mathbb{R}^{n_u}, \tag{2.19}$$

  where the inequalities in (2.19) are to be understood component-wise.[15]
  When it comes to represent control profiles over some time interval $[k, k+N]$ as discussed above, the constraints (2.19) obviously induces the following con-

[15] For instance if $u \in \mathbb{R}^2$, then $u \le \begin{bmatrix}1, 3\end{bmatrix}$ means that $u_1 \le 1$ and $u_2 \le 3$.

straints on the sequence of control inputs $\mathbf{u}_k$[16]:

$$\mathbf{u}^{min} := \begin{bmatrix} 1 \\ \vdots \\ 1 \end{bmatrix} \otimes u^{min} \le \mathbf{u}_k \le \begin{bmatrix} 1 \\ \vdots \\ 1 \end{bmatrix} \otimes u^{max} =: \mathbf{u}^{max}. \tag{2.20}$$

This obviously corresponds to $(2N \cdot n_u)$ constraints.

- Box-Constraints on the State.
  These constraints apply to the state components. They generally reflect safety or comfort requirements and/or requirement regarding the quality of the outcome that needs to meet some minimal values, etc.
  These constraints are denoted as follows:

$$x^{min} \le x \le x^{max} \quad \text{where} \quad x^{min} \in \mathbb{R}^{n_x} \quad \text{, and} \quad x^{max} \in \mathbb{R}^{n_x}, \tag{2.21}$$

  where the inequalities in (2.21) are to be understood component-wise as previously stated. Here again, when it comes to considering the state profile over some time interval $[k, k+N]$, the saturation constraint (2.21) induces the following constraints on the state profile $\mathbf{x}_{k+1}$:

$$\mathbf{x}^{min} := \begin{bmatrix} 1 \\ \vdots \\ 1 \end{bmatrix} \otimes x^{min} \le \mathbf{x}_{k+1} \le \begin{bmatrix} 1 \\ \vdots \\ 1 \end{bmatrix} \otimes x^{max} =: \mathbf{x}^{max}. \tag{2.22}$$

  This obviously corresponds to $(2N \cdot n_x)$ constraints.
  Notice that the previous two classes of constraints force the control and the state vector to remain in a bounded **hyper-boxes**. This is the reason why they are also referred to as **Box-constraints**.
  Box-constraints are not the only type of constraints that might be necessary to consider. More general constraints are sometimes encountered involving combinations of state/input components. The associated notation is mentioned in the next item.
- More General Constraints
  Beside the previous **Box-constraints** that need to be handled as such for computational efficiency reasons, there might be some constraints that involve more complicated expressions that combine different components of the control and the state vectors through possibly nonlinear expressions.
  These constraints can be written in a general condensed form as follows:

$$g^{min} \le g(x, u) \le g^{max} \in \mathbb{R}^{n_c}, \tag{2.23}$$

[16] Recall that the Kronecker product is defined by $\begin{bmatrix} a & b \\ c & d \end{bmatrix} \otimes A = \begin{bmatrix} aA & bA \\ cA & dA \end{bmatrix}$.

where the map $g : \mathbb{R}^{n_x} \times \mathbb{R}^{n_u} \rightarrow \mathbb{R}^{n_c}$ maps the pair (x, u) into a vector of n_c expressions each of which stands for a quantity that should be kept within the associated bounds during the lifetime of the controlled system.
Similarly, when considering a time interval $[k, k+N]$, the above constraints should be enforced for all the pairs of the form

$$(x_{k+i+1}, u_{k+i}) \qquad \text{for } i \in \{0, \dots, N-1\}.$$

The resulting set of constraints can therefore be written by first defining the augmented constraint map given by:

$$\mathbf{g}(\mathbf{x}_{k+1}, \mathbf{u}_k) := \begin{bmatrix} g(x_{k+1}, u_k) \\ \vdots \\ g(x_{k+N}, u_{k+N-1}) \end{bmatrix} \in [\mathbb{R}^{n_c}]^N, \tag{2.24}$$

on which the following saturation constraints have to be imposed:

$$\mathbf{g}^{min} := \begin{bmatrix} 1 \\ \vdots \\ 1 \end{bmatrix} \otimes g^{min} \leq \mathbf{g}(\mathbf{x}_{k+1}, \mathbf{u}_k) \leq \begin{bmatrix} 1 \\ \vdots \\ 1 \end{bmatrix} \otimes g^{max} =: \mathbf{g}^{max}. \tag{2.25}$$

This obviously corresponds to $(2N \cdot n_c)$ constraints.
The previous discussion on the constraints can be summarized as follows:

Constraints
There are three types of constraints that might be necessary to handle when designing a control law:

1. box-constraints on the control (2.19),
2. box-constraints on the state (2.21), and
3. combined constraints of the form (2.25).

An important difference between the first two box-constraints is that generally, the first constraints (control saturation) are to be considered as *hard constraints* while the second might (and often should) be *softened*.[17]

Now that we defined the dynamic model, the cost function and the constraints, we have all the ingredients we need in order to define the so-called *OCP*. This is the aim of the next section.

2.6 The Optimal Control Problem

An OCP encompasses the control objective and the associated constraints. Solving an OCP amounts at finding the best sequence of values to be taken by the manipulated variables so that the *controlled system behaves* in a way that minimizes the cost function while persistently respecting the constraints.

Obviously, the presence of uncertainties makes this task harder; both in terms of the complexity of the problem's formulation but also, and especially, in terms of the computational effort needed to get the appropriate *optimal* solution.

In what follows and in order to progressively introduce the associated formulations, the paradigm is presented in the deterministic setting where no uncertainties are present before the more complicated uncertain setting is discussed.

2.6.1 Deterministic Optimal Control Problems

Deterministic OCPs can be defined when one is looking for the best sequence of actions that minimizes some cost function under constraints over a time interval $[k, k+N]$ for a given (and supposedly known) realization of the uncertainty.

The ingredients of a deterministic OCP are the following:

- the initial state of the controlled system x_k,
- the realization $\mathbf{w}_k$ of the uncertainty vector over $[k, k+N]$, and
- the vector of parameters p_{cost} that are needed to define the cost function.

Once these items are defined, we already know that a cost function can be defined which takes the form

$$J(\mathbf{u}_k | x_k, \mathbf{w}_k, p_{\text{cost}}),$$

that associates to each sequence of control inputs $\mathbf{u}_k$ a score (a real scalar).

! Control Versus Decision-Variable

The manipulated variable u refers to the vector of variables that can be manipulated by the control designer: **it is the output of the control algorithm.** Traditionally, this variable refers to some physical actuators' set-points. However, in modern control context, this can be any vector of decision variables such as: set-points, threshold on investment, switching decision between different operational modes, etc.

Based on the above discussion, **if the realization $\mathbf{w}_k$ of the uncertainty over the time interval $[k, k+N]$ were known,**[18] an associated **deterministic** OCP can be stated as follows:

[18] This is impossible by the very definition of an uncertainty! but let us admit it temporarily.

$$\min_{\mathbf{u},\mathbf{x}} J(\mathbf{u}|x_k, \mathbf{w}_k, p_{\text{cost}}) \quad \text{under} \quad \left| \begin{array}{l} \mathbf{u}^{min} \leq \mathbf{u} \leq \mathbf{u}^{max}, \\ \mathbf{x}^{min} \leq \mathbf{x} \leq \mathbf{x}^{max}, \\ \mathbf{x} = \mathbf{F}(x_k, \mathbf{u}, \mathbf{w}_k), \\ \mathbf{g}^{min} \leq \mathbf{g}(\mathbf{x}, \mathbf{u}) \leq \mathbf{g}^{max}. \end{array} \right. \tag{2.26}$$

Note that in the above formulation, designated as the *multiple-shooting* formulation, the decision variable is defined to be the pair $(\mathbf{u}, \mathbf{x})$ of control and state profiles over the time interval $[k, k+N]$. This might be intriguing since the state trajectory $\mathbf{x}$ is uniquely determined by $(x_k, \mathbf{u}, \mathbf{w}_k)$ and can therefore be *eliminated* using the equality constraint $\mathbf{x} = \mathbf{F}(x_k, \mathbf{u}, \mathbf{w}_k)$ leading to what is called the *the single-shooting* formulation.

As a matter of fact, while the number of degrees of freedom in the single-shooting formulation, namely $(N \cdot n_u)$ is lower that the number of decision variables in the multiple-shooting formulation, namely $(N \cdot (n_u + n_x))$, it is commonly considered that the **multiple-shooting formulation leads to more efficiently solvable optimization problems** when the dimensions increase. This is because the resulting problems are structurally sparse[19] and such that the constraints on the state are easier to enforce. Again, this advantage in using multiple-shooting formulation is in particular true for high-dimensional problems and when the number of state constraints increases.

For the sake of compactness of notation, let us denote by s_k the set of parameters that define the scenario under interest, namely:

$$s_k := (x_k, \mathbf{w}_k, p_{\text{cost}}), \tag{2.27}$$

so that the cost function at instant k can be denoted by[20]:

$$J(\mathbf{u} \mid s_k).$$

In what follows, the optimal solution of the deterministic OCP (2.26) will be denoted as follows:

$$\mathbf{u}^\star(s_k) = \mathbf{u}^\star(x_k, \mathbf{w}_k, p_{\text{cost}}) \quad ; \quad \mathbf{x}^\star(s_k) = \mathbf{x}^\star(x_k, \mathbf{w}_k, p_{\text{cost}}), \tag{2.28}$$

where the profiles $\mathbf{u}^\star(s_k)$ and $\mathbf{x}^\star(s_k)$ are to be interpreted as trajectories over the time interval $[k, k+N]$, that is to say:

$$\mathbf{u}^\star(s_k) := \begin{bmatrix} u_k^\star(s_k) \\ \vdots \\ u_{k+N-1}^\star(s_k) \end{bmatrix} \quad ; \quad \mathbf{x}^\star(s_k) := \begin{bmatrix} x_{k+1}^\star(s_k) \\ \vdots \\ x_{k+N}^\star(s_k) \end{bmatrix}. \tag{2.29}$$

---

19 This is because in the multiple-shooting formulation, each decision variable is coupled with only the neighboring ones while in the single-shooting approach the first control does impact every single instants in the state trajectory.

20 Which reads: The cost function for **u given the context defined by** s_k.

This completes the definition of deterministic OCPs.

Chapter 5 discusses `Python`-based solution to the deterministic OCP defined above. Moreover, Chap. 7 shows a heuristic that provides a feedback control design that explicitly takes into account the **presence of uncertainties** might be obtained by solving a cloud of **deterministic** OCPs of the form (2.26).

But for the time being let us examine what would be the ideal formulation to solve in order to perfectly take into account the statistics of the uncertainty dispersion should the computational complexity-related issue be *temporarily* left aside. This is the object of the next section.

2.6.2 Stochastic Optimal Control Problems

Let us assume that the statistics of the uncertain profile $\mathbf{w}_k$ is available. By this, it is assumed that one disposes of a random sampling algorithm that reflects the probability distribution $\mathcal{W}$ of the realizations of $\mathbf{w}_k$.

The question is the following:

How to Get Rid of the Unknown $\mathbf{w}_k$?

How one should modify the formulation of the optimization problem (2.26) such that the explicit knowledge of $\mathbf{w}_k$ is no more required as it is the case in (2.26)?

The guiding principle in this transformation lies in the way the cost function and the constraints have to be modified [2]:

1. The deterministic value of the cost function should be replaced by the expected value associated to the probability distribution $\mathcal{W}$. This means that the cost function becomes:

$$\mathbb{E}_{\mathcal{W}}\Big[J(\mathbf{u}|x_k, \cdot, p_{\text{cost}})\Big] := \int_{\mathbf{w}} J(\mathbf{u}|x_k, \mathbf{w}, p_{\text{cost}}) f_{\mathcal{W}}(\mathbf{w}) d\mathbf{w}. \tag{2.30}$$

   Notice that with some abuse of notation this is simply a weighted average of all possible values of the w-profile weighted by the probability density function map $f_{\mathcal{W}}$.
2. The deterministic satisfaction of the constraints **involving the realization of the uncertainty** must be replaced by the probability of them being satisfied.[21] For

[21] There are different ways to perform this softening of the constraints. Only one version is presented here for the sake of clarity.

the formulation (2.26), this concerns the two following constraints:

$$\mathbf{F}(x_k, \mathbf{u}_k, \mathbf{w}_k) \in [\mathbf{x}^{min}, \mathbf{x}^{max}] \tag{2.31a}$$

$$\mathbf{g}(x_k, \mathbf{u}_k, \mathbf{w}_k) \in [\mathbf{g}^{min}, \mathbf{g}^{max}], \tag{2.31b}$$

which are no more required to hold **for all possible realizations of the uncertainty profile**[22] $\mathbf{w}_k$. Rather, they are required to be satisfied with a high probability. More precisely, the previous constraints might be replaced by the following softer constraints

$$\text{Pr}_{\mathcal{W}}\Big[\mathbf{F}(x_k, \mathbf{u}_k, \cdot) \notin [\mathbf{x}^{min}, \mathbf{x}^{max}]\Big] \leq \eta, \tag{2.32a}$$

$$\text{Pr}_{\mathcal{W}}\Big[\mathbf{g}(x_k, \mathbf{u}_k, \cdot) \notin [\mathbf{g}^{min}, \mathbf{g}^{max}]\Big] \leq \eta, \tag{2.32b}$$

for sufficiently small $\eta \in (0, 1)$.

Now since it is difficult to know what is the smallest achievable η and in order to come out with a necessarily feasible problem, a possible formulation of the stochastic OCP gathering the cost function (2.30) and the probabilistically defined constraints (2.32) together with the saturation constraints on the control profile, would be the one given by:

$$\min_{\mathbf{u},\mathbf{x},\eta} \; \mathbb{E}_{\mathcal{W}}\Big[J(\mathbf{u}|x_k, \cdot, p_{\text{cost}})\Big] + \rho\eta, \tag{2.33a}$$

$$\text{under} \quad \left|\begin{array}{l} \mathbf{u}^{min} \leq \mathbf{u} \leq \mathbf{u}^{max} \\ \text{Pr}_{\mathcal{W}}\Big[\mathbf{F}(x_k, \mathbf{u}_k, \cdot) \notin [\mathbf{x}^{min}, \mathbf{x}^{max}]\Big] \leq \eta, \\ \text{Pr}_{\mathcal{W}}\Big[\mathbf{g}(x_k, \mathbf{u}_k, \cdot) \notin [\mathbf{g}^{min}, \mathbf{g}^{max}]\Big] \leq \eta, \end{array}\right. \tag{2.33b}$$

where the **slack variable** η has been introduced in order to soften the two constraints (2.32) and render the so-formulated problem always feasible.

Notice that the penalty term $\rho > 0$ should be chosen high enough to promote smaller values for the probability of constraints violation. Consequently, η becomes a part of the set of decision variables that are to be determined by solving the stochastic OCP.

The resulting stochastic formulation involves the terms $\mathbb{E}_{\mathcal{W}}[J]$, $\text{Pr}_{\mathcal{W}}[\mathbf{F}]$ and $\text{Pr}_{\mathcal{W}}[\mathbf{g}]$ which are expectations and probabilities of quantities that might be defined through nonlinear relationships on rather high-dimensional spaces. The computation of such quantities is extremely difficult and can only be (hardly) approximated with mediocre precision.

We'll come back to this later on in this book. For the time being, the goal is to set clearly the basic notation and paradigms without digging into the computational part of the issues leading to appropriate solutions to the underlying problems.

[22] Which would be too conservative and intractable anyway.

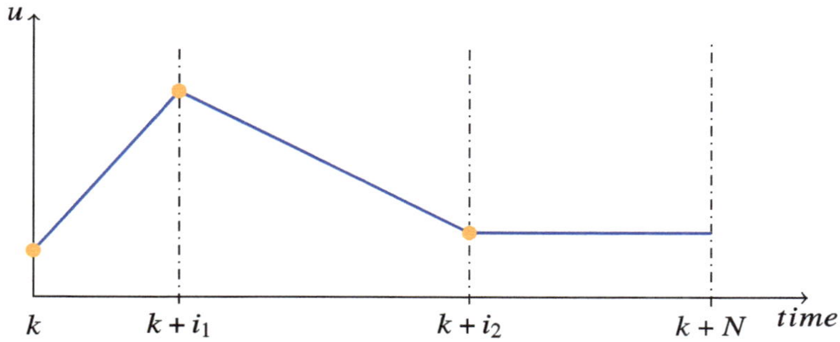

Fig. 2.3 Example of control parameterization with $n_u = 1$, $r = 3$. Note that by convention, when $r - 1 < N$, constant control is taken over the last interval $[k + r - 1, k + N]$

2.6.3 Control Parameterization

Control parameterization is sometimes used in the single-shooting approach in order to reduce the number of decision variables. Control parameterization amounts at describing the control input profile $\mathbf{u}_k$ which originally encompasses $(N \cdot n_u)$ degrees of freedom in terms of a lower number of degrees of freedom.[23]

More precisely recall that the future control input profile at instant k is given by:

$$\mathbf{u}_k := \begin{bmatrix} u_k \\ u_{k+1} \\ \vdots \\ u_{k+N-1} \end{bmatrix} \in [\mathbb{R}^{n_u}]^N. \tag{2.34}$$

The idea is to decide that only the values of the control vector at a smaller number r of future instants would be the set of free decision variables. The values at the remaining instants are then determined by linear interpolation as schematically shown in Fig. 2.3. More precisely, one would first choose the set of r instants (necessarily including the first one, denoted hereafter by κ:

$$\kappa = \left[k \; k + i_1 \; \dots \; k + i_{r-1}\right], \tag{2.35}$$

meaning that only the control input instances given by:

$$\mathbf{u}_k^r = \begin{bmatrix} u_k \\ u_{k+i_1} \\ \dots \\ u_{k+r-1} \end{bmatrix} \in \mathbb{R}^{r \cdot n_u}, \tag{2.36}$$

[23] The simplest form of control parameterization appeared together with the concept of *control horizon* $N_c < N$ which is the number of instants over the prediction horizon where the control input values are free to manipulate, the remaining values of the control over the remaining sampling periods over the prediction horizon are frozen to the last input value. In this section, a more general scheme is described that can induce a stronger reduction in the number of decision variables.

```
def compute_Pi(N, kappa, nu):
    # function that computes the matrix
    # of the reduced paramterization
    r = len(kappa)
    # head of the control profile
    Pi = np.zeros((N*nu, r*nu))
    Pi[0:nu, 0:nu] = np.eye(nu)
    Inu = np.eye(nu)
    for i in np.arange(1, N):
        iy = np.arange(i*nu, (i+1)*nu)
        # tail of the control profile
        if (i>=kappa[-1]):
            Pi[iy, (r-1)*nu:r*nu] = np.eye(nu)
        # intermediate values
        else:
            I = [j for j in range(r) if kappa[j]<=i]
            j = np.argmax(I)
            ik = (i-kappa[j])/(kappa[j+1]-kappa[j])
            Pi[iy, j*nu:(j+1)*nu] = (1-ik)*Inu
            Pi[iy, (j+1)*nu:(j+2)*nu] = ik*Inu

    return Pi
```

Fig. 2.4 Python script of the function `compute_Pi` which computes the matrix of the control profile's reduced dimensional parameterization. The output of the function is the matrix Π_κ involved in the parameterization map (2.37) while the input argument of the function is the triplet (N, κ, n_u)

are considered to be free to be manipulated by the optimization algorithm.

Consequently, once the sequence of instants κ such as (2.35) is defined, there is a matrix Π_κ that links the original control profile $\mathbf{u}_k$ to the new vector $\mathbf{u}_k^r$ of decision variables [see (2.36)] through the following linear relationship:

$$\mathbf{u}_k = \Pi_\kappa \cdot \mathbf{u}_k^r. \tag{2.37}$$

Based on this relationship any function $h(\mathbf{u}_k, \dots)$ involving the original decision variable $\mathbf{u}_k$ can now be expressed in terms of the reduced dimensional new decision variable $\mathbf{u}_k^r$ though the expression:

$$h\Big(\Pi_\kappa \mathbf{u}_k^r, \dots\Big).$$

This might be of a great help when formulating single-shooting optimization problems since it accelerates the computation when using the fast gradient method as it is shown later on in Chap. 5.

Figure 2.4 shows the python's script that implements the function that computes the parameterization matrix Π_κ invoked in (2.37). Note that this function takes the arguments (N, κ, n_u) defined earlier.

The following example shows how this function can be used to define a control profile with $n_u = 2$-dimensional input vector and $N = 20$ using a 4 decision instants

```
from learning_in_control import compute_Pi
N = 21
kappa = [0, 1, 5, 10]
nu = 2
ur = [1, 0, 3, 1, 2, -2, -1, 0]

Pi = compute_Pi(N, kappa, nu)
u_profile = (Pi @ ur).reshape(-1,2)
plt.plot(u_profile, '-o')
...
```

Fig. 2.5 The python scripts that calls the function `compute_Pi`, shown in Figure 2.4, in order to compute the whole control profile as a function of the value of the reduced-dimensional decision variable. The resulting control profile is shown in Fig. 2.6

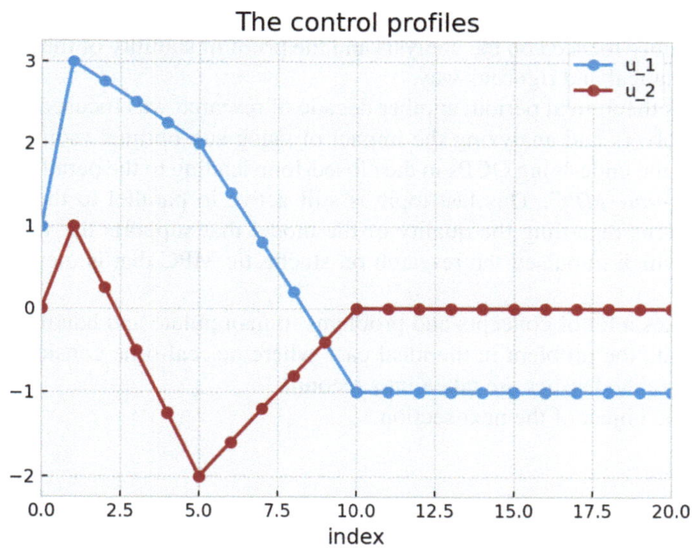

Fig. 2.6 The control profile resulting from the script shown in Fig. 2.5

$\kappa = [0, 1, 5, 10]$. The reduced dimensional degree of freedom vector value in the example is given by:

$$u^r_k := [1, 0, 3, 1, 2, -2, -1, 0], \tag{2.38}$$

which means that the first control takes the values $1, 3, 2, -1$ at instants $0, 1, 5$ and 10 respectively while the second control takes the values $0, 1, -2, 0$ at the same instants.

The resulting control profiles are given in Fig. 2.6.

The reason for which optimal control paradigm is invoked here is that solving such problems lies in the heart of the more advanced control design that catches increasing attention in the control community. This control design is called MPC and this is precisely the topic of the next section.

2.7 Model Predictive Control

MPC [3,4] is a feedback design methodology that is based on the repetitive on-line solution of OCPs. At each decision instant, a new OCP is defined taking into account the current value of the state vector and the process is repeated.

MPC development started in the late seventies [5] as a pragmatic solution to some industrial problems. The formulation was initially based on the concept of Dynamic Matrix Control (DMC) [6] which performs a local linearization of the dynamics of the system.

The successful application of this strategy finally attracted the attention of the nonlinear control community in the nineties [7–10] enhancing an intensive period of research mainly focused on the analysis and the proof of stability of the closed-loop in a mathematical and rigorous way.

After this theoretical period, another decade of research was focused on building dedicated solvers and analyzing the impact of using sub-optimal (non terminated) solutions to the underlying OCPs in the closed-loop leading to the paradigm denoted by *fast real-time MPC*. This last topic is still active in parallel to the emergence of the concerns regarding the quality of the model that supports the whole design procedure which impulsed the research on stochastic MPC that is mentioned later on.

This makes a lot of concepts and problems to manipulate and handle. But let us first formulate the problem in the ideal case where no real-time consideration, nor uncertainty-related issues are taken into account.

This is the object of the next section.

2.7.1 Ideal MPC Setting

The OCPs defined in the deterministic setting as in (2.26) or in the stochastic setting as in (2.33) enable to characterize the optimal sequences of control over some future time interval $[k, k+N]$ also called the prediction horizon.

Once either of these problems has been solved at some instant k, one disposes of an optimal sequence of actions over this interval that is denoted by[24]:

$$\mathbf{u}^\star(x_k) = \begin{bmatrix} u_0^\star(x_k) \\ \vdots \\ u_{N-1}^\star(x_k) \end{bmatrix} \in [\mathbb{R}^{n_u}]^N. \tag{2.39}$$

[24] Omitting the reference to the parameter vector p_{cost} for the sake of compactness of notation.

The question is: what next?

In other words, once one disposes of an optimal planning of actions over some future time horizon, what is the next move?

The obvious and rationale answer is to start applying the first action of the optimal sequence of actions, namely, apply the first action $u_0^\star(x_k)$ of the optimal sequence over the next sampling period $[k, k+1]$!

That is obviously what we all do in our everyday life when planning something that needs several steps to be fully achieved!

Assume that one takes this intuitive option and applies the first action $u_0^\star(x_k)$ of the optimal sequence of actions $\mathbf{u}^\star(x_k)$ over the next sampling period $[k, k+1]$ until the next sampling instant, namely $k+1$ has been reached. What to do next?

Here the answer is less obvious. Indeed, one can consider that the previous computation is still valid and the next step to do is to apply the next pre-computed actions, already contained in the sequence, namely to apply $u_1^\star(x_k)$ over the next sampling period $[k+1, k+2]$. This choice is based on the following implicit assumption:

> The remaining part of an optimal choice is optimal!

As a matter of fact, this seemingly simple statement is known in the control literature as the Bellman principle of optimality that can be stated as follows:

> **The Bellman Principle**
> An optimal policy has the property that whatever the initial state and the initial decision are, the remaining decision must constitute an optimal policy with regard to the state resulting from the first decision.

However, when digging into the assumptions under which this principle has been established, it comes out that it holds only under the following assumptions:

1. The initially computed optimal strategy is optimal with respect to an **infinite prediction horizon.** This is because with this condition, the computed strategy encompasses already the whole future without possible *surprises*. It can be easily understood that without this assumption, as the time goes, we get closer to *never-accounted-for situations* and the optimality has no reason to be satisfied.
2. The second condition that is required for the Bellman's principle to hold[25] is the absence of disturbances and the perfect knowledge of the system's model. This is again all about the absence of possible *surprises* as the time goes on.

[25] Used in the discussion of the previous action regarding the absence of *surprises* that would invalidate the optimality of the previous computation.

Bellman's principle is formulated in a perfect world without uncertainties and without disturbances.

Since for constrained nonlinear dynamical systems, the computation of the optimal control profiles can only be done for OCPs which are formulated over **finite prediction horizons**,[26] the first condition stated above is not satisfied. On the other hand, since disturbances and imperfect models are the rule rather that the exception in the real world, it comes out obviously that the second condition stated above for the Bellman principle to hold is also violated.

To make it short, applying the previously pre-computed action $u_1^\star(x_k)$ over the sampling period $[k+1, k+2]$ is almost always not optimal in a real world. If optimal decisions are to be taken, the ideal option is to consider the following road map:

- reformulate the OCP with the new value x_{k+1} of the state,
- solve the new optimization problem to get the new optimal control sequence denoted by $\mathbf{u}^\star(x_{k+1})$:

$$\mathbf{u}^\star(x_{k+1}) := \begin{bmatrix} u_0^\star(x_{k+1}) \\ \vdots \\ u_{N-1}^\star(x_{k+1}), \end{bmatrix}, \tag{2.40}$$

- and apply the first action $u_0^\star(x_{k+1})$ of the optimal sequence $\mathbf{u}^\star(x_{k+1})$ during the sampling period $[k+1, k+2]$.

This *wise* strategy that takes into account the imperfection of the world and the finite character of the prediction horizon leads to the implicit state feedback strategy given by:

$$u_k := K_{\text{MPC}}(x_k) := u_0^\star(x_k). \tag{2.41}$$

This feedback is referred to as MPC. Figure 2.7 summarizes the principle explained above by showing two successive updating instants.

MPC is today (and probably will remain for a while in the future) the more advanced and complete control design methodology. This is because of its ability to handle many aspects the real-world control problems such as:

- the presence of nonlinearities in the model,
- the need to handle input/state constraints, and
- the need to formulate complex control objectives.

[26] In the case of linear dynamics under unconstrained formulations, the fact that exponential terms sum over infinite horizons makes it possible to formulate unconstrained OCPs over infinite horizons leading to the feedback design commonly referred to as Linear Quadratic Regulator (LQR).

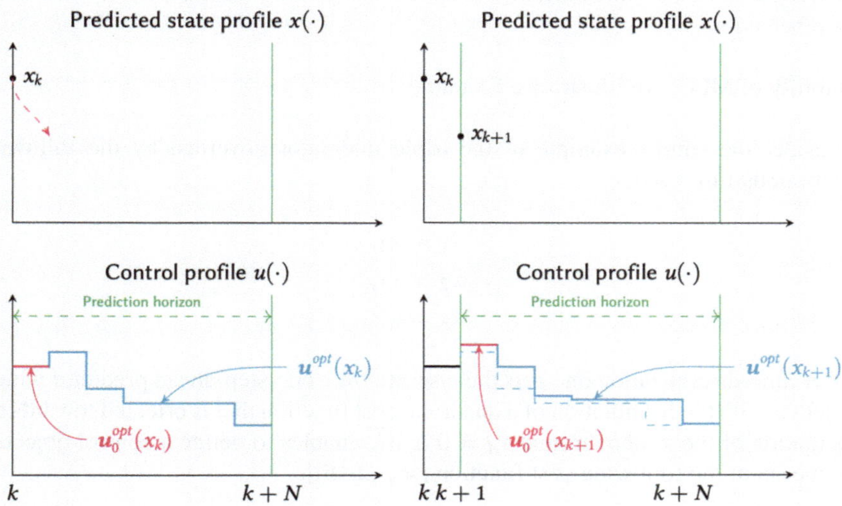

Fig. 2.7 Schematic representation of the MPC principle

Many big companies are promoting and selling MPC-based software products (ABB, Siemens, Schneider Electric, Honeywell, to cite a few of them) and whole dedicated academic workshops take place periodically to exchange the latest code developments, formulations and success stories of the MPC methodology.

Now assuming that the above mentioned road-map is applied, the following question remains to be addressed:

Closed-Loop Stability?

Is there any guarantee regarding the stability of the resulting closed-loop behavior?

This question is always in the heart of control-related studies since introducing feedback can destabilize even open-loop stable systems.

This question is studied in the next section.

2.7.2 The Stability Issue

Because of the intuitive definition of MPC as a control feedback design, it has been used within the linear control community during almost a decade [11] without formally questioning the stability of the resulting closed-loop system. This was probably due to the fact that in the early years, MPC was mainly applied to systems which are open-loop stable and hence, instability was not a recurrent issue in everyday practice.

Before giving some hints about the stability arguments and conditions, let us start by a simple example showing that a closed-loop systems under MPC might not be unconditionally stable and therefore, when applying MPC to intrinsically unstable dynamical systems some care should be taken regarding the stability of the closed-loop behavior.

Stability of MPC: An Illustrative Example

Consider the simple example of the triple integrator governed by the following continuous-time ODEs:

$$\dot{x}_1 = x_2,$$
$$\dot{x}_2 = x_3,$$
$$\dot{x}_3 = u.$$

After time-discretization one gets the discrete-time one-step-ahead predictor which, together with the formulation of a quadratic cost function that is oriented towards the regulation of the system around $x_d = 0 \in \mathbb{R}^3$, enables to define a control objective by means of the following cost function [see (2.15)]:

$$J(\mathbf{u}|x_k) := \sum_{i=0}^{N} \|x_{k+i+1}\|_{Q_x}^2 + |u_{k+i}|^2. \tag{2.42}$$

By minimizing this cost function in the decision variable $\mathbf{u}$, one can derive the resulting MPC feedback.

Now because of the following two properties:

- the quadratic structure of the cost function, and
- the absence of constraints,

it can be shown that the optimal control sequence $\mathbf{u}^*$ is a linear function of the state, namely (recall that $x_d = 0$):

$$\mathbf{u} = -[K_{\text{MPC}}(N)]x \quad \text{where} \quad K_{\text{MPC}}(N) \in \mathbb{R}^{N\times 3}. \tag{2.43}$$

Notice that in the previous expressions, it is recalled that the MPC feedback gain matrix, namely K_{MPC}, depends on the length N of the prediction horizon's length N. This is because N is involved in the definition of the cost function that determines the optimal solution and hence the resulting matrix K_{MPC}.

Now according to the principle of MPC which consists in applying the first control in the optimal sequence it comes that the feedback is given by[27]:

$$u = -[K^0_{\text{MPC}}(N)]x \quad \text{where} \quad K^0_{\text{MPC}}(N) = K_{\text{MPC}}(N)[0, :]. \tag{2.44}$$

[27] $K_{\text{MPC}}[0, :]$ denotes the first line of the feedback gain matrix K_{MPC}.

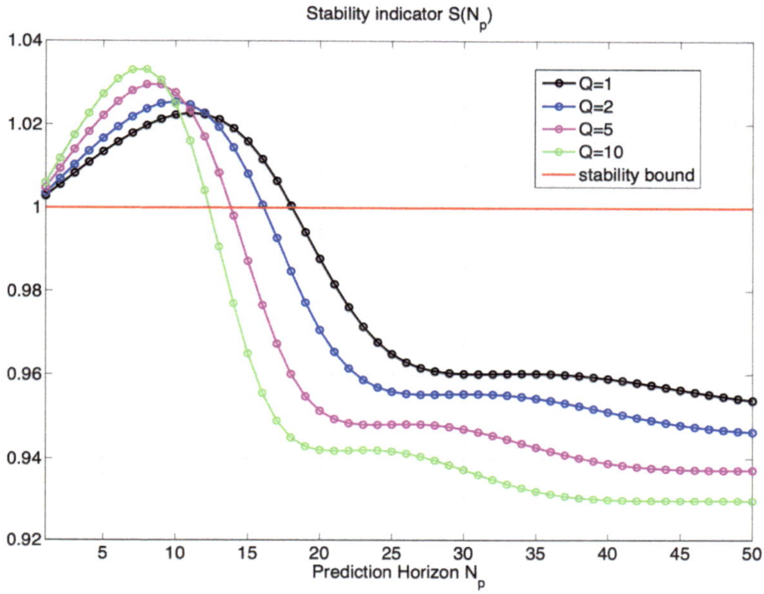

Fig. 2.8 Evolution of the stability indicator $S(N)$ as a function of the prediction horizon's length N for a penalty matrix $Q_x = \text{diag(Q, 0, 0)}$ for different values of the output penalty Q. Notice that only the distance of the regulated output x_1 to 0 is penalized while no penalty is imposed on non vanishing x_2 and x_3

By so-doing, the stability of the closed-loop system can be investigated by examining the eigenvalues of the closed-loop matrix given by[28]:

$$A_{\text{cl}}(N) := A - BK^0_{\text{MPC}}(N). \tag{2.45}$$

More precisely, the closed-loop asymptotic stability is equivalent to the following inequality being satisfied:

$$S(N) := \max_{i=1,2,3} |\lambda_i(A_{\text{cl}}(N))| < 1, \tag{2.46}$$

where $\lambda_i(A_{\text{cl}}(N))$ denotes the ith eigenvalues of the closed-loop matrix $A_{\text{cl}}(N)$.

Now we are ready to analyze the impact of the choice of the prediction horizon's length N on the stability of the closed-loop. Another parameters that might influence the stability of the closed-loop system is the choice of the penalty matrix Q_x.

Figure 2.8 shows the Evolution of the stability indicator $S(N)$ as a function of the prediction horizon's length N for a penalty matrix $Q_x = \text{diag(Q, 0, 0)}$ for different values of the output penalty Q. Notice that only the distance of the regulated output

[28] Assuming that (A, B) are the matrices of the system in discrete-time.

x_1 to 0 is penalized while no penalty is imposed on non vanishing x_2 and x_3. It comes out from this simple example that:

> **Closed-Loop Under MPC Is Not Necessarily Stable**
> The closed-loop under MPC feedback **is not always stable**. Something specific has to be done in order to guarantee the stability in closed-loop.

Beyond this warning, the plots in Fig. 2.8 also suggest that higher prediction horizons enhance the stability of the closed-loop under MPC as far as this example is concerned.

Let us now examine what happens when the whole state distance to the target is penalized. We can do this by using the following definition of the weighting matrix Q_x used in the definition of the cost function (2.42):

$$Q_x = \mathbb{I}_{3\times 3}.$$

The result is shown in Fig. 2.9 where one can see that this time and as far as this specific system is concerned, using a full penalty on the distances of all the components of the state to their targeted values 0 leads to a closed-loop stability **regardless of the prediction horizon length**.

This unconditional (w.r.t N) closed-loop stability is not always true for all controlled systems. Nevertheless, the following general rule can be considered to be always qualitatively true:

Conditions in Favor of MPC Stability
Using **longer** prediction horizons and using penalties on the **whole state** deviation to target generally enhances the closed-loop stability under MPC feedback laws.

Notice that the terms of the above statement are rather *fuzzy* and only lead to qualitative recommendation regarding the MPC formulation in terms of prediction horizon's length and penalties.

Over the last three decades, the nonlinear control community developed a rich literature on MPC formulations with guaranteed stability that we are going to briefly explore in the sect section.

Some Provably Stable Formulations

In this section, a quick informal review of some of MPC formulations that are associated to proofs of stability of the closed-loop are briefly mentioned with the associated references for those among the readers who would be interested in a completely rigorous statements and proofs.

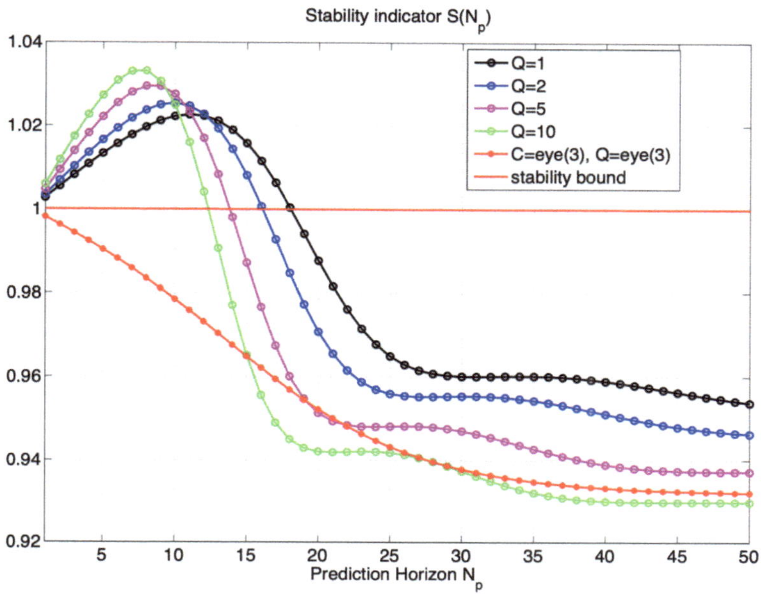

Fig. 2.9 Evolution of the stability indicator $S(N)$ as a function of the prediction horizon's length N for a penalty matrix $Q_x = \text{diag}(Q, 0, 0)$ for different values of the output penalty Q and for a full state penalty $Q_x = \mathbb{I}_{3\times 3}$

The objective of this presentation is to have the underlying success conditions in mind as guidelines in the heuristics provided here and there in the following chapters. The statements of closed-loop stability are generally composed of sufficient (and never necessary) conditions. Moreover, these conditions are almost rarely totally respected in the implementation of MPC on real-life problems. Nevertheless, being aware of the sufficient conditions enables to adequately *correct/improve* the formulation when things turn on wrong by incorporating some appropriate features and components in the formulation.

It is important to keep in mind that the stability results are generally derived for the deterministic setting (no unknown w in the formulation of the dynamics). In the case were uncertainties are present, the stability of the nominal design coupled with a reasonable assumption regarding the fact that the uncertainties are bounded, enable to conclude that the closed-loop system is rather safe.

Let us start by the first provable stability result in the context of nonlinear discrete-time systems which has been provided in [7]. The idea in this formulation is to enforce a terminal **equality constraint on the state at the end of the prediction horizon**. This leads to the following formulation of the cost function together with the stability-related terminal constraint:

$$\mathbf{u}^\star = \arg\min_{\mathbf{u}} \; J(\mathbf{u}|x_k) := \sum_{i=0}^{N} \ell(x_{k+i+1}, u_{k+i}) \qquad \text{under} \quad x_{k+N} = 0, \tag{2.47}$$

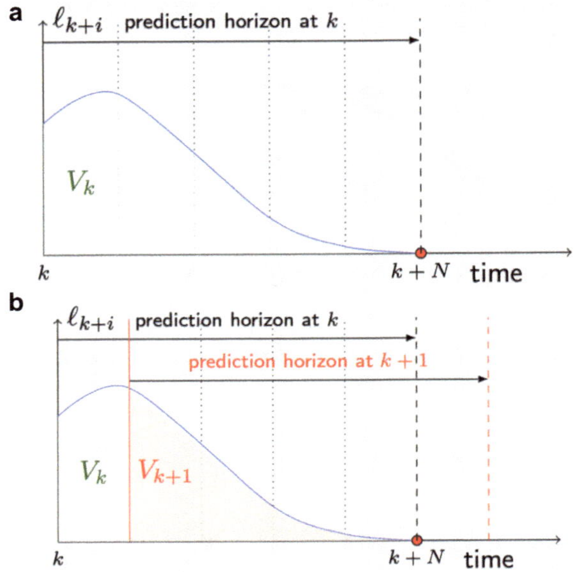

Fig. 2.10 Schematic view of the stability argument for MPC with terminal equality constraints on the state. **a** image of the optimal cost V_k at instant k. **b** image of a candidate sub-optimal cost at instant $k+1$

where ℓ denotes some stage cost that is assumed to be non negative.

Notice that, in all the stability proofs, it is assumed that the targeted desired state is $x_d = 0$ keeping in mind that otherwise, a shift in the coordinates defined by $\xi = x - x_d$ can be considered in order to derive the conditions associated to the non zero targets.

The stability argument is schematically depicted in Fig. 2.10. Here V_k stands for the optimal cost computed at instant k, namely:

$$V_k := J(\mathbf{u}_k^\star(x_k) \mid x_k).$$

This optimal cost is associated to a trajectory that ends precisely at the desired terminal state $x_d = 0$ by virtue of the equality constraint used in the formulation of the optimization problem (2.47).

Now since by assumption, 0 is a steady state of the dynamics, one has $0 = f(0, 0)$ which means that if the system is already at $x = 0$ it can be maintained at that position by applying $u = 0$. This means that when the beginning of the prediction horizon shifts from k to $k+1$, a candidate sequence to solve the new optimization problem might be given by:

$$\tilde{\mathbf{u}}_{k+1} := \begin{bmatrix} u_1^\star(x_k) \\ \vdots \\ u_{N-1}^\star(x_k) \\ 0 \end{bmatrix}, \tag{2.48}$$

in which the first action, namely $u_1^\star(x_k)$ is precisely **the second action in the previously computed optimal sequence** since:

$$\mathbf{u}_k^\star := \begin{bmatrix} u_0^\star(x_k) \\ u_1^\star(x_k) \\ \vdots \\ u_{N-1}^\star(x_k) \end{bmatrix}, \tag{2.49}$$

while the last action is simply $u = 0$ in order to maintain the state at 0 and hence satisfying the terminal equality constraint.

This results in the fact that the candidate sequence $\tilde{\mathbf{u}}_{k+1}$ is admissible (as it meets the terminal constraints) and therefore, the truly optimal solution $\mathbf{u}^\star(x_{k+1})$ can only be better in terms of the associated cost. This can be written as follows:

$$V_{k+1} := J(\mathbf{u}_{k+1}^\star(x_{k+1})\,|x_{k+1}) \leq J(\tilde{\mathbf{u}}_{k+1}|x_{k+1}) = V_k - \ell(x_{k+1}, u_k^\star), \tag{2.50}$$

and since $V_k \geq 0$, it cannot decrease indefinitely by a non vanishing rate (that would be $-\ell(x_{k+1}, u_k^\star)$). Consequently, the last equality leads to:

$$\lim_{k\to\infty} \ell(x_k, u_{k-1}) = 0. \tag{2.51}$$

Therefore, provided that the stage cost ℓ is defined such that the whole state is penalized, the last asymptotic property leads to the convergence of x_k to 0.

It is important to underline that adding the terminal equality constraint on the state $x(k+N) = x_d$ implicitly implies that the whole theory holds **only on the set of initial states that can be steered to the desired state in less than N steps under the control saturation limit.**[29]

Another way to enforce the terminal equality constraint on the state without explicitly add it to the formulation of the OCP is to adopt an infinite horizon.

Indeed, if the current state is inside the N-reachability set, we know that there is a control sequence (of finite length N) that steers the state to the desired value. This candidate sequence sets an upper bound $\bar{J}$ on the possible values of the optimal cost. This makes it impossible for an optimal sequence to remain infinitely long time at a distance ϵ of the target. This is because in this case, after some period of time during which the state remains ϵ-far from the desired state, the cumulated stage cost increases continuously at a rate ϵ and the corresponding cost ultimately goes beyond the upper bound $\bar{J}$. This would contradict the optimality of such a competing solution.

This argument can be exploited [12] to show that there always exists a **sufficiently long prediction horizon that induces the closed-loop stability under MPC without the terminal equality constraints on the state being explicitly required.** Here,

[29] This is sometimes referred to as the N-reachability subset.

the infinitely long prediction horizon plays the same role as the terminal equality constraint used in (2.47).

Obviously for nonlinear systems, the use of very long prediction horizon is rarely a good option due to the associated computational burden. This is the reason why other formulations that intended to **relax the equality constraint into an inclusion constraint** have been worked out by the control community during almost a decade before a seminal paper [4] showed that the *majority* of provably stable MPC formulations are specific instantiations of a general formulation that takes the following form:

$$\mathbf{u}^\star(x_k) = \arg\min_{\mathbf{u}} \left[\Phi(x_{k+N}) + \sum_{i=0}^{N-1} \ell(x_{k+i+1}, u_{k+i}) \right], \tag{2.52a}$$

$$\text{under} \quad x_{k+N} \in \mathcal{X}, \tag{2.52b}$$

which can be proved to induce closed-loop stability provided that the following conditions hold true:

1. There is a state feedback law[30] $\kappa(\cdot)$ that is defined on the terminal set $\mathcal{X}$ which makes $\mathcal{X}$ invariant. This means that if $x \in \mathcal{X}$, then $f(x, \kappa(x)) \in \mathcal{X}$.
2. The function Φ used in the terminal penalty involved in (2.52a) is a control Lyapunov function inside $\mathcal{X}$ under the local control law κ invoked in the previous item, more precisely[31] :

$$\forall x \in \mathcal{X}, \quad \Phi(f(x, \kappa(x)) - \Phi(x) \leq -\ell(x, \kappa(x)). \tag{2.53}$$

3. The targeted state is a steady state that belongs to the terminal set, namely $0 \in \mathcal{X}$, $f(0, 0) = 0$.
4. Some additional regularity assumptions regarding the maps involved in the model and the formulation.

Despite the fact that this formulation is less stringent (from a computational point of view) since $\mathcal{X}$ is typically much wider than $\{0\}$, it remains difficult to find the requested coupled pair $(\Phi, \mathcal{X})$ of terminal penalty and terminal set for general nonlinear systems encountered in real-life applications.

The consequence of this fact it that the above enumerated success conditions for the stability of the closed-loop dynamics are almost never checked or enforced in real-life applications and use-cases and remain exclusively used in academic papers about the stability of MPC schemes.

These limitations justified some non conventional formulations that can be gathered under the label *(terminal constraints)-free formulations*. This family gathers mainly two classes of formulations, namely:

[30] The knowledge of this feedback is not required, only its existence is needed for the proof.

[31] This implies that the value of Φ decreases by a rate faster than $-\ell$ over the closed-loop trajectory induced by the local control law κ.

- formulations based on the contraction property,
- formulation involving monotonically increasing penalty.

These two formulations are briefly discussed hereafter.

Formulation Based on the Contraction

In the contraction-based formulations such as the ones proposed in [13,14], an attempt is made to build a bridge between the traditional explicit analytical Lyapunov design approach where the Lyapunov function decreases at each sampling instant and the ideal MPC formulation where the Lyapunov function is never explicitly used. Bridging this gap is done by allowing for candidate Lyapunov functions that can be forced to decrease *on a multi-step basis*.

More clearly, in **traditional analytical Lyapunov design** which generally applies to unconstrained problems, one tries to find:

- a positive definite map V of the state,
- a state feedback control K, and
- a non-negative function ℓ,

such that the following dissipation inequality holds true over the subset of state space of interest:

$$V\big(f(x, K(x))\big) - V\big(x\big) \leq -\ell(x), \tag{2.54}$$

which is simply the inequality (2.53) invoked earlier inside the terminal set while here one would look to find the ingredient *in the large*. As it has been mentioned earlier, it is the difficulty in addressing this problem that justified the use of MPC.

Now contraction-based MPC tries to come back to this idea with a relaxation that is made possible by the conjunction of two *tricks*:

1. By using a multi-step formulation of the inequality (2.54) in which the map V is required to decrease not necessarily in one step but after a finite number, say N_c, of steps.
   This defines the contraction condition which requires that there always exists a sequence of actions of length N_c and a contraction factor $\gamma \in (0, 1)$ such that:
   $$V(x_{\mathbf{u}}^{(N_c)}(x_k)) \leq \gamma V(x_k), \tag{2.55}$$
   where $x_{\mathbf{u}}^{(N_c)}(x_k)$ is the state reached after N_c steps starting from the state x_k under the control sequence $\mathbf{u}$.
2. By taking N_c *sufficiently* large, there is no need to look for the positive definite map V since it can be proved that **any positive definite function tends to meet the contraction requirement as N_c increases**. This greatly facilitates the instantiation of the underlying MPC formulation.

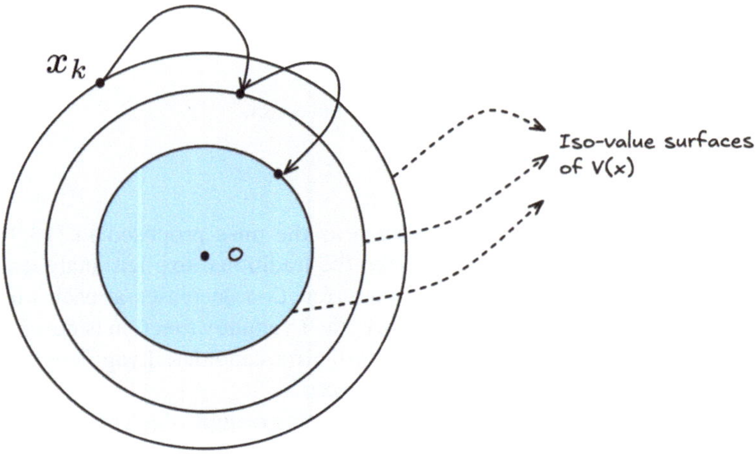

Fig. 2.11 Illustration of the contraction condition: There is always a control sequence of length N_c such that the map $V(x) = \|x\|$ (here) decreases with a contraction factor γ

When the contraction assumption mentioned above holds, a typical formulation of MPC in the deterministic setting can be given by:

$$\mathbf{u}^*(x_k, z_k) \leftarrow \min_{\mathbf{u},q}\Big[z_k \times \sum_{i=1}^{q}[\ell(x_{k+i}, u_{k+i-1})] + \alpha \min_{i=1}^{q} W(x_{k+i})\Big], \tag{2.56}$$

where

- ℓ is the standard stage cost usually used in MPC schemes,
- W is the map which satisfies the contraction assumption, and
- z is an internal state of the controller which is governed by the following dynamics:

$$z_{k+1} = \begin{cases} z_k & \text{if } W(x_k) > z_k, \\ \beta z_k & \textit{otherwise}, \end{cases} \tag{2.57}$$

for some parameter $\beta \in (0, 1)$.

Roughly speaking, the internal state z serves as the next target level for the contraction of W which is decreased by a β rate each time the contraction is achieved (Fig. 2.11).

The Prediction Horizon is a Decision Variable
Notice that in the contractive formulations, **the length of the prediction horizon (denoted by q) is used as a part of the decision variable** (together with the control sequence $\mathbf{u}$) as it can be observed in (2.56).

To summarize, in the contraction-based formulations, there is no terminal set to be reached each time the OCP is formulated and solved, rather, there is an intermediate objective corresponding to the contraction of a storage function. It is the repetitive success of the contraction that guarantees the asymptotic stability of the closed-loop.

Let us now move to the second formulation that is (terminal-constraint)-free.

Formulations with Time-Increasing Penalty.

Another possibility to derive provable stability results without using dedicated terminal constraints in the MPC formulation might be obtained by smoothly implement the effect of either terminal equality constraints or sufficiently long prediction horizon which both emphasize the penalty on the latest stages of the prediction horizon.

Indeed, by applying a monotonically **increasing penalty over the prediction horizon** and provided that the ratio between the terminal penalty and the initial penalty is high enough, a stability proof can be provided.

More precisely, the corresponding formulation takes the following form in the deterministic setting [15,16]:

$$\mathbf{u}^\star \leftarrow \min_{\mathbf{u}} \left[\sum_{i=1}^{N} \left[\frac{i}{N}\right]^m \left(\varphi_m \ell(x_{k+i-1}) + (1-\varphi_m) V(x_{k+i+1})\right) \right], \tag{2.58}$$

where

- φ_m is any positive weight that tends to 0 as $m \rightarrow \infty$,
- V is some positive definite map that can be computed using the linearized model at the targeted state.

This quick survey on the provably stable formulations intended only to give the big picture and not necessarily the details of the proof or an exhaustive list of the technical assumptions that are needed for the proofs to hold.

Obviously, as the reader might have already noticed, the stability analysis is a tricky technical issue in MPC design and assessment. This is the reason why this issue induced hundreds of research papers over a decade or so!

Now beyond the stability issue, one might legitimately raise the following intuitive question:

Is It Worthwhile to Tune the MPC Formulation?

Is the impact of the formulation on the quality of the resulting behavior and/or on the computation time worth the investigation?

The answer is definitively *yes!* although a more elusive one such as *it depends!* is quite tempting!

But nothing is better than a concrete real-life example to answer the above legitimate question.

2.7.3 Example Regarding the Impact of the Formulation

This section gives a real-life example that illustrates the impact of the formulation on the resulting closed-loop performance on one hand and on the underlying computation time of a single instance of the OCP on the other hand.

The example used for the illustration is a controlled **cryogenic refrigerator** which can be shortly defined as follows:

Cryogenic Refrigerator

A cryogenic refrigerator is a technological process that implement a specific thermodynamic cycle in which some amount of heat is extracted from a thermodynamic fluid:

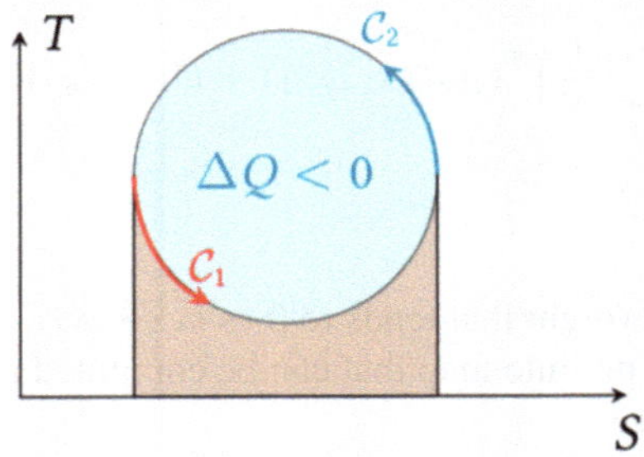

How?

Force a thermodynamic fluid to make a counter-clock cycle in the (S, T)=(Entropy, Temperature) plan.

$$\int dQ = \underbrace{\int_{\mathcal{C}_1} TdS}_{>0} + \underbrace{\int_{\mathcal{C}_2} TdS}_{<<0}$$

This resulting cold fluid can then be used, for instance, to cool-down the supraconducting devices that are used in the physical instruments in order to convey a high current that is used to create high magnetic fields. Such fields are necessary either to accelerate particles (like in the CERN facility) or to create plasma (like in the Tokamak used in nuclear fusion facilities an illustration of which can be examined at https://www.euro-fusion.org).

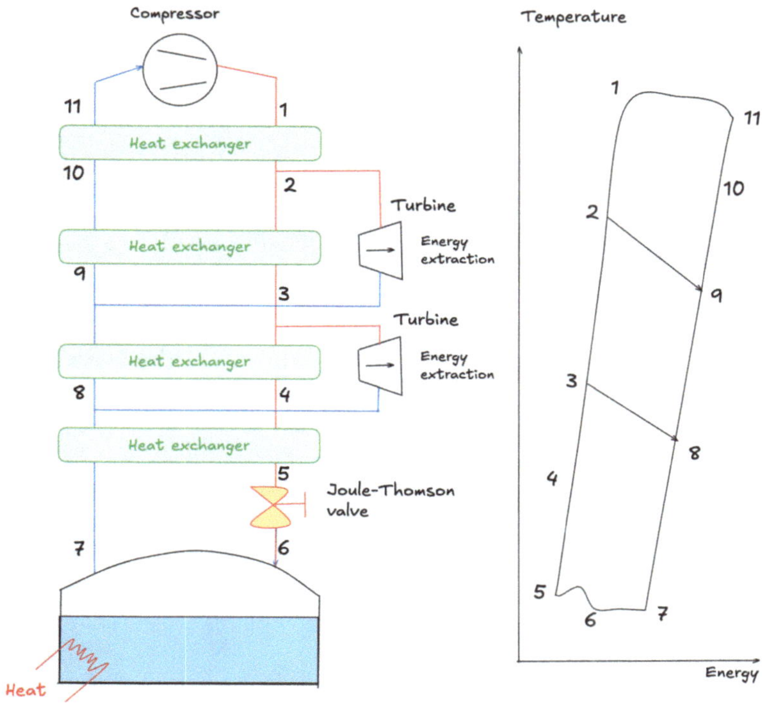

Fig. 2.12 Schematic view of the 400W experimental cryogenic refrigerator at CEA-Grenoble, France. Adapted from [17] [Bonne et al. Experimental investigation of control updating period monitoring in industrial PLC-based fast MPC: Application to the constrained control of a cryogenic refrigerator. in *Control Theory and Technology 15(2), 92–108, 2017.* copyright Springer-Nature 2017]; used with permission

While the figure shown in the above definition represents ideal thermodynamic refrigerating cycle, real technological realizations of this cycle show more *accidental* trajectories in the entropy-temperature plan (S, T). They nevertheless keep the essential task unchanged by implementing an anti-clockwise cycle corresponding to an overall negative balance of heat exchange (heat extraction) viewed by the thermodynamic fluid. Helium is commonly used as the thermodynamic fluid in refrigerators.

An example of such a technological realization is represented in Fig. 2.12 which shows a schematic view of the 400W-experimental facility which is available at the atomic and renewable energy commission (CEA)-Grenoble.

One can see the multiple heat-exchangers stages as well as the turbine and the compressor. On the right-hand side of this figure, the resulting anti-clockwise thermodynamic cycle is shown realizing the heat extraction task.

In order to get an idea of the real size of the process, Fig. 2.13 shows pictures representing the compressor (top image) and the so-called *cold box* representing the helium bath. Note that this two parts of the system are located in two different buildings!

Fig. 2.13 Images of two parts of the cryogenic refrigerator of the CEA-SBT Grenoble showing the compressor **a** and the helium bath **b** in which liquid helium regulated around 4K is maintained to serve in cooling down the supra-conductive devices. [Copyright CEA-INAC-SBT]; used with permission

From a control point of view, the objective is to regulate some key variables such as some temperatures and the level of liquid helium in the bath while respecting some safety constraints such as the minimum bound on the turbine's temperature that should be respected in order to avoid the formation of solid droplets that might definitively damage this high cost device.

This regulation task should be achieved in spite of unmeasured and unpredictable disturbances which are represented by the **heat pulses** induced by the electric current cycles that are not announced in advance and whose characteristics are not known to the control algorithm.

A typical realization of the heat pulsed profile (the disturbance affecting the controlled system) is shown in Fig. 2.14.

Typically, disturbance profiles are cyclic but neither the period of the cycles nor the details and the amplitude of the curves are known in advance nor are they easily reconstructible although some dedicated observers can be developed and used [18]. Notice however that this is not the approach adopted in the following results that simply aim at showing the impact of the choices of the MPC formulations on the regulation's quality and the corresponding computation time.

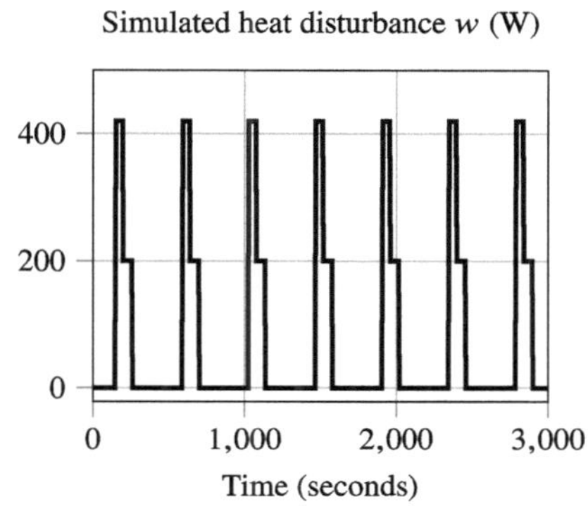

Fig. 2.14 A typical disturbance profile of the heat pulses that disturb the cryogenic refrigerator

A linear dynamic model can be obtained by linearizing the knowledge-based equations around the operation point of interest. This leads to a 24-dimensional model involving 3 control inputs and 3 regulated outputs (the details of which are not mandatory in this presentation):

$$x_{k+1} = f(x_k, u_k, w_k), \tag{2.59}$$

where $x \in \mathbb{R}^{24}$, $u \in \mathbb{R}^3$ and $w \in \mathbb{R}$.

Now that the system and the control problem are presented, let us compare three different MPC controllers that are defined with three different cost functions used in the formulation of three different deterministic MPC laws. More precisely, we consider the following formulations:

1. The first controller implements the following intuitive cost function defined by:

$$J(\mathbf{u}|x_k) := \sum_{i=0}^{N}\Big[\|y_{k+i+1}\|^2_{Q_y} + \epsilon\|x_{k+i+1}\|^2\Big]. \tag{2.60}$$

   This cost function involves standard uniform (in time) penalty but still add a **full state penalty** as suggested in the previous rules of thumb since such penalty enhances the stability of the resulting closed-loop. The resulting closed-loop behavior depicted in Fig. 2.15 shows that the average (over the periods) level of the liquid Helium in the bath is continuously decreasing which means that **the refrigerator cannot afford this disturbance for a long period of operation**.
2. The second formulation acknowledges that the previous formulation is not stabilizing the process and hence add a stabilizing item in the form of a **final penalty on the terminal state**, namely:

$$J(\mathbf{u}|x_k) := \epsilon_f\|x_{i+N}\|^2 + \sum_{i=0}^{N}\Big[\|y_{k+i+1}\|^2_{Q_y} + \epsilon\|x_{k+i+1}\|^2\Big]. \tag{2.61}$$

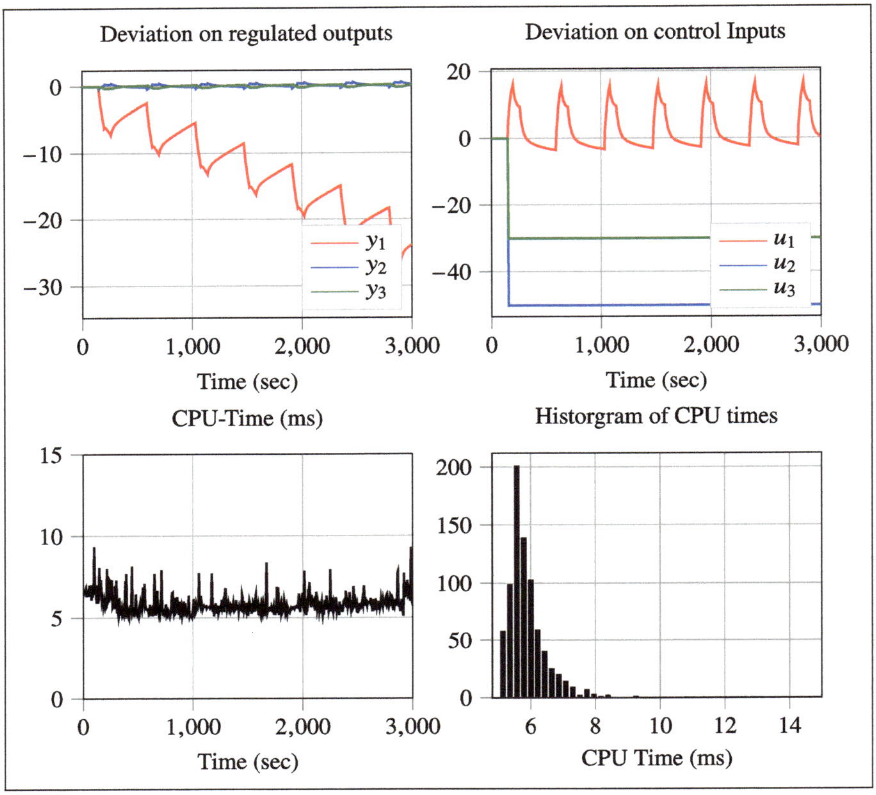

Fig. 2.15 Impact of MPC formulation: Closed-loop associated to the MPC formulation given by (2.60)

The resulting closed-loop behavior is shown in Fig. 2.16 where it clearly appears that the terminal penalty term does indeed induce better closed-loop stability as the helium level is stabilized in average making the situation sustainable in the long term for the considered disturbing pulses.
Notice however that the computation time associated to this second formulation is almost twice the one associated to the first unstable formulation.

3. The third formulation implements the monotonically increasing penalty over time in the stage cost without the terminal penalty introduced in the second formulation. More precisely, the same stage cost used in the first formulation (2.60) is used except that, instead of a uniform penalty over time, an exponentially increasing weight is implemented ($m = 2$ is used in the simulation):

$$J(\mathbf{u}|x_k) := \sum_{i=0}^{N}\Big[\frac{i}{N}\Big]^m\Big[\|y_{k+i+1}\|^2_{Q_y} + \epsilon\|x_{k+i+1}\|^2\Big]. \tag{2.62}$$

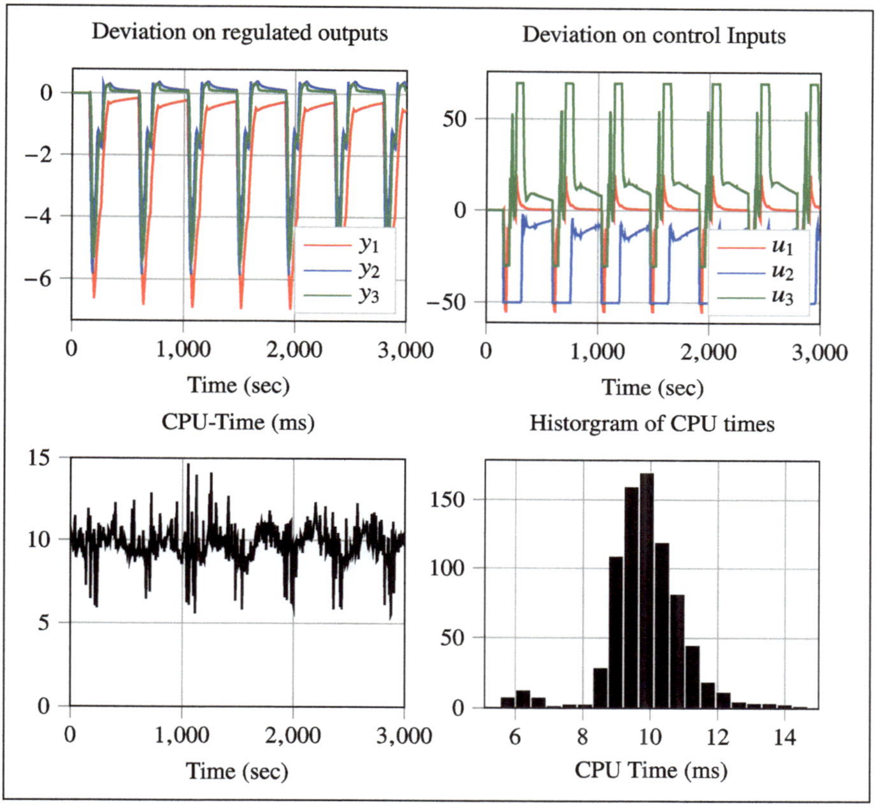

Fig. 2.16 Impact of MPC formulation: Closed-loop associated to the MPC formulation given by (2.61)

The behavior of the resulting closed-loop, depicted in Fig. 2.17 shows that not only the quality of the regulation is improved as it can be seen by a careful examination of the tracking error on the regulated outputs, but strangely enough, this is obtained with shorter computation times as it can be observed on the statistics regarding the computation time.

It is important to underline that this comparison holds only for the specific example used in the illustration. There is no proof that the comparison on both issues (computation time and the quality of the regulation) would be systematically in favor of such or such formulation.

The message of this section is the following:

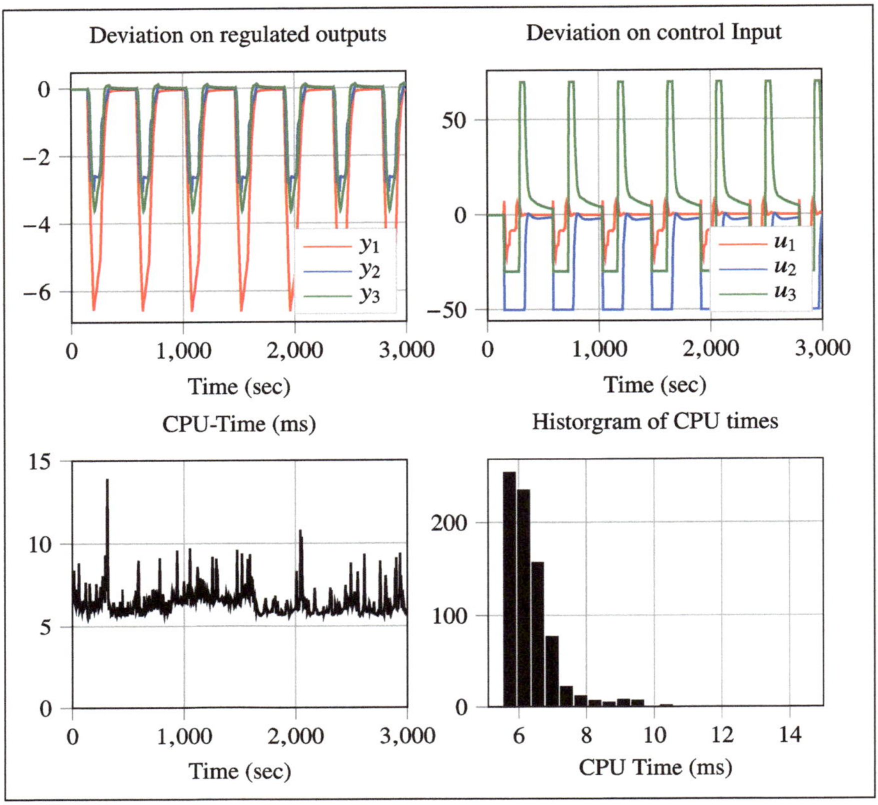

Fig. 2.17 Impact of MPC formulation: Closed-loop associated to the MPC formulation given by (2.62)

The Relevance of Tuning the MPC Formulation
The choice of the **formulation might generally have a strong impact** on the results in the wider sense and this is the reason why, before adopting the formulation to be implemented, some trials and essays are worth doing by simulating the different resulting closed-loop.

For the easiness of comparison, Fig. 2.18 shows the results for the two stabilizing formulations side-by-side for an easier comparison.

The previous presentation of the MPC principle is rather idealistic. Indeed, it assumes that at instant k, given the current state x_k, the optimal control sequence $\mathbf{u}^\star(x_k)$ can be made instantaneously available to be applied over the sampling period $[k, k+1]$. This implicitly assumes that the computation time, needed to solve the underlying OCP is negligible.

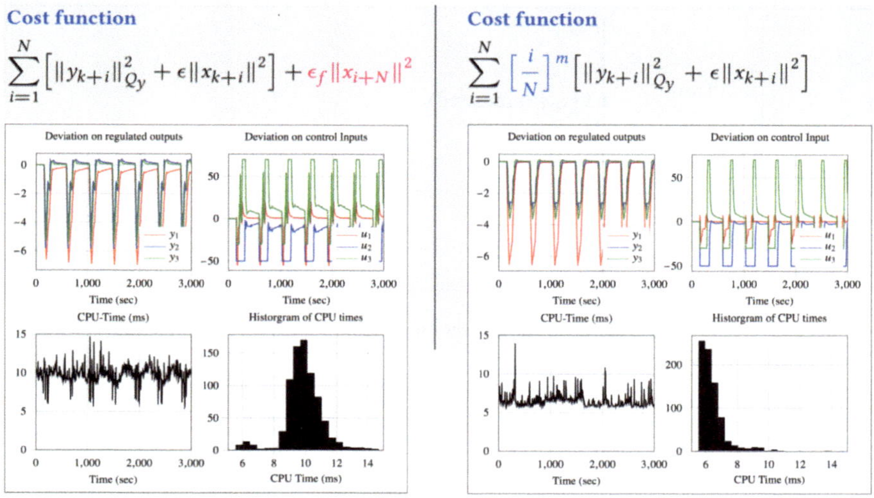

Fig. 2.18 Closed-loop behavior of the two stabilizing formulations

Given the complexity of the associated nonlinear constrained OCPs and the generally short sampling periods adopted for many real-life technological systems, this assumption regarding the computation time needed to deliver the control $u^\star(x_k)$ being lower than the sampling period, might be over-optimistic.

That is the reason why, explicit handling of the real-time issue associated to the computation time of the underlying OCPs needs to be undertaken.

This is the object of the next section.

2.7.4 Real-Time MPC Setting

Notice first of all that the discussion of this section assumes that the state is fully measured. For the reconstruction of the state via state observers, see Sect. 2.10 hereafter.

The first step towards a concrete real-time setting is to start by acknowledging that solving the OCPs (2.26) or (2.33b) requires a non negligible amount of time that cannot generally be asssumed to be a fraction of the sampling period.

This time is denoted hereafter by[32] $\tau_u > 0$.

This means that if one needs the optimal first action $u_0^\star(x_k)$ to be available at instant t_k, its computation has to start before $t_{k-1} = t_k - \tau_u$. But since the computation starts before the state of interest (x_k) is known, the computation of the optimal sequence has to be done **based on some predicted value $\tilde{x}_k$ of this true yet unavailable state**. This is shown in Fig. 2.19.

[32] u-for updating so that τ_u becomes the control updating period.

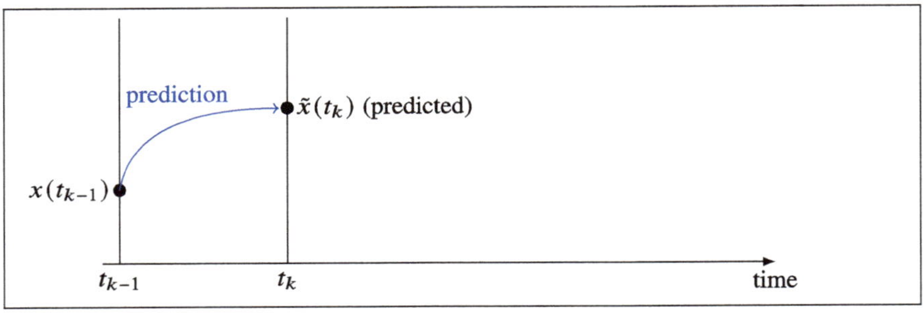

Fig. 2.19 In order to solve the OCP for the yet unavailable state $x(t_k)$, a prediction $\tilde{x}(t_k)$ of this future state is needed at instant $t_{k-1} = t_k - \tau_u$

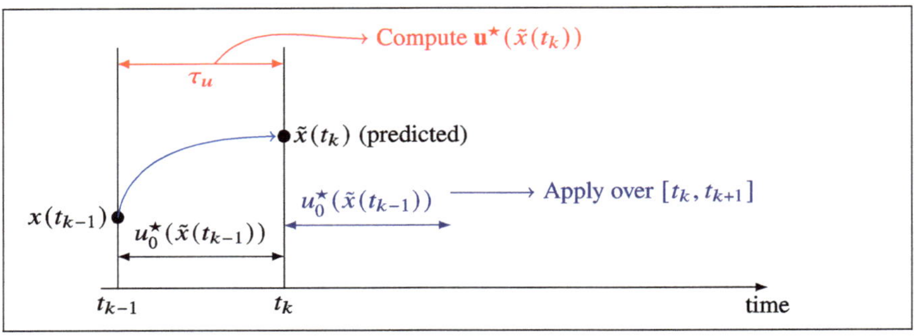

Fig. 2.20 During the control updating period $\tau_u = t_k - t_{k-1}$, the algorithm computes the optimal control sequence by solving the OCP parameterized by the predicted future state $\tilde{x}(t_k)$. During the computation, the previously computed control based on the predicted state $\tilde{x}(t_{k-1})$ is applied to the system

The prediction $\tilde{x}(t_k)$ of the next state is obtained using the simulation[33] of the model of the system **under the control $u_0^\star(\tilde{x}(t_{k-1}))$ that has been computed before in order to be applied over the computation period** $[t_{k-1}, t_k]$ (Fig. 2.20)**.**

Notice that in this presentation it is assumed that the first control $u_0^\star(\tilde{x}(t_k))$, which is based on the predicted value $\tilde{x}(t_k)$ of the state, is applied to the system.

In some real-time MPC settings, the solver delivers also the sensitivity of the solution to the error on the prediction of the state, namely[34]:

$$\frac{\partial u_0^\star}{\partial x}(\tilde{x}(t_k)).$$

[33] In case of uncertain systems, the **expected** value of $x(t_k)$ has to be computed, unless one uses the nominal value of the uncertain parameter w to perform the prediction task.

[34] Under some obvious regularity assumptions that are skipped here.

When this is the case, one can correct the first input of the presumably optimal sequence according to:

$$u_0^\star(x(t_k)) \approx \overbrace{u_0^\star(\tilde{x}(t_k))}^{\text{Prediction-based}} + \overbrace{\left[\frac{\partial u_0^\star}{\partial x}(\tilde{x}(t_k))\right] \cdot (x(t_k) - \tilde{x}(t_k))}^{\text{Measurement-based correction}}, \tag{2.63}$$

as soon as the *true* value of the state at instant t_k, namely $x(t_k)$, becomes available.[35]

Notice that the correction defined by (2.63) requires a simple matrix multiplication and its computation can be considered as instantaneous inducing almost no delay between the instant where the value of $x(t_k)$ is made available and the instant where the corrected value of the control is available for use.

In the literature dedicated to the real-time MPC issue [19], the previous two steps are referred to respectively as:

- the **preparation step** which is dedicated to solving the OCP based on the predicted value of the state $\tilde{x}(t_k)$, and
- the **feedback step** which is dedicated to correcting the previously computed control by taking into account the true value $x(t_k)$ of the state.

Before closing this brief section on the real-time aspects of MPC implementation, it is worth mentioning that the above formulation still assumes that the optimal solution $\mathbf{u}^\star(\tilde{x}(t_k))$ is possible to achieve within the updating computation time τ_u. Here again, this might be an over optimistic assumption. This is because during the computation time τ_u the system is left in open loop since a pre-computed control $u_0^\star(\tilde{x}(t_{k-1}))$ is applied without feedback during τ_u time units.

The problem is that for any real system, there is a maximum amount of time during which the system can be left without the slightest feedback that reacts to the collected measurements from the available sensors on the system.

Let us denote this maximum duration τ_u^{max}.

Based on this definition, any choice of the control updating period τ_u must be such that:

$$\tau_u \le \tau_u^{max}.$$

This upper bound depends only on the characteristic time of the system and its environment. and the way new set-points and disturbances might come into picture to make the previous computation irrelevant.

Now assuming that the ingredients of the implementation, namely:

- the formulated OCP,
- the hardware used to implement the solver, and
- the complexity of the optimization algorithm,

[35] Recall that the state is assumed here to be fully measured.

are such that the time needed to solve the underlying OCP is greater than τ_u^{max}, then the optimal solution cannot be computed in the allowable control updating period τ_u.

Under such circumstances, only a sub-optimal solution can be obtained which is provided by the finite number of iterations q that is defined by:

$$q = \left\lfloor \frac{\tau_u}{\tau_c} \right\rfloor := \text{integer part of } \tau_u/\tau_c, \tag{2.64}$$

where τ_c denotes the time needed, on the used computational device, to perform a single iteration of the algorithm being used to solve the underlying OCP. More precisely:

cpu for a Single Iteration τ_c
The notation τ_c denotes the time needed to perform a single iteration of the algorithm involved in the solution of the instances of the OCPs used in the MPC design. This time depends not only on the algorithm itself but also on the computing device (hardware).

When the available number of iterations q is not sufficient to completely solve the OCP, one is obliged to deliver the sub-optimal computed solution, denoted by[36] $\mathbf{u}(t_k)$ where

$$\mathbf{u}(t_k) = \mathcal{S}^{(q)}\Big(\mathbf{u}(t_{k-1}) \mid \tilde{x}(t_k)\Big) \quad ; \quad q = \left\lfloor \frac{\tau_u}{\tau_c} \right\rfloor, \tag{2.65}$$

where the right hand side of the equation denotes the solution obtained after q iterations of the used solver $\mathcal{S}$, starting from the previous update $\mathbf{u}(t_{k-1})$ and based on the new parameter $\tilde{x}(t_k)$ of the optimization problem.[37]

Now since $\tilde{x}(t_k)$ is obtained by simulating the model of the system, starting from the *true* previous state $x(t_{k-1})$ and applying the first control action in the previously computed sub-optimal control sequence $\mathbf{u}(t_{k-1})$ and the realization of the uncertainty $w(t_{k-1})$, it comes out that

$$\tilde{x}(t_k) := \text{Function}\Big(x(t_{k-1}), \mathbf{u}(t_{k-1}), w(t_{k-1})\Big),$$

which means that Eq. (2.65) can be re-written by overloading the notation as follows:

$$\mathbf{u}(t_k) = \mathcal{S}^{(q)}\Big(\mathbf{u}(t_{k-1}) \mid x(t_{k-1}), w(t_{k-1})\Big) \quad ; \quad q = \left\lfloor \frac{\tau_u}{\tau_c} \right\rfloor, \tag{2.66}$$

36 The $\star$-optimality indicator is dropped since the result is no more optimal!

37 Note that since one does not have the optimal solution of the OCP defined by the state $\tilde{x}(t_k)$, it is no more possible to write $\mathbf{u}^\star(\tilde{x}(t_k))$ since the sub-optimal obtained value after q iterations is no more a function of the state $\tilde{x}(t_k)$ but the consequence of a dynamic truncated iteration process that depends on its previous update $\mathbf{u}(t_{k-1})$.

or equivalently:

$$\mathbf{u}(t_{k+1}) = \mathcal{S}^{(q)}\big(\mathbf{u}(t_k) \mid x(t_k), w(t_k)\big) \quad ; \quad q = \left\lfloor \frac{\tau_u}{\tau_c} \right\rfloor. \tag{2.67}$$

On the other hand, by rewriting the dynamical state-space equation:

$$x(t_{k+1}) = F(x(t_k), u(t_k), w(t_k)), \tag{2.68}$$

it becomes obvious from (2.67) and (2.68) that in the case where the updating period τ_u is too short to solve the optimization problem, one should consider an *extended system* in which the extended state at instant t_k is given by:

$$z(t_k) := \begin{bmatrix} x(t_k) \\ \mathbf{u}(t_k) \end{bmatrix}, \tag{2.69}$$

and the extended state evolution is given by:

$$\begin{bmatrix} x(t_{k+1}) \\ \mathbf{u}(t_{k+1}) \end{bmatrix} = \begin{bmatrix} F(x(t_k), u(t_k), w(t_k)) \\ \mathcal{S}^{(q)}\big(\mathbf{u}(t_k) \mid x(t_k), w(t_k)\big) \end{bmatrix} \quad ; \quad q = \left\lfloor \frac{\tau_u}{\tau_c} \right\rfloor, \tag{2.70}$$

which can be written in a more compact form as follows[38]:

Extended State Dynamics and Output
The system with varied τ_u can be viewed as an extended system governed by the following dynamics:

$$z(t_{k+1}) = F_{ext}(z(t_k), \tau_u, w(t_k)) \tag{2.71a}$$
$$y(t_k) = h(z(t_k)) = J(\mathbf{u}(t_k) \mid x(t_k)), \tag{2.71b}$$

where $z(t_k) := (x(t_k), \mathbf{u}(t_k))$.

This enables to view the problem of choosing the *appropriate* control updating period as a **control problem** in which (see Fig. 2.21):

- the state is given by z,
- the measured *regulated* output y is given by the cost function J,
- the disturbance is given by w, and
- **the decision variable is given by** τ_u.

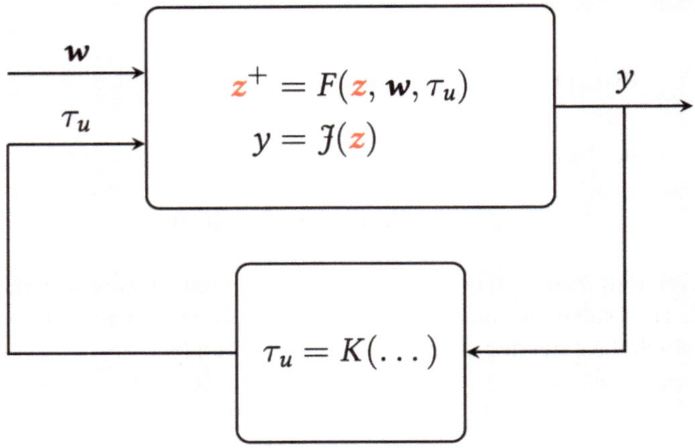

Fig. 2.21 Viewing the control updating period as a control variable for an extended uncertain system with state $z = (x, \mathbf{u})$ and a measured regulated variable represented by the cost function to be minimized J

In other words:

The Control Updating Period as a Control Input
Choosing dynamically the control updating period can be viewed as a control problem in which, the `control updating period` τ_u is the manipulated variable in a control problem in which the state is given by

$$z = (x, \mathbf{u}),$$

while the regulated output is the cost function.

Notice however that this results in a control problem with measured output but without a really known evolution model since the map $\mathcal{S}^{(q)}$ involved in (2.70) is not available in explicit, nor even in a black-box programmable form.

Nevertheless, viewing the rational choice of the control updating period as the outcome of a control problem's solution enables efficient heuristics [17,20] to be derived that exploit the previous measurements of the cost function in order to derive an adaptation mechanism to the number of iterations to perform at a given instant.

Typical results performed by a heuristic enabling adaptive control updating period on a triple integrator [20] based on the above described paradigm are given in Figs. 2.22 and 2.23.

[38] Recall that $u(t_k)$ is the first action in the sub-optimal control sequence $\mathbf{u}(t_k)$.

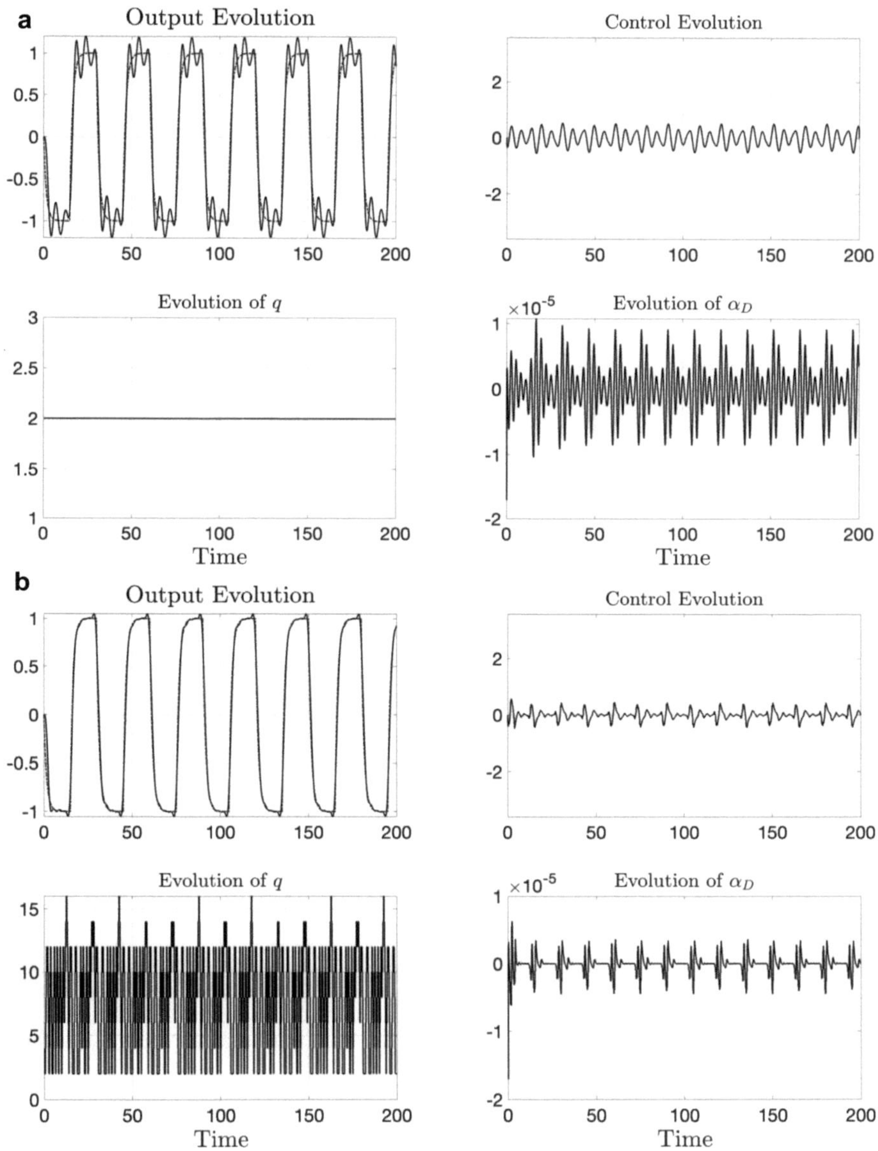

Fig. 2.22 Example of results performed by a heuristic enabling adaptive control updating period on a triple integrator. **a** Results with constant τ_u enabling $q = 2$ iterations of the fast gradient. **b** Results with active adaptation mechanism starting from $q = 2$

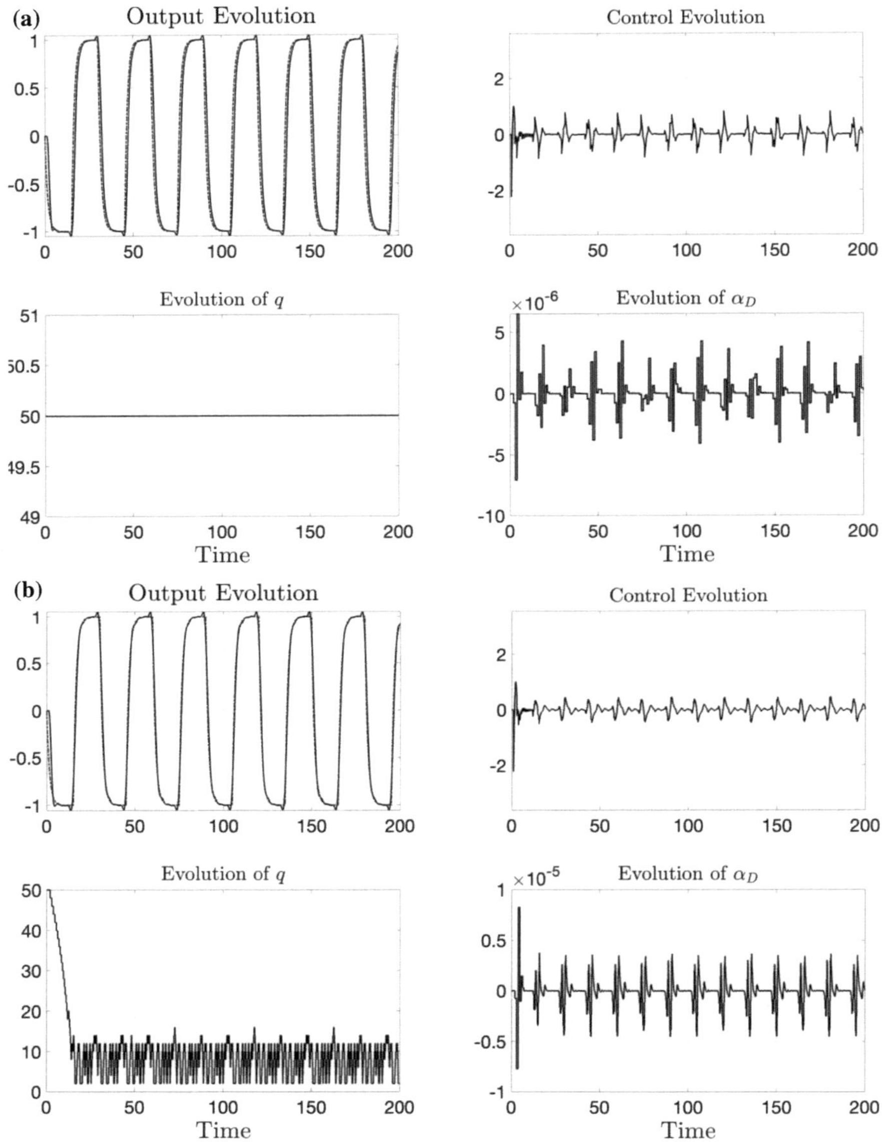

Fig. 2.23 Example of results performed by a heuristic enabling adaptive control updating period on a triple integrator. **a**: Results with constant τ_u enabling $q = 50$ iterations of the fast gradient. **b** Results with active adaptation mechanism starting from $q = 50$

By the time this book is written, the problem of real-time implementation of MPC feedback for general nonlinear systems is an active and very rich topics of investigation with many new results appearing in a monthly if not weekly basis.

The objective of the present section is simply to give the reader a hint regarding the issues being involved in this rather technical and problem-dependent issue.

It is also important to notice that the control updating period is not the only implementation parameter that can be investigated. Many other parameters show high impact on the computation time and hence the real-time implementability of the resulting MPC feedback. This is investigated in a deeper level in Chap. 6.

Now in spite of the huge progress in dealing with the real-time computation issue when using MPC-related implementation on real-life nonlinear systems, it cannot be excluded that under specific circumstances, the use of MPC and in particular, stochastic MPC can be impossible given the characteristic time of the system under interest.

In this case, and provided that the corresponding dimension of the problem (number of state and control components) remain moderate, it is possible to experiment some feedback design alternatives, in particular, the one that is based on the SDP principle might be of some help in some specific circumstances.

This is discussed in the following section.

2.8 Stochastic Dynamic Programming (Q-Learning)

A book dedicated to control under uncertainties can hardly avoid mentioning SDP as it is a quite popular control design methodology that fits the needs in some *specific circumstances*.

The limitation of the standard formulation of SDP stems from the non scalability of the associated solving algorithms when the dimension of the state/control space exceeds moderate values.

Nevertheless, the principle behind the approach and the associated algorithms are worth keeping in mind within the set of paradigms and tools that a good control designer should think off when facing a control problem in the presence of uncertainties.

2.8.1 Problem Statement

SDP is invoked when dealing with an uncertain dynamical system that is governed by a discrete-time state-space equation of the form:

$$x^+ = f(x, u, w), \tag{2.72}$$

with an unknown vector of parameters w for which we dispose of relevant random sampling procedure implementing an underlying statistics $\mathcal{W}$ as mentioned earlier in Sect. 2.6.2.

SDP aims at deriving an **explicit state feedback** that optimally solves, at each instant k, the following discounted cost-based infinite horizon stochastic OCP, expressed in the single-shooting form as follows:

$$\min_{\mathbf{u}\in\mathbb{U}^\infty} \quad (\mu+\alpha\sigma)\Big[\sum_{i=1}^{\infty}\gamma^i\ell(x_{k+i},u_{k+i-1},\cdot)\Big], \tag{2.73}$$

where

- $\mu[]$ and $\sigma[]$ stand for the expectation and the standard deviation of their arguments given the underlying statistics $\mathcal{W}$ on the uncertainty vector w,
- $\alpha > 0$ is the penalty on the standard deviation, and
- $\gamma \in (0, 1]$ is a discount factor that enables infinite horizon problems to be considered with finite cost.

About Penalizing the Standard Deviation

In many works, the coefficient

$$\alpha = 0,$$

is used when economic cost functions are considered where *losses might be compensated by gains*. This is the case, for instance, when the objective is to maximize the profit or minimize some cost.

This is not the appropriate view to adopt **when risk is involved** as in the context of cancer therapy for instance. Indeed, in such cases, it is not only the expectation that matters but also the probability of occurrence of quite bad scenarios. These have to be explicitly taken into account in the cost function through non vanishing penalty α on the standard deviation of the *augmented* cost.[39]

A complete example that underlines the relevance of using $\alpha > 0$ is provided in Chap. 17 that explicitly handles the problem of drug's injection rationalization in the context of combined therapy of cancer.

Notice that in this formulation, it is assumed that all the constraints other than the saturation constraints on the control input components are formulated as soft constraints by using appropriate exact penalty terms. More precisely, a constraint of the form

$$c(x,u) \leq 0,$$

induces an additive term in the stage cost ℓ in (2.73) of the form:

$$\rho\max\Big\{0, c(x_{k+i},u_{k+i-1})\Big\},$$

[39] The cost might be augmented to include penalties on the violation of constraints.

with some high weight $\rho > 0$.

Obviously, by taking quite large but still finite prediction horizons in the above formulation, one might put the problem in the framework of stochastic MPC discussed in the previous sections. However, the on-line computation of the statistical properties μ and σ might be challenging leaving aside the optimization problem itself.

In order to avoid this computational issue, the SDP approach proposes to perform the computation off-line to come out with an explicit feedback which is ready to use on-line while representing a good approximation of the solution of the above quite ambitious formulation.

This is done by examining a point-wise characterization of the optimal solution of the above problem. This characterization takes the form of a functional equation in which the unknown is precisely the the function that represents the optimal cost. The remaining task is to derive an approximate *tractable* solution to the associated functional equation.

This is more detailed in the following section.

2.8.2 The Stochastic Hamilton–Jacobi–Bellman Equation

SDP methods are based on the fact that the state feedback that solves (2.73) can be obtained through the following two steps:

1. First solve **off-line** the following functional equation in which, the unknown is the map $Q : \mathbb{R}^{n_x} \times \mathbb{R}^{n_u} \rightarrow \mathbb{R}$:

$$Q(x, u) = \ell(x, u) + \gamma \min_{v \in \mathbb{U}} \Big[(\mu + \alpha\sigma) \Big[Q(f(x, u, \cdot), v) \Big] \Big], \tag{2.74}$$

   which can be viewed as a stochastic version of the Bellman principle.
   Notice that the above equation should be satisfied at all pairs (x, u) which makes it a functional equation in the unknown Q (as a function) that appears in both side of (2.74). This underlines the fact that (2.74) is not a definition of $Q(x, u)$, rather it is a condition on Q.
2. Once the function Q is made available upon solving, **off-line**, the functional equation (2.74), the state feedback can be obtained by solving **on-line** the following optimization problem:

$$u_k = K(x_k) := \min_{u \in \mathbb{U}} Q(x_k, u), \tag{2.75}$$

   which is a deterministic **low-dimensional** optimization problem that is far less complicated to solve than (2.74).

Notice that since the map Q appears in both sides of (2.74), this suggests the use of a fixed-point iteration as it is shown in the next section.[40]

2.8.3 The Fixed-Point Iteration

The function Q involved in the HJB equation (2.74) is called the **value map**. This map has an argument (x, u) of dimension $n_x + n_u$. It is important to notice that the HJB (2.74) shows a **fixed-point** structure that might be written in a more condensed way as follows:

$$Q = \text{HJB}(Q). \tag{2.76}$$

Such form suggests to use the fixed-point iteration. Indeed, if any initial guess, say $Q^{(i)}$, is available as a map defined on $\mathbb{R}^{n_x} \times \mathbb{R}^{n_u}$, the r.h.s of (2.76) fully defines another map, say $Q^{(i+1)}$ on $\mathbb{R}^{n_x} \times \mathbb{R}^{n_u}$, meaning that one can formally define a fixed-point iteration of the form:

$$Q^{(i+1)} = \text{HJB}(Q^{(i)}). \tag{2.77}$$

The word *formally* is crucial here since indeed, the last equation is a functional equation that cannot be implemented in an exact way since this would mean that the equality should be computed for all pairs (x, u) which is obviously impossible.

In order to concretely implement the fixed point iteration (2.77), a finite-dimensional parameterization of Q is first chosen such as:

$$\hat{Q}(z|q) \quad ; \quad z := (x, u), \tag{2.78}$$

where $q \in \mathbb{R}^{n_q}$ is a finite dimensional parameter[41] which totally defines the approximated version of Q.

By so-doing, having a current guess $q^{(i)}$ of the parameter q, one obtains the current value of $\hat{Q}(z|q^{(i)})$ at any point in the z-space. Injecting this in the r.h.s of the HJB enables to derive the new cloud of values at any grid of values $\mathcal{Z}$ of z, namely

$$\mathcal{Q}(q^{(i)}) := \left\{\hat{Q}(z|q^{(i)})\right\}_{z\in\mathcal{Z}}. \tag{2.79}$$

This cloud can then be used to update the value of q by solving the following Machine Learning (ML) fitting problem defined by:

$$q^{(i+1)} \leftarrow \text{Fit}\Big(\underbrace{\mathcal{Z}}_{\texttt{features}}, \underbrace{\mathcal{Q}(q^{(i)})}_{\texttt{label}}\Big), \tag{2.80}$$

[40] This is not the only option although it is the one adopted in this book as far as SDP is concerned.

[41] One possible instantiation of q is to start by choosing a grid of points in the z-space, say $\mathcal{Z} := \{z^{(i)}\}_{i=1}^{n_q}$ and to consider the parameter q to be the vector of values of the candidate value-function Q at the grid-points, namely $q := \{Q(z^{(i)}\}_{i=1}^{n_q}$.

which obviously defines a tractable and *concrete* (no more formal) fixed-point iteration **in the unknown vector** q for a given choice of the supporting grid of points in the z.

! What Does Convergence of the Fixed-Point (FP) Iteration Mean?
The price to pay in order to get tractable and concrete implementation of the fixed-point iteration is that the convergence of the finite-dimensional fixed point iteration (2.80) concerns the approximated version $\hat{Q}$ of the value function. The distance between $\hat{Q}$ and the *exact* unknown Q depends on the density of the grid of points $\mathcal{Z}$ and the structure that has been chosen for the underlying approximating map $\hat{Q}(z|q)$.

The underlying argument is the following: Assuming that the fixed-point iteration converges, the distance between $\hat{Q}$ and Q is expected to converge to zero as the number of samples used in the grid, namely n_q, tends to infinity (assuming uniform distribution over the compact set of states of interest).

This explains why this approach is limited to problems with moderate values of $n_z = n_x + n_u$.

In other words, the convergence of the fixed-point iteration does not mean that the approximated map so-obtained, namely

$$\hat{Q}(\cdot\,|Q^{(\infty)}),$$

is a *good approximation* of the true solution (if any) of the stochastic HJB equation (2.74).

The quality of the state feedback that is based on the approximate solution $\hat{Q}$ upon convergence of the fixed-point iteration should be investigated separately. For instance, the *Probabilistic Certification* framework presented in the remainder of this chapter (see Sect. 2.11 hereafter) can be used for this purpose.

This being said, even if the convergence of the fixed-point does not guarantee the quality of the resulting solution, this convergence is a necessary condition to proceed and it is not guaranteed to take place.

The next section gives a possible set of sufficient conditions that guarantee the convergence of the fixed-point iteration.

2.8.4 Convergence of the Fixed-Point Iteration

The conditions for the convergence of the fixed-point iterations that are given in the literature are generally sufficient conditions which are not always necessary but this is a quite common feature when it comes to produce provable results.

An example of convergence results for the above fixed-point iteration on the parameters of the value function is given in [21]. These conditions require the following assumptions and definitions to be stated:

Conditions on the Dynamics and Uncertainties

Assumption 2.1 (**Continuity**) The map f describing the one-step ahead discrete-time state evolution (2.72) is **continuous** in its arguments.

Assumption 2.2 (**Finitely supported uncertainties**) There exists a compact set $\mathbb{W}$ to which belongs all possible realizations of the uncertainty vector w.

Notice that the second assumption seems to be a quite reasonable assumption when physically meaningful parameters uncertainties are involved. Still, this is rigorously not the case for Gaussian distributions for instance for which any arbitrarily high values are rigorously possible.

In order to formally state the convergence result, the following definition is needed:

Bounded Excursion Maps

Definition 2.1 A scalar map G defined on $\mathbb{X} \times \mathbb{U}$ is said to show B-bounded excursions if the following inequality holds true:

$$|G(f(x, u, w), v) - \mu[G(f(x, u, \cdot), v)]| \leq B, \tag{2.81}$$

for all $(x, u, v, w) \in \mathbb{X} \times \mathbb{U}^2 \times \mathbb{W}$ where $\mathbb{W}$ is the bounded set referred to in Assumption 2.2 above.

This definition simply states that the excursion of the realizations

$$G(f(x, u, w), v)),$$

away from the expected value

$$\mu[G(f(x, u, \cdot), v)],$$

is uniformly bounded by B over all possible values of the triplet (x, u, v).

It goes without saying that, under Assumption 2.1, such a bound B exists as soon as the map G is continuous since all the involved bounding sets are compact.

Using the above definitions and assumptions, the following result can be proved [21]:

Convergence of the Fixed-Point Iteration

Lemma 2.1 *Under Assumptions* 2.1 *and* 2.2, *if the fixed-point map* `Fit` *involved in* (2.80) *produces only* **continuous** *maps* $\hat{Q}^{(i)}$ *with B-bounded excursions for some* $B > 0$ *then the fixed-point iteration converges on*

$\mathbb{Z} := \mathbb{X} \times \mathbb{U}$ provided that the penalty $\alpha > 0$ on the variance term in (2.74) satisfies:

$$\alpha < \frac{1-\gamma}{2\gamma B}, \tag{2.82}$$

and provided that the sampling period used in the derivation of the discrete-time dynamics (2.72) is sufficiently small.

From the above Lemma it comes clearly that in the absence of a penalty on the standard deviation ($\alpha = 0$), the convergence of the fixed-point iteration is guaranteed provided that the other conditions of Lemma 2.1 are satisfied.[42]

It is worth mentioning that the convergence result in the previous statement is obtained provided that the successive iterates $Q^{(i)}$ produced by the fixed-point map (2.80) shows a B-bounded excursions. In [21], it is shown that this condition holds true for the setting associated to the following specific steps and choices that are used in the definition of the fixed-point iteration components:

- First of all, choose a fixed grid of instances of the z variable:

$$\mathcal{Z} := \{z^{(i)}\}_{i=1}^{n_q},$$

  that belong to the compact subset $\mathbb{Z}$ of interest.
- Denote the vector of values of the function Q at instances by

$$\mathbf{q} := \{q_i\}_{i=1}^{n_q},$$

  namely $q_i = Q(z^{(i)})$.
- Choose some regression model (Polynomial, Gaussian, Support Vector Machine Regressor (SVMR), etc.)[43] and denote it by $\hat{Q}$. For instance, in Chap. 17 where the framework is used to address the problem of drug's injection scheduling, a SVR with Radial basis is chosen with the default parameters used in the scikit-learn ML library [25].
- This choice enables to associate to each vector of values $\mathbf{q} \in \mathbb{R}^{n_q}$ a function $\hat{Q}(\cdot|\boldsymbol{q})$ defined on $\mathbb{Z}$ which is precisely the one identified (or *learned* to use a ML vocabulary) using the learning data $(\mathcal{Z}, \mathbf{q})$.

Based on the above items, it is now possible to state the following definition:

[42] Other sufficient conditions for the convergence of the fixed-point iteration in the absence of penalty α on the standard deviation can be found in [22–24].

[43] More about ML will come later on in this book. For the time being consider just that there is a library of structures that might be used.

Admissible Regression Model

Definition 2.2 A regression model is said to be admissible if and only if for all set $\mathcal{Z}$ of finite grid points in $\mathbb{Z}$ and any compact set of function values $\mathbb{Q} \subset \mathbb{R}^{n_q}$, there exists a positive real $B > 0$ such that all identified maps $\hat{Q}(\cdot|\boldsymbol{q})$ with $\boldsymbol{q} \in \mathbb{Q}$ shows B-bounded excursions in the sense of Definition 2.1.

Using the above preliminary results and definitions, the following main result regarding the convergence of the proposed FP iteration scheme can be stated as follows:

> **Convergence of the FP Iteration**

Proposition 2.1 *Under Assumptions* 2.1 *and* 2.2, *if an admissible regression model (in the sense of Definition* 2.2*) is used over some set of points* $\mathcal{Z} := \{z^{(i)}\}_{i=1}^{n_q}$ *with the initialization* $q_i = \ell(z^{(i)})$ *to solve the HJB equation* (2.74), *then the resulting fixed-point iteration is convergent provided that the variance related penalty* α *and the sampling period* τ *are taken sufficiently small.*

The only remaining task is to find a regression model that is admissible in the sense of Definition 2.2. Fortunately, there are many options. One of them is the **Support Vector Machine Regressor (SVMR) with Gaussian Kernel**[44] which shows the following structure:

$$\hat{Q}(z|\boldsymbol{q}) := \beta_0(\boldsymbol{q}) + \sum_{i=1}^{n_g} \beta_i(\boldsymbol{q}) G(z^{(i)}, z), \tag{2.83}$$

in which, the coefficients $\beta_i(\boldsymbol{q})$ admits uniform bound if $\boldsymbol{q}$ belongs to some compact set $\mathbb{Q}$. This last property comes from the continuity of the involved evolution maps that take bounded values over compact sets.

Recall also that **the use of fixed-point is not the only option** and other approaches can be used such as the gradient approach which is adopted when Neural Networks-based frameworks are adopted.

As a matter of fact, the framework of Q-learning via fixed-point iteration on the value function Q is one instantiation among many others of a more general framework that is referred to through the now very famous keyword of *Reinforcement Learning* (RL) that we shall recall and discuss in the next section.

[44] This is just an example and not an invitation to prefer this structure over other alternatives.

Links to Other Chapters
The Q-learning design is used in Chap. 17 to address the problem of designing a risk-averse uncertainty-aware drug's injection feedback law. In particular, it is shown that in the presence of risk, it might be crucial to use a non vanishing penalty α on the standard deviation-related term.

2.9 Reinforcement Learning

2.9.1 Definition

RL is one of the three main topics of ML which are:

- The **supervised learning** where one tries to find a map that matches a presumably existing relationship between a vector of features x and a target label y. Instances of both x and y are provided in the so-called learning data. When y takes continuous real-values, the problem is referred to as a *regression* problem while a *classification* problem is obtained when a finite number of values of y are present which can be viewed as the classes names.
- The **unsupervised learning** in which there is no labels that are provided and the aim is to cluster the available features data according to some similarity metric defined on the space of definition of the features.
- The **reinforcement learning** where a strategy delivering an appropriate action, say u, on an environment has to be learned in order to optimize some income which is represented by some performance index or equivalently in order to minimize some cost function that can be defined on a whole scenario called the *episode*.

More precisely, assuming that a discrete-time dynamical system with state vector $x \in \mathbb{R}^n$ and a decision vector $u \in \mathbb{R}^{n_u}$ are considered and that some *measurable stage cost*:

$$\text{(measurable stage cost)} \qquad \ell(x, u),$$

can be associated to each pair (x, u) being *visited* by the system, then the integration of the stage cost over some time interval is considered to be an indicator of the *quality* of the resulting system's trajectory over this time interval.

Reinforcement learning needs to consider a relevant sets of interest, namely $\mathbb{X} \subset \mathbb{R}^n$ and $\mathbb{U} \subset \mathbb{R}^{n_u}$, to be given which are assumed to contain all possible realizations of the state and the control vector values.

It goes without saying that choosing these sets is not an easy task[45] but we leave this difficulty aside here since it is a common and shared difficulty that affect the assessment of any control design framework.

[45] This is because whatever is the set $\mathbb{X}$ being chosen, one needs to be sure that if the initial state lies on the boundaries of the set $\mathbb{X}$, then any admissible trajectory is still inside the set which is obviously not easy to guarantee.

Reinforcement learning comes with many different *flavors*, we stick here to the so-called **deterministic-policy** proposed in [26] which is quite representative of the spirit of RL frameworks.[46]

In this framework, the objective of the RL is to determine an optimal state feedback *policy*:

$$\pi : \mathbb{X} \rightarrow \mathbb{U},$$

such that the following cost function is minimized:

$$J(\pi) := \int_{\mathbb{X}} \rho^{\pi}(x)\ell(x, \pi(x))dx, \tag{2.84}$$

where $\rho^{\pi}(x)$ is the *discounted* probability density function of the state under the control policy π, namely

$$\rho^{\pi}(x) := \int_{\mathbb{X}} \sum_{k=1}^{\infty} p_0(s)p(s \rightarrow x, k, \pi)ds, \tag{2.85}$$

in which:

- s represents an initial state that spans all the values in $\mathbb{X}$.
- x is the terminal state at which the l.h.s is computed. Notice that the set of values of k spans the set of instants at which the trajectories starting from the initial state s reach the terminal state x.
- $p_0(s)$ is the probability density function of the initial state s.
- $p(s \rightarrow x, k, \pi)$ is the probability that a trajectory starting at s reaches x after k sampling period under the policy π.

The consequence of this definition if that $\rho^{\pi}(x)$ sums all the probabilities of trajectories starting somewhere in $\mathbb{X}$ and reaching x and since the outcome of this operation is to *land* on the pair $(x, \pi(x))$ which costs $\ell(x, \pi(x))$, this results in a global stochastic evaluation of the expectation of the cost under the policy π that is denoted by $J(\pi)$.

It is important to underline that:

RL Applies to Uncertain Systems, at least Conceptually!

The simple use of the probability

$$p(s \rightarrow x, k, \pi),$$

[46] The subtile differences between the different candidate formulations are secondary as far as the aim of this chapter is concerned which is to give the main guidelines for the understanding of the method.

indicates that the case where the transition map is uncertain can be handled via the associated probabilities of transition.
In a deterministic setting, given the initial state s and the control law π, the outcome x would be perfectly known at each suture instant k so that $p(s \to x, k, \pi)$ can be represented via Dirac's distributions.

The ambition of RL frameworks is to provide a state-feedback policy $\pi^\star$ that is optimal considering all the possible realizations of the quantities that impact the outcome, namely, the initial state and the uncertainties. This is precisely why RL is (conceptually) viewed as the perfect ideal answer to the presence of uncertainties that affect the systems one is trying to control.

2.9.2 The RL Optimization Problem

It is obvious that the above defined problem is infinite dimensional as the decision variable is the policy π which is nothing but an **unknown function** defined on the state set $\mathbb{X}$ and having values inside the set $\mathbb{U}$ of admissible control inputs.

It is therefore not surprising that the first step in the solution is to choose some parameterization of the this map through a vector of parameters denoted hereafter by θ. The resulting policy is then denoted by π_θ and the decision variable becomes the finite dimensional vector of parameters θ paving the way to implementable algorithms which iterate on θ in order to minimize the cost function $J(\pi_\theta)$.

With the huge progress of Neural Networks (NN)-related softwares, it becomes a very common option to choose a NN structure for π_θ and to use a gradient-descent like strategy to perform the minimization.

As a matter of fact, regardless of the choice of NN structure, the gradient of the cost function w.r.t θ can be formally written as follows [26]:

$$\nabla_\theta J(\theta) = \int_{\mathbb{X}} \rho^{\pi_\theta}(x)\big[\nabla_\theta \pi_\theta(x)\big]\nabla_u Q^{\pi_\theta}(x, \pi_\theta(x))dx \tag{2.86a}$$

$$= \mathbb{E}\Big[\big[\nabla_\theta \pi_\theta(\cdot)\big]\nabla_u Q^{\pi_\theta}(\cdot, \pi_\theta(\cdot))\Big], \tag{2.86b}$$

where

$$Q^{\pi_\theta}(x, u) := \ell(x, u) + \mathbb{E}\Big[\sum_{k=0}^{\infty} \gamma^k \ell(x_{k+1}, \pi_\theta(x_{k+1}))\Big] \tag{2.87a}$$

$$\text{under} \quad (x_0, u_0) = (x, u), \tag{2.87b}$$

in which one recovers the value function $Q^{\pi_\theta}(x, u)$ that we studied in Sect. 2.8.

Notice that Eqs. (2.86a)–(2.86b) simply implement the gradient of the expectation-related cost function via the chained derivation rule in which the ultimate decision variable is θ. This gradient formulae is used to iterate over the decision variable θ which define the strategy one is looking for.

As a matter of fact, the Bellman principle enables to infer from (2.87a) the already encountered Bellman equation [see (2.74)] in the value function that should be satisfied at the optimal solution $\theta^\star$, namely:

$$Q^{\pi_{\theta^*}} = \ell(x, u) + \gamma \min_{v \in \mathbb{U}} \mathbb{E}\Big[Q^{\pi_{\theta^\star}}(x^+, v)\Big]. \tag{2.88}$$

Reinforcement learning attracted the attention of a large community of researchers mainly because of two reasons:

1. the recent impressive success stories of Deep reinforcement learning (DRL) in the game area (GO, chess) and since, in many fancy applications areas including painting and music,
2. the availability and the impressive easiness of use of some Deep Learning-related tools such as `pytorch` [27], `Keras` [28], `TensorFlow` [29] to cite but few examples.

The impressive success stories induced and are still inducing *unreasonably over-optimistic* expectation and regarding the extent to which RL frameworks might address complex problems **without the slightest knowledge regarding the system's model** and using only real-life data through experimentation and associated learning.

Indeed contrary to the fixed-point method described above, the attractive version of RL implementation is a kind of interactive learning experiment in which, the *actor/controller* trying to improve its algorithm updated the latter based on the evaluation (*reward*) of the current strategy returned by the *critic/environment*.

It turns out that the above described *iterative* principle might be adopted either in the presence of a model of the system (model-based RL) or in the absence of such a model leading to the so-called *model-free* RL.

We shall successively examine both contexts and discuss their relevance in the following sections.

2.9.3 RL Implementation in the Presence of an Uncertain Model

In this case, given the current value of the strategy's parameter θ, the expectation involved in (2.86b) can be approximated by averaging the values of the function over a cloud of realizations of the state.

Notice however that, for each value of x in the used grid of instances, the estimation of $Q^\theta(x, \pi_\theta(x))$ requires the estimation of the expectation of the integral cost (2.87a) which involves model-based simulations in which the sampled quantity is the realization of the uncertainties for the current sample of x.

This obviously leads to a rather huge number of simulations. This number can be determined, for instance, following the results of [30] that are explained at the end of this chapter (see Sect. 2.11.4) depending on the targeted quality of the approximation

Recall that the above cascaded evaluations are performed in order to only be able to get the evaluation of the gradient for the current value of the decision variable θ during the gradient-based optimization process.

! The Curse of Dimensionality

This means that the number of required simulations grows exponentially in the state dimension. This is the price to pay if one sticks to the ideal promises of RL, namely to deliver the stochastically optimal state feedback strategy defined as a function of the state over the set of interest $\mathbb{X}$.

This curse of dimensionality has been quite early identified by the practitioners of the Approximate Dynamic Programming and it induced a commonly shared conclusion stipulating that this kind of approaches is restricted to systems of moderate sizes in terms of the number of states.

To summarize, when models of the underlying dynamics and cost are available, what is to be done is quite well known but it is extremely expensive in terms of computation. On the other hand, when the model exists there are many other options that do not show explosion in the state dimension as it is shown in this book.

The reason for the current attractiveness of RL lies in its *presumed* ability to address the case where no model is available. The next section shows that in this case, the challenge is even stronger and the expectation should be seriously revised.

2.9.4 Model-Free Context

The presumed ability of RL frameworks to provide state-feedback policies in the absence of any knowledge regarding the underlying dynamics has a lot to do with the high expectation they raise in the minds of many decision makers, engineers and young researchers.

Before we discuss the solidity of such an expectation, we need to give some details in order to clarify *how on earth* can one manage to design a feedback without a dynamical model of the system.

In this section, the simplest version of model-free implementation of RL is briefly sketched before we can examine the extent to which the commonly shared optimism regarding RL is justified.

Notice first that in the absence of a model, the previous computational scheme that is based on model's simulation is no more an option. Even a single step enabling x^+ involved in (2.88) to be computed is impossible to execute due to the lack of a dynamical model providing such a one-step-ahead prediction. In other words, **Only real-life measurements-based updating** steps can be involved.

More precisely, the algorithm can only use the successive values of triplets of the form:

$$\Big\{z^{(i)}, \ell(z^{(i)})\Big\}_{i\in I} \qquad z^{(i)} := (x^{(i)}, u^{(i)}), \tag{2.89}$$

which are **effectively visited** by the real-life system during some data collection experiment or during the system life-time should a continuous adaptation be adopted.

Moreover, notice that this means that the quantity x involved in (2.89) is supposed to be measured. This suggests that the signification of x is not necessarily the same as before. Namely, it is not necessarily the physical state of the system should the state be only partially measured. Rather it might represent a vector of previous values of the measured output over some past window.

Since the algorithm needs not only a parametric representation of the control strategy but also a parametric representation of the value function Q which is no more a consequence of the strategy via the no more available dynamic model, a parametric representation of the form:

$$Q^{\psi}(\cdot, \cdot)$$

is considered where the vector of parameters ψ is to be found together with the control-related vector of parameters θ.

More precisely, having the parametric representation of Q^{ψ} and π_θ, an alternate updating steps of the two parameters ψ and θ are computed using gradient-based descent, namely:

$$\delta^{(i)} = \Big[\ell(z^{(i)}) + \gamma Q^{\psi^{(i)}}(z^{(i+1)})\Big] - Q^{\psi^{(i)}}(z^{(i)}) \tag{2.90a}$$

$$\psi^{(i+1)} = \psi^{(i)} - \alpha_{\psi}\Big[\nabla_{\psi} Q^{\psi^{(i)}} z^{(i)}\Big] \tag{2.90b}$$

$$\theta^{(i+1)} = \theta^{(i)} - \alpha_{\theta}\nabla_{\theta}\pi_{\theta}(x^{(i)})\nabla_u Q^{\psi^{(i)}}(z^{(i)}), \tag{2.90c}$$

in which:

- $\delta^{(i)}$ is the error on the equality (2.88) related to Q at the iteration i.
- The Eq. (2.90b) implements a gradient-based correction of the vector of parameters ψ based on the error $\delta^{(i)}$ computed in the previous step.
- The Eq. (2.90c) implements a gradient-based correction of the vector of parameters θ aiming at decreasing the current value of the cost function at the pair $z^{(i)}$ which is *visited* at iteration i.

This is about the principle of the method which is based on alternate improvement of the strategy π_θ and the value function Q^{ψ} using the gradient descent approach.

This seems to be quite intuitive and attractive. In particular thanks to the automatic gradient handling and efficient computation performed by freely available tools,[47] this kind of quite involved task is now an option that can be undertaken by almost anybody by means of relatively few lines of `python` code.

[47] Such as the `pytorch` package which can now be executed on a distant and freely available `GPU` machines through the `google-colab` notebooks environment for instance.

Nevertheless, these nice and attractive features tends to hide some inherent difficulties and challenges. Indeed, a deeper examination of the details of the method reveals some bottlenecks and highly challenging issues that are discussed in the next section.

2.9.5 Some Challenging Issues in Model-Free RL

The first challenging issue is related to the exploration of the set of values of the pair (x, u). Indeed, it is obvious that the relevance of any resulting computation of the value function Q^ψ is conditioned by a representative set of *visited* values of (x, u) as this is a condition for the resulting estimated map to be valid over the domain of possible values.

This issue was only computational **in the presence of mathematical model** of the system due to the high number of system's simulations to be performed. The representativity issue was conceptually addressed in the model-based framework via the cloud generation step. In other words, by including a sufficiently rich number of pairs

$$(x^{(i)}, u^{(i)}) \quad i = 1, \ldots, n_{\text{grid}},$$

in the cloud of possibilities, it was only a matter of time for the computation to be done and the algorithm to converge.

In the model-free context however, only the effectively and experimentally visited pairs can be included while the system is effectively operated by the progressively improved feedback strategy.

But in order to explore in a sufficiently relevant manner the space of (x, u) pairs so that the expectation can be correctly estimated, actions u_k that are different from the ones suggested by the expression $\pi_\theta(x)$ should be regularly applied to enrich the explored domain. This is referred to as the *exploration step*. The way and the frequency at which the exploration should be done remains an open and largely unsolved problem and will probably remain as such forever.

This is because only random-like variations can be applied around the current value of the control such as:

$$u^{(i)}_{\text{exploration}} = \pi_\theta(x^{(i)}) + \nu^{(i)}, \tag{2.91}$$

since in the absence of an underlying dynamical model, there is strictly no way to know what is the correction $\nu^{(i)}$ to be added in order to *force* the system to visit *still not-yet-visited* regions of the state/action space.

This first challenge can be summarized as follows:

! Challenge 1: Domain Exploration

Without an underlying dynamical model, it is extremely difficult to choose the actions that force the system to visit *not-yet-explored* regions of the space of pairs (x, u) while this is mandatory to derive a relevant estimation of the value map Q^w that might hold over the space of interest $\mathbb{X} \times \mathbb{U}$.

Notice that the above discussed challenge of exploration holds even in the absence of uncertainties, that is to say, even in a nominal design in the sense that the context of the experiments is the one fixed by the control designer and all sources of uncertainties are carefully neutralized or suppressed.

When one consider dynamical systems that are subject to uncertainties and/or disturbances, another challenging aspect comes into picture, namely, the difficulty of exploring sufficiently relevant set of uncertainty realizations.

Indeed, since in the absence of model, no simulations are possible with different instantiations (realizations) of the uncertainties or parameters changes, only purely experimental data collection can be involved. Therefore, one can only encounter the *realizations of the uncertainties* that are *decided by the fate* during the specific interval of time during which the data is collected.

Now one might argue that in some cases, it is possible to artificially change the parameters of the system during the experimentation so that a more representative and intentionally chosen combinations of the system's parameters. This might be extremely expensive and can never explore the real set of possibilities on one hand, on the other hand, this applies to only a specific set of situations and systems.

In other words:

! **Challenge 2: Fate Is Not Controllable**
The outcome of iterations in model-free RL depends on an unavoidably limited set of realizations of the uncertainties that might have no statistical relevance given the true set of possible realizations that the original statement of the RL was intended to explicitly account for.

The above mentioned two challenges are already quite serious but let us assume for the sake of argumentation, that *magically*, one gets a nice representative set of visited pairs (x, u) and moreover, the random uncertainties encountered during the experimental data collection process were miraculously representative so that the statistics can be approximated to derive a correct formulation.

The next challenge we shall discuss is the one linked to the convergence of the gradient-based iterations involved in the very definition of the updating laws (2.90a)–(2.90c).

Indeed, every experimented person in optimization still recall these *ancient times* where the problem of the presence of local minima associated to the gradient descent methods were largely acknowledged and largely experimented. It seems however that the impressive success of the Stochastic Gradient Descent (SGD) algorithms [31] in the context of Deep Learning (DL) induced a collective forgetting of this simple and solidly assessed fact.

The recent better understanding of the behavior of SGD iterations in the context of DL [32] suggests that the convergence is enhanced by the combination of two factors:

1. the use of a *sufficiently large* number of hidden layers leading to an **over-parameterization** of the underlying DNN on one hand, and

2. the use of the SGD on the other hand which reduces the risk of the iterations being captured inside a local minimum.

Unfortunately enough, the use of over-parameterization in the context of model-free RL computation which is necessary to increase the probability of avoiding the local minima comes at the price of much larger required number samples used in the expression of the cost function. This increases the number of iterations and hence the training duration (since samples are acquired via real-life experiments) that would be required to avoid the issue of over-fitting.

Recall that *over-fitting* arises when the number of degrees of freedom (which increases with the number of hidden layers) is not small enough compared to the number of samples participating in the least squares problem's formulation.

This makes the framework inappropriate in many control-related situations if not in the majority of them.

This is summarized in the following statement[48] :

! **Challenge 3: The Vicious Circle of Deepening the Neural Networks**
Avoiding local minima in Deep Learning via SGD requires deepening the NN which requires increasing the number of samples which might increase the duration of the learning experiment beyond reasonable values.

It is important to distinguish between two concepts that are involved in the previous discussion, namely: the issue of local minima and the issue of over-fitting. These two concepts might appear to be close as they are both related to the problem of having bad solutions to an optimization problem. Without a rigorous understanding of the difference between these concepts, one might think that the previous discussion induces a somehow self-contradictory conclusion that can be stated as follows:

Deepening the network is allegedly used to reduce the risk of local minima but using it induces the need for more data to avoid over fitting!

So obviously, putting an equal sign between *local minima* and *over-fitting* renders the above statement self contradictory.

To grasp the difference, recall that:

[48] As an example, several recently published works regarding the use of RL frameworks in order to perform model-free RL regarding energy management strategies in smart buildings require a number of *batches* that last over 10–15 years which is obviously a nonsense from a business or economic point of view.

> **Local Minima Versus Over-Fitting Issues**
> The local minima issue is faced by an optimization algorithm that tries to solve a specific optimization problem. It is associated to the risk of the algorithm being trapped at a value of the decision variable that is a *local minimum* which is not a *global* one **for a given instance** of the optimization problem.
> By opposition, the over-fitting issue concerns problems whose formulation depends on a given instance of data and is defined to be the risk of the **true global minimum** of a problem, formulated using some *training data*, to be significantly different from the true global minimum of the problem under **different evaluation data**.

having these differences in mind, the above statement might be rightly understood: Increasing the number of samples reduces the risk of the different formulated problems showing significantly different global solutions while deepening the NN reduces the risk of missing the associated region of common solutions to all these problems.

The last, but not least, challenge that is associated to the model-free RL paradigm is the one related to the safety of the learning process at its early stages as well as the economic cost (in time, material and resources) of reinventing the knowledge that we accumulated regarding the physical laws governing the phenomena.

Indeed, when the model is totally unknown, it is very difficult to derive exploration strategies that are safe since we have no prediction capability by simulating a model of the underlying process in order to be sure that the currently applied action has no harmful consequences on the process we are controlling.

This is the reason why many recent works started to emerge on this topics with the keyword *safe learning* somewhere in the titles [33,34]. But unfortunately, the majority of them are tautological and hide under the mathematically involved formulations some used, while presumably unknown, knowledge regarding the system's behavior.

Now given all the above listed challenges, one might ask

Are RL's Success Stories Fake?

Given the above list of challenging issues, how is it that RL is associated to so many successful and so impressive use-cases? What is the specificity of the class of problems that makes RL successful?

Answering this question is hazardous as it is always possible to find examples where only a partial use of model-free reinforcement problem is involved. More precisely, RL might be used to solve a sub-problem of a wider one. This together with some available models of the remaining phenomena enables the whole scheme to be viable.

Nevertheless, it is possible to take the risk of mentioning some characteristics of the problems that make them eligible to a successful and even impressive handling by means of model-free RL, namely:

Conditions for a Successful Model-Free RL Design
Model-free RL might be impressively successful if the following conditions hold true:

1. It is possible to run a *huge* number of data-collecting experiments without risk nor exorbitant financial cost!
2. The success is expressed through rather *qualitative* criteria by opposition to tight control and high precision-like metrics.

Tow typical examples of such situations might be:

- building a strategy for the Chess game,
- learning feedback control enabling a rough directional movement of a multi-body spider-like robot.

Indeed, nothing prevents a huge number of GPU devices to play fictitious chess-games during months involving an almost humanly unimaginable number of matches in order to explore the discrete space of state combinations.

Similarly, we can leave the toy spider-like robot during days and weeks if not months trying to go almost smoothly from a point A to point B without much damages and with moderate energy cost.

At the same time, it appears very difficult to get a faithful model for any of these two examples. This makes the use of model-free RL not only a possibility but probably the only reasonable remaining choice.

On the other hand, everyone might easily agree that the above stated conditions are not satisfied for the majority of industrial systems that cannot be made accessible for random experimentation for obvious safety and economic reasons if not for a simple chronological reasons.[49]

2.9.6 RL: The Revival of an Old Dream!

To close the discussion of this section, it is worth to take a step back to see the big picture in the appropriate time scale!

Recall that in the 90s, the nonlinear control design via analytic Lyapunov methods [1] was the dominant option. The emergent nonlinear MPC (NMPC) [4] faced some initial opponents who were claiming that it was not really a *state feedback* because of its implicit nature (no explicit expression of the control as a function of

[49] The time needed for the iterations to converge in case they really do!

the measured information, rather, the input is the outcome of some underlying and hidden optimization algorithm).[50]

The difficult, if not impossible derivation of a Lyapunov-based solution to general constrained nonlinear systems incited some authors [35] to classify systems for which, modeling involving high nonlinearities is necessary, as legitimate candidates for data-driven control design (this was in 2013!).

The success of NMPC in handling the unavoidable presence of constraints in control system design was detrimental in achieving its acceptance in the wide control community as a finally not so bad an option.

However, the old dream of disposing of an explicit representation of the control feedback law that can be computed once for all resisted in the linear case leading to the design of **explicit linear MPC** computation tools [36].

Several years were necessary to the control community before acknowledging that the associated algorithms are not scalable and that the computational complexity of the on-line computation which might be used to deliver standard NMPC on-line version were comparable [37] to the off-line promoted explicit solutions.[51] Moreover, contrary to the off-line solution which remained limited to linear and moderate size problems, the on-line version generalizes to nonlinear and high-dimensional cases.

Having this historical reminder in mind, it is hard not to see in the euphoric fervor around reinforcement learning a new *avatar* of this sadly buried dream consisting in being able to compute the control as a strategy (pre-computed function of the state) that would be available to the designer against the uncertainties in an explicit, guaranteed, pre-computed and on the top of it, optimal manner.

It can be thought that this position is somehow *oriented* by personal strong history of commitment to the development of Nonlinear MPC. So let us examine Y. LeCun's say on the subject from one of his 2024 tweets[52]:

> *Indeed, I do favor MPC over RL. I've been making that point since at least 2016. RL requires ridiculously large numbers of trials to lean any new task. In contrast MPC is zero shot: If you have a good world model and a good task objective, MPC can solve new tasks without any task-specific learning. That's the magic of planning. It doesn't mean that RL is useless, but its use should be a last resort.*
>
> *Yann LeCun*

This section suggests that the possibility to achieve the RL-induced dreams is strongly questionable, at least in its general scope of its original claim and for the majority of real-life industrial problems.

[50] I may understand some readers having a difficulty to believe this but you should believe me here as it is a personal experience.

[51] Because of the computation time needed to determine the region in the state space to which the current state belongs and hence in order to determine which explicit expression to use.

[52] By the time he delivered this opinion, Y. LeCun, the inventor of *Convolutional* NN and one of the recognized fathers of Artificial Intelligence (AI), was Chief AI Scientist at Meta.

On the other hand, it is always good to keep in mind the possibility to use the paradigm of RL and its associated algorithms and tools to address specific problems in a way that exploits its specificity and possible angles of attack. In a broader sense, this is the state of mind exposed in the solutions proposed in this book. In particular, some of such existing attempts in the literature are presented or recalled in [38].

It is worth recalling that in all the previous discussion regarding MPC, SDP and RL, it was assumed that all the components of the state vector were available via dedicated sensors. Unfortunately, this is rarely the case in real-life as only a *small* subsets of the components of the state can be accessible to measurement.

When this is the case, one needs to reconstruct the useful part of the state vector that is necessary for the implementation of the control law. The associated paradigm is referred to as the problem of state estimation.

As far as this quick overview of concepts is concerned, this is the object of the next section. More detailed study is proposed in Chaps. 9 and 11.

2.10 State Estimation

Traditionally, state estimation algorithms aim at reconstructing all the components of the state vector (full state estimation). It is under this paradigm that the concept of state observability has been introduced. That is the reason why this section starts by addressing the full state estimation problem. Later on, partial state estimation versions are introduced and discussed.

2.10.1 Full State Estimation

Recall that the state vector $x(t_k) \in \mathbb{R}^{n_x}$ of a controlled system contains the information that totally describes the system itself at some instant k (leaving aside the exogenous signals that might affect its behavior).

For a system that can be described by a set of ordinary differential equations (ODEs), the number of components of the state vector is equal to the total number of derivations that are needed to express the laws governing the dynamics of the system.

The state estimation paradigm stems from the fact that it is generally not possible to have a dedicated physical sensor that is associated to each component of the state vector. Only some of the components and/or sometimes some combinations of them are measured through a set of sensors.

But even if they are not measured, some components of the state or some non measured combinations of them might be necessary to know in order to express some items of the control requirements that the control needs to enforce. This can be related to some safety constraints on non measured components of the state or it might be related to the quality of the resulting closed-loop.

Example

As an example, imagine that the system is a car equipped with a sensor measuring the longitudinal speed and that the energy consumption is related to the traction force being applied by the engine. The latter is directly related to the acceleration of the car. Consequently a proper handling of the control objective necessarily requires the reconstruction of the acceleration via some algorithm that accepts the speed measurement as input.

In order to handle the state estimation issue, it is important to know what are the available sensors on board and how they are related to the state, the control and the system's parameters.

Therefore, in addition to the dynamics governing the state vector expressed by (2.6) the information regarding the measured quantities is also required. This is referred to as the *measurement equation* and takes generally the following form:

$$y = h(x, u) = \begin{bmatrix} u \\ h_s(x) \end{bmatrix} \in \mathbb{R}^{n_y}, \tag{2.92}$$

which explicitly indicates that the control input u is supposed to be measured and hence is a part of the measured output vector. On the other hand, the map h_s indicates the other measurements which are dependent on the state vector itself.

Notice that in standard textbooks, the input vector is also supposed to be measured, nevertheless the notation y is generally used to refer to the measured variables except u. This section stipulates that the notation y includes the input vector u since this leads to simpler notation and because as far as the estimation issue is concerned, there is no need to distinguish between u and the other measured variables.

! The Control Input u Is a Part of the Measurement Vector y

The commonly used notation in the control community separates the input from the *other measurements*. In this section, we include the control input inside the measurement vector since this control input is supposed to be measured for the sake of control-related needs! Therefore in addition to be coherent with this fact, this notation enables shorter expressions in the formulation of the state estimation problem and algorithms.

The *full state estimation* paradigm is depicted in Fig. 2.24.

The question is whether it is possible to reconstruct the state vector x_k from the knowledge of the **past measured quantities** (including the control input as mentioned above) over some sufficiently high number N_O of sampling periods[53] :

[53] $[k - N_O, k]$ is sometimes called the observation window.

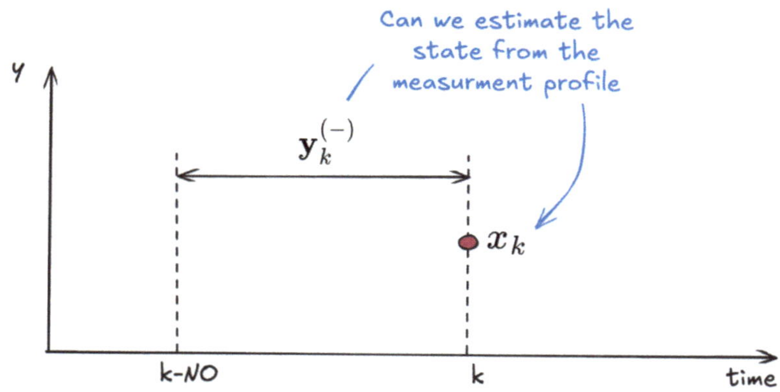

Fig. 2.24 The full state estimation paradigm: can we reconstruct the value of the state vector at instant k, namely x_k, by using the previous measurement (including the input) over some sufficient number N_O of sampling instants?

$$\mathbf{y}_k^{(-)} := \begin{bmatrix} y_k \\ \vdots \\ y_{k-(N_O-1)} \end{bmatrix} \in [\mathbb{R}^{n_y}]^{N_O}. \tag{2.93}$$

While the previous paradigm is oriented towards the full state estimation, let us now consider the case where one is not necessarily interested in the reconstruction of the whole state vector but a less complete information that would nevertheless be sufficient to achieve a specific task.

This leads to the concept of *observation target* that is presented in the next two sections where the case of deterministic systems and then that of uncertain systems are successively discussed.

2.10.2 Reconstruction of an Observation Target for Deterministic Systems

As a matter of fact, the full state estimation paradigm addresses the question of the reconstruction of the whole state vector which might be too ambitious or simply not necessary.

An example that helps illustrating this fact is the situation where the control designer starts by deriving a *state feedback* of the form:

$$u_k = K(x_k),$$

that achieves some task **assuming in a first stage that the state vector is available** and then builds a state observer that provides an estimation of the state, commonly denoted by:

$$\hat{x}_k = O(\mathbf{y}_k^{(-)}),$$

so that the truly implemented feedback control is defined by[54] :

$$u_k = K(\hat{x}_k) := \big[K \circ O\big](\mathbf{y}_k^{(-)}),$$

which bowls down to be an implementable dynamic output feedback control as it needs only available measured quantities over the past observation horizon.

The previous presentation used the common practice consisting in replacing the unavailable quantity x_k by the *reconstructed* quantity $\hat{x}_k$ **inside** the map K.

The problem in the previous approach is that it can be conducted **only if the full state is reconstructible.** which might be a strong invalid assumption.

The other option is to define the expression:

$$z = K(x) \in \mathbb{R}^{n_u}, \tag{2.94}$$

to be **the target of the estimation** scheme [39], namely the quantity one would like to reconstruct.

Notice that the ability to reconstruct z is generally less stringent than the full state reconstruction. One simple way to see it is by simply acknowledging that if x is known then z will be also known while the inverse is not necessarily true. Moreover, the dimension of K and hence the dimension of z, namely n_u is generally lower than the dimension n_x of the state.

Obviously, the paradigm of the **partial estimation of an observation target** z is a more general paradigm that encompasses the commonly used full estimation paradigm since the latter can be recovered in the special case where $z = x$ is used.

More generally, one can consider the following definition:

Deterministic Partial Observability

Definition 2.3 Consider a nonlinear deterministic dynamical system governed by the state-space evolution and the measurement map defined by:

$$x^+ = f(x, u), \tag{2.95a}$$

$$y = h(x, u), \tag{2.95b}$$

and consider an observation target defined by

$$z = T(x, u). \tag{2.96}$$

[54] Here $\circ$ stands for map composition operator namely $\big[K \circ O\big](\cdot) = K(O(\cdot))$.

We say that z is N_O-observable for some integer N_O if and only if, there exists a map O such that the following equality holds true:

$$z_k = O(\mathbf{y}_k^{(-)}) \quad \texttt{where} \quad y_k^{(-)} := \begin{bmatrix} y_k \\ \vdots \\ y_{k-n_O-1} \end{bmatrix}, \tag{2.97}$$

for any k and any realization of the trajectories of (2.95).

The relevance of this paradigm of partial observability is even more crucial when one consider uncertain dynamical systems.

2.10.3 Reconstruction of an Observation Target for Uncertain Systems

Let us first revisit the full-state estimation paradigm in the presence of uncertainties. To do so, let us consider an uncertain system that is governed by the following discrete-time dynamics:

$$x^+ = f(x, u, w),$$

where w is an uncertain vector. Assume that one disposes of a way to derive an *ideal* feedback

$$u = K(x, w) \in \mathbb{R}^{n_u}, \tag{2.98}$$

should the vector of uncertainties w be known.

Now going back to real-life where w is not available by measurement, the commonly used approach is to extend the dynamics by considering the following extended state evolution equation:

$$x^+ = f(x, u, w), \tag{2.99a}$$

$$w^+ = w. \tag{2.99b}$$

Notice that despite the fact that a constant dynamic is presumed in the model's extension, it can be shown that when an observer[55] is designed using this model and provided that the observation window is not too long, the resulting estimation leads to a dynamic behavior of the estimated w despite the fact that the underlying model expresses constant w.

[55] Or an extended state estimator which is an inter-changeable denomination.

Another way to look at this is to imagine that the observer equation involves evolution law of the estimated vector of uncertainties of the form: $\hat{w}^+ = \hat{w} +$ correction term meaning that a dynamic evolution of the correction term is used.

The implicit assumption that underlines the above solution is that **full observability holds for the extended system** meaning that there exists a map O such that:

$$\begin{bmatrix} \hat{x}_k \\ \hat{w}_k \end{bmatrix} = O(\mathbf{y}_k^{(-)}), \tag{2.100}$$

enabling an implementable estimation-based version of the ideal feedback law (2.98) mentioned above which takes the following form:

$$u_k = K(\hat{x}_k, \hat{w}_k) := \Big[K \circ O\Big](\mathbf{y}_k^{(-)}).$$

The full observability assumption that lies behind this approach might become extremely hard to rigorously meet when the dimension n_w of the uncertainty vector increases. Indeed, this approach supposes that $n_x + n_w$ components of the extended state (x, w) can be fully recovered by using the previous measurements profile $\mathbf{y}_k^{(-)}$.

Here is where the concept of observation target and partial observability becomes extremely useful. Indeed, regardless of the number n_w of components of w, the number of variables one tries to reconstruct in this use-case remains equal to the number n_u of control input components.

This does not mean that the problem would always have a solution but simply shows that the domain of success is generally much wider than the full extended observability case which is not even needed in the first place, at least in this context.

Based on the above discussion, the following definition extends Definition 2.3 to the case of uncertain dynamical systems:

Partial Observability of Uncertain Systems
Definition 2.4 Consider a nonlinear uncertain dynamical system governed by the following uncertain dynamics and measurement equations:

$$x^+ = f(x, u, w), \tag{2.101a}$$
$$y = h(x, u, w), \tag{2.101b}$$

and consider an observation target defined by:

$$z = T(x, u, w). \tag{2.102}$$

We say that z is N_O-observable for some integer N_O if and only if there exists a map O such that the following equality holds true:

$$z_k = \mathcal{O}(\mathbf{y}_k^{(-)}), \tag{2.103}$$

for any k and any realization of the system's trajectory.

It goes without saying that the conditions of this definition might rarely be satisfied for a real-life process. In particular the equality (2.103) might certainly not be satisfied fully rigorously. Nevertheless, the definition gives the ground of intuition that enables to understand the rationale of some of the heuristics given in the remainder of this book.

Referring to the conditions of this definition enables to quickly understand the kind of assumptions one might have in mind when doing such or such thing while keeping in mind that these conditions can be only approximately satisfied but sufficiently for a learning process to capture a good level of precision.

When this partial observability property is satisfied, given any measurement profile, there is only a unique value of the observation target z that is compatible with both the measurement profile and the Eq. (2.101) governing the dynamics and the sensors.

Links to Other Chapters

The ideas developed in this section regarding the partial estimation in the presence of uncertainties provide the foundation of the uncertainty-aware dynamic output feedback design studied in Chap. 7 and then applied in Chap. 16 to solve the real-life problem of power management in hybrid vehicles.

Anticipating the content of Chaps. 7 and 16 where this concept of partial observability is extensively used to design uncertainty-aware output dynamic feedback, let us simply mention that ML tools are applied to *learn* the map (2.103) from learning data-sets that are built by appropriately simulating a cloud of open-loop optimal control trajectories of the uncertain model corresponding to a cloud of uncertainty realizations.

For the time being, let us achieve the presentation of the (extended)-state estimation problem towards the definition of the corresponding dynamical system that is called (extended)-state estimator.

The need for some extra work comes from the fact that the quantity denoted in the previous discussion by $\hat{z}_k$ is uniquely determined from the previous measurement profile $\mathbf{y}_k^-$ through the presumed relationship $\hat{z}_k = \mathcal{O}(\mathbf{y}_k^-)$. This totally discards the previous estimated values and leaves the estimation precision uniquely dependent on the measurement that might be affected by a possible high level of noise.

This issues is discussed and handled in the next section.

2.10.4 The Observer Dynamics

The starting point is to notice that the inverse map given by (2.103) whether it is defined in the deterministic or stochastic setting is a measurement-related quantity and hence is prone to measurement noise effects. These effects can partially be compensated for by taking into account the dynamics of the system.

More precisely, the behavior of the observed quantity should always account for two sources of priors:

✓ The first is the one derived from the previous measurements over the observation horizon of length N_O (z should explain, to some extent, the recent measurements!).

✓ The second is the one derived from the previously estimated value z_{k-1}. This information should impact how far the newly estimated z_k might be from the last estimated value given our knowledge (even if this knowledge is only partial) on the underlying dynamics.

This recalls the use of the correlation matrices in the design of the optimal Kalman filter that enable to determine a balance between the confidence in the knowledge of the dynamics and the confidence in the precision of the measurement in order to come out with an optimal mitigation of the sources of uncertainties induced by the state evolution and the measurement noises.

Links to Other Chapters

In the case of **full extended state estimation**, the total handling of the issue as well the precise implementation of the extended state Moving-Horizon Estimator (MHE) is studied in depth in Chap. 9. Moreover some hints enabling a real-time implementation of the underlying on-line optimization process is also examined.

As for **partial target's observation**, an example is given in Chap. 11 where the sparse identification algorithm developed in Chap. 10 is used to design partial state estimator in the presence of uncertainties and measurement noise.

In the context of **partial estimation** of an observation target z, one can implement such consideration by acknowledging that given the previous estimation $\hat{z}_{k-1}$ of the target at instant $k-1$ and because there is no corresponding previous estimation of the whole state that provides a model based estimation of z_k from the knowledge of z_{k-1}, one can still introduce an amount of trade-of between the two potentially contradictory requirements by considering the following dynamics for $\hat{z}_k$:

$$\hat{z}_k = \varepsilon \hat{z}_{k-1} + (1-\varepsilon) O(\mathbf{y}_k^{(-)}), \tag{2.104}$$

for some $\varepsilon \in (0, 1)$. Obviously, matrix weighting can also be used in case relevant information on specific components of z are available.

Let us summarize the previous discussion regarding the estimation paradigm:

> **The Estimation Paradigm**
> The estimation paradigm is about reconstructing the value of some quantity
>
> $$z_k = T(x_k, u_k, w_k),$$
>
> from the previous measurement profile $\mathbf{y}_k^{(-)}$.
> Rigorously speaking, the success of the estimation relies on the existence of a map
>
> $$z_k = O(\mathbf{y}_k^{(-)}),$$
>
> meaning that the targeted quantity is reconstructible from the only measurements.

As mentioned in the *Links to Other Chapters* sections, this topic is extended in Chaps. 9, 7 and 16.

For the time being, let us resume our overview of the main concepts that helps tackling the uncertainty issue. A major one is linked to the probabilistic certification paradigm that is discussed in the next section and that is extensively used in the forthcoming chapters, in particular, the one dedicated to applications.

2.11 Probabilistic Certification

The certification paradigm emerged quite recently in the control literature from the acknowledgement of the fact that in many situations, it is quite impossible to come out with a rigorous proof of a statement according to which *a property holds for all possible realizations of the uncertainties*.

This difficulty simply stems from the fact that a proof cannot be provided for a fact that does not hold in reality unless a set of conditions that would be **too conservative** is supposed to hold for the sake of the needed proof. In such circumstances, the associated proofs take the form of a *tautology*.

As a matter of fact, the intuitive idea behind the probabilistic certification was in the air for a long time although a definitive acceptance of the qualitatively different kind of associated assertions took a while to be considered as valid and solid results in the control-related literature.

The idea can be simply presented as follows: Consider a property that depends on a unknown vector of parameters that takes values in a known bounded set and following some known (or possible to approximate) statistical law. Then if we need to prove that this property holds with a high probability then we can check this property over a high number, say N, of random realizations of the uncertain vector and then compute the proportion of samples for which the property is not satisfied. This would give an idea about the probability of its satisfaction.

This intuitive idea is the same that was involved in the famous *Monte-Carlo* simulation framework that was first suggested by the polish Mathematician Ulam [40,41] in the early years of the emergence of computers.

Another instantiation of this *counting samples* paradigm is the computation of the statistical expectation of a quantity as its average computed over a large set of N random realizations.

But the following question remained: *How to formulate the link between the number of random simulations to be performed and some probabilistic measure of the solidity of the resulting statement about the satisfaction of a property.*

More precisely, the probabilistic certification paradigm can be stated as follows:

> **Probabilistic Certification**
> The probabilistic certification framework is about determining the number of samples to be drawn (following the known statistical law of uncertainty) in order to get a prescribed level of confidence regarding the assertion according to which the property is satisfied with a given probability.

Let us take an example of a control objective and a control design and see how this paradigm applies to the assessment of control feedback design. Beside its role in illustrating the probabilistic certification paradigm, this example is also used later on in the book to illustrate some control design heuristics.

2.11.1 Illustrative Example: The Combined Therapy of Cancer

Problem Statement

Let us consider a general form of a dynamic system representing the evolution of the tumor and the number of circulating lymphocytes among other necessary quantities under a combined action of several drugs injection rates $u \in \mathbb{R}^{n_u}$:

$$\dot{x} = f(x, u, w), \tag{2.105}$$

where $x \in \mathbb{R}^n$ is the state of the model while $w \in \mathbb{R}^{n_w}$ stands for the vector of parameters involved in the mathematical expressions.

Assume that the state x is defined such that:

- x_1 stands for the tumor size (to be reduced).
- $x_2 = C$ stands for the amount of circulating lymphocytes that is commonly viewed as an indicator of the patient health/resistance. Therefore, any drugs' injection

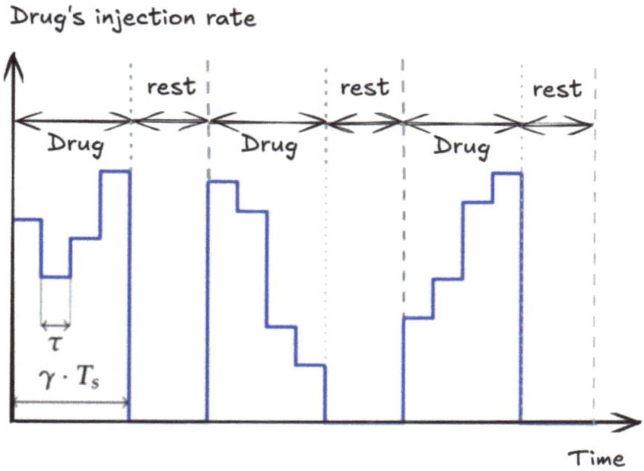

Fig. 2.25 Temporal structure of the multi-drug therapy. Example of a treatment period consisting of $N_T = 3$ sub-periods of a duty cycle γ. The periods of rest are periods during which no treatment of any kind is applied

strategy has to be defined such that the following constraint is satisfied[56]:

$$(\forall t \geq 0) \qquad x_2(t) := C(t) \geq C_{min}. \tag{2.106}$$

Other state components may be necessary to complete the description of the model (namely $n \geq 2$), in particular, the description of targeted therapies might need the dynamic of auxiliary agents/antibodies to be described by dedicating some components of the state vector x to represent corresponding sub-populations.

It is assumed that the dynamic model (2.105) describes the evolution of the system under the combined effect of n_u different drugs such as chemotherapy, immunotherapy, anti-angiogenesis and so on.

Let us consider a feedback-based therapy of duration T consisting of N_T sub-periods (of duration $T_s = T/N_T$) each of which involving a treatment phase and a rest phase as shown in Fig. 2.25.

Notice that in Fig. 2.25 the injection curves have to be interpreted as a multi-variable signals when several drugs are combined.

Let us consider that some *parameterized*[57] feedback strategy is designed which delivers, at each decision instant, the values of the injection rates. Assume however that the parameters of the feedback strategy are not totally fixed yet. Rather, they should be optimized so that some properties of the outcome of the therapy can be certified despite of the high uncertainties on the model's parameters.

Assume that the above mentioned *parameterized* feedback strategy to be applied during the *drug-delivery* periods depicted in Fig. 2.25 takes the following form:

$$u_k = K(x_k, p_c), \tag{2.107}$$

---

[56] Any drug delivery strategy (feedback law) has to guarantee with a very high probability that this health constraint is fulfilled all the time. In other word, a probabilistic certification is needed on this basis before the associated drug delivery protocol is accepted.

[57] Meaning that the feedback strategy is defined up-to the choice of some parameters such as: gains, thresholds, etc.

where p_c is a vector of feedback **design parameters**. Namely, these are parameters that are used in the very definition of the feedback. This can include thresholds, gains, delays, slopes, etc. In other words these are tunable parameters that does not change the structure of the feedback but changes the resulting drug delivery quantitatively and hence changes the overall results of the therapy over the therapy's duration.[58]

As a matter of fact, the parameters of the injection time structure shown in Fig. 2.25 are also tunable parameters that can be (and should be) chosen in order to improve the outcome of the treatment and the probability of success of the whole therapy of duration T and/or to improve the patient acceptability in terms of treatment/rest periods balance.

More precisely, the tunable parameters are:

- The duty cycle γ representing the ratio between the drug's injection sub-periods and the rest sub-periods.
- The duration of the basic period T_s over which the previous duty-cycle is defined.

Consequently, the vector of design parameters can be defined by the concatenation of the feedback law's vector of parameters p_c invoked in (2.107) and the above two time-schedule related parameters.

This leads to the following vector of parameters to be optimized in order to get acceptable certification of the outcome of the therapy:

$$\theta = \begin{bmatrix} p_c \\ \gamma \\ T_s \end{bmatrix}. \tag{2.108}$$

Now the question is the following:

Evaluating the Outcome?

How to evaluate the performance associated to a given choice of the design parameter θ defined by (2.108)?

How to evaluate the probability of a successful therapy when using a feedback law that is defined by the components of the design vector θ?

Answering this question is mandatory should one try to *optimize* the control design parameters.

One way to rigorously state the problem is to define a set of constraints on the overall therapy such that when these constraints are satisfied for a choice θ of the design parameter, this choice is declared to be **admissible**.

[58] The details are skipped intentionally since the objective here is to present the ingredients of the probabilistic certification tools and not to show how to concretely solve the specific combined therapy of cancer. For a complete example of such *parameterized* feedback strategy in the combined therapy of cancer, the reader can consult [42].

Moreover, in order to rank all the admissible choices, a cost function should be defined so that the best choice $\theta^\star$ can be taken to be the one, **among all admissible choices** that minimizes the cost function.

For instance, in the specific example of combined therapy of cancer, two constraints can be considered, namely:

1. **The health constraint** stating that (2.106) must hold true all the time and for all realizations of the uncertainty vector w[59]:

$$(\forall w) \qquad \max_{k\in\mathcal{K}}\Big[C_{min} - x_2(k|w,\theta,x_0)\Big] \le 0, \tag{2.109}$$

   where

   - $x_2(k|w,\theta,x_0)$ is the value of the health indicator x_2 at future instants $t_k = k\tau$ starting from the initial state x_0 under the control defined by the design parameter θ and the realization w of the uncertain parameters.
   - $\mathcal{K} := \{0,\dots,N\}$ where N is the number of sampling periods that covers the total duration T of the therapy.

2. **The tumor contraction constraint** stating that the size of the tumor at the end of the therapy duration is contracted compared to its initial size by a pre-defined target factor $\gamma_c \in (0,1)$ which is again to be enforced for any realization of the uncertainty vector w. This constraint can be formally written as follows:

$$(\forall w) \qquad \left[\frac{x_1(N|w,\theta,x_0)}{x_1(0)}\right] \le \gamma_c. \tag{2.110}$$

By defining the two constraints (2.109) and (2.110), one disposes of a criterion to separate *admissible* from *non admissible* values of the design parameters θ.

But one question remains unanswered in the above formulation:

Achievable Contraction Rates?

How is it possible to be sure that there is an admissible solution that contracts the tumor size with a predefined ratio γ_c? What if this is simply impossible given the initial state x_0 and the worse possible parameters value?

This question is about the feasibility issue and is commonly encountered in constrained optimization problems.

[59] This simply means that $\forall(t_k, w)$, one has: $C(t_k) \ge C_{min}$.

The solution is to *free* the parameter γ_c and add it to the design parameter θ that hence becomes:

$$\theta = \begin{bmatrix} \gamma_c \\ p_c \\ \gamma \\ T_s \end{bmatrix}, \tag{2.111}$$

and to set the cost function to be equal to $\theta_1 = \gamma_c$.

By so-doing, we are looking for a design parameter that maximizes the tumor contraction (by minimizing γ_c) while satisfying the health constraints.

This kind of choices and *tricks* always show up when setting the items of a probabilistic certification problem. There is never a single way of formulating the problem and that is why it is mandatory to address this kind of *problem-specific* step in the design process.

Based on the above discussion and notation, the problem can be stated in the following terms[60]:

> **Optimal Robust Constraints Satisfaction Problem**
> *For the given initial state* x_0, Find a feedback design parameter $\theta^\star$ that solves the following **optimal robust constraint satisfaction problem** given by:
>
> $$\min_\theta [\theta_1] \quad \text{under} \quad (\forall w) \quad g(\theta, w) = 0, \tag{2.112}$$
>
> where $g(\theta, w)$ is *failure*'s counting map defined by:
>
> $$g(\theta, w) := \begin{cases} 0 \text{ if (2.109)–(2.110) are satisfied,} \\ 1 \text{ otherwise,} \end{cases} \tag{2.113}$$
>
> and where θ is the design parameter defined by (2.111).

Formally speaking, the problem is well defined and seems quite sound. Unfortunately, this formulation shows many drawbacks and bottlenecks that are discussed in the next section.

[60] In the formulation and for the sake of the general presentation that follows, the explicit reference to the initial state is omitted.

2.11.2 Why Robust Constraints Satisfaction is Problematic?

Keeping in mind the above discussed illustrative example, a general optimal robust constraints satisfaction problem takes the following form:

$$\min_{\theta\in\Theta} J(\theta) \quad \text{under} \quad (\forall w) \quad g(\theta, w) = 0, \tag{2.114}$$

where $g(\theta, w)$ is defined by:

$$g(\theta, w) := \begin{cases} 0 \text{ if the specification constraints are satisfied,} \\ 1 \text{ otherwise,} \end{cases} \tag{2.115}$$

and where the statistics $\mathcal{W}$ of realizations of w is supposed to be available.

Notice that the word ***robust*** in the above formulation is associated to the fact that the constraint $g(\theta, w) = 0$ is required to hold for **any possible realization of the uncertain vector** w.

This requirement comes with two drawbacks:

1. The first is related to the **computational aspect**. Indeed, showing that a constraint is satisfied for all w is impossible to do for general nonlinear settings. One might say that it suffices to show that

$$\max_{w} |g(\theta, w)|^2 = 0.$$

   But this is generally a **non-convex** optimization problem and for such problems,[61] only local maxima can be found meaning that it is impossible to exclude that even if the result of the computation is 0, there might still exist values of w for which the constraint is violated.
   Moreover, it is even unlikely that there is no such w that induces constraints violation since we are not looking at the irrelevant character of such realizations. So not only is it possible to miss a non vanishing maximum of the above maximization problem from a computational point of view, it is even highly probable that the statement we would be trying to prove is generically wrong.
2. Leave alone the computational difficulty in addressing the problem, the formulation itself is uselessly pessimistic. This is because it asks for zero risk which is known to be contradictory with real-life problem's instances. There is always even a tiny risk of having a configuration of uncertainties that leads to harmful results, we just need to **check that the associated probability is sufficiently small to be accepted**.

[61] Not withstanding the fact that the function to be maximized is boolean and hence many efficient available solvers would not be eligible.

These are the reasons why alternative formulations incorporating risk assessment were proposed and are now widely used in the control community under the keyword: probabilistic certification and randomized optimization. These are briefly presented in the next section in a specific setting that is used in the remainder of the book.

2.11.3 Risk Assessment via Probabilistic Certification

The above mentioned robust constraints satisfaction problem can be softened through two steps:

1. In the first, the original $(\forall w)$-based robust formulation is replaced by the probability of constraints violation being sufficiently small, namely:

$$\min_{\theta \in \Theta} J(\theta) \quad \text{under} \quad \Pr_{\mathcal{W}}\Big\{g(\theta, w) = 1\Big\} \leq \eta. \tag{2.116}$$

   This formulation is conceptually less stringent than the original robust one as it accepts values of θ for which some residual subset of low probability realizations of w induces a violation of the constraints.
   Notice that the previously examined robust formulation can be obtained as a particular case by using $\eta = 0$. Therefore, η stands for a deterioration in the precision from 0 to $\eta \neq 0$. This is the reason why the parameter η **is referred to as the precision parameter** of the probabilistic certification.
   However, from the computational side, the formulation remains difficult to implement because of the multi-dimensional probability term that is not easy to compute. The second softening step comes into play in order to tackle this computational issue.
2. The second steps implements the intuitive averaging idea recalled in the beginning of this section which approximates the probability of occurrence by a counting events process. More precisely, the randomized methods [43] simplify (2.116) by replacing the probability by the mean value over N drawn independent identically distributed (i.i.d) samples of the uncertainty vector w according to the statistics defined by the presumably known $\mathcal{W}$.
   This leads to the new more tractable formulation given by:

$$\min_{\theta \in \Theta} J(\theta) \quad \text{under} \quad \frac{1}{N}\sum_{s=1}^{N} g(\theta, w^{(s)}) \leq \frac{m}{N} \leq \eta, \tag{2.117}$$

   which simply replaces the constraint on the probability by a different constraint stating that the mean value of $g(\theta, w^{(s)})$ over N random trials to be lower than $m/N \leq \eta$, or to state it differently that at most m between the total number N of samples lead to the violation of the specifications.

It comes therefore that, for each choice of the number m of samples showing constraints violation, N must be such that:

$$\frac{m}{N} \leq \eta, \tag{2.118}$$

which is obviously **only a necessary condition**.
Indeed, in addition to the condition (2.118) which is necessary for the approximated expectation to be lower than the prescribed precision level η, the number of sampled scenarios N must also be sufficiently large for the averaging-based approximation of the expectation to be sufficiently precise.
In other words, N must be sufficiently high in order to be allowed to legitimately infer from the fulfillment of (2.117) that the condition (2.116) on the probability is satisfied with a probability greater than $1 - \delta$ with a pre-specified small value δ.
That is the reason why the minimum value of N that makes the above-described two-step simplification relevant involves the following items:

- the level of precision specified by η,
- the level of confidence specified by δ, and
- the number of accepted instances of constraints violation defined by m.

In [43] and [44], several expressions of the minimal value of N are given under different settings. In this book, we'll restrict ourselves to the simple case where the set of candidate values of the design vector θ is **a discrete set** Θ of cardinality n_Θ. Note that this cardinality is to be distinguished from the dimension n_θ of the vector θ itself.[62]

In this specific case, the following proposition holds true:

> **Certification Rule**
> **Proposition 2.2** (N samples for probabilistic certification) *Let $m \in \mathbb{N}$ be any integer representing the admissible number of unsuccessful samples. Let $\delta \in (0, 1)$ be a targeted confidence parameter for the certification and let $\eta \in (0, 1)$ be the targeted precision parameter.*
> *Take N satisfying the following inequality:*
>
> $$N \geq \frac{1}{\eta}\left(m + \ln(\frac{n_\Theta}{\delta}) + \left(2m\ln(\frac{n_\Theta}{\delta})\right)^{1/2}\right), \tag{2.119}$$

[62] As an example, the set:

$$\Theta := \left\{\begin{bmatrix}1\\2\end{bmatrix}, \begin{bmatrix}-1\\0\end{bmatrix}, \begin{bmatrix}3\\-6\end{bmatrix}\right\}$$

is a set of cardinality 3 while its elements lie in $\mathbb{R}^2$. Therefore, $n_\theta = 2$ and $n_\Theta = 3$.

where $n_\Theta := \text{card}(\Theta)$ *then any solution to the optimization problem:*

$$\min_{\theta \in \Theta} J(\theta) \quad \text{under} \quad \frac{1}{N}\sum_{s=1}^{N} g(\theta, w^{(s)}) \le \frac{m}{N} \le \eta, \tag{2.120}$$

in which the realizations

$$\left\{w^{(s)}\right\}_{s=1}^{N},$$

are randomly i.i.d drawn using the probability measure $\mathcal{W}$ *satisfies the constraint in* (2.116) *with a probability* $\ge 1 - \delta$*, provided that the true probability of dispersion of the uncertainties* w *is governed by* $\mathcal{W}$*.*

A remarkable property of the expression (2.119) enabling the computation of N is that it is totally independent of the dimension of w[63] (the number of parameters involved in the dynamic model).

This property is of a tremendous importance in the context of application examples such as the certified therapy of cancer since there are typically a quite high number of parameters.[64] It is a rather good news that this does not influence the number of samples that are needed to define the constraint in the optimization problem (2.117). This is a rather counter intuitive property for a first simple examination.

Another interesting feature of Proposition 2.2 is that the confidence parameter δ appears through a logarithmic term which means that **it is possible to seek highly confident assertions without dramatic increase in the number of trials**.

Table 2.1 shows the value of N for different precision levels (η) when the confidence level $\delta = 10^{-3}$ is used and the number of failures $m = 1$.

Table 2.1 Typical values of the sample size N (number of samples needed to achieve the certification) satisfying (2.119) as a function of the precision η and the cardinality n_Θ of the design parameter set Θ (confidence parameter $\delta = 10^{-3}$ and $m = 1$ are used)

n_Θ	$\eta = 0.1$	$\eta = 0.05$	$\eta = 0.01$	$\eta = 0.001$
1	132	264	1317	13164
5	154	308	1536	15354
10	163	326	1628	16280
100	193	386	1930	19299
1000	223	445	2225	22249
10000	252	503	2515	25148

[63] The number of needed samples N does not depend on the dimension of the uncertainty vector w.
[64] These are typically the gains associated to each functional term in the compartmental model.

An example of information delivered by this table is precisely formulated as follows:

Reading Table 2.1
When 1000 options for the setting of the design parameter θ are considered (n_Θ=1000), if one of these values, say $\theta^\star$, lead to no more than a single failure (m=1) among (N=2225) samples of the uncertainty vector, one can assert with 99.9% of confidence (δ=0.001) that when using $\theta^\star$ the requirements expressed through g will be satisfied in 99% of the cases ($\eta = 0.01$).

In the next section, other probabilistic certification frameworks addressing specific problems that might be helpful in formulating and solving some other control-related issues under uncertainties are briefly presented.

Links to Other Chapters
The probabilistic certification is used many times in different chapters:

- In Chap. 6, the framework is used to design a certifiable choice of the NMPC implementation parameters.
- Chapter 14 shows the application of the framework to the certification of the Quality of Service (QoS) of an Electric Vehicle Charging Station (EVCS).
- Chapter 15 shows its role in the design of Propofol injection during anesthesia that provides certifiable result despite of large discrepancies between patients.

2.11.4 Other Probabilistic Certification Frameworks

In this section, some definition and results suggested by [30] are presented that might be useful to address some interesting problems in the presence of uncertainties.

More precisely, we are interested in the solution of the following class of optimization problems whose formulation is defined through some badly known parameters:

$$\theta^\star \leftarrow \min_{\theta\in\Theta} \quad \mathbb{E}_{\mathcal{W}}\Big[J(\theta|\cdot)\Big], \tag{2.121}$$

where

- θ is some decision/design variable that is supposed to belong to some bounded set $\Theta \in R^{n_\theta}$. This set is not related to the set mentioned in the previous section and needs not to be discrete.
- w is some uncertain vector that is governed by the statistics defined by $\mathcal{W}$.

- $J(\theta, w)$ is the value of some cost function at the pair (θ, w) of decision/design variable θ and the realization w of the uncertain vector. **It is assumed in this section that the cost function values lie inside the interval** $[0, 1]$.[65]
- $\mathbb{E}_{\mathcal{W}}[J(\theta|\cdot]$ is the expectation of the cost function at the decision variable value θ over possible realizations of the uncertainty w.

Recall that we already encountered an instantiation of this problem in Sect. 2.6.2 when invoking the formulation of stochastic OCPs. In this context, θ is represented by the sequence of control inputs $\mathbf{u}$, the uncertainty vector w is represented by the uncertainty profile $\mathbf{w}$ over the prediction horizon while the cost function is simply the cost function to be minimized whic expresses the control objective.

In order to state the probabilistic certification result, we need to state and illustrate the following definition [see also Fig. 2.26]:

Probably Approximate Near Minimum
Definition 2.5 Let the following be defined:

- $f : \Theta \rightarrow [0, 1]$ is a function to be minimized.
- $f^\star$ the true minimum of f.
- P_θ is a given probability measure on Θ.
- Let $\varepsilon, \alpha > 0$ be given numbers.

A number f_0 is said to be a **probably approximate near minimum** of $f(\cdot)$ to accuracy ε and level α, if the following conditions hold true:

1. $f^\star \leq f_0 + \varepsilon$,
2. $P_\theta \Big\{\theta \in \Theta : f(\theta) < f_0 - \varepsilon\Big\} \leq \alpha$.

Figure 2.26 helps illustrating this rather complicated statement. Let us try to better examine it.

Keep in mind that f_0 is a candidate approximation of $f^\star$. The statement suggests that this approximation would be quite accurate provided that ε and α are quite small. In other words, we need to show that $f^\star$ cannot go *far from* f_0 in whatever direction. Regarding the upward direction, we know that $f^\star$ cannot be greater than f_0 by more than ε because of the first condition that reads: $f^\star \leq f_0 + \varepsilon$. In the other direction ($f^\star \leq f_0$), things are less strict but the second condition stipulates that the probability that $f^\star$ lies below $f_0 - \varepsilon$ is associated to a probability that is lower than α.

[65] This is always formally possible to meet by applying a strictly monotonic shaping function to any originally defined cost function. It should be checked however whether this change can induce a change in the optimal solution because of the expectation operator.

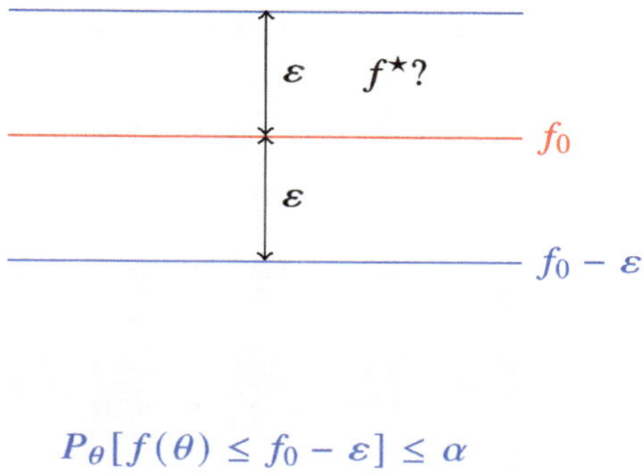

Fig. 2.26 The concept of probably approximate near minimum: the computed value is f_0, the true minimum is $f^\star$. The consequence of definition 2.5 is that the probability of $f^\star$ lying outside the green region is lower that α

Now that we *fully* understand this concept of probably approximate near minimum, let us examine an algorithm that enables to compute such approximation that applies to the problem stated by (2.121).

Obviously the function f here is defined to be the expectation of J for the given value of the decision variable θ, therefore, we shall need to sample some representative

$$w^{(s)},\, s = 1, \ldots, n_s,$$

of the uncertainties in order to reconstruct an approximation of the expectation.

On the other hand, since we seek a sampled based computation of the optimal solution, a set of samples $\theta^{(i)}$, $i = 1, \ldots, n$ of the decision variables is used.

These two populations are involved in the following algorithm [30]:

Algorithm 2.1

Choose a confidence level $\delta \in (0, 1)$ and $\alpha \in (0, 1)$.
Compute n and n_s according to:

$$n \geq \frac{\log(2/\delta)}{\log(1/(1-\alpha))} \quad \text{and} \quad n_s \geq \frac{1}{2\varepsilon^2}\ln(\frac{4n}{\delta}). \tag{2.122}$$

Generate i.i.d samples $\theta^{(1)}, \ldots, \theta^{(n)} \in \Theta$ according to P_θ.
Generate i.i.d samples $w^{(1)}, \ldots, w^{(n_s)}$ according to $\mathcal{W}$.
Compute:

$$\hat{J}_i := \frac{1}{n_s}\sum_{s=1}^{n_s} J(\theta^{(i)}, w^{(s)}) \quad ; \quad \hat{J}_0 := \min_{1\leq i\leq n} \hat{J}_i. \tag{2.123}$$

Then with confidence $1 - \delta$, it can be stated that $\hat{f}_0$ is a probably approximate near minimum of the cost function invoked in (2.121) to accuracy ε and level α.

[!] Recall that an underlying assumption here is that

$$J \in [0, 1].$$

This can be done using normalization step or by using `tanh`-based transformation to cite but some possibilities.

Figure 2.27 shows typical values of the pairs (n, n_s) for a confidence parameter $\delta = 10^{-2}$ and for different choices of the parameters ε and α.

Notice that the last column suggests some boundary effect of the results since it is assumed as a preliminary assumption that the cost function lies in [0, 1] meaning that when considering a precision $\epsilon = 1$ one can be certain that probability of lying outside the interval $[f_0 - 1, f_0 + 1]$ is literally equal to 0.

It is important to underline that Algorithm 2.1 involves three tuning parameters, namely:

- The confidence level $\delta \in (0, 1)$ which sets the level of confidence at which stands the whole statement of the results associated to the algorithm.
- The precision $\epsilon > 0$ associated to the quality of the approximation of the minimum (see Fig. 2.26), namely, the distance between the computed minimum and the real minimum which is guaranteed only to a certain level of probability α.
- The level of probability $\alpha \in (0, 1)$ to which the relevance of the precision ε holds.

Notice that in order to perform the computation needed by Algorithm 2.1, the number of evaluation instances of the cost function is equal to the product $(n \cdot n_s)$ which can be quite an large and dissuasive number.

	eps=0.01	eps=0.05	eps=0.1	eps=0.5	eps=1
alpha=0.005	(1058, 64779)	(1058, 2592)	(1058, 648)	(1058, 26)	(1058, 7)
alpha=0.01	(528, 61303)	(528, 2453)	(528, 614)	(528, 25)	(528, 7)
alpha=0.05	(104, 53180)	(104, 2128)	(104, 532)	(104, 22)	(104, 6)
alpha=0.1	(51, 49617)	(51, 1985)	(51, 497)	(51, 20)	(51, 5)

Fig. 2.27 Typical values of the pairs of integer (n, n_s) invoked in (2.122) which are used in the sampling process for different values of the parameters (ε, α) when the confidence parameter is taken equal to $\delta = 10^{-2}$

2.12 Conclusion

In this chapter some main concepts in control under uncertainties are recalled in a rather informal way. The presentation of these concepts and results serve as a basis for the understanding of the ideas and heuristics examinedmail in the next chapters.

As such, this chapter is not an official course in either of the explored topics as each one of these deserves an academic book for its own sake given the available underlying results over many years.

Consequently, it is obvious that this chapter condenses too many concepts and frameworks to be *absorbed* in a single first reading. That is the reason why the reader is encouraged to come back to the present chapter as soon as a topics explored in the next chapters show strong links with one of the concepts described in this chapter that might be forgotten.

References

1. Khalil, H.: Nonlinear systems. Pearson Education. Prentice Hall (2002). https://books.google.fr/books?id=t_d1QgAACAAJ
2. Mesbah, A.: Stochastic model predictive control: An overview and perspective for future research. IEEE Control Syst. Mag. **36**(6), 30–44 (2016)
3. Morari, M., H. Lee, J.: Model predictive control: past, present and future. Comput. Chem. Eng. **23**(4), 667–682 (1999). https://doi.org/10.1016/S0098-1354(98)00301-9. https://www.sciencedirect.com/science/article/pii/S0098135498003019
4. Mayne, D.Q., Rawlings, J., Rao, C.V., Scokaert, P.O.M.: Constrained model predictive control: Stability and optimality. Automatica **36**, 789–814 (2000)
5. Richalet, J., Rault, A., Testud, J., Papon, J.: Model predictive heuristic control: Applications to industrial processes. Automatica **14**(5), 413–428 (1978) https://doi.org/10.1016/0005-1098(78)90001-8. https://www.sciencedirect.com/science/article/pii/0005109878900018
6. Cutler, C.R., Ramaker, B.L.: Dynamic matrix control?? a computer control algorithm. In: Joint Automatic Control Conference, 17, p. 72 (1980)
7. Keerthi, S.S., Gilbert, E.G.: Optimal infinite horizon feedback laws for a general class of constrained discrete-time systems: Stability and moving horizon approximations. J. Optim. Theory Appl. **57**, 265–293 (1988)
8. Mayne, D.Q., Michalska, H.: Model predictive control of nonlinear systems. In: 1991 American Control Conference, pp. 2343–2348. IEEE (1991)
9. Alamir, M., Bornard, G.: On the stability of receding horizon control of nonlinear discrete-time systems. Syst. Control Lett. **23**(4), 291–296 (1994)
10. Alamir, M., Bornard, G.: Stability of a truncated infinite constrained receding horizon scheme: the general discrete nonlinear case. Automatica **31**(9), 1353–1356 (1995)
11. Richalet, J., Rault, A., Testud, J.L., Papon, J.: Algorithmic control of industrial processes. In: Proceedings of the 4th IFAC symposium on indentification and plant parameter estimation, pp. 1119–1167 (1976)
12. Alamir, M., Bornard, G.: Stability of a truncated infinite constrained receding horizon scheme: the general discrete nonlinear case. Automatica **31**(9), 1353–1356 (1995). https://doi.org/10.1016/0005-1098(95)00042-U, https://www.sciencedirect.com/science/article/pii/000510989500042U
13. Kothare, S., de Oliveira, L., Morari, M.: Contractive model predictive control for constrained nonlinear systems. IEEE Trans. Autom. Control **45**, 1053–1071 (2000)

14. Alamir, M.: Contraction-based nonlinear model predictive control formulation without stability-related terminal constraints. Automatica **75**, 288–292 (2017)
15. Alamir, M.: Stability proof for nonlinear mpc design using monotonically increasing weighting profiles without terminal constraints. Automatica **87**, 455–459 (2018)
16. Alamir, M.: Numerical investigation regarding an mpc scheme with non uniformly weighted stage cost without terminal constraints: Application to the control of a real-life cryogenic plant. IFAC-PapersOnLine **51**(20), 418–423 (2018). 6th IFAC Conference on Nonlinear Model Predictive Control NMPC 2018
17. Bonne, F., Alamir, M., Bonnay, P.: Experimental investigation of control updating period monitoring in industrial PLC-based fast MPC: Application to the constrained control of a cryogenic refrigerator. Control Theory Technol. **15**(2), 92–108 (2017)
18. Bonne, F., Alamir, M., Bonnay, P.: Nonlinear observer of the thermal loads applied to the helium bath of a cryogenic joule-thompson cycle. J. Process Control **24**(3), 73–80 (2014)
19. Bock, H.G., Diehl, M., Leineweber, D.B., Schlöder, J.P.: A direct multiple shooting method for real-time optimization of nonlinear dae processes. In: Allgöwer, F., Zheng, A. (eds.) Nonlinear Model Predictive Control, pp. 245–267. Birkhäuser Basel, Basel (2000)
20. Alamir, M.: Monitoring control updating period in fast gradient based nmpc. In: 2013 European Control Conference (ECC), pp. 3621–3626 (2013). https://doi.org/10.23919/ECC.2013.6669122
21. Alamir, M.: Explicit approximation of stochastic optimal feedback control for combined therapy of cancer (2019). https://doi.org/10.48550/ARXIV.1905.04937. https://arxiv.org/abs/1905.04937
22. Jaakoola, T., Jordan, M.I., Singh, S.P.: On the convergence of stochastic iterative dynamic programming algorithms. J. Mach. Learn. Res. **6**(6), 1185–1201 (1994)
23. Bertsekas, D.P.: Dynamic programming and optimal control. Athena Sci. (2017)
24. Kirkby, R.: Convergence of discretized value function iteration. Comput. Econ. **49**(1), 117–153 (2017)
25. Pedregosa, F., Varoquaux, G., Gramfort, A., Michel, V., Thirion, B., Grisel, O., Blondel, M., Prettenhofer, P., Weiss, R., Dubourg, V., Vanderplas, J., Passos, A., Cournapeau, D., Brucher, M., Perrot, M., Duchesnay, E.: Scikit-learn: Machine learning in Python. J. Mach. Learn. Res. **12**, 2825–2830 (2011)
26. Silver, D., Lever, G., Heess, N., Degris, T., Wierstra, D., Riedmiller, M.: Deterministic policy gradient algorithms. In: International conference on machine learning, pp. 387–395. Pmlr (2014)
27. Stevens, E., Antiga, L., Viehmann, T.: Deep learning with PyTorch. Manning Publications (2020)
28. Ketkar, N., Ketkar, N.: Introduction to keras. Deep learning with python: A hands-on introduction, pp. 97–111 (2017)
29. Abadi, M., Barham, P., Chen, J., Chen, Z., Davis, A., Dean, J., Devin, M., Ghemawat, S., Irving, G., Isard, M., et al.: {TensorFlow}: a system for {Large-Scale} machine learning. In: 12th USENIX symposium on operating systems design and implementation (OSDI 16), pp. 265–283 (2016)
30. Vidyasagar, M.: Randomized algorithms for robust controller synthesis using statistical learning theory. Automatica **37**, 1515–1528 (2001)
31. Bottou, L.: Stochastic gradient descent tricks. In: Neural networks: tricks of the trade: second edition, pp. 421–436. Springer (2012)
32. Allen-Zhu, Z., Li, Y., Song, Z.: A convergence theory for deep learning via over-parameterization. In: International conference on machine learning, pp. 242–252. PMLR (2019)
33. Garcıa, J., Fernández, F.: A comprehensive survey on safe reinforcement learning. J. Mach. Learn. Res. **16**(1), 1437–1480 (2015)
34. Gu, S., Yang, L., Du, Y., Chen, G., Walter, F., Wang, J., Knoll, A.: A review of safe reinforcement learning: Methods, theory and applications. arXiv preprint arXiv:2205.10330 (2022)

35. Hou, Z.S., Wang, Z.: From model-based control to data-driven control: Survey, classification and perspective. Inf. Sci. **235**, 3–35 (2013)
36. Tøndel, P., Johansen, T.A., Bemporad, A.: An algorithm for multi-parametric quadratic programming and explicit mpc solutions. Automatica **39**(3), 489–497 (2003)
37. Borrelli, F., Baotić, M., Pekar, J., Stewart, G.: On the complexity of explicit mpc laws. In: 2009 European Control Conference (ECC), pp. 2408–2413. IEEE (2009)
38. Alamir, M.: Learning against uncertainty in control engineering. Ann. Rev. Control **53**, 19–29 (2022)
39. Alamir, M.: Partial extended observability certification and optimal design of moving-horizon estimators. IEEE Trans. Autom. Control **67**(7), 3663–3669 (2022). https://doi.org/10.1109/TAC.2021.3108490
40. Eckhardt, R.: Stan ulam, john von neumann. Los Alamos Science, p. 131 (1987)
41. Metropolis, N., Ulam, S.: The monte carlo method. J. Am. Stat. Assoc. **44**(247), 335–341 (1949)
42. Alamir, M.: On probabilistic certification of combined cancer therapies using strongly uncertain models. J. Theoret. Biol. **384**, 59–69 (2015)
43. Alamo, T., Tempo, R., Camacho, E.F.: Randomized strategies for probabilistic solutions of uncertain feasibility and optimization problems. Autom. Control IEEE Trans. **54**(11), 2545–2559 (2009)
44. Alamo, T., Tempo, R., Luque, A., R., R.D.: Randomized methods for design of uncertain systems: sample complexity and sequential algorithms. Automatica **52**, 160–172 (2015)

Quick Python Reminders and Key Modules

3

M. Alamir, *Nonlinear Control of Uncertain Systems*,
https://doi.org/10.1007/978-3-031-93287-8_3

3.1 Chapter Content at a Glance

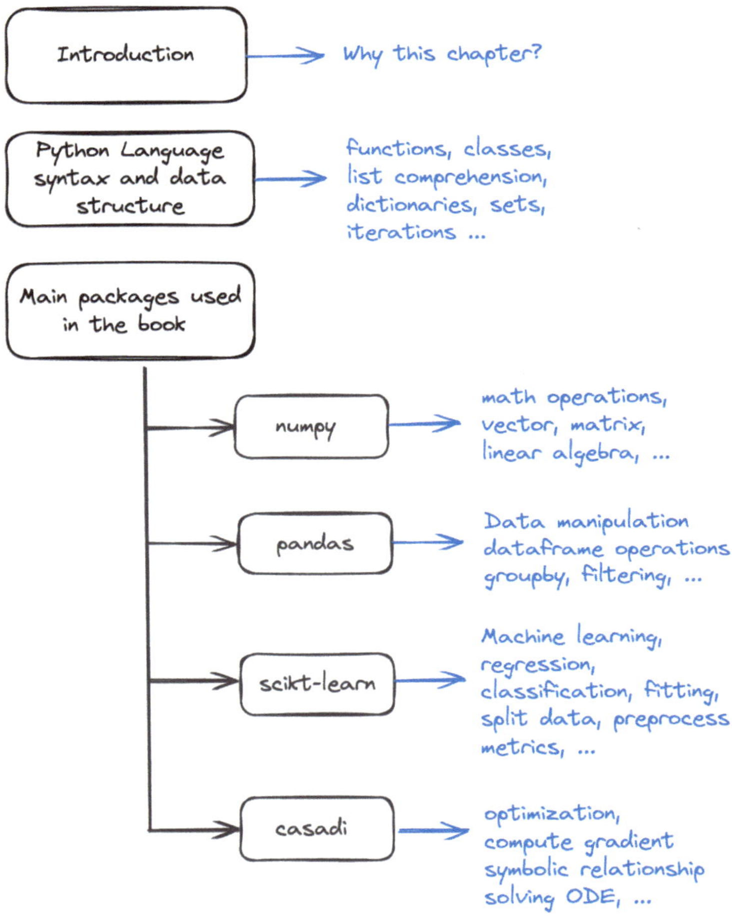

3.2 Introduction

This chapter is not a Python tutorial!

It is a collection of reminders and hints that enable an easier and more profitable reading of the code given here and there throughout the next chapters.

The content of this chapter could have been placed in an appendix at the end of the book. But as a reader, I would prefer to have it here in order to be aware of the different notions and modules that are extensively used in the book. These are obviously a small part of what can be known about Python. So anyway, should it be reported in a late appendix, I would read it before proceeding to the next chapters, so what is the point in putting it later?

As I mentioned earlier, *I am anything but a Geek* and in particular, I am not a *native*-`Python` programmer. I learned this beautiful language 6 years before I wrote this book. Consequently, the pieces of code included in this book are not necessarily optimal, or, as geeks would say, *not pythonistic* enough!

Nevertheless, the items or specific `Python` instructions and elements of syntax mentioned in this chapter are somehow an image of what I extensively used to produce the results. Some of them are specifically mentioned because I found them *tricky* and were very repetitively the source of my coding errors. So I thought that mentioning them would spare some of an average reader's own precious debugging time.

This chapter quickly covers the following items:

✓ First of all, **general information** are given regarding the `Python` *files, functions, classes, dictionaries, lists* that are the building bricks of the language. Moreover, the principles underlying the `Python` *modules* and *packages* are also recalled.

✓ **The main packages** used in this book are successively introduced together with the main calls that come very often in the codes of this book.

More precisely, the following packages are discussed:

`numpy`: The main `Python` package for manipulating numbers, arrays, and matrices. It also enables to use linear algebra's very efficient utilities.

`pandas`: The main `Python` package for manipulating data through the so-called *dataframes* used in the data-science-related computation and manipulation.

`scikit-learn`: The main `Python` package for data-science algorithms performing all possible tasks such as preprocessing, features selection and generation, classification, clustering, and regression.

`casadi`: This is a nice, easy-to-use, and powerful `Python` framework for the formulation and the efficient solution of many engineering problems such as ODE, DAE integration, symbolic differentiation, and constrained optimization.

`matplotlib`: The main `Python`'s visualization package.

This simple enumeration of the packages alone immediately shows that the chapter can only be an entry point to these excellent resources.

Notice however that this chapter has been written *recursively* by looking at the codes and identifying what would be needed to remind or to comment for the reader to smoothly understand why things are written the way they are.

Consequently, it might be necessary to go back regularly to this chapter when some chunks of code seem weird or difficult to understand.

Regarding the Python installation procedure, please have a look on the official Python website depending on your operating system:

www.python.org.

It is also possible to temporarily skip the installation step by using on-line Python environments such as:

https://www.online-python.com,
https://colab.research.google.com,

in order to try small *snippets* and get familiarized with the language. Nevertheless, this might become quite limited very quickly and I advise to make the effort to get a proper local installation.

3.3 General Overview of the Python-Related Concepts and Syntax

1. A Python file shows the py extension: `filename.py`.
   The file might contain a list of functions, classes, instructions for the import of needed packages together with executable commands in almost any order.
2. Assume a first file `file1.py` containing the following code implementing two functions named add and `multiply` respectively:

   A File is a Module

   ```
   # file1 (this is just a comment)

   def add(x,y):
       return x+y

   def mult(x,y):
       return x * y
   ```

   then another file `file2.py` **living in the same folder** can import the two functions and use them as follows:

   Importing Modules

   ```
   from file1 import add, mult

   x1, x2 = 2.3, -2
   ```

```
s = add(x1, x2)
m = mult(x1, x2)

print(s, m)

>> 0.3, -4.6
```

Using this `import` feature, one can gather the *utilities* in different files (organized by topic, leading to a so-called `Python module`) and use them in different other files which *import* these utilities from the module.

3. **There is no mandatory type declaration in Python**. The following screenshot shows different types of variables:

Variables Types

```
a = 1.2 # float
b = int(a) # integer
c = "hello"     # string
d = [1, "Oh my god", 2, 3] #list
e = d[1] # e="Oh my god"
f = d[0] # d=1
g = {1, 2, 3, "a"} # non ordered set -> s[0] is illegal
```

---

4. Indexing in `Python` start with 0. This is a typical loop:

Loops on Ranges

```
for i in range(5):
print(i)

>> 0, 1, 2, 3, 4 # notice that 5 is not included.
```

---

5. It is possible to loop over sequences and sets among many other `iterables`:

Loops on Lists

```
L = [1, "C", 1.3]
for e in L:
print(e)
```

```
>> 1, "C", 1.3
```

6. There are many operations on lists, in particular the **append** operation

Operations on Lists

```
L1 = [1, 2, 3]
L2 = [5, 19]
L3 = L1 + L2
L4 = L2.append(-1)

>> L3 = [1, 2, 3, 5, 19]
>> L4 = [15, 19, -1]
```

7. This is how lists can be sliced and specific elements can be chosen:

Slicing Lists

```
L = [1, 2, 3, 4, 5, 6]

print(L[-1]) # last element of the list
print(L[1:4]) # elements of index 1, 2, 3

>> 6
>> [2, 3, 4] # notice again that the last index is missing
```

8. **List Comprehension** is extensively used to build lists from iterators and selection rules, here are some examples:

Lists Comprehension

```
L = [i for i in range(10) if np.mod(i,2)==0]
# Even integer lower than 9
print(L)

>> [0, 2, 4, 6, 8]

R = ['c1', 'c3', 'a1', 'b23', 'abc6']
Lc = [x for x in R if 'c' in x] # elements containing 'c'.
print(Lc)
```

```
>> ['c1', 'c3', 'abc6']
```

---

9. Dictionaries are composed of `(key, value)` pairs

Dictionaries

```
d = {"k1":5, "k2":-3.4, 5:"Good", (1,2):[1, 2, 3]}
print(d["k1"], d[5], d[(1,2)], d[(1,2)][2])

>> 5, "Good", [1,2,3], 3
```

---

10. The elements (key and values) of a dictionary can be iterated over simultaneously using the dictionary `items()` iterator as it is shown below:

Iterations on Dictionaries

```
d = {"k1":5, "k2":-3.4, 5:"Good", (1,2):[1, 2, 3]}

for k, v in d.items():
 print(k, v)

>> "k1", 5
>> "k2", -3.4
>> 5, "Good"
>> (1,2), [1, 2, 3]
```

---

11. A Python class encapsulates attributes and methods that are associated to an object (instances of the class). This is a very important concept that cannot be fully introduced in this book, the syntax is the following:

Classes

```
#----- definition of the class -------
class Cat:

 # function to create an instance of the class
 def __init__(self, age, sex):
    # attributes of the class
  self.age = age
```

```
        self.sex = "M"
    # Methods exposed by the class
    # method to print info
    def print_info(self):
        print(f'I am a {self.sex} cat aged {self.age}')

    # method to talk
    def make_noise(self, sentence):
        message = sentence + " Miaaaaou!"
        return message

# ---- Use of the class -----

Minou = Cat(age=2, sex="M")
print(Minou.age)
print(Minou.sex)
print(Cat.make_noise("Hello"))

>> 2
>> "M"
>> "Hello Miaaaaou!"
```

Note that inside a method, new attributes of the class might be created to enrich the list of its attributes although this is not recommended. This might be the case when some attributes are not entirely defined unless a method is called with some exogenous parameters that are not available at the class instantiation step.[1]

12. A very helpful `fstring` formatting is worth considering when printing some expressions that depends on some computed arguments:

fstrings

```
x = 5
print(f'The value of x = {x}')

>> The value of x = 5
```

[1] Even in this case, it is recommended to define all the possible future attributes at the instance creation (inside the `__init__` function with the generic `None` value that can later be replaced by the outcome of some method.

13. **List comprehension versus iterator**: Assume that a list L of matrices is created via list comprehension and the resulting elements of the list are used to successively update the value of some variable K as in the following code:

List Comprehension vs Iterators (1)

```
def some_function(x)
    ...
    return M # some matrix computed using x

L =  [some_function(x) for x in X]
K = ...
for M in L:
    # process using M
    K = K + function(M)
    if condition(K):
        break
```

Notice that the iterations stops if some condition on K is encountered. Notice that in the above code the list is created first and then used. This means that the function `some_function` will be called for all the item in the set `X` regardless of what will happen later when the items of the list will be used.
So if X contains 10^6 elements and even if each matrix `M(x)` lies in $\mathbb{R}^{1000\times 1000}$, the list L created via list comprehension contained 10^{12} elements. This might induce two kind of problems:

a. the first is obviously a memory issue;
b. the second might also be a computational issue. Indeed, what if the execution of `some_function` is computationally expensive and what if the computational loop is interrupted by the `condition(K)` being validated after only using the five first elements of the list?
   In this case, a million-elements list L would have been computed (and stored!) while only five of them where necessary.

Here where the concept of `iterator` becomes useful and at almost no additional complication. Indeed, we have only to modify the previous code as follows:

List Comprehension vs Iterators (2)

```
def some_function(x)
    ...
    yield M #            x <--- (yield instead of return)

L =  (M(x) for x in X)   #<------- () instead of []
K = ...

for M in L:
    # process using M
    K = K + function(M)
    if condition(K):
        break
```

where only the two modifications have been introduced, namely:

a. The `return` word has been replaced by `yield`.
b. The parentheses `()` are used instead of the brackets `[]` in the definition of L.

By doing so, L is no more a list, but what is called an *iterator*. The difference is that the elements of the iterator **are not computed before the iteration**. Rather, they are computed on the fly as the iteration proceeds.
This obviously remove the two drawbacks regarding the memory and the computation of useless terms.

This gives a very brief overview of the main syntax that will be encountered in the forthcoming scripts.

In the remainder of this chapter, some of the main Python packages used in the book are presented and their main syntax is presented.

3.4 The Main Packages Used in the Book

3.4.1 The numpy Package

This is the main numerical manipulation package. It handles numbers, multidimensional arrays, linear algebra manipulation (eigenvalues, eigenvectors, rank, kernel, solving least-squares problems, perform singular values decomposition to cite but few examples).

These are the features that are worth recalling regarding this package.

1. When importing this package, the common practice is to import it while renaming it using the short name np. This is done using the following call:

   `import numpy as np.`

2. When matrices are concerned, a row-based approach is adopted, the row are represented using lists and the matrix is represented as a list of rows[2]:

Matrix Creation and Lines/Columns Selection

```
import numpy as np

M = np.array([[1, 2, 3], [4, 5, 6]]) # this is a 2x3 matrix
MT = M.T # transpose of M
M[0,:] # first line of M
M[:,1] # second column of M
```

This leads to the matrix M given by

$$M = \begin{bmatrix} 1 & 2 & 3 \\ 4 & 5 & 6 \end{bmatrix} \in \mathbb{R}^{2\times 3}.$$

3. **Matrix multiplication** can be performed using either the `np.matmul` function or the @ operator:

Matrix Multiplication

```
M1 = np.array([[1, 2, 3], [4, 5, 6]]) # 2 x 3 matrix
M2 = np.array([[1, 2], [2, 3], [0,1]])  # 3 x 2 matrix

P1 = np.matmul(M1, M2)
P2 = M1 @ M2

print(P1)
print(P2)

>> array([[5, 11], [14, 29]]) # 2 x 2 matrix
>> array([[5, 11], [14, 29]]) # 2 x 2 matrix
```

[2] Note that the argument of `np.array()` here is the list of list given by `[[1, 2, 3], [4, 5, 6]]` which is a list of two elements: the first one is the list [1, 2, 3] representing the first line and the second element is the list [4, 5, 6] representing the second line.

4. **Element-wise multiplication and division** can be performed using `np.multiply` and `np.divide` functions or simply by using the (*) and (/) operators as shown below:

Element-wise Matrix Multiplication

```
import numpy as np

M1 = np.array([[1,2], [3,4]])
M2 = np.array([[2,3], [-1,1]])

print(M1*M2)

>>
[[ 2  6]
 [-3  4]]
```

which shows the results of multiplying element-wise the two matrices

$$M_1 = \begin{bmatrix} 1 & 2 \\ 3 & 4 \end{bmatrix} \quad \text{and} \quad M_2 = \begin{bmatrix} 2 & 3 \\ -1 & 1 \end{bmatrix}.$$

5. **Concatenation of matrices** can be obtained using the functions `np.hstack` (horizontal) and `np.vstack` (vertical). The argument of these function is a list matrices to be concatenated horizontally or vertically. Here are some examples:

Matrix Concatenation

```
M1 = np.array([[1, 2, 3], [4, 5, 6]])
Mh = np.array([[0], [-1]])
Mv = np.array([[0, 1, -1], [2, 2, 6], [-2, -3, -4]])

print(np.hstack([M1, Mh]))
print(np.vstack([M1, Mv]))

>> array([[ 1,  2,  3,  0],
       [ 4,  5,  6, -1]])

>> array([[ 1,  2,  3],
       [ 4,  5,  6],
       [ 0,  1, -1],
       [ 2,  2,  6],
       [-2, -3, -4]])
```

6. The **shape** of a numpy object can be investigated using its shape **attribute** as shown below:

Matrix Reshaping

```
M1 = np.array([[1, 2, 3], [4, 5, 6]])
b = np.array([1, 2, 3])
b_modified = b.reshape(-1,1)
M3 = M1.reshape(3,2)

print(M1.shape, b.shape, M1.shape[0], b_modified.shape, M3)

>> (2,3)
>> (3,)
>> 2
>> (3, 1)
>> array([[1, 2],
       [3, 4],
       [5, 6]])
```

Note how the reshape method reshapes the numpy object to new shape on a row first basis. when –1 is used in one of the two dimensions, the method set the remaining dimension as required and the dimension associated to –1 inherits the resulting dimension.

7. The module numpy.random offers many utilities for the generation of random objects that might be scalar, arrays or matrices with float or integer type. In particular, the numpy.random.permutation enables a random permutation of the elements of a list given as argument.
   Notice that the "." in the expression np.random.randn means that the package numpy contains a sub-module called random which exports a function called randn which generates random matrix normally distributed with unity variance.

3.4.2 The Pandas Package

The pandas package is a highly efficient and widely used package in Data-Science. It enables to handle efficiently very large datasets on which it offers an impressive number of operations and transforming actions.

The pandas package operates on so-called dataframe objects which take the form shown Fig. 3.1. Notice that this is not the simplest way to construct the dataframe of Fig. 3.1. The construction is intentionally divided into several steps in order to illustrate some features used in pandas.

Creating DataFrames

```
import numpy as np
import pandas as pd

X = 10 * np.random.randn(8,4)
# setting the index names
I = [f'#{i+1}' for i in range(len(X))]
# setting the columns names
col = ['A', 'B', 'C', 'D']
# define the dataframe with the named index and list of columns
df = pd.DataFrame(X, index=I, columns=col)
n = len(df)
# A new column is created on the fly
df['E'] = [f'a_{i}' for i in range(n-1)]+[5]
```

Fig. 3.1 An example of a pandas dataframe **a** and the Python code used in its construction **b**

Regarding the previous script, the following comments help understanding some facts and techniques that of great help:

1. First a random matrix of dimension (8×4) is created using the `numpy.random.randn` function.
2. A list `I` is created that will serve as the value of the index argument for the creation of the `dataframe`. The index can be viewed as the list of tags of the different lines in the `dataframe`.
3. A list `col` is created that will serve as the value of the columns argument for the creation of the `dataframe`; this is the list of tags for the columns in the `dataframe`.
4. A `dataframe` called `df` is created using the `pandas` function `DataFrame` which creates an instance of the `dataframe` class.
5. At this stage, the `dataframe` contains only four columns.
6. A column named `E` is added to the `dataframe` that contains a list of strings **Except for the last element in the new column which is an integer = 5**. This means that the elements inside a `dataframe` need not to be of the same type
7. Note also that the instruction `len(df)` provides the number of lines in the `dataframe`.
   Slicing the `Dataframe`
8. Different ways can be used to access the elements of the `dataframe`: the result shown in the following code refer to the `dataframe` shown in Fig. 3.1.

Slicing Dataframes

```
print(df.iloc[0]['B'])      # first line, column = 'B'
print(df.loc['#1']['B'])  # line indexed by '#1', column = 'B'
print(df.loc['#2', 'E'])

>> -13.546743
>> -13.546743
>> 'a_1'
```

Notice the difference between using `df.iloc` and `df.loc`: The `iloc` refers to the line number regardless of the index, while the `loc` refers to the content of the index column.

9. By using the ":", one can take all the lines of a list of columns, for instance, the following selection:

```
df.loc[:, ['E', 'A']]
```

provides the following `dataframe`[3]:

	E	A
#1	a_0	11.835849
#2	a_1	0.466508
#3	a_2	-10.229831
#4	a_3	-7.782419
#5	a_4	-9.144093
#6	a_5	-5.386116
#7	a_6	5.831405
#8	5	-9.733600

Similarly, the following selection:

```
df.loc[['#2', '#8'], ['E', 'A']]
```

[3] Note how the order of the columns' names in the selection list is used to reorder the resulting dataframe columns.

gives the following dataframe:

	E	A
#2	a_1	0.466508
#8	5	-9.733600

10. It is possible to extract the underlying matrix (to apply mathematical operations by numpy functions should all the elements of the dataframe be numbers) by using the values attribute as in the following code[4]:

Extracting the Values

```
Xr = df.iloc[0:3][['A', 'B', 'C']].values

print(type(Xr))
print(Xr.shape)
print(Xr)

>> <class 'numpy.ndarray'>
(3, 3)
[[ 11.83584899 -13.54674258  -4.44447143]
 [  0.46650837   6.26992632   0.76166182]
 [-10.22983112   8.14047239   2.22168065]]
```

11. The info method of a dataframe gives useful information about it. Executing df.info() on the above dataframe produces the following output:

Summarizing Information on Dataframes

```
print(df.info())

>> <class 'pandas.core.frame.DataFrame'>
Index: 8 entries, #1 to #8
Data columns (total 5 columns):
 #   Column  Non-Null Count  Dtype
---  ------  --------------  -----
 0   A       8 non-null      float64
 1   B       8 non-null      float64
 2   C       8 non-null      float64
 3   D       8 non-null      float64
```

[4] Notice how the function type() gives the Python type of the argument.

```
 4   E        8 non-null      object
dtypes: float64(4), object(1)
memory usage: 684.0+ bytes
```

Notice in particular that one can see that the column "E" contains a non-numerical element and there is no `None` value in the dataframe. It gives also the number of lines and columns and their names.

12. One of the most interesting feature consists in applying maps on the lines or columns of a `dataframe` (or on a selection of them), here are some examples that apply to the `dataframe` shown on Fig. 3.2:

```
df.mean(axis=0)
c1     1.00
c2     3.00
c3     6.25
dtype: float64
```

```
df.mean(axis=1)
0     2.000000
1     5.000000
2     2.000000
3     4.666667
dtype: float64
```

```
abs(df).mean(axis=1)
0     2.000000
1     5.000000
2     3.333333
3     4.666667
dtype: float64
```

```
df.min(axis=0)
c1   -2
c2    2
c3    3
dtype: int64
```

```
df.std(axis=0)
c1     2.449490
c2     1.414214
c3     2.872281
dtype: float64
```

```
abs(df).std(axis=1)
0     1.000000
1     1.000000
2     2.309401
3     4.725816
dtype: float64
```

Notice how the use of `axis=0` induces the application of the operation over the lines of the `dataframe`, while the use of `axis=1` performs the operation on the columns.

13. It is also possible to apply custom functions defined by the user, here is an example that defines a function that computes the excursion (max–min) of the columns or the lines of the dataframe.
Note here that the size of the dummy argument v used in the function depends on whether the function is applied using the `axis=0` or `axis=1` option.

The Apply Method on Dataframes

```
def compute_exursion(v):
    return v.max()-v.min()

print(df.apply(compute_exursion, axis=0).values) # over lines
print(df.apply(compute_exursion, axis=1).values) # over columns

>> [6 3 7]
[2 2 8 9]
```

14. Operations over a **Moving window** are also useful in some cases to get filtered properties, the following example computes averaged versions of the `dataframe` of Fig. 3.2 over a window of size 2 for the sake of illustration:

	c1	c2	c3
0	1	2	3
1	4	5	6
2	-2	2	6
3	1	3	10

Operation on Rolling Windows

```
dfa = df.rolling(window=2).mean()
print(dfa)

>>
     c1   c2   c3
0   NaN  NaN  NaN
1   2.5  3.5  4.5
2   1.0  3.5  6.0
3  -0.5  2.5  8.0
```

Obviously, this leaves one line undetermined as the computation formula for line i is `(v[i]+v[i-1])/2` and there is no line indexed by –1, the first line is undetermined and should be removed from any later use of the filtered `dataframe`. The first line can be removed using the `drop` function as follows:

```
dfa = dfa.drop(index=0).
```

15. The **groupby** function enables to group the elements of a `dataframe` based on the values of a given column so that some computation can be done over all the lines that share the same values in the selected column.
In order to illustrate this utility, consider the `dataframe` of Fig. 3.3 that is used for the sake of illustration. This is an example showing the `groupby` method into action:

```
df_grouped = df.groupby(by='A').sum().
```

Fig. 3.2 `dataframe` used to illustrate the application of maps on the components of a dataframe

	c1	c2	c3
0	1	2	3
1	4	5	6
2	-2	2	6
3	1	3	10

Fig. 3.3 The `dataframe` serving to illustrate the `groupby` method of a `dataframe`

	A	B	C	D	E
0	v1	1	4	4	2
1	v2	1	5	3	5
2	v1	7	0	5	5
3	v2	1	4	1	1
4	v1	5	6	4	0
5	v2	1	0	1	6
6	v1	2	2	5	4
7	v2	0	7	1	3
8	v1	1	4	5	0
9	v2	6	6	4	5

This results in the `dataframe` shown below:

	B	C	D	E
A				
v1	16	16	23	11
v2	9	22	10	20

Notice that since the column `A` used in the aggregation through the `groupby` function contains only two distinct values, namely `v1` and `v2`, the resulting `dataframe` contains two lines and since there are four remaining columns, one get a `dataframe` with four columns.

Now each element in the resulting `dataframe` contains the sum over the associated columns of all the elements of this column positioned in lines sharing the same value of the column `A`.

For instance, the number 11 that is situated in the line indexed by `v1` and the column `E` represents the sum of all the elements in the column `E` corresponding to the lines where the column `A` has the value `v1`, that is to say:

$$11 = 2 + 5 + 0 + 4 + 0.$$

The `groupby` feature is extremely useful and show multiple useful options. Notice that the `by` argument which has the value `A` in the preceding example, might take a list such as `['B', 'C']` which leads to a bi-dimensional index. Moreover, only the `sum` function has been chosen above but it is possible to apply a sequence of maps, including user-defined custom maps making the treatment potentially infinitely rich and useful in gathering deep understanding of the properties of the `dataframe` for any specific grouping rationale.

The selection of possibilities shown here is only an illustration of how powerful is the `pandas` package. Interested readers are strongly encouraged to consult the user manual of this handy and efficient package:

https://pandas.pydata.org/docs/,

in order to grasp the full extent to its power in manipulating and viewing many properties of large datasets in few lines of code.

! The Polar Package

While writing of this book, another package, namely `polar`, is seemingly becoming increasingly popular among some data-scientists I know but which I never used. By the time this book will be available, this might be confirmed or not. Please, consult the website:

https://pola.rs,

for more updated details.
This is just an example of how dynamics is the world of `Python` and data-science thanks to the open and very active community.

3.4.3 The `scikit-learn` Package

The `scikit-learn` [1] `Python` package, imported using the package name `sklearn`, is probably the most commonly used `Python` package for data-science.

It enables to solve the major ML problems, namely

- *classification*,
- *regression*, and
- *clustering*,

in a unified and easy-to-use way.

The first two problems, namely *classification* and *regression* amount at finding a function F that captures the relationship between an argument vector x and a target **label** y. The vector x is also called the **vector of features**:

$$y \approx F(x). \tag{3.1}$$

About the Notation x and y
In this section, the notation x and y are used in the specific realm of machine learning. They should not be confused with the same notation used in Chap. 1 where x and y denoted respectively the state vector of a dynamical system and the measured output obtained from physical sensors.

When the values taken by y are restricted to a limited number of possible values, the underlying problem is called a *classification problem.*

On the contrary, when y can take a priori any real value, then the underlying problem is called a *regression* problem.

The *Clustering* problem involves no labels and consists in splitting the set of values taken by x in the so-called *learning* set of data into *clusters* according to some rational that depends on the algorithm being used. Since this is done without any label, the problem is called *unsupervised.*

Regardless of the underlying problem, the learning process needs a set of occurrences of the features vector and the label (if any). This set is called the **learning dataset**:

$$\mathcal{D} := \Big\{(x^{(i)}, y^{(i)}\Big\}_{i=1}^{n_D} \quad ; \quad x^{(i)} \in \mathbb{R}^{n_F} \quad y^{(i)} \in \mathbb{Y}, \tag{3.2}$$

in which each of the pairs $(x^{(i)}, y^{(i)})$ is called a **sample**.

Notice that in the above notation, n_F is the dimension of the vector of features $x^{(i)}$, hence n_F is the number of features used in the ML problem statement.

The domain of the definition of the label, namely $\mathbb{Y}$ is the set to which belong the values of the label in the dataset. In the specific case of classification problems, this set might be, for instance, $\mathbb{Y} := \{A, B, C\}$, while in a regression problem, $\mathbb{Y}$ might be $\mathbb{Y} = \mathbb{R}$ or $\mathbb{Y} = [0, 1]$, for instance.

Machine learning tools generally gather the features in a matrix X where a **sample is represented by a line**, while all the corresponding label's values are gathered in a column vector:

$$X = \begin{bmatrix} x^{(1)} \\ \vdots \\ x^{(n_D)} \end{bmatrix} \in \mathbb{R}^{n_D \times n_F} \quad ; \quad y = \begin{bmatrix} y^{(1)} \\ \vdots \\ y^{(n_D)} \end{bmatrix} \in \mathbb{Y}^{n_D}. \tag{3.3}$$

Based on the above discussions and notation, it comes out that:

Reading the Data X and y
Referring to the notation:

$$X = \begin{bmatrix} x^{(1)} \\ \vdots \\ x^{(n_D)} \end{bmatrix} \in \mathbb{R}^{n_D \times n_F} \quad ; \quad y = \begin{bmatrix} y^{(1)} \\ \vdots \\ y^{(n_D)} \end{bmatrix} \in \mathbb{Y}^{n_D},$$

the element $x_j^{(i)}$ represents the value of the feature number j in the i-th samples of the dataset, while $y^{(i)}$ is the value of the label corresponding to the same sample.

In the following section, an illustrative example of a typical use of `scikit-learn` utilities in solving a regression problem is given before the main instructions and syntax of this package are briefly explored. This enables to grasp the whole picture showing a pipeline of processing before we dig in deeper details.

An Illustrative Example

In this example, a cloud of samples in the $\mathbb{R}^3$ is first created. Then an *arbitrary* function [see (3.4)] that associates to each $x \in \mathbb{R}^3$ a real number is chosen.

This function is then used to create a vector of labels corresponding to the previously created cloud of samples. Then `sklearn` utilities will be used to solve the regression problem consisting in *finding approximately* the previously mentioned function which is supposed to be unknown to the *learner* (the `sklearn`'s user).

Beyond the presentation of the regression tools of `sklearn`, this example will help introducing some important concepts and *tricks* that are of general scope in the machine learning context such as learning data splitting, data completeness, overfitting, shuffling, and some other items.

These steps are successively implemented hereafter.

Creation of the Learning Dataset
Let us consider the following function of three variables:

$$y = \begin{cases} x_1 x_2 + x_3^2 \text{ if } x_1 \leq 0, \\ \sin(x_1 x_3) \text{ otherwise,} \end{cases} \tag{3.4}$$

and consider the following set of interest where the values of the features vector x lie:

$$\mathbb{X} = [-2, +2]^3. \tag{3.5}$$

Let us first create the features matrix `X` containing 10000 samples of x and a label vector `y` containing the corresponding values of the function (3.4). This can be done using the following script:

Creating the Dataset

```
def func(x):
    if x[0]<=0:
        return x[0]*x[1]+x[2]**2
    else:
        return np.sin(x[0]*x[2])

Nsamples = 10000
X  = np.array([-2 + 4*np.random.rand(3) for _ in range(Nsamples)])
y = np.array([func(x) for x in X])
```

Note how in the last instruction, the list comprehension mechanism is used together with the iteration over the lines of the matrix `X` which is viewed as a list of rows.

Splitting the Dataset into Training and Test Datasets
This is a typical and mandatory step in any ML design. Indeed, we need to learn on a part of the learning data and test the *fitted* function on another part which has never been seen in the training step. This enables to check the capability of the model to *generalize* to new unseen instances of data.

There are lots of details and dedicated tools to properly address this major point in ML design. It is strongly recommended that the readers refer to the excellent references [2].

To perform the split of the learning dataset into training and test sub-datasets, the function `train_test_split` of the `model_selection` package of `scikit-learn` is used in the following script:

Splitting the Dateset into Train/Test Subsets

```
# split the data into training and test
from sklearn.model_selection import train_test_split

Xtrain, Xtest, ytrain, ytest = train_test_split(X, y,
                                        test_size=0.33)
print(f'shape of Xtrain = {Xtrain.shape}')
print(f'shape of Xtest = {Xtest.shape}')

>> shape of Xtrain = (6700, 3)
>> shape of Xtest = (3300, 3)
```

Notice how the fraction of the targeted test dataset has to be given through the argument `test_size`. The resulting matrices `Xtrain` and `Xtest` corresponds to the chosen size as shown in the output of the previous script.

Fitting a Model Using the Training Dataset and Evaluating its Quality on the Test Data

The next major step is to choose a model structure among the various possibilities offered by the `scikit-learn` libraries and to identify the corresponding model using the pair `(Xtrain,ytrain)` resulting from the splitting operation described in the previous step.

The following script uses the `RandomForestRegressor` structure to fit the data. Once the model is fitted, it is tested on the test pair `(Xtest, ytest)` in order to evaluate its precision.

The result is shown as a scatter plot comparing the ideal predictions and the predictions provided by the so-fitted `RandomForest model`. This is done using the following script[5]:

Fit a Model and Using it to Predict

```
# Choose a model structure and fit the data

from sklearn.ensemble import RandomForestRegressor
reg = RandomForestRegressor()
```

[5] Note how easy it is to fit a complex structure such as the random forest model from the user point of view!

```
# here is where the model is fitted
reg.fit(Xtrain, ytrain)
# fit ended

ytest_pred = reg.predict(Xtest)
ytrain_pred = reg.predict(Xtrain)

plt.rc('font', size=18)
fig, ax = plt.subplots(ncols=2, nrows=1)
ax[0].plot(ytrain, ytrain_pred, 'ro', ytrain, ytrain, 'k-')
ax[0].grid(True)
ax[0].set_xlabel('True values')
ax[0].set_ylabel('Predicted values')
ax[0].set_title('Results on training data')

ax[1].plot(ytest, ytest_pred, 'ro', ytest, ytest, 'k-')
ax[1].grid(True)
ax[1].set_xlabel('True values')
ax[1].set_ylabel('Predicted values')
ax[1].set_title('Results on test data')
```

The prediction results corresponding to the fitted `RandomForest` map are shown in Fig. 3.4 where it can be seen that the prediction on the training data is slightly better that those on the test data.

It is important to note however that the quality of the prediction on the test data **is probably over-optimistic**. This is because when splitting the dataset in training and test data using the `train_test_split` function, the by-default behavior is to *shuffle* the lines randomly. The consequence of this is that **for almost any sample in the presumably unseen test data, there is a sample that is very close to it in the learning data** which means that the lines in the test data are almost all *already seen* in the training data. This phenomenon is termed *data leakage* in the machine learning context.

Understanding the `Shuffle` Parameter When Splitting the Data

In order to better understand this `shuffling` feature and then examine the `realistic` way to check the quality of the obtained model, let us apply the `train_test_split` function on a fictitious pair `(X,z)` having the same number of lines as before but where `z` is a vector such that:

$$z = (1, \ldots, 10000),$$

in order to see the difference in the splitting choices when the argument `shuffle` of the function `train_test_split` is set to `True` or `False`.

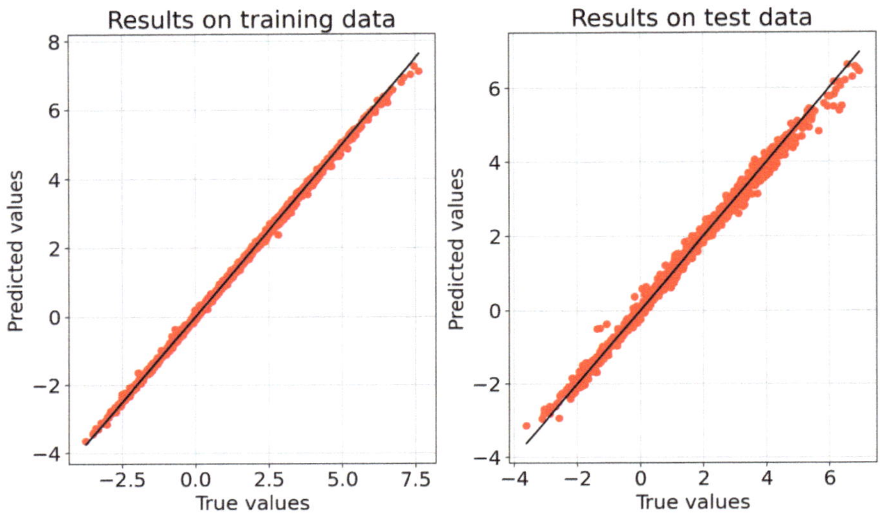

Fig. 3.4 Results of the predictions provided by the fitted Random Forest (RF) model on the data representing the non-smooth function defined by (3.4). Notice that the splitting process used the by default option: `shuffle=True`

The following script implements this idea and show the histograms of selected indices (through z) in both cases.

The Impact of Shuffling the Samples

```
# compare shuffle = True/False in train_test_split
z = np.arange(1, Nsamples+1)

_, _, ztrain_tr, ztest_tr = train_test_split(X, z,
                                        test_size=0.33,
                                        shuffle=True)

_, _, ztrain_fa, ztest_fa = train_test_split(X, z,
                                        test_size=0.33,
                                        shuffle=False)
#---------
fig = plt.figure()
plt.hist(ztrain_tr, bins=1000, alpha=0.4)
plt.hist(ztrain_fa, bins=1000, alpha=0.4)
plt.title('Comparison shuffle=True/False upon splitting!')
plt.xlabel('Selected indices bins')
plt.ylabel('number of selections')
plt.grid(True)
```

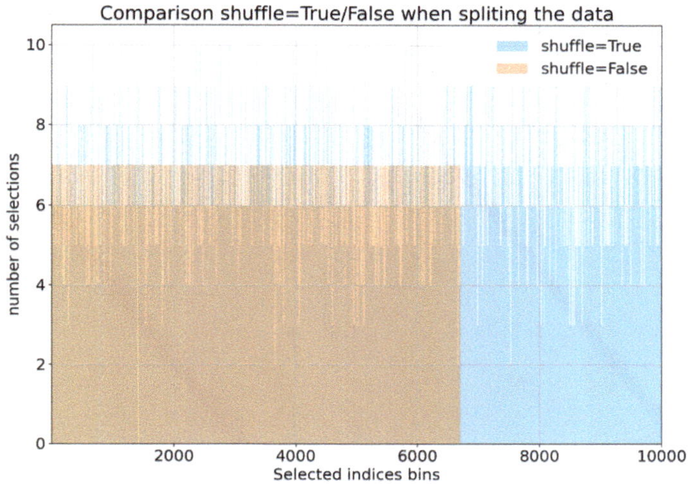

Fig. 3.5 Comparison of the line indices selected while splitting the data when the argument `shuffle` of the `train_test_split` is set to `False` or `True`. Recall that the default value of this parameter is `True`

```
plt.xlim([1, 10000])
plt.legend(['shuffle=True', 'shuffle=False'])
```

The histograms given by the previous script are shown in Fig. 3.5. One can clearly see that when `shuffle=False` is used, the first (1 − `test_size`) * 100% of the data are selected as the training lines and the remaining `test_size` * 100% of the lines are used for the test. This obviously reduces the probability to find close lines in the training data for each line in the test data. On the other side, when `shuffle=True` is used (which is the option by default), the lines are shuffled before the choice is done. This leads to over-optimistic results.

So let us repeat the learning and validation steps using the option:

`shuffle=False`,

and compare the resulting precision on the train vs test datasets. The results are shown in Fig. 3.6.

The results seem to be quite correct. But *wait a little bit!* This might come from the fact that the lines in the `X` matrix has originally been created using a random process which contains inherently a sort of shuffling:

```
X  = np.array([-2 + 4*np.random.rand(3)
                                   for _ in range(Nsamples)]).
```

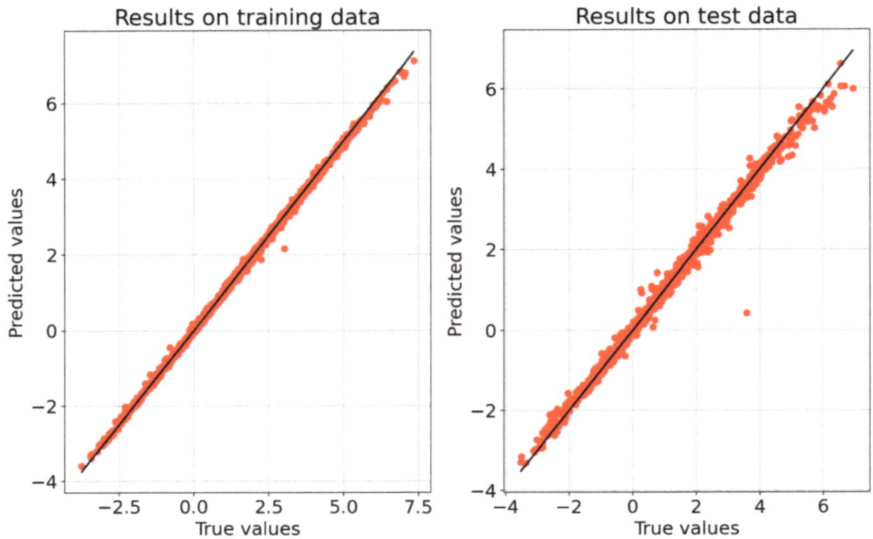

Fig. 3.6 Results of the predictions provided by the fitted RF model on the data representing the non-smooth function defined by (3.4) when `shuffle=False` is used in the data splitting step

So let us imagine that, for the sake of illustration, the features matrix X has been created using the following *triky* script:

Creation of a Non-Homogeneous Train/Test Data

```
Nsamples = 10000
X1 = np.linspace(-2, 2, 10000).reshape(-1,1)
X23 = np.array([-2 + 4*np.random.rand(2)
                for _ in range(Nsamples)])
X = np.hstack([X1, X23])
```

Indeed, the `tricky malicious` side in this script is that it exploits the fact that the very definition of the function (3.4) is very sensitive to the first feature x_1 so if we create the data such that the values of x_1 are monotonically increasing from –2 to 2 then the `shuffle=False` version will have an unbalanced representation with high values of x_1 that are not present in the training data while in the `shuffle=True` version, this unbalance would be corrected.

The regression results are shown in Fig. 3.7. It comes clearly that when the argument `shuffle=False` is used in the split process, the quality of the generalization on *unseen* data is much worse than in the shuffled case.

This was a quite long digression! But it was worth the effort because the topic is very important and understanding the impact of the way the data is collected and

(a)

(b)

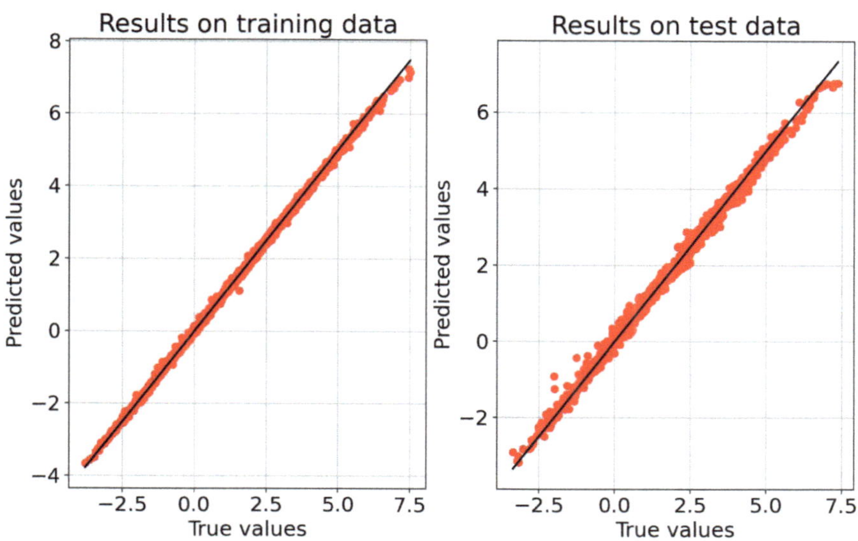

Fig. 3.7 Regression result under maliciously created features matrix `X`. **a** with `shuffle=False` **b** with `shuffle=True`

the way the data is used in the learning is one of the most important topics when it comes to use machine learning tools in solving real-life problems.

So although this book is not about machine learning, it is important that one keeps in mind the easiness with which it is possible to draw over-optimistic results from bad data preparation or from forgetting some elementary precaution rules regarding the data processing and the risk of over-fitting.

> **About Shuffling the Data During the Split Process**
> It is easy to get over-optimistic assessment of the quality of a fitted model in machine learning if the splitting data step is done while randomly shuffling the samples. This is because in this case, to each sample in the test data it is likely that close samples is present in the training data.

Let us now briefly overview some other interesting utilities and functions provided in the `scikit-learn` package.

- **Pre-processing Functions.** Many problems might be better solved if the features matrix is first scaled. `StandardScaler` and `Normalizer` functions (as well as some robust variants of them) can be used to this mean (these functions lie inside the `sklearn.preprocessing` module).
- **Principal Component Analysis (PCA) and Extraction.** PCA tools can be used to extract the combination of the features (columns of the features matrix `X`) that contains the most variance in the data. This might help reducing the number of features and hence the risk of over-fitting. The corresponding function is `sklearn.decomposition.PCA`.
- **Polynomial Features Generation.** Starting from a features matrix `X`, the

  `sklearn.preprocessing.PolynomialFeatures`,

  utility helps creating a matrix with as many columns as necessary to get all the monomials (up to some desired degree) contained in the polynomial expansion of the original features vector represented in the original features matrix `X`.
- **Regression and Classification Tools.** In the previous illustrative example, the `RandomForestRegressor` has been used but there are so many other functions that can be used. Some of them are used in the forthcoming chapters with exactly the same calling syntax including `fit` and `predict` methods.
  The same syntax is used when classification is required (fitting labels with low number of specific non-necessarily numerical values). Each of these models has its own vector of parameters that might limit its complexity in order to reduce the risk of over-fitting (number of branches of a forest, degree of polynomials in support vector machine polynomial model, regularization coefficient in linear `Ridge` regressor, etc.)
  Moreover, for some of the available models, it is possible to ask for auto-tuning of the regularization coefficients in order to get almost optimal setting. This can be done via the so-called `cross-validation` procedure in which the data is

split into a number cv=3 (for instance) sub-sets. Then successively, each of these sub-sets is used as a test sub-set for the model that is fitted using the other sub-sets for a grid of values of the regularization coefficients. The retained regularization parameter is the one that leads to the statistically best results on test sub-sets. Such utilities are generally called with the same name of the original structure with the name ending by CV. For instance, the cross-validation version of the linear model Ridge is called RidgeCV, the one associated to the ℓ_1-regularization function LassoLars is called LassoLarsCV and so on.

- **Evaluation Metrics.** Having at hand the predicted label, say ypred that is to be compared to the true label y, there are many evaluation metrics that enable to compute different indicators of the quality of the results.
  For instance, the precision of the regression on the training and the test datasets involved in the results shown in Fig. 3.7 can be obtained using the following script which uses the r2_score metrics that is available in the sklearn.metrics module. Other specific metrics are available to quantify the accuracy of a classifier such as the confusion_matrix and roc_curve.

Computing the Regression Scores

```
from sklearn.metrics import r2_score
print(r2_score(ytest, ytest_pred))
print(r2_score(ytrain, ytrain_pred))

>> 0.7162568623565579
>> 0.9992918584814008
```

- **Cascaded Instructions in scikit-learn.** Cascaded execution of operators is sometimes extensively used in Python programs in general but this is particularly true when using scikit-learn.
  For instance, the following instruction starts by standardizing the features matrix X, then transforms the result by creating polynomial features of degree 3, then fit a regressor reg using the resulting pair of new features matrix and the original label y:

The Polynomial Features

```
sc = StandardScaler()
pol = PolynomialFeatures(degree=3)
Xsc = StandardScaler.fit_transform(X)
Xpol = Pol.fit_transform(Xsc)
reg.fit(Xpol, y)
```

The same results can be obtained by using the following cascaded version if one is only interested in the resulting regressor being fitted:

```
reg.fit(PolynomialFeatures(degree=3).fit_transform(
          StandardScaler().fit_transform(X)), y).
```

The advantage of the first detailed version is that the fitted standard scaler `sc` and polynomial features transformer `pol` can be reused to transform new matrices (with the same fitted parameters), while in the condensed above shown formulation, there is no traces that are kept of these objects. Moreover, the first version is more clear which is to be preferred should the result and the computation time be equivalent.

Notice also that the `sklearn` package offers a very nice feature through the use of the `Pipeline` object which explicitly defines the process of successively applying an ordered list of transformation and prediction that can be very useful in the `Grid Search` process in the space of parameters. This feature is not used in the present book but interested readers are advised to consult the appropriate chapter on *Pipelines* in [2].

About Machine Learning

Machine learning tools and possibilities can never be totally mastered a priori because the topic is very large and involves so many disciplines (optimization, statistics, computer science, signal processing, linear algebra to cite but few ones) but also because new tools appear on a monthly basis that do better or faster what previously available tools were able to do. It results that the only reasonable option is to keep informed and to learn the different aspects as soon as they show useful for the problem at hand.

3.4.4 The CasaDi Package

The `CasaDi` package[6] proposes a framework [3] for the formulation and the solution of many engineering problems such as:

- Solving Ordinary Differential Equations.
- Solving Differential Algebraic Equations.
- Computing exact sensitivities of general multi-variable nonlinear functions.
- Solving a wide class of optimization problems: Nonlinear Programming, Quadratic Programming, Least squares, and much more.

[6] https://web.casadi.org/.

For each of the above utilities, different algorithms developed by different research groups are integrated and made ready to use through a unified syntax.
This represented a huge step for many researchers sparing them the need to learn in details the syntax and the requirements of each of these algorithms individually. Notice that for MATLAB® users, `casadi` is also available through a simple installation procedure.

It is worth understanding that `CasaDi` is based on building tractable **symbolic relationships** between variables which enables to derive exact sensitivity functions, for instance.

The price to pay is that the expressions that can be used in the definition of the variables are *restricted* to mathematical expressions that belong to a set of predefined and known-to-`Casadi` list of functions. This does not mean that this list is too restrictive but it simply means that some expressions such as those involving `look-up-tables`[7] or `Python-build-in` libraries cannot be used.

As an example, the use of terms in the expressions that are the outputs of linear system's solver, for instance, is not allowed. Note that even some common functions such as the Python function `abs` cannot be used as it is. Instead, the build-in `CasaDi` version `fabs` should be used.

So the easiest way to proceed is to look at the documentation and list of functions when needed.

An example of such *not-easy-to-guess* expressions is given in the following script which simply implements the relationship involved in the r.h.s. of (3.4):

Declaring Symbolic Variables and Computing Symbolic Functions

```
from casadi import MX, Function, if_else

# declare a casadi variable called 'x' of dim 3
x = MX.sym('x', 3)

# compute the associated output delivering y as output
y = if_else(x[0]<=0, x[0]*x[1]+x[2]*x[2], np.sin(x[0]*x[2]))

# define casadi function having x as argument and
F = Function('F', [x], [y])
```

Let us go through this script to introduce some `CasaDi`-related concepts and syntax:

[7] In this case, explicit formulas that interpolate the table should be explicitly derived before use in building symbolic relationships via `CasaDi`.

- First, some utilities (classes, methods, etc.) are imported, namely: `MX`, `Function`, and `if_else` which are used in the next lines.
- A `CasaDi` vector named `x` of dimension 3 is declared. From here, any variable that is defined in an expression which involves the components of `x` becomes itself a `CasaDi` object and its dependency to `x` is tracked-over for possible mathematical manipulations.
- This is the case with the variable `y` since its definition involves the components of `x`. This definition, according to (3.4) involves a conditional branch depending on the sign of $x_1 =$ `x[0]`. Here is where the utility `if_else` enters into action: if its first argument is `True` then the output is defined by its second argument, otherwise it takes the value of the third argument.
- The last instruction builds a `CasaDi` function that takes the value of the first `CasaDi` object `x` as argument and delivers the value of the corresponding `y` according to the relationship that is defined in the previous line.
- Note also that the `numpy` function `sin` is accepted by `CasaDi` to build the relationship between x and y although a dedicated `sin` and `cos` function from the `casadi` package are also available.
- Note also that if `y` is built using two `CasaDi` variables `x` and `z`, the instruction that creates the map would have been:

  ```
  F = Function('F', [x,z], [y]),
  ```

  and if another variable `q` was defined using `x` and `z`, one can create a function with multiple outputs as follows:

  ```
  F = Function('F', [x,z], [y, q]).
  ```

Now let us see how the function `F`, created in the above script, can be concretely used in a numerical and no more symbolic way.

Using the Created Functions

```
x_1 = np.array([-1, 1, 2]) # First branch of the definition
x_2 = np.array([+1, 1, 2]) # Second branch of the definition

y_1 = F(x_1)
y_2 = F(x_2)

print(f'y_1 = {y_1}')
print(f'y_2 = {y_2}')
```

```
>> y_1 = 3
>> y_2 = 0.909297
```

Defining Relationships Using Algorithms

One of the most interesting features of `CasaDi` is that the symbolic relationships can be tracked through complicated set of instructions that might involve loop and iterations. To say it shortly, **relationships can be defined through algorithms**.

To see an example of this feature, let us consider the recurrent equation involved in the *The Syracuse conjecture*[8]:

$$u_{n+1} = \begin{cases} \dfrac{u_n}{2} & \textit{if } u_n \text{ is even,} \\ 3u_n + 1 & \text{if } u_n \text{ is odd.} \end{cases} \tag{3.6}$$

Assume that we want to build a `CasaDi` function that takes u_0 as argument and delivers the value of the 20 next terms. This can be done using the following script[9]:

Defining Function Through Sequence of Instructions

```
# syracuse sequence using CasaDi.

from casadi import MX, Function, if_else, vertcat, fabs

# initialize the sequence
u0 = MX.sym('u0')

# initialize the list U containing the final result
m, u, U = 20, u0, u0

# iterate to compute the successive terms in U
for i in range(m):
    u_next = if_else(fabs(np.floor(u/2)*2-u)<1e-6, u/2, 3*u+1)
    U = vertcat(U, u_next)
    u = u_next

syrac = Function('syrac', [u0], [U])
```

[8] The Syracuse conjecture, elaborated in 1928, can be stated as follows: *For any integer N used to initialize the sequence ($u_0 = N$), there is an integer n such that $u_n = 1$. This simple statement has never been formally proved!*

[9] Notice that there is no way integer-specific manipulation in casadi. That is way, in the proposed version below, float manipulations are used to emulated the `Syracus` rules definition.

Notice the use of the concatenating utility `vertcat` that is imported at the beginning of the script and which is extensively used later on. This utility plays the role of the `Python's append` list method that enables to iteratively construct the lists.

The resulting `CasaDi` function is used by the following script to compute the sequence for two different initial values, namely 8 and 13.

Using the Created Functions

```
for u0 in [8, 13]:
    Lsyrac = syrac(u0)  # Using the function
    print(f'results for u0={u0} --> {Lsyrac}')
```

```
Results for u0=8 --> [8, 4, 2, 1, 4, 2, 1, 4, 2, 1,
                         4, 2, 1, 4, 2, 1, 4, 2, 1, 4, 2]

Results for u0=13 --> [13, 40, 20, 10, 5, 16, 8, 4,
                         2, 1, 4, 2, 1, 4, 2, 1, 4, 2, 1, 4, 2]
```

The statement of the conjecture seems true for these two initial values.

Billions of such initial conditions have been already checked and proved to meet the requirements (hitting the 1 after a number of iterations, called the flight time) but this is obviously not a formal proof of the conjecture!

Computing Exact Sensitivities Using Casadi

One of the main advantages of using and tracking symbolic relationships is the ability to symbolically compute the *exact*[10] sensitivities or the partial derivatives of an expression with respect to the arguments used in its definition.

Computing the sensitivities is of a great importance in applied mathematics as they can be used to appreciate the relevance of the different parameters impacting an outcome but also to get the gradient that can be used in various optimization algorithms.

This can be done in `CasaDi` using the function `jacobian`. The following example pursues the developments involving the `CasaDi` implementation of the relationship defined by (3.4) by computing the gradient of `y` w.r.t. `x`.

[10] By opposition to numerical evaluation via finite differences.

Jacobian Computation

```
# computing sensitivities using CasaDi

import numpy as np
from casadi import MX, Function, if_else, jacobian

x = MX.sym('x', 3)
y = if_else(x[0]<=0, x[0]*x[1]+x[2]*x[2], np.sin(x[0]*x[2]))
F = Function('F', [x], [y])

jac = jacobian(y, x)
G = Function('G', [x], [jac])
```

Note that the definition of the gradient function named `G` is done in two steps:

1. In the first, a `CasaDi` variable named `jac` is first computed using the function `jacobian` which takes two arguments: the quantity whose sensitivity is to be computed, namely `y` here and the variable with respect to which the sensitivity is to be evaluated, namely `x` in our example.
2. Once the gradient is computed as a `CasaDi` object, the previously examined `Function` utility is used to provide the sensitivity map that can be used with different values of the argument `x`.

Let us now check the resulting gradient function on two values of `x` involving both branches of the definition.

Checking the Computed Gradient

```
x_1 = np.array([-1, 1, 5])
x_2 = np.array([1, 0, np.pi])

print(f'Gradient at {list(x_1)} = {G(x_1)}')
print(f'Gradient at {list(x_2)} = {G(x_2)}')

Gradient at [-1, 1, 5] = [[1, -1, 10]]
Gradient at [1.0, 0.0, 3.141592653589793] = [[-3.14159, 0, -1]]
```

Simple check shows that the results are correct!

Obviously, this is a simple example showing how simple it is to compute exact sensitivities using this nice framework. The approach is totally scalable and the efficiency of this symbolically computed sensitivities is of great interest in the forthcoming developments when it comes to use gradient-based descent approaches that are sometimes needed to solve optimization problems in real time.

Formulating/Solving Optimization Problems in CasaDi

As an illustrative example and in order to exploit the already written steps, let us consider the function y=F(x) defined by (3.4) and the following optimization problem:

$$\min_{x\in\mathbb{X}=[-2,2]^3} F(x) \quad \text{under} \quad \left| \begin{array}{ll} x_1^2 x_2 & \leq 1, \\ \sin(x_1 x_3) & \leq -0.5. \end{array} \right. \tag{3.7}$$

Declaring this optimization problem under CasaDi is done by the following script which uses the previous definition of the function F which is used in (3.7) to represent the cost function to be minimized:

Declaring the Optimization Problem

```
from casadi import MX, Function, if_else, jacobian, nlpsol

# define the decision variables and its bounds
x = MX.sym('x', 3)
lbx, ubx = [-2]*3, [2]*3

# define the cost funciton
y = F(x)

# define the list of constrained terms and their bounds
g = vertcat(x[0]**2 * x[1], np.sin(x[0]*x[2]))
lbg, ubg = [-1e9]*2, [1, -0.5]

# define the optimization problem
prob = {'f':y, 'x':x, 'g':g}

# define the solver
solver = nlpsol("solver", "ipopt", prob)

# define the initial guess and call the solver
x0 = np.random.randn(3)
solution = solver(x0=x0, lbx=lbx, ubx=ubx, lbg=lbg, ubg=ubg)
```

Let us go through the lines of the previous script to introduce and comment the associated instructions and tools:

- The new imported utility here is `nlpsol` for the class of *nonlinear programming solvers*.
- Note that in Python it is possible to easily create long lists containing the same elements through the syntax `[-2]*3` which leads to the list `[-2, -2, -2]`. This syntax is used to define the lower and upper bounds on the decision variable `x` and later, the lower bound `lbg` on the constrained expressions.
- It is worth mentioning that the saturation bounds on the decision variable `x` introduced through `lbx` and `ubx` are treated differently from the coupled constraints for efficiency reasons.
- Note also that the definition of the problem **does not include the bounds**. Only the definitions of the cost function, the constrained variables is required. An additional argument representing the parameters that are required to totally define the cost and the constraints can be included, in this case the field is named `p` and the definition of the problem would take the following form[11]:

```
prob = {'f':y,
        'x':x,
        'g':g,
       'p':vertcat(...)}.
```

- The declaration of the solve:

```
solver = nlpsol("solver", "ipopt", prob)
```

  involves the choice of the algorithm to be used (here the interior point solver `ipopt`) and the problem to solve which is here the previously defined problem.
- The last instruction

```
solution = solver(x0=x0,
                  lbx=lbx,
                  ubx=ubx,
                  lbg=lbg,
                  ubg=ubg)
```

[11] Recall that the notation {} denotes a Python dictionary that is recalled in the beginning of the present chapter. It is important to note that in this dictionary, the named fields `f, x, g, p` are fixed by `Casadi`. However the true values associated to these keys need not to have the same name. This is already the case for the field with keyword `f` which takes the value of the `Casadi` variable `y`. It goes similarly for the field with the keyword `x` which could have been attributed the value of a variable named `z`, for instance.

calls the solver with the initial guess `x0` for the iterations as well as the lower and upper bounds `lbx, ubx` to be affected to the decision variable and the lower and upper bounds `lbg, ubg` on the constraints defined by g.
This call runs the iterations for the computation of the optimal solution of the previously defined problem `prob` using the chosen interior-point algorithm `ipopt`.

- Once this last instruction is executed, the resulting *presumably optimal solution* `x` and the corresponding presumably optimal value `F(x)` are available in the corresponding **dictionary solution** fields, namely:

```
solution['x'], solution['f']
```

```
print(f'optimal solution found = {solution["x"]}')
print(f'optimal value found = {solution["f"]}')

>> optimal solution found = [1.43719, -1.00562, -1.09297]
>> optimal value found = -1.
```

Figure 3.8 shows a typical output delivered by the `ipopt` solver used in the previous script to define the `nlpsol` object. This output gives valuable information regarding the cost function, the constraints fulfillment, the constraint-related co-state variables,[12] the number of iterations, the exit status and the CPU time among many others.

The level of details shown in the output can be modified through dedicated options or the whole logout can be oriented towards a file to be saved and later analyzed.

3.4.5 Other Packages

So far, the packages that are extensively used in the following chapters have been presented. As it has been mentioned at the end of the previous section dedicated to machine learning packages, there are some packages that deserve to be mentioned in addition to all those which might appear in the forthcoming months and years. Some of them would have been used instead of the ones that are used in the book.

Among them, the following are intimately linked to the control topic we are focusing on in the present book. For each of them, a glimpse of the scope and spirit are mentioned together with the appropriate links for deeper examination by the interested readers.

Pyomo

The **Pyomo** package [4] is intended to be a *Flexible modeling framework for optimization problems in Python.*

[12] These variables represent the sensitivity of the optimal solution to the bounds on the constraints.

```
This is Ipopt version 3.12.3, running with linear solver mumps.
NOTE: Other linear solvers might be more efficient (see Ipopt documentation).

Number of nonzeros in equality constraint Jacobian...:        0
Number of nonzeros in inequality constraint Jacobian.:        4
Number of nonzeros in Lagrangian Hessian.............:        4

Total number of variables............................:        3
                     variables with only lower bounds:        0
                variables with lower and upper bounds:        3
                     variables with only upper bounds:        0
Total number of equality constraints.................:        0
Total number of inequality constraints...............:        2
        inequality constraints with only lower bounds:        0
   inequality constraints with lower and upper bounds:        2
        inequality constraints with only upper bounds:        0

iter    objective    inf_pr   inf_du lg(mu)  ||d||  lg(rg) alpha_du alpha_pr  ls
   0  1.3624097e+00 9.95e-01 1.70e+00  -1.0 0.00e+00    -  0.00e+00 0.00e+00   0
   1 -9.6443655e-02 4.04e-01 9.96e-01  -1.0 8.77e-01    -  5.98e-01 1.00e+00f  1
   2 -9.9420830e-01 0.00e+00 4.17e+00  -1.0 1.09e+00   0.0 4.38e-01 1.00e+00h  1
   3 -9.7994201e-01 0.00e+00 1.10e+00  -1.0 3.04e-01   0.4 5.93e-01 1.00e+00h  1
   4 -9.9215129e-01 0.00e+00 7.41e-01  -1.0 4.21e-01  -0.1 9.90e-01 5.00e-01h  2
   5 -9.9909904e-01 0.00e+00 5.36e-01  -1.0 8.21e-01  -0.5 9.90e-01 2.50e-01h  3
   6 -9.9938014e-01 0.00e+00 7.26e-02  -1.0 9.29e-02  -0.1 9.90e-01 1.00e+00h  1
   7 -9.9210731e-01 0.00e+00 1.70e-01  -1.0 8.18e-01    -  9.91e-01 1.00e+00h  1
   8 -9.9968381e-01 0.00e+00 1.84e-02  -1.7 2.33e-01    -  1.00e+00 1.00e+00h  1
   9 -9.9999948e-01 0.00e+00 1.01e-03  -3.8 1.98e-02    -  1.00e+00 1.00e+00h  1
iter    objective    inf_pr   inf_du lg(mu)  ||d||  lg(rg) alpha_du alpha_pr  ls
  10 -1.0000000e+00 0.00e+00 3.22e-07  -5.7 1.11e-03    -  1.00e+00 1.00e+00h  1
  11 -1.0000000e+00 0.00e+00 1.09e-10  -8.6 3.32e-05    -  1.00e+00 1.00e+00h  1

Number of Iterations....: 11

                                   (scaled)                 (unscaled)
Objective...............:  -1.0000000000000000e+00   -1.0000000000000000e+00
Dual infeasibility......:   1.0943222817874796e-10    1.0943222817874796e-10
Constraint violation....:   0.0000000000000000e+00    0.0000000000000000e+00
Complementarity.........:   2.5466617630358012e-09    2.5466617630358012e-09
Overall NLP error.......:   2.5466617630358012e-09    2.5466617630358012e-09


Number of objective function evaluations             = 18
Number of objective gradient evaluations             = 12
Number of equality constraint evaluations            = 0
Number of inequality constraint evaluations          = 18
Number of equality constraint Jacobian evaluations   = 0
Number of inequality constraint Jacobian evaluations = 12
Number of Lagrangian Hessian evaluations             = 11
Total CPU secs in IPOPT (w/o function evaluations)   =      0.005
Total CPU secs in NLP function evaluations           =      0.000

EXIT: Optimal Solution Found.
      solver  :   t_proc      (avg)   t_wall       (avg)    n_eval
       nlp_f  |  68.00us (  3.78us)  76.27us (  4.24us)        18
       nlp_g  |  66.00us (  3.67us)  64.11us (  3.56us)        18
  nlp_grad_f  | 100.00us (  7.69us)  85.56us (  6.58us)        13
  nlp_hess_l  | 133.00us ( 12.09us) 132.50us ( 12.05us)        11
   nlp_jac_g  |  48.00us (  3.69us)  47.63us (  3.66us)        13
       total  |  10.74ms ( 10.74ms)   9.51ms (  9.51ms)         1
```

Fig. 3.8 Typical output of the `ipopt` solver used in `CasaDi`

http://www.pyomo.org/.

Similar to CasaDi, Pyomo defines an optimization problem and let the user choose the appropriate solver among some possible options. There is a community of users providing examples of use that might be inspiring. After a careful examination of Pyomo, no clear evidence supporting a preference of Pyomo over CasaDi appeared to me although some reasons might exist that I am not aware of.

Gekko

The **Gekko** package [5] is a Python package for machine learning and optimization of mixed-integer decision variables as well as for the integration of differential algebraic equations. It provides easiness of use of large-scale solvers for linear, quadratic, non-linear, and mixed-integer programming (LP, QP, NLP, MILP, MINLP). This enables to solve problems like parametric regression, data reconciliation, real-time optimization, dynamic simulation, and nonlinear predictive control.

https://gekko.readthedocs.io/en/latest/.

The link below:

https://gekko.readthedocs.io/en/latest/examples.html,

contains an exhaustive list of use-cases including control-related ones such as the one implementing a Moving-Horizon Estimator (MHE), a Nonlinear Model Predictive Controller (NMPC) among many other interesting engineering problems.

Gekko is certainly under-estimated in the present book and I advise a careful reading of the documentation and probably an extensive use of the Gekko package, including by trying to implement Gekko-flavored versions of some of the codes provided in the next chapters.

Pytorch

The **Pytorch** package [6] is the reference package for learning Deep Neural Networks (DNN) models. It is made freely available by the Meta company for a large use by the computer science community to learn (fit) large models for regression or classification, including an easy implementation of parallel computing via Graphical Processing Units (GPU).

While neural networks are not specifically invoked in the present book, if one chooses to use DNN to address some regression or classification tasks invoked in the book, this would be the package to use. On the other hand, Pytorch also offers the possibility to compute exact sensitivity which can be used instead of the CasaDi's function Jacobian invoked earlier in this chapter.

`TensorFlow`

The **`Tensorflow`** package [7] is the `Google` equivalent of `Pytorch` that is also widely used by a part of the computer science community as a preferred framework for DNN training.

https://www.tensorflow.org.

Beyond DNN, tensor flow is a framework for creating and manipulating high-dimensional numerical object in an efficient and parallelizable way.

`Scipy`

This is a `numpy`-based efficiently implemented signal processing Python package that is worth examining when some specific signal processing and linear algebra tasks are to be achieved. The `scipy` package includes ODE solvers, general purpose optimization routines, linear algebra dedicated functions (singular value decomposition, eigenvalues computation, matrix condition number computation, and so on), Fast Fourier Transformation (FFT), wavelets-related utilities, and polynomial/spline interpolation facilities to cite but few of the large number of utilities provided by `scipy`.

3.5 Conclusion

In this chapter, some `Python`-related main elementary topics have been recalled and the main packages that are extensively used in the forthcoming chapters are introduced for an easy reading of the provided `Python` scripts.

It goes without saying that the `Python` language itself as well as any of the discussed packages (`numpy`, `pandas`, `sklearn`, and `CasaDi`) deserves a book for its own sake. The remainders and introduced topics included in this chapter should be considered as a starting point for a deeper and endless investigation and learning using the provided links and references.

Keep an Eye on Novelties Keep in mind that the `Python` world is extremely dynamics and there are valuable and game-changing packages that are made available on a monthly basis.This is the reason why a continuous and systematic checkout should be adopted in order to be informed of the last advances on the many algorithmic fields that can be of great help in improving the control-related solutions.

References

1. Pedregosa, F., Varoquaux, G., Gramfort, A., Michel, V., Thirion, B., Grisel, O., Blondel, M., Prettenhofer, P., Weiss, R., Dubourg, V., Vanderplas, J., Passos, A., Cournapeau, D., Brucher, M., Perrot, M., Duchesnay, E.: Scikit-learn: Machine learning in Python. J. Mach. Learn. Res. **12**, 2825–2830 (2011)
2. Raschka, S.: Python machine learning. PACKT, Birmingham-Mumbai (2015)
3. Andersson, J.A.E., Gillis, J., Horn, G., Rawlings, J.B., Diehl, M.: CasADi - A software framework for nonlinear optimization and optimal control. Math. Program. Comput. **11**(1), 1–36 (2019). https://doi.org/10.1007/s12532-018-0139-4
4. Hart, W.E., Laird, C.D., Watson, J.P., Woodruff, D.L., Hackebeil, G.A., Nicholson, B.L., Siirola, J.D.: Pyomo–Optimization modeling in Python, 2 edn. Springer Optimization and Its Applications. Springer (2017). https://doi.org/10.1007/978-3-319-58821-6, http://www.pyomo.org
5. Beal, L.D., Hill, D.J., Martin, R.S., Hedengren, J.D.: Gekko optimization suite. Processes **6**(8), 106 (2018). https://doi.org/10.3390/pr6080106, https://www.mdpi.com/2227-9717/6/8/106
6. Paszke, A., Gross, S., Massa, F., Lerer, A., Bradbury, J., Chanan, G., Killeen, T., Lin, Z., Gimelshein, N., Antiga, L., Desmaison, A., Kopf, A., Yang, E., DeVito, Z., Raison, M., Tejani, A., Chilamkurthy, S., Steiner, B., Fang, L., Bai, J., Chintala, S.: Pytorch: An imperative style, high-performance deep learning library. In: Advances in Neural Information Processing Systems 32, pp. 8024–8035. Curran Associates, Inc. (2019). http://papers.neurips.cc/paper/9015-pytorch-an-imperative-style-high-performance-deep-learning-library.pdf
7. Abadi, M., Agarwal, A., Barham, P., Brevdo, E., Chen, Z., Citro, C., Corrado, G.S., Davis, A., Dean, J., Devin, M., Ghemawat, S., Goodfellow, I., Harp, A., Irving, G., Isard, M., Jia, Y., Jozefowicz, R., Kaiser, L., Kudlur, M., Levenberg, J., Mane, D., Monga, R., Moore, S., Murray, D., Olah, C., Schuster, M., Shlens, J., Steiner, B., Sutskever, I., Talwar, K., Tucker, P., Vanhoucke, V., Vasudevan, V., Viégas, F., Vinyals, O., Warden, P., Wattenberg, M., Wicke, M., Yu, Y., Zheng, X.: TensorFlow: Large-scale machine learning on heterogeneous systems. https://www.tensorflow.org/ (2015)

Part II
Methodologies and Algorithms

This Part II of the book discusses different frameworks and algorithms that enable to handle different instances of the problems introduced in Chap. 1.

Some of these frameworks use rather conventional control approaches, although modern probabilistic certification-based aspects are incorporated. Other frameworks make use of recent learning-based modules when appropriate.

Some of the chapters focus on generic problems whose solution can be used as a part of a whole in order to address complex problems formulation. The last chapter of this part gives an overview of existing uncertainty-oriented solutions that are not discussed in detail in the present book and can be viewed as *further reading* suggestions.

The content of this part is later used in Part III, dedicated to the examination of use-cases that are strongly impacted by the presence of uncertainties.

Handling Uncertainties via Standard Methods

4

M. Alamir, *Nonlinear Control of Uncertain Systems*,
https://doi.org/10.1007/978-3-031-93287-8_4

4.1 The Chapter's Content at a Glance

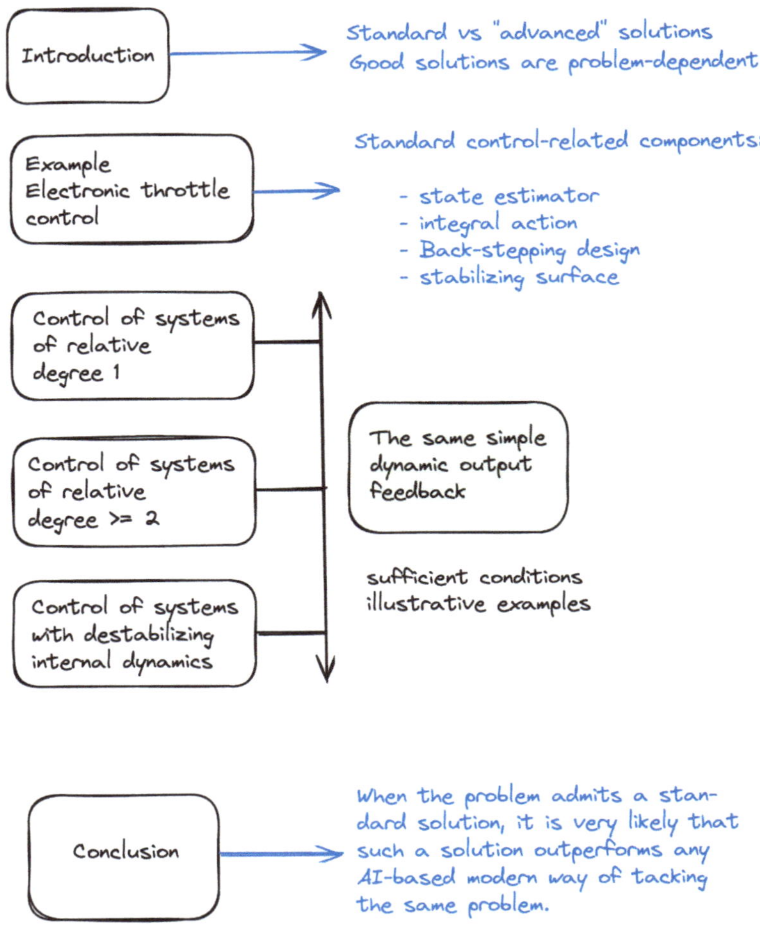

4.2 Introduction

Many of the recent learning-based approaches discussed in this book are intimately linked to the emergence and the increasing availability of new tools coupled with an increasing computational power. The latter is in particular enhanced by increasingly efficient tools and workstations available to *ordinary actors* also by the development of Graphical Processing Unit (GPU) devices.

Sometimes, these approaches truly enlarge the domain of application of control and estimation algorithms. But sometimes, they simply makes it easier for everyone to forget the very rich set of concepts and design principles that made the success of the control theory over decades.

As such, this represents a real threat because:

! New Is Not Necessarily Better

The new tools and methods do not necessarily provide *better* solutions to problems that were previously solvable (for such problems, standard approaches are generally more suitable). Rather, they sometimes help extending the set of solvable problems by addressing problems that found no solution using the standard approaches.

Being unaware of this subtle positioning might be harmful since better *old* solutions will be increasingly forgotten at the price of having *poor solutions with expensive computational burden while cheap previously mastered solutions would have provided more successful and cheaper-to-get outcomes.*

Examples of such situations are proposed in this chapter.

This is done in order to clarify the rather general arguments invoked above and in order for the reader to be aware of the corpus of simple and powerful *control-inspired* solutions that one should keep in mind when it comes to addressing control-related *real-life* problems and objectives.

Obviously, only some instances of such *standard control* solutions are proposed in this chapter in order to stimulate the unaware reader's curiosity about a whole discipline which is nowadays quite abusively considered as an *old-fashioned* subcategory of the corpus of Artificial Intelligence (AI).

While some generic tools and concepts are in action here, the underlying implemented solutions almost always involve a part of **problem-dependent** instantiation. This seriously excludes the possibility of building *push-button* ready-to-use packages that provide *control-inspired* solutions for any underlying engineering problem.

This viewpoint may be challenged by a part of the control community and the community of researchers in general. Indeed, one might argue that the *glory* of the *scientific method* is intimately linked to its ability to produce general concepts and tools that *fly highly above* specific real-life instantiations of the problems.

This point of view is an ideal that does not resist my personal (*rather long*) experience in solving real-life problems. This experience can be summarized through the following statement:

Any Good Design Is Problem-Dependent

Notice however that this is not a separation line between *Standard control-inspired* solutions and the *AI-inspired* ones. Rather, it is the *presumably* generic and *solve-any-problem* capability that is commonly attributed to the AI-inspired solutions that appears to me as a wishful thinking that is fed by a not yet mature analysis of the situation.

It is important to underline that this is an active undergoing debate in the control scientific community that we should silence here, at least temporarily, in order to address more concrete issues.

It is very important to keep in mind that the intention of this chapter is not to provide an overview of traditional control-related approaches. This is because such a task would require several books on one hand; on the other hand, the present book focuses on learning-related new alternatives and topics.

The only aim of the present chapter is to remind the reader that there is an existing corpus of concepts and tools that addressed the presence of uncertainties in dynamical systems and that it is wise to be aware of their very existence so that one can choose the appropriate tools and solutions for the specific problem under study.

This chapter is divided into two parts:

1. In the first part, a concrete example is given of a problem that should be addressed using standard control-related approaches. This problem is given together with a possible standard solution among many other possibilities.
2. In the second part, a very specific set of control design solutions are given because they are used in some of the future chapters as parts of the complete solutions to some of the use-cases considered in the book.

Let us start by a concrete example.

4.3 The Electronic Throttle Controlled System

Figure 4.1 shows a schematic of the Electronic Throttle Controlled (ETC) system that is used in this chapter for the sake of illustration. It is also used in **Chap.** 9 in order to illustrate the `MHE_solver Python` package that is intended to implement full extended-state Moving-Horizon Estimators (MHE) that is discussed briefly in **Chap.** 2.

This system describes the control of the angle

$$x_1 = \theta,$$

that moderates the admission of the airflow entering the combustion engine.

Two parts are clearly distinguishable in Fig. 4.1, namely the mechanical part and the electric part.

The role of the electric part is to act on the mechanical part through an electro-motor force:

$$x_3 := e_a,$$

that is controlled through the current:

$$i_a = u,$$

which directly controls the electro-motor force $x_3 := e_a$ by producing a torque acting on the rate of change of the angular velocity:

$$x_2 = \dot{x}_1 = \dot{\theta},$$

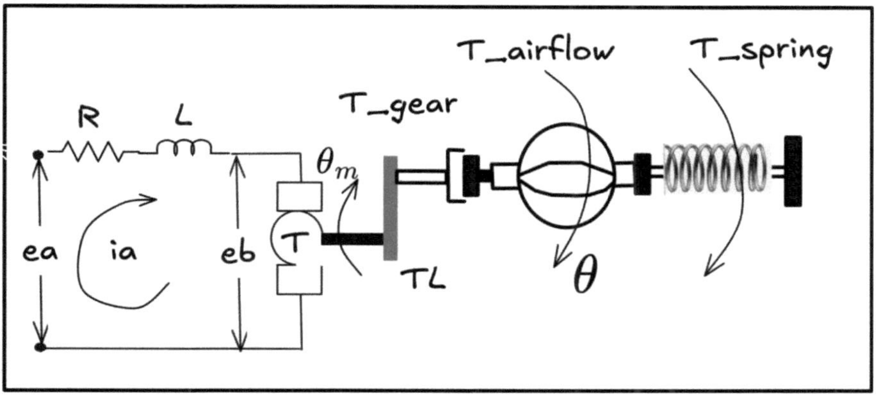

Fig. 4.1 The ETC automotive unit

so that the latter follows some reference value. This reference value is determined by another sub-system that we are not studying here.

This system is used hereafter to define a control problem in which a high level of uncertainty is present and which admits an excellent solution that no AI-inspired methodology can equal in terms of both performance level and computational burden.

Let us first examine the equations that govern the system. This is done in the next section.

4.3.1 The Dynamic Model of the ETC

The Ordinary Differential Equations (ODEs) that govern the ETC involve the state vector defined by

$$x := \begin{bmatrix} \theta \\ \dot{\theta} \\ e_a \end{bmatrix} \in \mathbb{R}^3,$$

and the folloxing vector of parameters:

$$p := \big[N_m, J_m, J_g, b_m, b_t, K_{sp}, K_t, R_p, R_{af}, L_a, K_b, R_a\big] \in \mathbb{R}^{12}. \tag{4.1}$$

These parameters are used within the following system of ODEs:

$$\dot{x}_1 = x_2, \tag{4.2a}$$

$$\dot{x}_2 = \frac{1}{N_m^2 J_m + J_g}\Big[\phi(x, p) + N_m K_t x_3\Big], \tag{4.2b}$$

$$\dot{x}_3 = \frac{1}{L_a}\Big[-N_m K_b x_2 - R_a x_3 + i_a\Big], \tag{4.2c}$$

in which the map $\phi(x, p)$ is given by

$$\phi(x, p) := -K_{sp}(x_1 - \pi/2) - (N_m^2 b_m + b_t)x_2 - \pi R_p^2 R_{af}\Delta(x_1)\cos^2(x_1), \tag{4.3}$$

where

$$\Delta(x_1) := 2P_{\text{atm}} \times (1 - x_1/\pi),$$

is the over-pressure depending on the opening angle $x_1 = \theta$ of the device.

The nominal values of the parameters p are given in Table 4.1.

The significance of all the parameters and the equations involved in the previous equation is not mandatory here. Nevertheless, the following comments are worth to be provided for a better understanding of the functioning of this *engineered* system:

- Equation (4.2a) follows from the definition of the second state $x_2 = \dot{\theta}$ being the derivative of the first state $x_1 := \theta$.
- Equation (4.2b) expresses the second law of mechanics in its angular version where, as explained above, $x_3 = e_a$ is proportional to the torque applied on the device while the denominator in the right-hand side represent the equivalent moment of inertia.
- The function ϕ gathers three terms that represent:
  - the stiffness of the rotational spring leading which produces a mechanical torque that is proportional to the distance $(x_1 - \pi/2)$ to the rest position $\pi/2$,
  - the friction terms that is proportional to $(-x_2)$, and
  - the last term representing the torque that is induced by the difference in pressure on both sides of the device.

Obviously, the true values of the parameters might differ from the nominal values shown in Table 4.1 so that one might think that *advanced approaches* that are dedicated to handling uncertainties on the model's parameters are necessary here in order to meet the following control objective:

Problem Statement

Design a feedback law that takes into account the uncertainty on the values of the vector of parameters p so that the angle θ tracks an exogenous reference signal θ_r.

It will be shown in the sequel that this problem can be almost perfectly solved using *standard approaches* such that the resulting performance can hardly be equaled or even approached by any *modern so-called intelligent or data-driven solutions such as the one developed elsewhere in the book.*

Table 4.1 Nominal values of the ETC system

Parameter	Value	Parameter	Value
$p_1 = N_m$	4	$p_7 = K_t$	0.1045
$p_2 = J_m$	0.0004	$p_8 = R_p$	0.0015
$p_3 = J_g$	0.005	$p_9 = R_{af}$	0.002
$p_4 = b_m$	0.03	$p_{10} = L_a$	0.003
$p_5 = b_t$	3.4×10^{-3}	$p_{11} = K_b$	0.1051
$p_6 = K_{sp}$	0.4316	$p_{12} = R_a$	1.9

However, as mentioned above, the price to pay is to look carefully at the principles that underline the functioning of the system which has been engineered in order to achieve its control task.

Once this is done and based on this understanding of the system, it is shown how some standard reflexes and *approaches* from the control theory regarding the stability of dynamically varying quantities can be put together in order to address the *tight* control problem as stated above despite the presence of uncertainties.

This is explained step-by step in the following section.

4.3.2 The Parameterized Feedback Design

Along the design steps explained in the present section, some design parameters are progressively introduced whose value can be tuned by hand or in a systematic way through probabilistic certification (see Sect. 2.11.3). That is the reason why this section is entitled *Parameterized feedback design*.

The **ETC** system is a human-engineered system that is designed in order to achieve a specific task by acting on a specific manipulated input which is here precisely the current $u = i_a$.

Consequently, the engineers behind the design of this technological system *hopefully* worked hard in order to ensure that by changing the value of the current i_a, it is possible to force the regulated variable, namely the angular position θ, to follow any future possibly required trajectory θ_r. They even most likely sized the actuator maximum achievable torque so that the targeted response time can be achieved.

As a matter of fact, by looking at the system's equations (4.2), it is possible to *follow* a *path of causality* going from i_a to θ. Namely, increasing i_a ends by increasing x_3 which then ends by making x_2 increase leading ultimately to the increase of the angular position x_1. The same sentence can be said using the word decreasing. This chain of causality can be represented schematically by the following graph:

$$u = i_a \longrightarrow x_3 \longrightarrow x_2 \longrightarrow x_1 = \theta.$$

The question is precisely:

Question?

When to do what? and with which rate of change?

Let us start by the simple fact according to which, if the following relationship is satisfied over some period[1]:

$$\dot{\theta} = \lambda(\theta_r - \theta), \quad \lambda > 0, \tag{4.4}$$

then the angle θ_1 will asymptotically follow θ_r should the latter be a constant. When it is not the case, higher values of λ would induce a tracking error that is inversely proportional to λ.

These are basic dynamic systems considerations.

Consequently, since by definition, $x_2 = \dot{\theta}$, one would be interested in forcing the following equality which is nothing but a simple re-writing of (4.4):

$$S_1(x, \theta_r, \lambda) := x_2 - \lambda(\theta_r - x_1) = 0. \tag{4.5}$$

This can be done by following the simple rule:

If $S_1 < 0 \longrightarrow$ increase x_2 **otherwise** decrease x_2.

But increasing [resp. decreasing] x_2 is done by increasing [resp. decreasing] x_3 according to the above discussion.

Therefore, a way to summarize the above rule is to consider the quantity S_3 defined by (4.6), that takes the opposite sign to S_1 as shown in Fig. 4.2, as a reference value for x_3[2]:

$$S_3(S_1, \alpha_1, \beta) := -\beta \times \tanh\Big(\alpha_1 S_1\Big). \tag{4.6}$$

Obviously, high values of β induce higher amplitudes for x_3, while higher values of α_1 induces higher dynamics on x_3 for the same *distance* of S_1 to its ideal value 0.

Now again, ideally, we would like to have $S_3 = 0$ because when this happens, one has according to (4.6), $S_1 = 0$ which is the intermediate objective explained above.

This means that as long as S_3 is different from 0, the control $u = i_a$ should take the opposite sign! Here again, we can use the same principle as before and get the final expression of the feedback law which can therefore be given by the following simple expression:

$$u = i_a = u_{\max} \times \tanh(\alpha_2 S_3), \tag{4.7}$$

where $u_{\max}$ is the maximal allowable value of the current.

---

[1] λ is our first control design parameter! Its impact on the resulting behavior is as follows: High values of λ induce rapid convergence, while low values induce a slow convergence. The price to pay for quick convergence (high λ's) is a higher dynamic on the current signal that should be compatible with the underlying electric hardware providing the needed electric torque.

[2] α_1 and β are two other control design parameters!

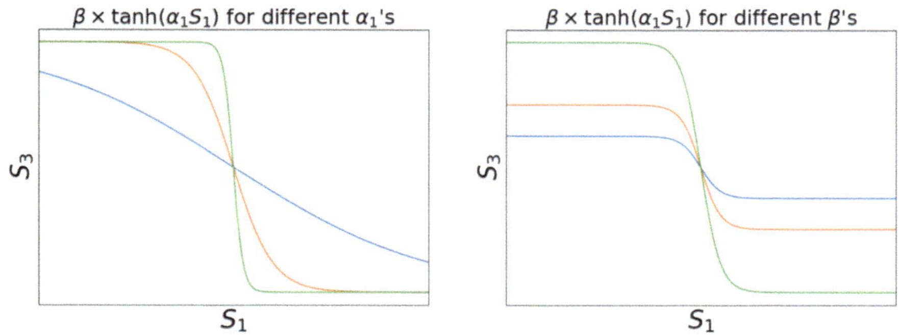

Fig. 4.2 Schematic view of the implementation of the relationship (4.6) expressing the fact that the reference on x_3, namely S_3 should take an opposite sign to the one of S_1 defined by (4.5)

Based on the above discussion, the parameterized feedback can be summarized as follows[3]:

Parameterized Feedback—Version 1

Definition 4.1 Given the design parameters $\lambda, \alpha_1, \alpha_2 > 0$ and the saturation level $u_{\max}$ on the current, the control-inspired, *old-fashioned*, feedback would be given by

$$i_a = u = u_{\max} \times \tanh\Big(\alpha_2 \tanh\big(\alpha_1 (x_2 - \lambda(\theta_r - x_1))\big)\Big), \tag{4.8}$$

where θ_r is the reference value on the angular position $x_1 := \theta$.

Now this version is not bad but it might lead to small tracking errors even for a constant reference signal θ_r. Although the error level can be made as small as desired by increasing the gain λ, a *respectful control designer* would not be fully satisfied with a control loop that allows permanent tracking errors for constant reference signals.

Here is where a second control design principle enters into action, namely the use of *integral action*. The integral action enables to asymptotically remove persistent errors which might be induced by constant model mismatches that lead to bias on the regulated variable.

This is done by adding a corrective term δ_θ to the reference in the expression of the feedback law (4.8) which becomes

$$i_a = u = u_{\max} \times \tanh\Big(\alpha_2 \tanh\big(\alpha_1 (x_2 - \lambda(\theta_r + \delta_\theta - x_1))\big)\Big). \tag{4.9}$$

[3] Since α_2 is multiplied by β, they are merged in a single parameter α_2.

The correction term is a dynamical internal state of the controller that is governed by the following equations:

$$\delta_\theta^+ = \delta_\theta + \lambda_I \times (\theta_r - x_1), \tag{4.10}$$

where $\lambda_I > 0$ is the integrator gain that adds to the vector of control design parameters. Notice that this last definition means that the correction term is the result of integrating the error between the reference value θ_r and the current value of the regulated variable $x_1 = \theta$ multiplied by the gain λ_I.

As for the choice of the integrator gain λ_I, since the role of the integrator is to recover asymptotically the regulation error, small values should be used in order to avoid overshooting issues that are generally due to the integral correction term δ_θ taking high values because of the unavoidable error during transient regimen.

Now in order to avoid this high sensitivity in tuning this integrator gain, we can update the value of the internal state of the integrator only when the angular value θ is sufficiently close to the targeted value, say, when

$$|\theta - \theta_r| \le \texttt{th} \approx 0.05.$$

Based on the above, we obtain the following second version of the parameterized feedback law:

Parameterized Feedback—Version 2

Definition 4.2 Given the positive design parameters $\lambda, \alpha_1, \alpha_2, \lambda_I > 0$ and the saturation level $u_{\max}$ on the current, the control-inspired feedback would be given by

$$\delta_\theta^+ = \begin{cases} \delta_\theta + \lambda_I \times (\theta_r - x_1) & \text{if } |x_1 - \theta_r| \le \texttt{th} \\ \delta_\theta & \text{otherwise,} \end{cases} \tag{4.11a}$$

$$u = u_{\max} \times \tanh\Big(\alpha_2 \tanh\Big(\alpha_1\big(x_2 - \lambda(\theta_r + \delta_\theta - x_1)\big)\Big)\Big), \tag{4.11b}$$

where θ_r is the reference value on the angular position $x_1 := \theta$.

Regarding the above feedback design, the following comments are worth giving:

- Notice that the expression of the feedback law makes no use of any presumed value for the vector of parameters p involved in the definition of the dynamic model (4.2) governing the system. This is because we simply used qualitative causality relationships that one can capture by looking at the model from a structural point of view and obviously by giving some credit to the engineers who designed the equipment.

- The counter-part of this seemingly parameter-independent design is that there is at this design stage no formal proof regarding the stability of the closed-loop behavior (convergence of the regulated θ to its reference). We shall therefore consider a certification step with a cloud of parameters that is designed knowing the statistics of the parameters and based on the principle of probabilistic certification explained in **Chap.** 1.
- Notice that the derivation of this proposal for the parameterized control feedback is first based on the comprehension of the operation principle of this engineered system on one hand, but also, it relies on some control design concepts and tools. For instance, the fact that the constraint $S_1(x, \theta_r, \lambda) = 0$ asymptotically leads[4] to $x_1 = \theta_r$ and the concept of eliminating persistent errors, under some circumstances, via integrator action represented in the above design by the role played by the integrator state λ_θ.
- The definition of the parameterized feedback uses only the measurement of $\theta := x_1$ and $\dot{\theta} = x_2$ which is quite reasonable since generally the angular position is measured using high-precision incremental encoders that provide a very accurate and almost noise-free measurement of the angular position. This enables a high-precision estimation $\hat{x}_2$ of the angular speed via simple numerical differentiation although dedicated filtering estimator can also be deployed.

In the next section, a `Python` code for the implementation of the above-designed parameterized controller is first shown before its efficiency in meeting the control objective as stated above is examined over a cloud of possible vectors of parameters.

4.3.3 `Python`-Based Implementation

We shall use the `CasaDi` framework in order to accelerate the computation of the one-step-ahead map. This is not mandatory since we do not use any specific utilities such as optimization or gradient computation. Nevertheless, the use of `CasaDi` accelerates the simulation of the closed-loop behavior and since we might need to simulate the closed-loop system for a large number of instantiations of the parameters in a possible certification step, this might be interesting to do.

The following script shows the implementation that yields the `CasaDi` map `F(x,u,p)` that implements a one-step-ahead computation for the ETC dynamical system. Let us start by importing the needed modules and setting some parameters:

```
import numpy as np
from casadi import MX, Function
from casadi import vertcat, power, cos
import matplotlib.pyplot as plt
```

[4] I agree that this is not rocket science but it is likely that a standard expert in deep learning might not think about it immediately!

```
nx = 3
nu = 1
# Atmospheric pressure
P_atm = 1.01325e5
# sampling period for the control
dt = 1e-3
# Number of internal steps to integrate over dt
nsub_step = 5
```

Then the following script implements a one-step-ahead map for the ETC dynamical system:

The One-Step-Ahead Map

```
bm = MX.sym('bm',1)
bt = MX.sym('bt',1)
Jm = MX.sym('Jm',1)
Jg = MX.sym('Jg',1)
Kb = MX.sym('Kb',1)
Ksp= MX.sym('Ksp',1)
Kt = MX.sym('Kt',1)
La = MX.sym('La',1)
N = MX.sym('N',1)
Ra = MX.sym('Ra',1)
Raf = MX.sym('Raf',1)
Rp = MX.sym('Rp',1)
#-----
J = N * N * Jm + Jg
Kf = N * N * bm + bt
p = vertcat(N, Jm, Jg, bm, bt, Ksp,
            Kt, Rp, Raf, La, Kb, Ra)
#----- xdot
x = MX.sym('x', nx)
u = MX.sym('u',nu)
term = 2*np.pi*Raf*power(Rp*cos(x[0]),2)
term *= P_atm*(1-x[0]/np.pi)
xdot1 = x[1]
xdot2 = 1.0/J*(-Ksp*(x[0]-np.pi/2)-Kf*x[1]+N*Kt*x[2]-term)
xdot3 = 1/La*(-N*Kb*x[1]-Ra*x[2]+u[0])
xdot = vertcat(xdot1,xdot2,xdot3)
f = Function('f',[x,u,p],[xdot])
#------ one step dt
k1 = f(x,u,p)
k2 = f(x+0.5*k1*dt,u,p)
k3 = f(x+0.5*k2*dt,u,p)
k4 = f(x+1.0*k3*dt,u,p)
x_plus = x + dt / 6 * (k1 + 2 * (k2 + k3) + k4)
Finter = Function('Finter', [x,u,p], [x_plus])
#------ one large step = nsub_step * dt
xact = 1.0 * x
for _ in range(nsub_step):
```

```
    xact = Finter(xact,u,p)
F = Function('F', [x,u,p], [xact])
```

The following `ControlParam` `Python` class implements the feedback control defined above with the four parameters discussed previously, namely:

$$\alpha_1, \alpha_2, \lambda \text{ and } u_{max}.$$

Standard control textbooks consider that the upper bounds on the control input level are given parameters that depend on the chosen actuator. As a matter of fact, we can use the closed-loop simulator to choose the optimal dimension of the actuator so that some required performance levels can be achieved.

This is why these parameters are left here as degrees of freedom that might be computed during the design step.

The ControlParam Class

```
class ControlParam:

    def __init__(self, alpha1, alpha2, lam, lamI, umax):

        self.alpha1 = alpha1
        self.alpha2 = alpha2
        self.lam = lam
        self.lamI = lamI
        self.umax = umax
        # the internal state of the integrator
        self.xI = 0.0

    def update(self,x, theta_ref):

        # update the integrator state

        if abs(theta_ref-x[0])<0.05:
            self.xI += lamI*(theta_ref-x[0])

    def feedback(self, x, theta_ref):

        self.update(x, theta_ref)
        S1 = self.lam * (theta_ref + self.xI - x[0]) - x[1]
        S_inter = np.tanh(self.alpha1 * S1)
        u = self.umax * np.tanh(self.alpha2 * S_inter)

        return u
```

Notice how the `update` method responsible of the evolution of the value of the integrator's internal state is called inside the main `feedback` method so that one has to just call the latter which will automatically takes care of the integrator state updating.

Once an instance of the `ControlParam` class is created, it can be used to simulate a closed-loop scenario that is defined by the following items:

✓ An initial state $x_0 \in \mathbb{R}^3$.
✓ A vector of model's parameters p.
✓ A simulation's duration `Nsim` expressed as a multiple of the basic sampling period `dt` used in the definition of the discrete-time one-step-ahead map.
✓ A function `theta_ref(t)` that gives for any instant t the value of the reference angular position to be tracked.
✓ A control setting's instance `cp`.

This leads to the `simulate_cl` function that is given in the following script:

```
def simulate_cl(x0, p, cp, Nsim, theta_ref):

    # simulate the closed loop using the cp settings

    tcl = np.array([i*dt for i in range(Nsim+1)])
    Xcl = [x0]
    Ucl = [0.0]
    xi = 0
    for i in range(Nsim):
        u = cp.feedback(Xcl[-1], theta_ref(tcl[i])+xi)
        Ucl.append(u)
        xplus = F(Xcl[-1], u, p).full().flatten()
        Xcl.append(1.0 * xplus)

    Xcl = np.array(Xcl)
    Ucl = np.array(Ucl)
    return tcl, Xcl, Ucl
```

Let us use the above script to simulate some closed-loop scenarios for different values of the parameters of the model regarding which **the controller knows absolutely nothing**.

Let us first use the nominal values of the parameters (see Table 4.1) and the following set of controllers parameters:

$$\alpha_1 = \alpha_2 = 1 \quad \lambda = 20 \quad \lambda_I = 0.1 \quad u_{\max} = 200. \tag{4.12}$$

The key parameters here are λ and $u_{\max}$. The choice of λ depends on the highest bandwidth of the reference θ_{ref} which is taken here to be $\omega_{\max} = 10$ rad/sec. This correspond to a fastest response time of 0.12 s ($t_r = \frac{2\pi}{5\omega_{\max}} = 0.12$) leading to a $\lambda = 3/t_r \approx 20$.

As for the saturation level $u_{\max}$, it is linked to the feasibility issue. Practically, one can start using very high values to ensure feasibility (for the given bandwidth) and then decrease progressively until the quality of the closed loop is considered to be no more satisfactory.

It is obviously possible to determine the best choice of the control parameters settings via optimization but this is not done here since the objective of the current chapter is simply to provide a strong reminder regarding the possibility of using standard control-related solutions and not to go the deepest possible on this specific example.

So the idea here is to simulate the closed loop that uses the above controller settings over a cloud of instances of the model's parameters in order to show that the controller leads to *convenient* closed-loop behavior regardless of the true values of the parameters that are totally unknown to the controller.

The behavior of the closed-loop system using the control setting (4.12) and the nominal parameters of the system is shown on Fig. 4.3 for two kinds of trajectories for the reference angular position.

More precisely, the function's argument `theta_ref` used when calling the closed-loop simulator mentioned above are successively given by

```
theta_ref = lambda t: np.pi/2 + 0.5 * np.sign
(np.sin(omega*t)),
```

and

```
theta_ref = lambda t: np.pi/2 + 0.5 * np.sin(omega*t),
```

where $\omega = 10$ rad/sec.

Now in order to check the robustness of the so-designed and *hand-tuned* feedback control against the unavoidable dispersion of the true values of the system's parameters, a cloud of 200 realizations of the parameter vector p is created using the script below:

Checking Robustness

```
nu_p=0.1
Nsamples = 200
P = np.array([np.array([p[i]*(1+np.random.normal(0,nu_p))
              for i in range(len(p))])
             for _ in range(Nsamples)])
```

This leads to the dispersion showing the histograms of relative values depicted on Fig. 4.4.

The evolution of the 200 closed-loop simulations are shown in Fig. 4.5 for the two kinds of reference angle profiles. Notice in particular how the tracking error is

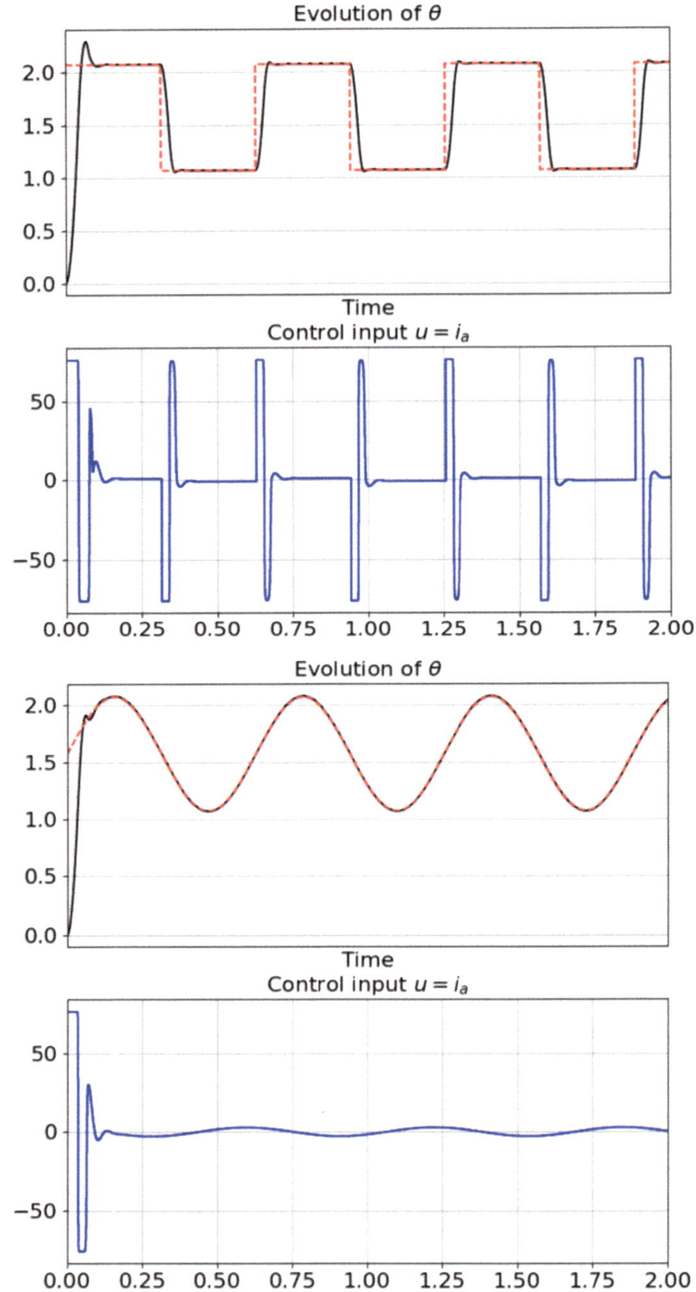

Fig. 4.3 ETC-example: Closed-loop performance using the control design parameter setting (4.12) and the nominal model's parameter of Table 4.1

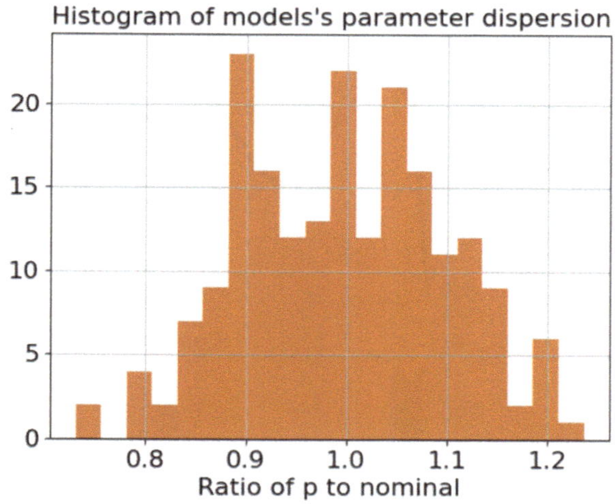

Fig. 4.4 ETC-feedback design: Histogram of the ratio of parameter values to nominal values for the validation of the robustness of the designed feedback law

vanishing whatever is the true unknown set of parameters **thanks to the integral action** used in the controller's design.

It is worth noticing there is no way that such closed-loop quality can be achieved using the so-called advanced learning-based control design approaches that takes explicitly into account the presence of uncertainties such as stochastic MPC, Q-learning, model-free reinforcement learning and so.

Only system-specific design together with some standard control-related tricks and concepts can achieve the kind of results shown in Fig. 4.5.

Moreover, when appropriately done, the design is easy with no heavy computational burden for both on-line and off-line side.

Finally, such a design is comprehensible and easily acceptable by practitioners thanks to its simplicity, explainability, and effectiveness.

The lesson from this example can be summarized as follows:

> **Take-Home Message**
> Intuitive and efficient control design can **sometimes** be designed by exploiting the fact that an engineering system is generally designed so as to achieve a specific task using the actuators involved in the design and through a tractable graph of causality and actions.
> Using the latter together with a bunch of simple control-related tricks and conventional practices **should be the first option to try when facing uncertainties on the controlled system's parameters.**

Recall that this was a simple but hopefully *convincing* example of the use of *standard* control-related concepts and tricks. It is not the objective of this chapter to go through

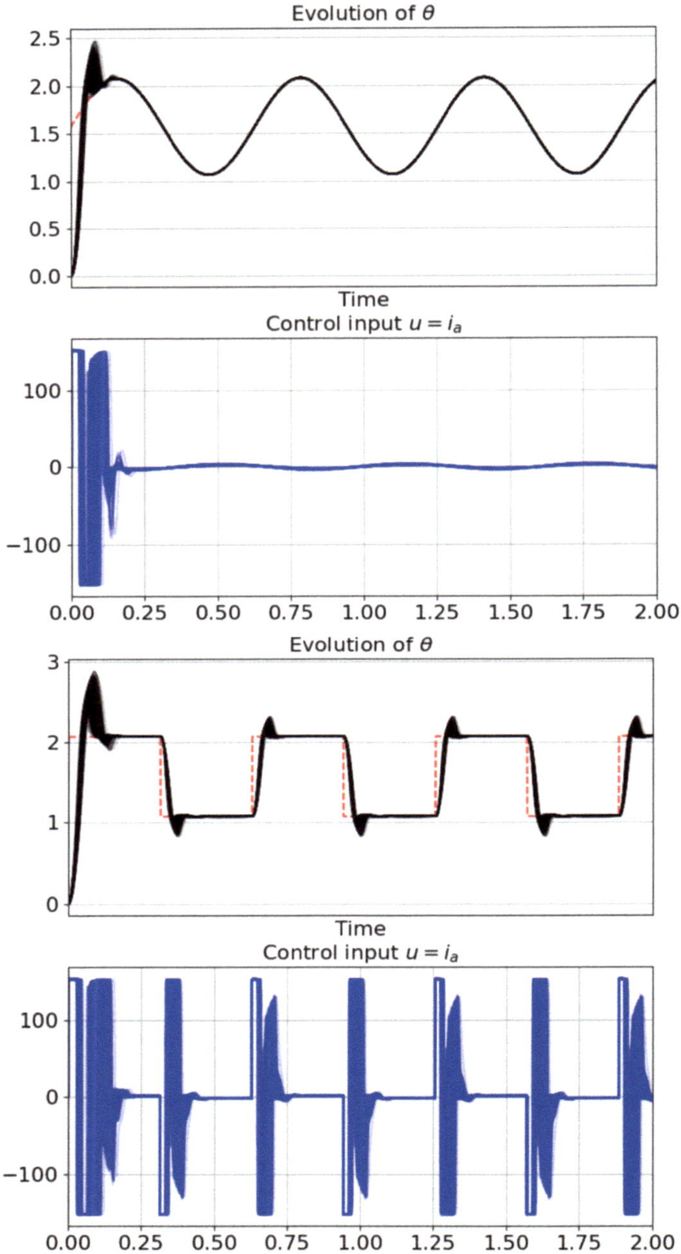

Fig. 4.5 ETC-feedback design: Behavior of the closed loop using the same designed feedback for 200 different instantiations of the model's vector of parameters (see the histogram of relative dispersion on Fig. 4.4)

the impressive list of such tools but to simply attract the reader's attention that *standard* control concepts are not a subset of AI solutions and that they deserve a specific attention and study should one be seriously involved in the *business* of control design for industrial and human-engineered systems.

Nevertheless, some specific control-related theory is given in the remainder of this chapter since some of them are used in some of the applied use-cases that are discussed in Part III of this book.

4.4 Control of Unknown Scalar Systems of Relative Degree 1

In the previous section, we tackled a specific example of control design in the presence of uncertainties. In this section, a generic setting is studied which fits a specific class of **scalar dynamical systems** for which, we rather ignore almost everything except the fact that the regulated output is of degree 1.

Obviously, we are not only interested in scalar systems for their own sake, but it can be shown that some control problems involving dynamical systems of higher state dimension can be decomposed into several sub-problems in each of which the problem boils down into the problem of controlling a set of smaller scalar systems of the generic form described below.

Let us first briefly recall the concept of relative degree that is defined for general nonlinear systems which are governed by ODEs:

> **Relative Degree of a Nonlinear System**
> **Definition 4.3** A nonlinear controlled system is said to be of relative degree r **with respect to some output** y if and only if the integer r is the lowest order of derivation of y that depends explicitly on the control action u.
> More precisely:
>
> $$\frac{\partial^\nu}{\partial u^\nu}[y] = 0 \quad (\forall \nu < r), \tag{4.13a}$$
> $$\frac{\partial^r}{\partial u^r}[y] \neq 0. \tag{4.13b}$$

The results of this section are extended in the next sections to more general systems (of higher dimensions and higher relative degrees) building on the results of the current section.

4.4.1 Working Assumptions and Problem Statement

While many similar results are available in the control literature, the results of the current section and the ones that follow are extensively used in some of the real-life applications dedicated part of this book. That is the reason why we focus on this class of specific so-called *standard* approaches.

In this section, we are interested in controlling the following class of scalar systems:

$$\dot{x} = \alpha[u - h] \quad (\alpha, h(\cdot)) \quad \text{unknown}, \tag{4.14}$$

where

- $\alpha \in [\alpha_{\min}, \alpha_{\max}]$ is an unknown parameter with known bounds $\alpha_{\min} > 0$ and $\alpha_{\max} \geq \alpha_{\min}$;
- u is the control input;
- h is an unknown expression that is **not necessarily constant** during the system's operation; and
- x is the scalar output of interest we would like to force to track some exogenous reference signal x_d.

It is also assumed that bounds on the possible excursions of the unknown term h are supposed to be known, namely, the following assumption is assumed to hold:

Known Bounds on h
Assumption 4.1 There is a known pair of lower and upper bounds $h_{\min}$ and $h_{\max}$ such that for all t, the following inclusion holds true:

$$h(t) \in [h_{\min}, h_{\max}]. \tag{4.15}$$

In order for the regulation problem to be feasible, it is necessary that the input control u can always ultimately *dominate* the term h. This is expressed through the following assumption which states that u is surrounded by lower and upper bounds that make this domination possible:

The Control Input u Is Authoritative Enough
Assumption 4.2 There exists lower and upper bounds $u_{\min}$ and $u_{\max}$ such that for all time t, the following inclusion holds true:

$$u(t) \in [u_{\min}, u_{\max}]. \tag{4.16}$$

Moreover, these bounds relate to the bounds on h, expressed in Assumption 4.1, through the following inequalities:

$$u_{\max} - h_{\max} \geq \varrho_+ > 0, \tag{4.17a}$$

$$h_{\min} - u_{\min} \geq \varrho_- > 0. \tag{4.17b}$$

In simpler terms, this assumption states that the highest value of u is strictly higher (with margin ϱ_+) than the highest value of h and the lowest value of u is strictly lower (with margin ϱ_-) than the lowest value of h.

Notice that if this assumption is not satisfied, one cannot guarantee that the sign of $\dot{x}$ can be ultimately enforceable by the control regardless of the underlying algorithm.[5]

The problem we are interest in can therefore be stated as follows:

Problem Statement
Design a dynamic output feedback law of the form:

$$\dot{z} = f(z, x, x_d), \tag{4.18}$$

$$u = K(z, x, x_d), \tag{4.19}$$

such that the tracking error $e = x - x_d$ is stabilized to a neighborhood that can be made as small as requested.

Links to Other Chapters
The feedback design studied in this section is applied in **Chap.** 15 as a part of the solution to the problem of rationalizing the Propofol injection during anesthesia in the presence of high dispersion of the model among populations of patients.

Before we examine the proposed solution to the above-stated problem, a discussion regarding some alternative solutions that might come to the mind of control-specialized readers is proposed in the next section.

4.4.2 Discussion Regarding Alternative Solutions

There are obviously two intuitive approaches to address the problem stated above, namely:

- the high-gain control solution, and
- the observer-based solution.

Notice that beyond the positioning of the proposed solution, it is interesting to examine the above mentioned alternatives that are definitively parts of the targeted and so-called *standard* control-related approaches. These alternatives can be useful in addressing different problems than the one we are interested in here.

[5] As a matter of fact, these assumptions are, kind of, necessary conditions for the very possibility of having any control which can guarantee the overall achievement of the regulation and/or tracking objective.

These alternatives are sketched in the following discussion.

Elementary High-Gain Feedback Option

The idea here is to *compensate for the effects of the unknown term h* by a high-gain control.

This is done using a sufficiently high value of λ that appears in the expression of the feedback law:

$$u = \lambda(x_d - x).$$

Indeed, by so doing, the equation of the error:

$$e = x_d - x,$$

becomes, assuming constant reference signal x_d:

$$\dot{e} = -\alpha\lambda e - \alpha h,$$

which steers the error towards the asymptotic set defined by

$$\mathcal{E} := \left\{ e \mid e \leq \frac{\max\{|h_{\min}|\,|h_{\max}|\}}{\lambda} \right\},$$

which can be made as small as required by taking high values of λ.

The problem with this solution is that **the transient of the control input variable** u **cannot be guaranteed to meet the saturation constraints** (4.16).

Observer-Based Feedback Option

In this option, the idea is not to *blindly compensate for* the unknown term but to estimate it through a dynamic observer using the measurement of both u and x.

Unfortunately, the fact that α is supposed to be unknown makes the quantity h rigorously non-observable. Indeed, α and h have to be both reconstructed and a simple examination of Eq. (4.14) leads to the fact that a pair of possibilities:

$$(\alpha_1, h_1) \quad , \quad (\alpha_2, h_2),$$

are rigorously non-distinguishable as soon as they are linked through the following relationships[6] for some real number β:

$$\alpha_1 = \beta\alpha_2 \quad \texttt{and} \quad h_2 = (1-\beta)u + \beta h_1,$$

[6] This can be established by writing that:

$$\alpha_1(u - h_1) = \alpha_2(u - h_2),$$

which look at the possibility of having two different explanation of the dynamic of x.

which means that the observability is rigorously broken, at least for when u is constant which is never a good news.

Nevertheless, there are standard ways to overcome these difficulties as it is always the case in smart solutions.

Let us simply say that the solution proposed in the forthcoming section is simply a different option that might be interesting to consider as a general and elegant solution to the stated problem.

4.4.3 The Proposed Solution

The expression of the feedback involves the following saturation map that depends on the bounds of the control input term:

The Saturation Map

$$S(v) := \begin{cases} u_{\max} & \texttt{if} \quad v \geq u_{\max}, \\ u_{\min} & \texttt{if} \quad v \leq u_{\min}, \\ v & \texttt{otherwise}. \end{cases} \tag{4.20}$$

Using this notation, the proposed feedback is given by the following dynamical system which obviously matches the required structure of the solution expressed by (4.18)–(4.19) invoked in the statement of the control problem:

The Dynamic Output Feedback

Definition 4.4 The structure of the dynamic feedback is given by the following equations in which z is the *the internal state of the controller*:

$$\dot{z} = \lambda_f \Big[S\Big(\lambda(x_d - x) + z\Big) - z \Big], \tag{4.21a}$$

$$u = S\Big(\lambda(x_d - x) + z\Big), \tag{4.21b}$$

where $\lambda > 0$ and $\lambda_f > 0$ are two design parameters to be appropriately chosen.

Now for a control designer's expert eyes, it appears quite immediately that the structure above is *a sort of saturated* ***Proportional–Integral*** controller.

Indeed, the term $\lambda(x_d - x)$ depends on the tracking error via the proportional gain λ. The adopted control adds to this proportional term the term z before saturating the sum. But the value of z is the result of the filtered integration of the control input u which is itself an image of the tracking error.

OK! it is not rigorously a PI controller and that is why it is interesting and less prone to some of the PI-related difficulties.

Now the question is the following:

Any Formal Analysis?

What about the resulting closed-loop behavior? Is it possible to come out with some theoretical analysis of this behavior or a kind of insight regarding the appropriate choice of the involved parameters λ_f and λ?

This is precisely what is meticulously done in [1]. Here, only the main result is stated for the sake of brevity. For those readers aiming at deeply understanding the associated proof, the examination of the above reference is recommended.

The analysis conducted in the above cited reference enables the following result to be stated:

> **Condition on Successful Parameters Settings**
> **Proposition 4.1** *For any $\lambda > 0$, **if**, the following conditions hold true:*
>
> 1. *λ_f is such that:*
>
> $$\lambda_f < \left[\min\left\{\frac{min\{\varrho_+, \varrho_-\}}{u_{\max} - u_{\min}}, \frac{\alpha_{\min}}{4}\right\}\right] \times \lambda. \tag{4.22}$$
>
> 2. *x_d is constant.*
> 3. *The dynamic of the unknown term h satisfies*
>
> $$\left|\frac{dh}{dt}\right| \leq \delta, \tag{4.23}$$
>
> ***then** the tracking error $e_x := x - x_d$ and the estimation error $e_h := z - h$ asymptotically satisfy the following inequalities:*
>
> $$\lim_{t\to\infty} |x(t) - x_d| \leq \frac{\delta}{\lambda\lambda_f}, \tag{4.24a}$$
>
> $$\lim_{t\to\infty} |z(t) - h(t)| \leq \frac{\delta}{\lambda_f}. \tag{4.24b}$$

Regarding the above convergence results, the following comments enable to grasp some of the properties of the proposed solution:

- First of all, since (4.22) implies that by increasing λ, one can increase λ_f also, the inequalities (4.24a)–(4.24b) clearly show that the asymptotic errors can be made as small as desired.

- When the unknown term h is constant, $\delta = 0$ is eligible in (4.23) and therefore the last two inequalities show that the errors tend asymptotically to 0.

4.4.4 An Illustrative Example

In this section, a simple example is provided in order to illustrate the theoretical results invoked in the preceding section.

Let us consider a system of the form (4.14) in which the unknown term h takes the following form:

$$h(t) = 5 - 1.3\cos(1.2t) - 1.6\sin(2t + \pi/6). \tag{4.25}$$

Notice that h can be a function of the state also. This doesn't matter as far as h is bounded since the controller has no idea about the origin for the variations of the unknown term h. By examining Fig. 4.6, it comes out that the following bounds are appropriate:

$$h_{\text{min}} = 2, \qquad h_{\text{max}} = 8, \qquad |\frac{dh}{dt}| \leq \delta = 4.76.$$

As for the unknown parameter α, the following unknown function is considered in the simulation:

$$\alpha(t) = 0.2 + |\sin(\frac{1}{2}\cos(3t)|, \tag{4.26}$$

leading to a known lower bound

$$\alpha_{\text{min}} = 0.2.$$

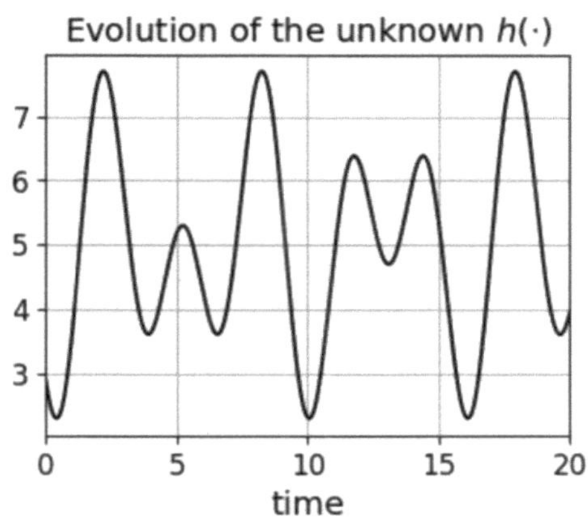

Fig. 4.6 Evolution of the unknown term h versus time

These choices are implemented using the following script:

```
h = lambda t: 5-1.3*np.cos(1.2*t)-1.6*np.sin(2*t+np.pi/6)
alpha = lambda t: 0.2+abs(0.5*np.cos(3*t))
delta = abs(np.diff(h(t))).max()/dt
```

Therefore, in order to meet the requirements of Assumption 4.2, the following bounds are used on the control input u:

$$u_{\min} = 1 \qquad u_{\max} = 10,$$

leading to the values $\varrho_- = 1$ and $\varrho_+ = 2$ regarding the margin parameters involved in (4.17).

Injecting the previous information in (4.22) leads to the following condition on the eligible choice of λ_f once a value of λ is adopted:

$$\lambda_f \leq \left[\min\left\{\frac{min\{2, 1\}}{10-1}, \frac{0.2}{4}\right\}\right] \times \lambda, \tag{4.27}$$

$$\leq 0.05\lambda. \tag{4.28}$$

The resulting closed-loop behavior can be simulated using the following script:

Simulating the Closed-Loop

```
xref = lambda t: 50 + 0 * np.sin(0.2*t)
umin, umax = 1, 10
S = lambda v: min(umax, max(umin, v))
K = lambda x, z, xd: S(lam * (xd-x)+z)

def syst(y,t,lam, lamf):
    x = y[0]
    z = y[1]
    u = K(x,z,xref(t))
    xdot = alpha(t)*(u-h(t))
    zdot = lamf * (u-z)
    return [xdot, zdot]

x0 = 40
z0 = 0.0
lam = 4000
lamf = 0.05 * lam
y0 = [x0, z0]
sol = odeint(syst, y0, t, (lam, lamf))
```

The resulting closed-loop behavior is shown in Fig. 4.7. This figure shows the evolution of the regulated variable x alongside with its reference value.

The second plots shows the evolution of the controller's internal state z which estimates the unknown term h.

The remaining two plots show the evolution of the two errors

$$e_x = |x - x_d|,$$

and

$$e_z = |z - h|,$$

together with their associated **theoretical bounds** given by (4.24a) and (4.24b). It can be observed that the amplitudes of the errors lie asymptotically below their theoretically derived bounds.

It is important to recall that while the controller still works for slowly varying reference signal x_d, the theoretical bounds are no more valid since they have been computed assuming that the derivative of the reference signal is 0.

Note also that in the case of time-varying signals, the success conditions might be amended also. Here is where probabilistic certification might be necessary to get certifiable closed-loop properties.

Figure 4.8 shows the behavior of the closed loop under a much lower value $\lambda = 50$ showing that while the bounds are always very much respected (following the theory), the levels of errors are much higher.

In the following section, the feedback design proposed in the current section is extended to the more general case that includes a higher order model for the internal dynamics of the actuator. This enables us to enlarge the domain of application of the proposed methodology.

4.4.5 Controlling Unknown Systems of Relative Degree ≥ 2

This section uses the results of the preceding section to enlarge the class of highly uncertain systems for which the same feedback expression (4.21), in which x is simply replaced by the regulated output y can be used, namely[7]:

$$\dot{z} = \lambda_f \Big[S\Big(\lambda(y_d - y) + z\Big) - z \Big], \tag{4.29a}$$

$$u = S\Big(\lambda(y_d - y) + z\Big). \tag{4.29b}$$

Notice however that, while the control definition remains unchanged, the difference in the structure of the controlled system (with a higher relative degree) induces some

[7] This is because x now stands for the non-scalar vector of state's components leaving y and u to be the only measured quantities.

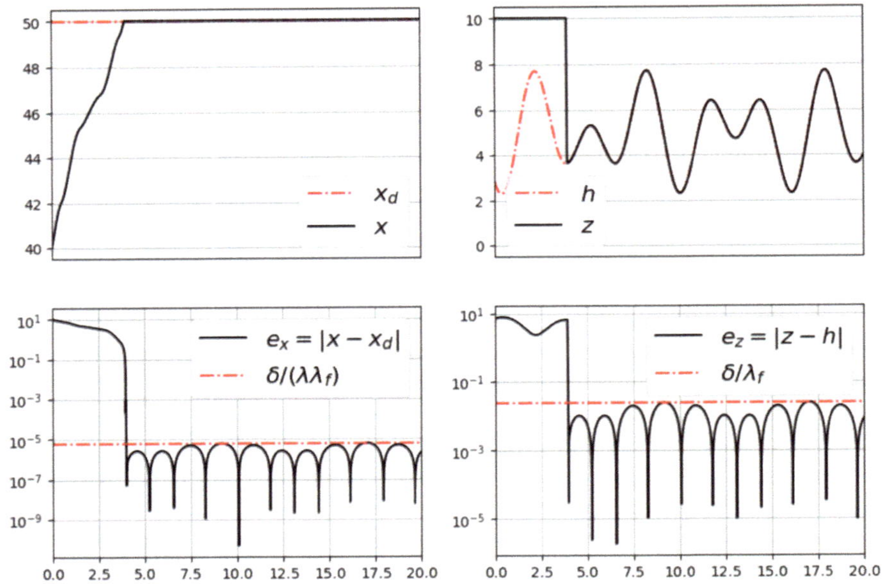

Fig. 4.7 The closed-loop behavior under $\lambda = 4000$ and $\lambda_f = 0.05\lambda$ under the control law that ignores almost everything about the dynamical system

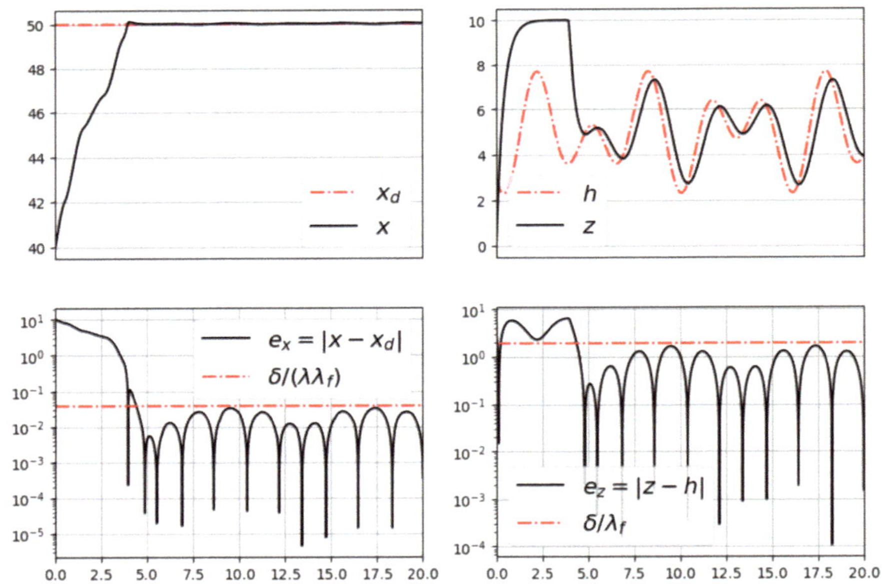

Fig. 4.8 The closed-loop behavior under $\lambda = 50$ and $\lambda_f = 0.05\lambda$ under the control law that ignores almost everything about the dynamical system

differences in the set of sufficient conditions that are, as expected, more stringent and complex to check.

As in the preceding section, the objective here is to state the results in a comprehensive way without digging deeply in the details of the proofs that can be found in [2].

Nevertheless, it is worth noticing that the underlying proof is based on the use of the spirit of the results of the previous section which makes the sufficient conditions more easy to grasp.

So let us consider the class of systems under interest[8]:

$$\dot{y}(t) = \alpha(t)\Big[q_1(t) - h(t)\Big], \tag{4.30a}$$

$$\dot{q}(t) = G(q(t), u(t)), \tag{4.30b}$$

where contrary to the case of the previous section, the control does not affect directly the derivative of the regulated output y. Rather, it acts on the derivative of some vector $q \in \mathbb{R}^{n_q}$ in which, the first components plays the role played by u in the previous section.

In other words, u has to control q so that q_1 acts as u was acting previously in the control scheme.

Notice that the relative degree of the dynamical system (4.30) with respect to y depends on the dimension of the sub-state q. More precisely, if $q \in \mathbb{R}^{n_q}$ then the relative degree might be as high as $n_q + 1$.

This obviously introduces some *delay* in the *action path* going from u and ending at $\dot{y}$. The presence of such a delay imposes additional conditions for the framework to be successful.

As a matter of fact the dynamic on q has to be viewed as the dynamic of some controlled actuator with unitary gain, this is expressed in the following assumption:

Assumption on the Actuator Dynamics

Assumption 4.3 The dynamic system (4.30b) is an asymptotically stable system with unitary gain, more precisely, for any reference signal u on q_1 and any $\rho > 0$ such that:

$$|\dot{u}| \le \rho, \tag{4.31}$$

there exist positive reals σ, π_1 such that the following properties hold true **after a finite time**:

$$|u(t) - q_1(t)| \le \sigma\rho, \tag{4.32a}$$

$$|\dot{u}(t) - \dot{q}_1(t)| \le \pi_1\rho. \tag{4.32b}$$

[8] Note that we changed the notation used to designate the regulated variable from x to y since the state vector now is $x = (y, q) \in \mathbb{R}^{1+n_q}$ and is no more equal to $x = y$ as in the previous section.

This assumption simply states that the control inside the actuator is such that the tracking errors on both the values and the derivative between u and q_1 are bounded by terms that are proportional to the limitation ρ on the dynamics $\dot{u}$ of the true control input. Notice that u serves as a reference delivered by the control algorithm to the internal actuator's control loop.

About the Parameters σ and π_1

In the remainder of the discussion, the parameters σ and π_1 involved in (4.32) are supposed to be two **known characteristics** of the controlled actuator dynamics.

The success conditions are expressed in terms of these parameters. Later on, an example of a candidate procedure for the identification of these parameters is given.

It is important to underline here that the general form of the actuator dynamics (4.30b) might be of any order, it can be linear or nonlinear depending on the actuator model and the way it is controlled.

What does really matter for the forthcoming discussion is the consequences of the actuator dynamics and its internal controller on the relationship between the first state q_1 and the reference trajectories u expressed through (4.32). This makes the forthcoming results of rather wide scope of applicability to many situations as it is shown in Part III.

The trick that is used in [2] in order to capitalize on the results of the preceding section is to view the dynamical system (4.30) as a system of relative degree 1 with a different unknown term denoted by h' that will play the role played by h in the previous section.

This is done by including the difference between the true control input u and the term q_1 inside a newly defined unknown term h'.

More precisely, the dynamics of the regulated variable y can be rewritten in a new form that is equivalent to the form used in the previous section with the only difference being that h' is used that depends on the applied control:

$$\dot{y} = \alpha\Big[q_1 - h\Big], \tag{4.33}$$

$$= \alpha\Big[u - h'\Big] \quad \text{with} \quad h' := h + (u - q_1). \tag{4.34}$$

By so doing, the results of the previous section can be applied **provided that one can adapt the conditions to the new definition of the additional term h'.**

Again, it is important to underline that the major difference here is that the possible values taken by the unknown term h' **as well as by its derivative**[9] **depend on the control** u since that latter is now a part of the definition of the unknown additional term h'.

[9] Which also plays a role in the set of the success conditions as it is stated by Proposition 4.1 through the role of the parameter δ.

Therefore, in order to keep the evolution of the unknown term under control, we need a bound on the term $|\dot{u}|$ for a given choice of the controller's pair of parameters (λ, λ_f).

This is provided by the following intermediate results [2]:

Bound on $|\dot{u}|$

Lemma 4.1 *For all $\lambda > 0$ such that:*

$$0 < \lambda < \frac{1}{\bar{\alpha}\sigma}, \tag{4.35}$$

the following inequality holds true:

$$|\dot{u}| \leq \rho(\lambda, \lambda_f) := \frac{\lambda}{1 - \lambda\bar{\alpha}\sigma}\Big[\bar{\alpha}\bar{\rho} + \frac{\lambda_f}{\lambda}\Delta_u\Big], \tag{4.36}$$

where the following definitions are used:

$$\Delta_u := u_{\max} - u_{\min}, \tag{4.37a}$$

$$\bar{\rho} := \max\Big\{u_{\max} - h_{\min}, h_{\max} - u_{\min}\Big\}, \tag{4.37b}$$

while $\bar{\alpha}$ is the upper bound on α.

Now that we have a bound on the excursion of the term $|\dot{u}|$, we can invoke (4.32) to revise the bounds on the input u which have to be enlarged so that the authority conditions expressed by Assumption 4.2 still holds for an unknown term that shows the following bounds:

$$h'_{\min} = h_{\min} - \sigma\rho(\lambda, \lambda_f), \qquad h'_{\max} = h_{\max} + \sigma\rho(\lambda, \lambda_f), \tag{4.38}$$

where $\rho(\lambda, \lambda_f)$ is defined by (4.36).

Figure 4.9 shows the conditions on the bounds $u_{\min}$ and $u_{\max}$ for a given setting of the parameters λ, λ_f.

This condition simply requires that the bounds on the control input u should be large enough in order to include the original domain of h enlarged by the potential excursions that might be induced by the tracking error term $(q_1 - u)$ in the newly defined unknown term h'.

Before going further, it is important to understand that including the tracking error $(u - q_1)$ in the unknown terms and the need for some bound on the excursion of this term introduces an implicit limitation on the dynamics of the internal actuator control loop.[10]

[10] Recall that it is the conditions (4.31) that introduces the ρ-related terms in the game!

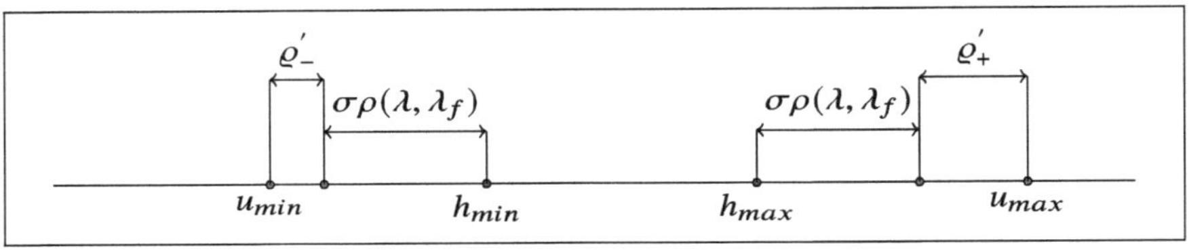

Fig. 4.9 Schematic view of the bounds involved in the condition on the control input's bounds

Indeed, if we increase the amplitude of the control dynamic, namely $|\dot{u}|$ (which arises when trying to get fast control with high bandwidth), one increases the tracking error and hence increases the excursion of the unknown term h' and this might end by the loss of controllability of the regulated output.

This is a major difference compared to the case of the previous section. This explains why in the results of the present section, there is an upper bound on the design parameter λ which is expressed by (4.35) while in the results of the previous section, candidate values of λ were not upper bounded.

The fact that the result stated in the next Proposition is inherited from the result of the previous section leads to the need for the following definition that comes directly from (4.22) with the only difference that the expression of ϱ_+ and ϱ_- are now made dependent on the controller's parameters λ and λ_f:

$$\varphi(\lambda,\lambda_f) := \left[\min\left\{\frac{min\{\varrho_+(\lambda,\lambda_f),\varrho_-(\lambda,\lambda_f)\}}{u_{\max}-u_{\min}}, \frac{\alpha_{\min}}{4}\right\}\right]\times\lambda, \tag{4.39}$$

where, in reference to Fig. 4.9, $\varrho_-(\lambda,\lambda_f)$ and $\varrho_+(\lambda,\lambda_f)$ are given by

$$\varrho_-(\lambda,\lambda_f) := \sigma\rho(\lambda,\lambda_f) + \varrho_-', \tag{4.40a}$$

$$\varrho_+(\lambda,\lambda_f) := \sigma\rho(\lambda,\lambda_f) + \varrho_+', \tag{4.40b}$$

in which $\rho(\lambda,\lambda_f)$ is given by the r.h.s. of (4.36).

Now we have all that we need to state the main result of the current section regarding the behavior of the class of system given by (4.30) under the parameterized control law given by (4.29):

> **Control of Unknown Systems of Relative Degree ≥ 2**

Proposition 4.2 *If the pair (λ,λ_f) is chosen such that the following inequalities are satisfied for some $\beta \in (0,1)$:*

$$\lambda \leq \frac{\beta}{\sigma\bar{\alpha}} \quad ; \quad \lambda_f \leq \varphi(\lambda,\lambda_f), \tag{4.41}$$

then the following property holds regarding the tracking error of a constant reference value y_d:

$$\lim_{t\to\infty} |y_d - y(t)| \le \frac{\delta + \pi_1 \rho(\lambda, \lambda_f)}{\lambda\lambda_f}, \tag{4.42}$$

where $\delta > 0$ *is an upper bound on the dynamic of* h*, namely* $|\dot{h}| \le \delta$.

Proposition 4.2 clearly shows that there is a trade-off in the choice of (λ, λ_f) since higher values are desired in order to reduce the asymptotic error despite the varying unknown term on one hand, but on the other hand, the inequalities expressed by (4.41) shows that there is a limitation in the choice of how high the values of λ and λ_f are allowed to be.

In such a case, the use of constrained optimization is the natural option. Indeed, the controller gains λ and λ_f might be determined by solving the following optimization problem:

Determining λ and λ_f via Optimization

The pair of parameters (λ, λ_f) are computed by solving the following constrained optimization problem:

$$(\lambda^\star, \lambda_f^\star) \leftarrow \text{arg} \min_{(\lambda,\lambda_f)} \left[\frac{\delta + \pi_1 \rho(\lambda, \lambda_f)}{\lambda\lambda_f}\right], \tag{4.43a}$$

$$\text{under the constraints} \left| \begin{array}{l} 0 \le \lambda \le \beta/(\sigma\bar{\alpha}), \\ 0 \le \lambda_f \le \beta\varphi(\lambda, \lambda_f), \\ 0 \le \rho(\lambda, \lambda_f) \le \rho_\ell, \end{array} \right. \tag{4.43b}$$

where $\varphi(\lambda, \lambda_f)$ and $\rho(\lambda, \lambda_f)$ are respectively defined by (4.39) and (4.36) while $\bar{\alpha}$ stands for an upper bound on α.

Notice that in the above formulation, $\beta \in (0, 1)$ is arbitrarily chosen while ρ_ℓ is a bound on the possible values of $\rho(\lambda, \lambda_f)$ that might be necessary for some special cases where too high gains are made possible theoretically while the designer needs to keep moderate gain for whatever reasons.

In the next section, a simple example is used to illustrate the rather complicated sufficient conditions involved in the statement of the previous result.

4.4.6 Illustrative Example

Let us consider the following system:

$$\dot{y} = \alpha[q_1 - h(t)], \tag{4.44a}$$

$$\dot{x}_a = AS(x_a) + Bu \quad ; \quad q_1 = Cx_a, \tag{4.44b}$$

where the unknown term $h(t)$ is given by

$$h(t) = 5 - 1.1\cos(1.2t) - 0.4\sin(2t + \pi/6), \tag{4.45}$$

and where (4.44b) describes the dynamic of the actuator that is supposed to be unknown to the controller.

Based on the expression of h, the bounds

$$h_{\min} = 3.5,\ h_{\max} = 6.5,\ \delta = 2.2,$$

represent valid bounds on the lower and upper values of h and on the module of its derivative respectively and can therefore be used in the validation of the results for this example.[11]

Moreover, $\alpha_{\min} = 0.8$ and $\bar{\alpha} = 1.2$ are used while the truly applied unknown value is $\alpha = 1$.

The actuator in use, which is totally unknown to the controller (except for its associated bounds), is given by (4.44b) in which:

$$x_a := (q_1, \dot{q}_1) \in \mathbb{R}^2,$$

stands for the internal state of the actuator while the term $S(x_a)$ introduces saturation on the actuator's internal state which represents a nonlinearity in the dynamics of the actuator.

To be more specific, explicit saturation mechanisms are used so as to enforce the following limitations on the value of q_1 and its derivative. More precisely, the following constraints are structurally imposed:

$$q_1 \in [0, 10],\quad \dot{q}_1 \in [-60, +60],$$

while the saturation bounds on u are taken such that:

$$u \in [0, 10].$$

Identification of σ and π_1

The determination of σ and π_1 can be obtained through an *identification experiment* conducted on the actuator.

More precisely, using an input signal $u = u_{id}$ given by[12]:

$$u_{id}(t) := \begin{cases} u_{\max} & \text{if } t \le t_s \\ c_0 + c_1\cos(c_2 t) & \text{otherwise,} \end{cases} \tag{4.46}$$

where $t_s = 10$ and c_0, c_1, c_2 are computed recursively as shown in the following algorithm:

---

[11] These bounds can be computed as shown in the example of the preceding section.

[12] The selection of the identification signal is based on the need to have $y = y_d$ at steady state, to achieve this, u_{id} is set to $u_{\max}$ for t_s seconds then oscillates in a way that mimics h' making use of the available knowledge on the excursion of this term.

Algorithm 4.1 Actuator bounds identification

Require: $\rho_\ell > 0$
Initialize $\sigma \leftarrow 0, \pi_1 \leftarrow 0$;
while $k < 200$ **do**
  $s_+ \leftarrow h_{max} + \sigma\rho_\ell$; $s_- \leftarrow h_{max} - \sigma\rho_\ell$ ▷ Bounds on h'
  $c_0 \leftarrow (s_+ + s_-)/2$ ▷ Mean value of h'
  $c_1 \leftarrow (s_+ - s_-)/2$ ▷ amplitude of excursion of h'
  $c_2 \leftarrow (\delta + \pi_1\rho_\ell)/c_1$ ▷ Make that c_1c_2 simulates derivative of h'
  $u_{id} \leftarrow$ `identsignal`$(u_{max}, c_0, c_1, c_2))$ ▷ Generate input signal
  $q_1, \dot{q}_1 \leftarrow$ `actuator`(u_{id}) ▷ Simulate the actuator dynamics
  $\sigma \leftarrow |u_{id} - q_1|/\rho_\ell$ ▷ Update σ
  $\pi_1 \leftarrow |\dot{u}_{id} - \dot{q}_1|/\rho_\ell$ ▷ Update π_1
  $k \leftarrow k + 1$
end while

More precisely, the signal u_{id} is injected in the dynamics (4.44b) and by simulating this dynamics (simulation here is used to imitate the real-life experiment), the profile of $|u_{id} - q_1|$ and $|\dot{u}_{id} - \dot{q}_1|$ are obtained. The functions `identsignal` and `actuator` invoked in the algorithm incorporate the computation of the input signal defined by (4.46) and the simulation of the actuator dynamics respectively.

Using the resulting values, the values of σ and π_1 are computed according to (4.32) where the upper bound on the allowable rate of change of the control input ρ is chosen to be equal to the maximum possible value of $\dot{q}_1$, namely $\rho_\ell = 60$ in accordance with the presumed bound mentioned above.

Using the procedure described above, we get the following values:

$$\sigma = 0.0107 \quad ; \quad \pi_1 = 0.0152,$$

which are used in the remainder of the presentation.

Optimal Pair $(\lambda^\star, \lambda_f^\star)$ and Resulting Closed-Loop Behavior

We can now solve the optimization problem given by (4.43) using $\beta = 0.99$.

Figure 4.10 shows the iso-values of the cost function together with the feasible values that lie in the set of points, being considered.

The figure suggests that a good approximation of the optimal solution is given by

$$\lambda^\star = 5.6 \quad ; \quad \lambda_f^\star = 1.2, \tag{4.47}$$

and that this optimal choice corresponds to the asymptotic error bound given by

$$\lim_{t\to\infty} |y_d - y(t)| \leq 0.5. \tag{4.48}$$

Figure 4.10 shows the closed-loop behavior when using the optimal values mentioned above showing a good tracking performance that is achieved after finite time. The expected theoretical bounds on the tracking error are obviously respected (Fig. 4.11).

Moreover, the differences between u and q_1 on one hand and between their derivatives meet also the theoretical bounds corresponding to the computed values of σ and π_1.

Moreover Fig. 4.12 shows that when using $\lambda = 10$ and $\lambda_f = 1.1$ which do not meet the theoretical conditions, higher tracking errors can be observed and a too *highly nervous* behavior of the control dynamics is observed. In particular, the bound on the tracking error that is predicted for the appropriate choice of the parameter is now violated due to the use of values that do not respect the theoretical prescription and the resulting behavior is obviously not satisfactory.

The take-away message from this quite technically involved section is that highly unknown dynamics regarding both the system dynamics and the controller dynamics can be rationally handled at the price of some experimentation on one hand and using some bounds that can be computed in real-life situations (regarding the bounds that some unknown terms might have given the context of use and some commonly known orders of magnitude).

Now it might be true that the rigorous use of the computational framework exposed in this section, in particular, might not be an option on a daily basis. Nevertheless the results of this section should incite the practitioner to use **randomized certification** of the underlying control structure in some situations since the results suggest that *there exist values of the pair* (λ, λ_f) *that lead to satisfactory results.*

The remaining task would be to define the probabilistic certification-related precise definition of the success/failure of an experiment in order to come out with the appropriate values to be used in the controller.

The last class of highly uncertain systems that we shall address before ending this chapter is the one where the dynamics of the control might excite some internal modes in the system making the control task no more achievable. This is handled in the following section.

Addressing these situations is technically involved but it occurred to me that since the readers already invested in the previous sections and probably start to grasp the mechanisms at work, it is interesting to capitalize on this understanding in order to address this last important class of *highly uncertain* systems.

4.5 Systems with Destabilizing Internal Dynamics

This last section addresses the problem of output feedback design for highly uncertain dynamical systems showing the following features:

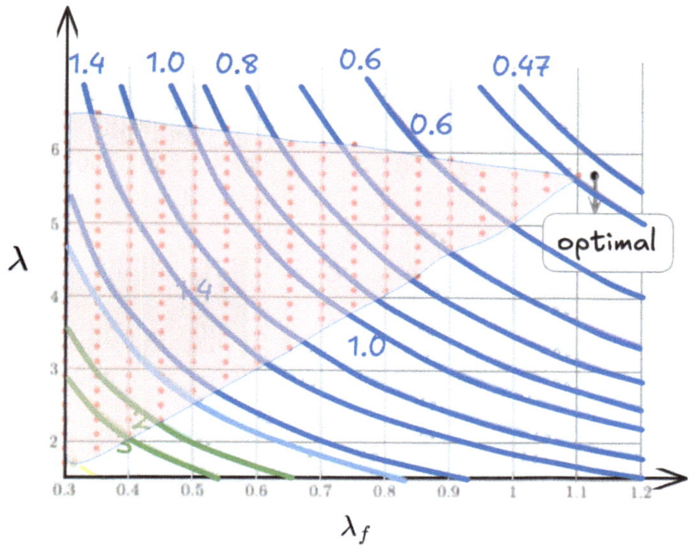

Fig. 4.10 Contour plot of the cost function where dots are the feasible admissible pairs given the constraints while the optimal solution is indicated by the arrow basis

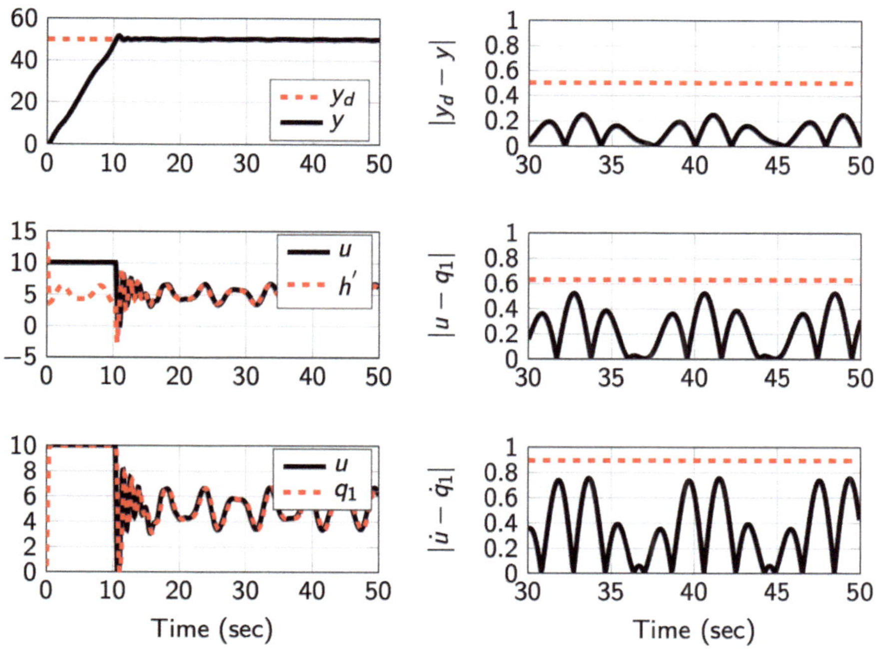

Fig. 4.11 Closed-loop simulation of the unknown system using the optimal pair $(\lambda^\star, \lambda_f^\star) = (5.6, 1.1)$. The right-hand plots compare the simulated terms to their theoretically computed bounds

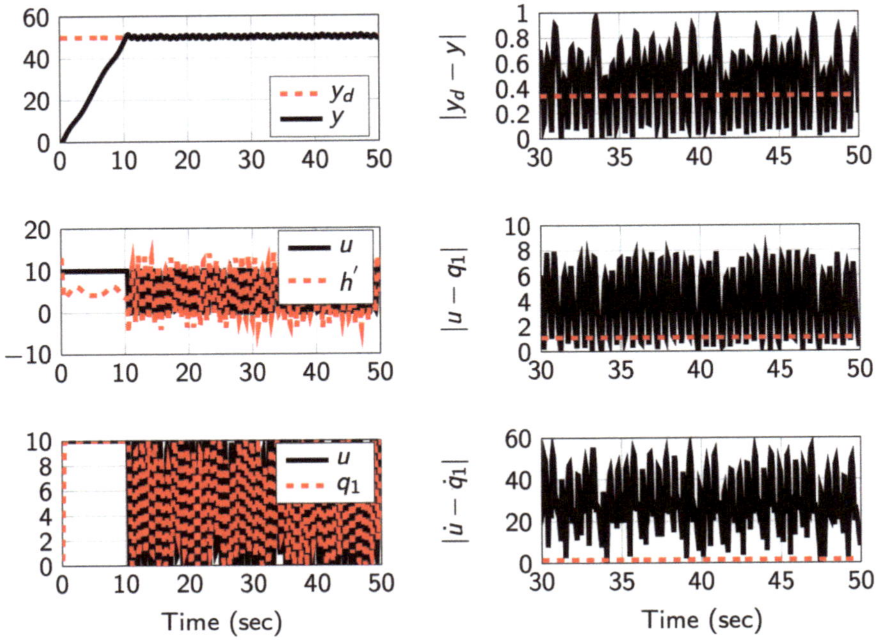

Fig. 4.12 Closed-loop simulation of the unknown system using $(\lambda^\star, \lambda_f^\star) = (10, 1.1)$ that **does not meet the theoretical requirements**. The right-hand plots compare the simulated terms to their theoretically computed bounds

1. There is a simple although uncertain relationship between the input u and the regulated variable y (that is of relative degree 1). This simplicity suggests the possible use of appealingly simple feedback design.
2. However, an `incautious` design of such a simple feedback law might *excite* some internal dynamics of the system leading to oscillations and/or even irreversible lack of controllability.
3. In spite of this issue, the use of *advanced* observer-based nonlinear controllers that would be designed using the entire system's dynamics including the dangerous internal dynamics is not welcomed by practitioners because of its high complexity and the need to derive a tight model involving complex behaviors that are not always well mastered. Moreover, the design of an observer is not always easy nor is it feasible for non-observable systems.

Once the structure of the system is defined, we shall give two real-life examples of such systems in order to motivate the rather technical study that follows.

4.5.1 The Class of Systems Considered

Let us consider the family of uncertain dynamical systems that are governed by the following equations:

$$\dot{y}(t) = \alpha(t)\Big[u(t) - g(t) + \ell(t, \eta(t))\Big], \tag{4.49a}$$

$$\dot{\eta}(t) = E\Big(\eta^{[t-\tau,t]}, \dot{u}^{[t-\tau,t]}, w^{[t-\tau,t]}\Big), \tag{4.49b}$$

where

- y is the regulated variable.
- $u \in [\underline{u}, \bar{u}]$ is the control input that is saturated with known bounds regarding which the following notation is used:

$$\Delta_u = \bar{u} - \underline{u}.$$

- $g(\cdot)$ is an unknown (possibly time-varying) term that is supposed to lie in a bounded set defined by

$$g(t) \in [\underline{g}, \bar{g}],$$

  for **known bounds** $\underline{g}$ **and** $\bar{g}$.
  The notation:

$$\Delta_g := \bar{g} - \underline{g},$$

  is also used in the sequel.
- $\eta \in \mathbb{R}^{n_\eta}$ represents a vector of internal states that are governed by (4.49b) in which $E(\cdot, \cdot, \cdot)$ is a possibly **unknown map**.
- $w \in \mathbb{R}^{n_w}$ is a vector of exogenous disturbances.
- $\alpha(t) \in [\underline{\alpha}, \bar{\alpha}]$ is an unknown possibly time-varying scalar with known bounds.

It is important to underline the fundamental difference between the class of systems given by (4.49) and the one studied in the previous section, namely (4.44). Indeed, for a system governed by (4.49) the control acts directly on the regulated output y (relative degree 1) while in the previous section, the control acts inside the actuator through multiple levels (at least one) before reaching the regulated variable y (relative degree ≥ 2).

This might suggest that the system we study in the present section should be easier to control. This is not necessarily the case because of the other difference between the two classes of systems which is related to the following two facts:

- The presence, in the definition of $\dot{y}$ of the term $\ell(t, \eta)$ that depends on the internal state η of the remaining dynamics.
- The fact that the internal dynamic $\dot{\eta}$ depends on the **the derivative** of the control input, namely $\dot{u}$.

The conjunction of the two above characteristic induces the need for a specific care regarding the choice of the parameters of the feedback law.

Recall that the feedback law is always the same, namely:

$$u = S(\lambda(y_d - y) + z), \tag{4.50a}$$

$$\dot{z} = \lambda_f(u - z), \tag{4.50b}$$

with the control design problem that can be stated as follows:

Problem Statement
Find a high level set of characterizations of the uncertainties and maps involved in (4.49) and associated *sufficient* conditions on the controller parameters (λ, λ_f) such that the behavior of the closed-loop system can be conveniently/quantifiably assessed.

Links to Other Chapters
The feedback design studied in the present section is used in **Chap.** 18 in order to address the problem of decentralized control of the frequency in micro-grids.

But before to go into the technical details, let us first assess the relevance of the framework and the class of systems being considered by showing two real-life industrial examples that totally fit the working framework depicted above.

This seems important because the above formulation and mathematics might induce the feeling that this is a toy hypothetical setting that correspond to no real-life instances.

4.5.2 Real-Life Examples

The class of systems described by (4.49) fits at least two important industrial problems that are briefly discussed hereafter:

Control of Hydraulic Turbines
Efficient control of hydraulic turbines in the context of Power Storage Plant (PSP) is gaining a renewed interest. In this context, the hydraulic plant is used as a mean to compensate for the intermittency of renewable sources of energy. As such, the response time of hydraulic turbines' speed control is a key parameter that determines the success of the whole power management system and beyond, the rate of integration of renewable which is obviously a major challenge of the coming decades.
In this context, the regulated variable $y = \Omega$ is the rotational speed of the turbine while the control input is the guiding vane's opening γ. The internal state η is the

state of the penstock (flow rate and pressure values along the penstock) that drives the water flow from the water reservoir to the turbine's level (see Fig. 4.13).

Practitioners know that the major obstacle that limits high rotational acceleration of the turbine lies in the induced oscillating pressure waves that might result in the penstock [3].

When referring to the internal dynamics involved in (4.49b), **the time laps τ represents twice the time needed for the pressure wave to travel the penstock's length**.

The potential pressure-wave-induced oscillations generate disturbing torque variations at the turbine level. If the control dynamics are not monitored in order to avoid these oscillations' settling in the penstock, the oscillations can exceed the control's ability to compensate for them.

That is the reason why simple control laws that are not carefully tuned against this phenomenon fail to produce satisfactory results in terms of response time of the hydraulic turbines.

Distributed/Decentralized Frequency Control in Isolated Micro-Grids

This is another example that falls into the framework studied in this section. Here the regulated variable is the micro-grid frequency denoted hereafter by f.[13]

The control input $u = P$ consists here in the sum of the produced powers by the Diesel generators (commonly referred to as *Gensets*) and the renewable sources while the uncertainty w is the derivative $\dot{P}_L$ of the unknown load power consumption P_L which results from the sum of the individual powers required by the users of the micro-grid.

The internal dynamics is induced by the flexibility of the power transmission between Gensets following sudden changes in the load power demand.

In this context, the need for a distributed/decentralized control design (in which only the frequency measurement can be used) emphasizes the need for simple control laws that cannot afford the task of modeling-based reconstruction of the internal dynamic which would require both observability and total knowledge of the model's details which might be simply impossible to dispose of by the local distributed controller of Gensets and renewable energy producers.

The internal dynamic disturbs the commonly used power balance frequency equation usually used in this context (see [4] and [5]). This equation is given by

$$\dot{f} = \alpha(P - P_L),$$

which simply expresses that the dynamic of the frequency is induced by the power *unbalance* between the production and the consumption.

A recent study [6] showed that in order to take into account the internal dynamic, it is necessary to introduce additional term to the above simple equation so that the

[13] The definition of this frequency and its concrete measurement is a quite *tricky* problem that is not studied here.

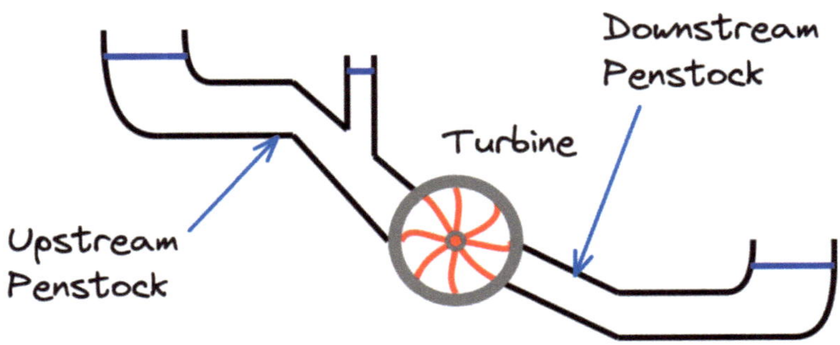

Fig. 4.13 Schematic view of a hydraulic turbine with its upstream and downstream penstocks

appropriate dynamic becomes

$$\dot{f} = \alpha(u - P_L + C\eta), \tag{4.51a}$$

$$\dot{\eta} = A\eta + B_1\dot{u} + B_2\dot{P}_L, \tag{4.51b}$$

which obviously takes the form of (4.49a)–(4.49b) with the following correspondences:

$$w = \dot{P}_L \quad , \quad g = P_L \quad \text{and} \quad \ell(\eta) := C\eta.$$

Notice that in this specific example, a 5-th order dynamics (in the model governing η) is shown to be required in order to correctly represent the phenomena when only two Gensets are considered [6].

Now that the relevance of the class of systems considered is assessed through the previously described relevant examples, let us examine the working assumptions under which it is possible to provide an assessment of the properties of the closed-loop system when controlling the class of systems given by (4.49a)–(4.49b) using the control feedback law given by (4.50).

4.5.3 Working Assumptions

We shall follow a path that is similar to the one followed in the previous section, namely, we shall considered a specific modified definition of the unknown term h that is suitable for the new problem at hand. This will enable us to capitalize on the result of Sect. 4.4 in order to derive a modified set of necessary conditions that apply to system defined by (4.49).

More precisely, a careful examination of (4.49) clearly shows that the unknown term h that has been used in Sect. 4.4 dedicated to the control of unknown scalar

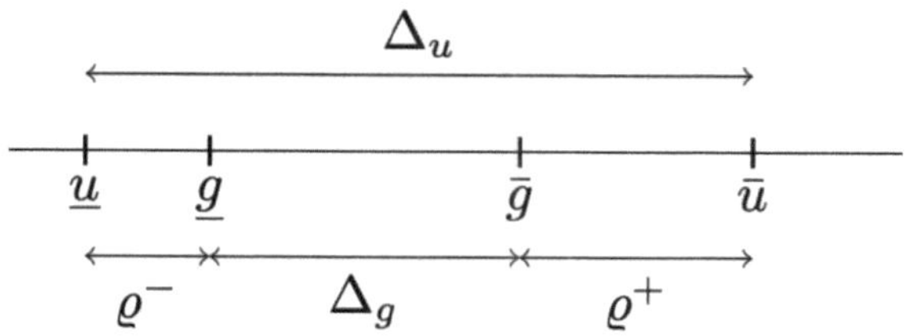

Fig. 4.14 Illustration of the control authority margin-related assumption

systems of relative degree 1 is now given by[14]:

$$h := g - \ell(\eta), \tag{4.52}$$

and this implies that a first **necessary condition** for the success of any control design is what we already referred to as the *the control authority condition* on u which stipulated that the bounds on the control input, namely, $\underline{u}$ and $\bar{u}$ should be such that the control is able to ultimately dominate any possible value of the above newly defined h.

This leads to the first working assumption of this section (see Fig. 4.14):

Control Authority Margin
Assumption 4.4 There exist two positive scalars, namely $\varrho_- > 0$ and $\varrho_+ > 0$ such that the following inequalities hold true:

$$\bar{u} \geq \bar{g} + \varrho_+ \quad \text{and} \quad \underline{u} \leq \underline{g} - \varrho_-. \tag{4.53}$$

The next working assumption characterizes the behavior of the additional term ℓ involved in (4.49a) under the dynamics (4.49b) for all possible disturbance profiles $\mathbf{w} \in \mathbb{W}$ and all possible control profile $\mathbf{u}$:

Bounds on the Internal Dynamics
Assumption 4.5 There exist positive scalars $c_0, d_0 > 0$ and $c_1, d_1 > 0$ such that for all $\mathbf{w} \in \mathbb{W}$ and all possible input profiles $\mathbf{u}$ (inside $[\underline{u}, \bar{u}]$), the following inequalities hold true:

$$|\ell(t, \eta(t))| \leq c_0 + c_1 \|\dot{\mathbf{u}}^{[t-\tau,t]}\|_\infty, \tag{4.54a}$$
$$|\dot{\ell}(t, \eta(t))| \leq d_0 + d_1 \|\dot{\mathbf{u}}^{[t-\tau,t]}\|_\infty. \tag{4.54b}$$

Moreover, the evolution of the state η remains bounded for any bounded profiles of $\|\mathbf{u}\|_\infty$ and $\|\dot{\mathbf{u}}\|_\infty$.

[14] Notice that u appears in the expression of $\dot{y}$ which means that while the uncertain term is more complicated, the system remains of relative degree 1 with respect to the regulated output.

Notice that inequalities (4.54) simply stipulate a structured bounds on the growth that can be expressed by affine functions of a specific measure of the amplitude of the control trajectories. More precisely, The specific measure is expressed in infinity norm on the control derivative following the intuition that comes from the example of the hydraulic turbines mentioned above.

Indeed, as soon as a highly dynamic action has been used at some previous time, this condemns the future of the behavior regardless of the actions that one can undertake later on because of the delay impacting the relationships.

Now since the relations become a little tricky, let us examine the special simple case where the dynamics involved in (4.49b) is affine as in (4.51b), namely:

$$\dot{\eta} = A\eta + B_1\dot{u} + B_2 w, \tag{4.55}$$

with a stable transition matrix A (open-loop stable internal dynamics as in the case of hydraulic turbine's related penstock damped dynamics invoked earlier).

In this specific case which holds for both industrial context described above the inequalities (4.54) become true after a finite time with the constants c_0, c_1, d_0 and d_1 given by the following expressions which involve the matrices A, B_1 and B_2 which appear in the internal dynamic (4.55)[15]:

$$c_0 := \epsilon + \sup_{\mathbf{w}\in\mathbb{W}} \sup_{t\geq 0} \left| C \int_0^t e^{A(t-\sigma)} B_2 \mathbf{w}_\sigma d\sigma \right|, \tag{4.56a}$$

$$d_0 := \epsilon + \sup_{\mathbf{w}\in\mathbb{W}} \sup_{t\geq 0} \left| CA \int_0^t e^{A(t-\sigma)} B_2 \mathbf{w}_\sigma d\sigma \right|, \tag{4.56b}$$

$$c_1 := \sup_{t\geq 0} \left| C \int_0^t e^{A(t-\sigma)} B_1 d\sigma \right|, \tag{4.56c}$$

$$d_1 := |CB_1| + \sup_{t\geq 0} \left| CA \int_0^t e^{A(t-\sigma)} B_1 d\sigma \right|. \tag{4.56d}$$

Regarding the Stability of Matrix A

In the case of hydraulic turbines, the asymptotic stability of the penstock dynamics comes from the fact that this system is theoretically marginally stable but becomes asymptotically stable by introducing a pre-compensator in the definition of the flow-rate feedback control. The control u discussed in the present section represents the definition of the remaining feedforward term. As for the micro-grid case, the asymptotic stability comes from the use of the droop frequency control as a first rapid feedback loop on the top of which, the control we are interested in amounts at translating the droop curve vertically. This is equivalent to an action on the total produced power.

[15] Beyond which the influence of the initial state that prevailed at the beginning of the system's first operation time becomes lower than ϵ.

For the sake of simplicity, the following notation is used in the forthcoming sequel:

Notation
In the remainder of this section, the short notation $\|\dot{\mathbf{u}}\|_\infty$ is used for $\|\dot{\mathbf{u}}^{[t-\tau,t]}\|_\infty$

Since the technicalities leading to the proof of the main results are quite tedious but especially since the spirit of the derivation methodology is hopefully understood in its main lines, the details of the proof is skipped here. Interested readers can refer to [7] for a detailed derivation of the result stated hereafter. Nevertheless, for the sake of completeness, all the terms that are used in the statement of the results are first explicitly defined.

But before let us summarize the quantities that are supposed to be given as a macro-descriptors of the bounds on the uncertainties and the control variables:

> **The Problem's Data**

- $\underline{u}$, $\bar{u}$: Lower and upper bounds on the control input. These define $\Delta_u = \bar{u} - \underline{u}$.
- $\underline{\alpha}$, $\bar{\alpha}$: Lower and upper bounds on the unknown gain $\alpha(t)$ appearing in (4.49a).
- $\underline{g}$ and $\bar{g}$: Lower and upper bounds on the unknown term g. These define $\Delta_g := \bar{g} - \underline{g}$.
- ϱ_- and ϱ_+ margins on the control input authority compared to the unknown term g.
- Coefficients c_0, c_1, d_0 and d_1 that defines the bounds on the internal dynamic-related function $\ell(\eta)$ as expressed in (4.54).

Using these data the following derived expressions can be computed for later use in the statement of the main result:

$$\underline{\varrho} := \min\Big\{\varrho_-, \varrho_+\Big\}, \tag{4.57a}$$

$$\beta := \Delta_g + \max\Big\{\varrho_-, \varrho_+\Big\}, \tag{4.57b}$$

$$\delta_u(\lambda, \lambda_f) := \frac{\lambda\bar{\alpha}(\beta + c_0) + \lambda_f \Delta_u}{1 - \lambda\bar{\alpha}c_1}, \tag{4.57c}$$

$$\varphi(\lambda) := \left[\Delta_u(1 + \frac{c_1\underline{\alpha}\lambda}{1 - \lambda\bar{\alpha}c_1})\right]^{-1}\left[\underline{\varrho} - c_0 - \lambda\frac{(\beta + c_0)c_1\bar{\alpha}}{1 - \lambda\bar{\alpha}c_1}\right]\underline{\alpha}, \tag{4.57d}$$

$$\lambda^* := \sup\Big\{\sigma \le \frac{1}{\bar{\alpha}c_1} \quad \text{such that} \quad \inf_{\lambda\in(0,\sigma)} \varphi(\lambda) \ge 0\Big\}. \tag{4.57e}$$

Note that the definition of λ^* is consistent[16] as soon as $c_0 \le \underline{\varrho}$ by the very definition of φ since $\varphi(0)$ becomes ≥ 0.

[16] This means that set over which sup operator is applied is not empty!

Having at hand all these *nasty* definitions and notation, we have all that we need to state the main results of this section:

> **Main Result**
> **Proposition 4.3** ***If*** *the following conditions are satisfied:*
>
> 1. *assumptions 4.4 and 4.5 hold true;*
> 2. $c_0 \leq \varrho, \lambda \in (0, \lambda^*)$ *and* $\lambda_f < \lambda\varphi(\lambda)$; and
> 3. *the unknown term* g *satisfies* $|\dot{g}| \leq \delta_g$,
>
> ***then*** *the tracking error* e_y *satisfies the following asymptotic inequality:*
>
> $$\lim_{t\to\infty} |y(t) - y_d| \leq \frac{\delta_g + d_0 + d_1\delta_u(\lambda, \lambda_f)}{\lambda\lambda_f}. \tag{4.58}$$

Notice that since $c_0 < \varrho$, as mentioned before, the term λ^* is well defined. Moreover, since $\lambda \leq \lambda^*$ all the expressions in (4.57) involving the denominator $1 - \lambda\bar{\alpha}c_1$ are well defined.

OK, all this seems very complicated!

Nevertheless, it might be the price to pay in order to design a simple feedback that achieves a quite critical task.

Beside this, a careful look at the expressions and conditions involved in the above statement enables to understand the different sources of trouble that might prevent the success of the proposed simple feedback law.

Analyzing these sources can also be a first step towards a control-oriented tuning of the joint design of all the pieces of items that impact the final success of the closed-loop design.

In the next section, an illustrative example is examined to show how these bounds can be computed and used to get predictable bounds on the quality of the closed-loop behavior.

4.5.4 An Illustrative Example

Let us consider the uncertain system given by (4.49) in which the internal dynamic (4.49b) is given by

$$\dot{\eta}(t) = A\eta(t) + B_1\|\mathbf{u}^{[t-\tau,t]}\|_\infty + B_2 w(t), \tag{4.59a}$$

$$\ell(\eta) = C\eta, \tag{4.59b}$$

where the matrices are given by

$$A = \begin{bmatrix} -3.4 & -1 & -0.5 \\ 0.25 & -1.7 & 0.5 \\ 1.2 & 2.75 & -3.6 \end{bmatrix} \; ; \; B_1 = \begin{bmatrix} 0.1 \\ 0 \\ 0 \end{bmatrix} \; ; \; B_2 = \begin{bmatrix} 0 \\ 0.2 \\ 0 \end{bmatrix}, \tag{4.60a}$$

$$C = \begin{bmatrix} 1 & 0 & 1 \end{bmatrix}, \tag{4.60b}$$

while the delay $\tau = 10$ is used.

The unknown signal $g(t)$ used in the simulation is given by

$$g(t) = 1 - 0.3\cos(0.1t) - 0.1\sin(0.2t + \pi/6), \tag{4.61}$$

for which the bounds are given by

$$g \in [0.5, 1.5],$$

can be used leading to $\Delta_g = 1$.

Moreover the bound:

$$\delta_g = 0.05,$$

on the amplitude of the derivative is g is used in the verification of the theoretical results.

Let us consider the saturation on the control defined by

$$u \in [-1, 3].$$

Given the previous elements, this leads to $\Delta_u = 4$ and the margins $\varrho_- = 1.5$ and $\varrho_+ = 1.5$ and hence $\underline{\varrho} = 1.5$.

Let us assume that the bounds on the gain α are $\underline{\alpha} = 0.1$ and $\bar{\alpha} = 0.3$ with the true unknown value $\alpha = 0.2$ used in the simulation.

As for the exogenous disturbance w, the upper bound $\bar{w} = 0.1$ and use the following time series $w(t) = \bar{w}\sin(t)$ in the simulation.

The computation of the quantities involved in (4.57) gives the following coefficients:

$$c_0 = 0.05 \; , \; c_1 = 0.07 \; , \; d_0 = 0.2 \; , \; d_1 = 0.42. \tag{4.62}$$

Using these values in the definition of $\varphi(\lambda)$ and λ^* leads to the bound on the asymptotic tracking error and the bound $\lambda\varphi(\lambda)$ on the admissible values of λ_f that are depicted in Fig. 4.15.

From this Figure, it can be inferred that

$$\lambda^* = 17.18.$$

Note that the upper bound:

$$1/(c_1\bar{\alpha}) \approx 47.4,$$

in the example meaning that in this example, λ^* is limited by the need for a positive value of φ.

Figure 4.16 shows the behavior of the closed-loop system when the value $\lambda^{(1)} = 17 \in (0, \lambda^*)$ is adopted together with $\lambda_f = 0.95\lambda^{(1)}\varphi(\lambda^{(1)}) \approx 0.0102$ satisfying the theoretical conditions.

This Figure shows quite good closed-loop behavior with the tracking error asymptotically lying below its theoretical bound a while after the set-point gets constants.

On the other hand, Fig. 4.17 shows the behavior of the closed-loop system when $\lambda^{(2)} = 10\lambda^{(1)}$ is used (with the same value of λ_f). The closed loop clearly shows that the trajectories are no more attracted towards a small neighborhood of the desired values.

This is mainly due to the high excursion of the internal state η which leads to high values of ℓ that breaks the attractiveness of the tight neighborhood of the desired value y_d. This excursion has to be compared to the rather moderate one which can be observed on Fig. 4.16.

This clearly demonstrates that a rationalized choice of the pair (λ, λ_f) is a key step in the success of the simple control law (4.50) for this particular and obviously relevant class of highly uncertain systems.

4.6 Conclusion

This chapter started by an example illustrating the power of *standard* control methods in addressing the problem of control in presence of high uncertainties in the dynamics. Then some provable results have been given which enable to address general problems after possible decomposition of the problem at hand.

This chapter is by no means exhaustive in terms of the provable results that the control community elaborated during decades. It is just a reminder of a whole tradition of addressing the uncertainty issue by mean of well-designed feedback laws.

The message of this chapter is that when the problem admits a standard solution (which might be the case on a strict subset of real-life problems!), it is very likely that such a solution outperforms any AI-based *modern* way of tackling the same problem. This is the reason why, these methods, concepts, and solutions need to be present in the mind of any control designer.

Notice that, even for control problems that do not lie in this category, the *conventional* control-inspired methods such as the ones discussed in this chapter, among many others, can be used as a part of *larger* solutions, including those that might be AI-inspired. For instance, one can imagine a problem where a subset of the regulated variables can be controlled via *conventional* methods to make their behavior less sensitive to model mismatches hence simplifying the space of search over which the learning process has to operate in order to address the remaining part of the solution.

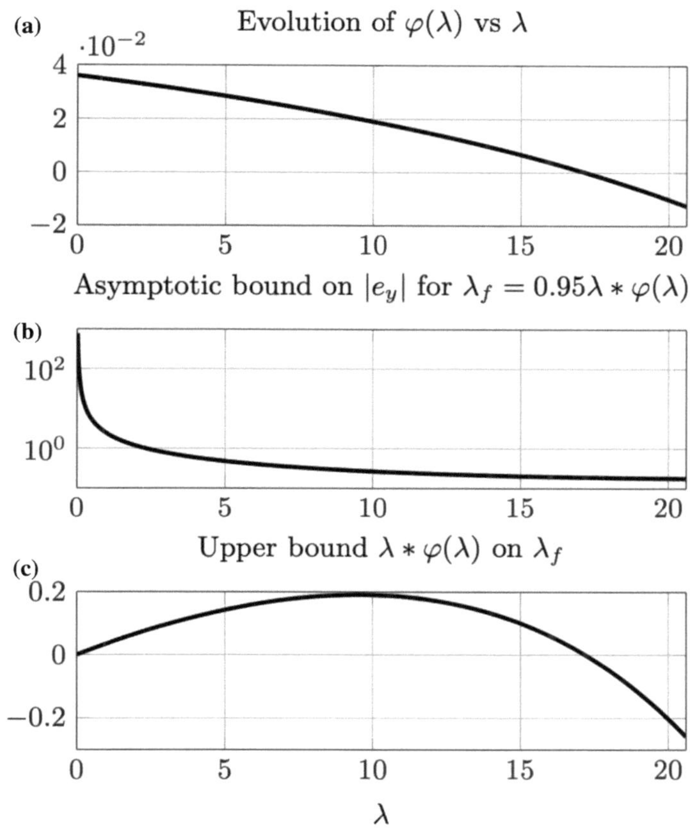

Fig. 4.15 **a** Evolution of $\varphi(\lambda)$. **b** Asymptotic bounds on the output tracking error. **c** Upper bound $\lambda\varphi(\lambda)$ on the control parameter λ_f

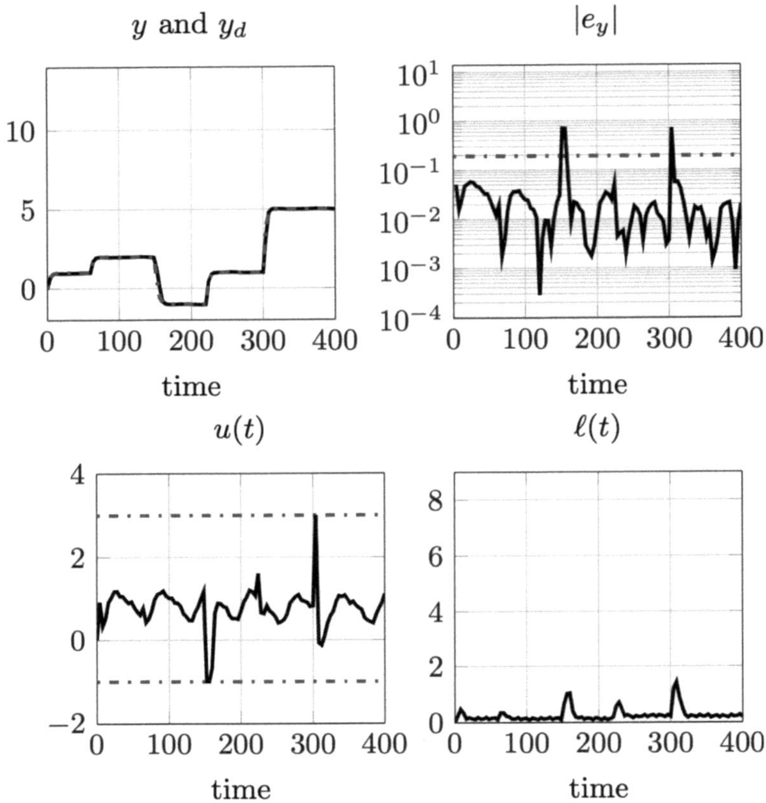

Fig. 4.16 Behavior of the closed-loop system for the choice $(\lambda^{(1)}, \lambda_f) = (17, 0.0102)$ meeting the theoretical conditions of success as expressed in Proposition 4.3

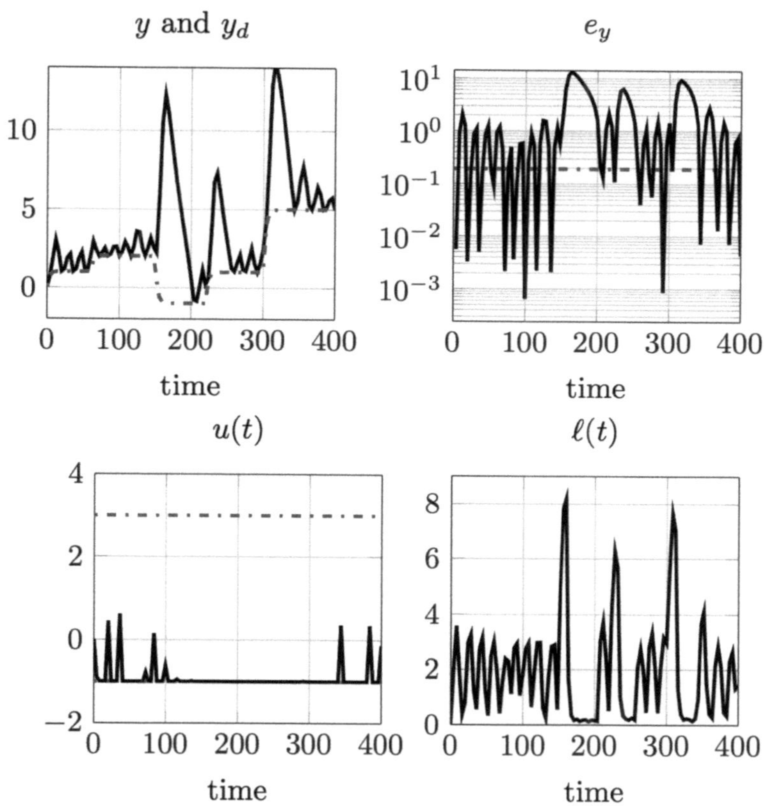

Fig. 4.17 Behavior of the closed-loop system for the choice $(\lambda^{(1)}, \lambda_f) = (170, 0.0102)$ that violates the theoretical conditions

References

1. Alamir, M.: On controlling unknown scalar systems with low gain feedback. Tech. rep., GIPSA-lab (2015). https://hal.archives-ouvertes.fr/hal-01215808/document
2. Mohamed, A., Alamir, M.: Robust output feedback controller for a class of nonlinear systems with actuator dynamics. IFAC-PapersOnLine **51**(25), 275–280 (2018)
3. Zhou, J., Karney, B., Hu, M., Xu, J.: Analytical study on possible self-excited oscillation in s-shaped regions of pump-turbines. Proceed Inst Mech Eng Part A: J Power Energy **225**(8), 1132–1142 (2011)
4. Zhao, J., Lyu, X., Fu, Y., Hu, X., Li, F.: Coordinated microgrid frequency regulation based on dfig variable coefficient using virtual inertia and primary frequency control. IEEE Trans Energy Conver **31**(3), 833–845 (2016)
5. Xialing, X., Jianghui, X., Yong, L., Youping, X., Ruyu, B., Hui, L., Tao, L., Jing, Y.: Research on the dynamic frequency characteristic of large-scale power grid considering the action of power system splitting and load shedding. In: 2014 International Conference on Power System Technology, pp. 103–108. IEEE (2014)

6. Dobrowolski, J., Alamir, M., Bacha, S., Gualino, D., Wang, M.X.: On higher order dynamics in the fundamental equation of frequency in islanded microgrids. IFAC-PapersOnLine **50**(1), 6636–6641 (2017)
7. Alamir, M., Dobrowolski, J., Mohammed, A.: Output feedback control of a class of uncertain systems under control-derivative dependent disturbances. IFAC-PapersOnLine **50**(1), 9973–9978 (2017)

Solving Deterministic Model Predictive Control Problems

5

M. Alamir, *Nonlinear Control of Uncertain Systems*,
https://doi.org/10.1007/978-3-031-93287-8_5

5.1 Chapter's Content at a Glance

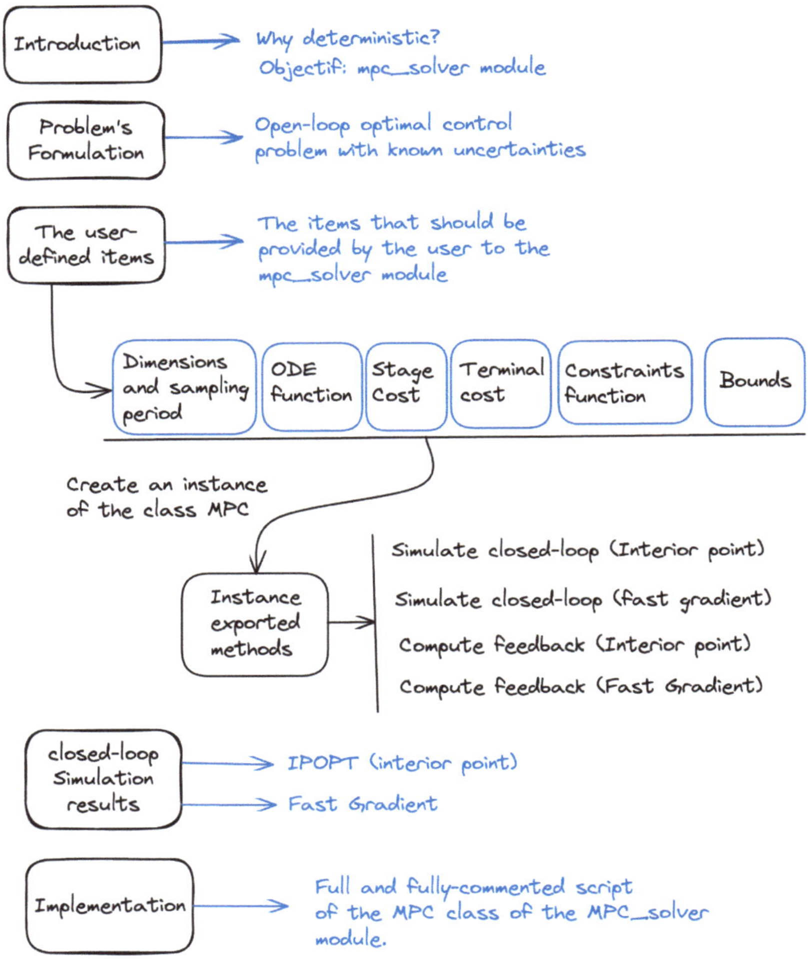

5.2 Introduction

In this chapter, a Python module (the MPC_solver) that enables to solve deterministic MPC problems as stated in Sect. 2.6.1 and to simulate the resulting closed-loop trajectories is presented and illustrated using the PVTOL example introduced in Sect. 2.2.

As it is mentioned at the end of Chap. 2 where some Python packages have been discussed, there are some available NMPC dedicated solvers and modeling frameworks, so one might ask, what is the point in presenting a new Python MPC-related package.

As a matter of fact, the answer is twofold:

- Firstly, the module `MPC_solver` serves here as a support to visit the different aspects of NMPC problem's definition and implementation so that the reader gets a better understanding of the large picture. Learning the general concepts is easier when a concrete example is used as a support together with a concrete implementation of all the details of the solver.
- A second reason in favor of this detailed presentation that progressively shows how the solver is built from scratch is that **available frameworks do not leave all the degrees of freedom that are available for the NMPC designer to manipulate**. Some of these degrees of freedom are explicitly used in **Chap.** 6 which is dedicated to the problem of tuning NMPC implementation options in order to cope with real-time implementation, constraints satisfaction and closed-loop stability. Consequently, understanding the detailed steps and parameters of the NMPC implementation is a crucial step towards understanding the tuning framework that is studied in **Chap.** 6.

This being said, another legitimate question remains:

Why Deterministic MPC?

Why are we interested in solving deterministic MPC problems in a book dedicated to controlling nonlinear systems in presence of uncertainties?

The reason lies in the fact that solving deterministic MPC problems is a key step in some of the solutions proposed in this book in order to explicitly handle the presence of uncertainties/disturbances in the dynamic models. More precisely:

Links to Other Chapters

The solver studied in this chapter is used in **Chap.** 7 to implement the uncertainty-aware dynamic output feedback strategy that approximates the solution to the stochastic NMPC problem. Indeed, it is shown in Chap. 7 that an approximate solution to the stochastic NMPC design problem can be obtained by means of learning data that involves the solution of a *cloud* of **deterministic** optimal control problems (OCPs).

The module `mpc_solver`, introduced in the present chapter, extensively uses the `Python` version of the `Casadi` package introduced in Sect. 3.4.4. The reader might take a second look on that section for a smoother reading of the present chapter.

As mentioned in Sect. 2.6, the OCP underlying the definition of the MPC design might be formulated in a **Single-Shooting (SS)** version where the decision variable is the control profile or in a **Multiple shooting (MS)** version where the vector of decision variables gathers the state and the control profiles over the prediction horizon.

For the sake of simplicity, only SS version is implemented here. The reader can easily derive the MS version following the simple examples proposed in the documentation of the `CasaDi` package.

It is important to keep in mind that `CasaDi`, as well as all available frameworks are generally in permanent evolution and the present material might become out-dated in quite a short time. Nevertheless, the concepts and the way the resulting solution can be used as a step in a disturbance/uncertainty-aware feedback control design framework should remain valid over time regardless of the way and the speed at which open-loop optimal solution will be obtained.

This chapter is organized as follows:

- First, the problem is stated by recalling the formulation of the open-loop OCP and the definition of the MPC feedback (Sect. 5.3).
- Before digging into the code of the `mpc_solver` module, we shall start by showing how it is used. In order to show this, Sect. 5.4 shows the user-defined functions and parameters that are needed in the call of the solver.
  An instantiation using the PVTOL example of Sect. 2.2 is proposed for the sake of concrete illustration.
- The behavior of the resulting closed-loop dynamics is examined in Sect. 5.5 for two possible solutions, namely:
  - using the interior-point solver `Ipopt` implemented via `CasaDi` (Sect. 5.6);
  - using variants of the fast-gradient algorithm in which the gradient function is computed using the `Jacobian` function of `CasaDi` (Sect. 5.7).
- Finally, the details of the `Python` module `mpc_solver` are presented and commented for the sake of completeness and in order to give the readers willing to implement their own MS version a clear overview of the steps that are be required. This is given in Sect. 5.8.

5.3 Problem Formulation

The `Python` module `mpc_solver`, developed in the present chapter, addresses the NMPC control of general nonlinear dynamics governed by a system of Ordinary Differential Equations (ODEs) that take the following general form:

$$\dot{x} = f(x, u, w, p), \tag{5.1}$$

where

- $x \in \mathbb{R}^{n_x}$ is the state vector;
- $u \in \mathbb{R}^{n_u}$ is the vector of manipulated variables (also referred to as the vector of control inputs);

- $w \in \mathbb{R}^{n_x}$ is the vector of uncertainty/disturbance; and
- $p \in \mathbb{R}^{n_p}$ is the vector of **known constant parameters**.

As it is shown later, the same vector p will also incorporate any constant parameter that might be needed to completely define the optimization problem such as the bounds on some variables, the weighting matrices, and the desired targeted state.

It is assumed that some prediction horizon length N is fixed over which the OCP is defined with the control profile:

$$\mathbf{u} \in [\mathbb{R}^{n_u}]^N,$$

as decision variable and the initial state x_0 and the **uncertainty profile w as given parameters**[1]:

> **Deterministic Optimal Control Problem**
> The deterministic OCP is the following constrained optimization problem:
>
> $$\min_{\mathbf{u}} J(\mathbf{u}|x_0, \mathbf{w}) := \Psi(x_N) + \sum_{k=0}^{N-1} \ell(x_{k+1}, u_k), \tag{5.2a}$$
>
> $$\textit{under} \quad \left| \begin{array}{l} u_k \in [u^{min}, u^{max}], \\ g_j(x_k, u_k) \in [g_j^{min}, g_j^{min}] \ j = 1, \ldots, n_g, \\ x_{k+1} = F(x_k, u_k, w_k, p) \ k = 0, \ldots, N-1, \end{array} \right. \tag{5.2b}$$

where

- ℓ stands for the so-called **stage cost**;
- Ψ is the **terminal penalty** applied to the terminal state;
- the inequalities in (5.2b) are to be interpreted component-wise with $u^{min} \in \mathbb{R}^{n_u}$ and $u^{max} \in \mathbb{R}^{n_u}$ being the lower and upper bounds respectively on the components of the control input vector; and
- $g^{min} \in \mathbb{R}^{n_g}$ and $g^{max} \in \mathbb{R}^{n_g}$ are respectively the lower and upper bounds on the vector of constraints.

[1] This is the reason why we refer to the problem as deterministic since the uncertainty are supposed to be known each time we solve the problem. We'll see later how this helps in addressing the unknown uncertainties case through learning!

The ideal MPC formulation assumes that there is an optimal solution, say $\mathbf{u}^\star$ to the above problem[2]:

$$\mathbf{u}^\star(x_0, \mathbf{w}) := \begin{bmatrix} u_0^\star(x_0, \mathbf{w}) \\ \vdots \\ u_{N-1}^\star(x_0, \mathbf{w}) \end{bmatrix} \in [R^{n_u}]^N, \tag{5.3}$$

and according to the receding-horizon principle, the corresponding MPC (deterministic and uncertainty-perfectly-aware) feedback would be given by

$$K_{\text{MPC}}(x) := u_0^\star(x, \mathbf{w}). \tag{5.4}$$

The module `mpc_solver` makes it easier to implement the above defined feedback by rendering transparent the repetitive and generic steps that are needed to implement the above formulation requiring the user to only focus on the **problem-dependent** items of the formulation.

More precisely, the module needs the user to define the functions that are specific to its current problem. The list of these functions are detailed in the following section. Moreover, the instantiation of these elements are given for the PVTOL example which is introduced in Sect. 2.2.

5.4 The User-Defined Parameters and Functions

The `mpc_solver` module implements the generic and common operations towards the derivation of the MPC feedback given by (5.2). For this to be possible, the user needs to provide the problem-specific information and functions. These are the following[3]:

- the different dimensions of the vectors involved,
- the prediction horizon length,
- the sampling period,
- the ode function,[4]
- the stage cost function ℓ,
- the terminal cost function Ψ,
- the constraints function g, and
- the bounds on the control and the constrained expressions.

[2] The solution also depends on the parameter vector p but this dependence is omitted for the sake of brevity.

[3] See (5.2a)–(5.2b).

[4] Models are generally available in ODE form. The transformation to discrete-time version is done automatically to get the corresponding one-step-ahead map for the provided sampling period.

The syntax of these elements as well as their instantiation in the specific case of the PVTOL example are given in the following sections.

Connect to the PVTOL Section
In this section, the instantiation of the problem's definition uses the presentation of the PVTOL example that has been introduced in Sect. 2.2. Please refer to this presentation in order to connect with the following code's snippets.

5.4.1 The Dimensions, Prediction Horizon, and Sampling Time

The dimensions of the different vectors, the length of the prediction horizon and the sampling period are denoted by[5] :

`nx, nu, nw, npar, N, dt.`

In the case of the PVTOL, this is given by the following script:

```
nx, nu, nw, npar, dt, N = 6, 2, 2, 6, 0.1, 20.
```

These elements are given as arguments of the instantiation of the main class `MPC` of the `mpc_solver` module as shown later on.

5.4.2 The ODE Function

The function that defines the ODEs governing the dynamics of the system as shown in (5.1) is to be defined using the following syntax[6]:

Structure of the ODE

```
def ode(x, u, w, p):
  ...
  # System's equations

  xdot = vertcat( ...)
  return xdot
```

In the specific case of the PVTOL, this function takes the following form:

[5] Notice that the use of the notation `np` is avoided because `np` is generally used as a short symbol for the `numpy-Python package through the import command (import numpy as np)`.
[6] It is important to use the `vertcat` after importing it through the command
`from casadi import vertcat.`
This is necessary for the compatibility with the `casadi` part of the program.

Instantiation of the ODEs for the PVTOL

```
def ode_pvtol(x, u, w, p):

    eps, gain = w[0], w[1]
    xdot = vertcat(
        x[3], x[4], x[5],
        -u[0] * np.sin(x[2]) + eps * gain * u[1] * np.cos(x[2]),
        u[0] * np.cos(x[2]) + eps * gain * u[1] * np.sin(x[2])-1,
        gain * u[1]
    )
    return xdot
```

Notice that in this example, the uncertainty vector comprises the two model's parameters `eps` and `gain` that are used in the sequel as uncertain parameters.

Recall that since `mpc_solver` is a generic module, all the variables should be present in the list of arguments even if they are not explicitly used in the function.[7] This is the case here for the parameter `p`.

5.4.3 The Stage Cost

The stage cost is the function ℓ invoked in the formal definition of the open-loop OCP (5.2a). It should be defined using the following syntax:

```
def stage_cost(x, u, w, p):
  ...
  # Definition of the stage cost

  ell = ...
  return ell
```

In the specific case of the PVTOL example, the stage cost includes a quadratic penalty on the distance to some desired state x^d given by

$$x^d = \left[y^d, z^d, 0, 0, 0, 0\right]^T,$$

as well as a quadratic penalty on the distance to the steady control given by

$$u^d := \left[1, 0\right]^T.$$

[7] There might be some advanced *pythonistic* way to avoid this but let us keep things simple!

This leads to the stage cost function defined by[8]:

$$\ell(x, u) = \|x - x^d\|^2_{\text{diag}(q_y, q_z, q_\theta, 1, 1, 1)} + \|u - u^d\|^2. \tag{5.5}$$

Consequently, the definition of the stage cost requires 5 parameters which are:

`yd, zd, qy, qz, qtheta,`

which are the first five components of the parameter vector p (the last argument of the stage cost according to the syntax defined above).

This leads to the following specific definition of the stage cost function in the case of the PVTOL example and for the specific choice of the penalties configuration[9]:

```
def stage_cost_pvtol(x, u, w, p):

    # p = vertcat(yd, zd, qy, qz, qtheta, rho_f)

    yd, zd = p[0], p[1]
    q = vertcat(p[2], p[3], p[4], 1, 1, 1)
    xd = vertcat(yd, zd, 0, 0, 0, 0)
    ud = np.array([1, 0])
    ex = x - xd
    eu = u - ud
    ell = 0
    for i in range(6):
        ell += q[i] * ex[i] ** 2
    for i in range(2):
        ell += eu[i] ** 2
    return ell
```

5.4.4 The Terminal Cost

The terminal cost refers to the term $\Psi(X_N)$ involved in the definition of the OCP (5.2a). The syntax of this function takes the following form:

[8] This is obviously a specific chosen parameterization of the stage cost. The penalty on the derivatives of y, z and θ need not to be necessarily equal to 1.

[9] The last component, named `rho_f` of the parameter vector p is used later on in the definition of the terminal penalty Ψ used in the definition of the cost function (5.2a).

```
def terminal_cost(x, w, p):
  ...
  # expression of the terminal cost
  psi =
  return psi
```

Notice that the argument u is absent from this function's list of arguments. This is because the penalty involves the terminal state regardless of the next control to be applied. However, the quality of the terminal state might well depend on the quantities w and p.

In the specific case of the PVTOL example, the terminal penalty function is chosen to be the same as the stage cost but with some higher penalty thanks to the weighting scalar rho_f, namely:

$$\Psi(x) := \rho_f \|x - x^d\|^2_{\text{diag}(q_y, q_z, q_\theta, 1, 1, 1)}, \tag{5.6}$$

which is implemented using the following function that respects the syntax given above:

```
def terminal_cost_pvtol(x, w, p):

    # par = vertcat(yd, zd, qy, qz, qtheta, rho_f)

    yd, zd = p[0], p[1]
    rho_f = p[5]
    q = vertcat(p[2], p[3], p[4], 1, 1, 1)
    xd = vertcat(yd, zd, 0, 0, 0, 0)
    psi = 0
    ex = x-xd
    for i in range(6):
        psi += q[i] * ex[i] ** 2
    psi *= rho_f
    return psi
```

5.4.5 The Constraint Function

This function defines the vector of quantities on which there are constraints to be enforced over the whole prediction horizon as it is expressed by the second line of the constraints definition (5.2b), namely:

$$g_j(x_i, u_i) \in [g_j^{min}, g_j^{min}] \quad j = 1, \ldots, n_g.$$

Now the function we are about to give the syntax is the one that defines the left-hand side of the above expression, namely the one that provides the vector g as a function of the state, control, disturbance and the parameters. The concrete values of the bounds on these constrained expressions are given elsewhere as parts of the list of parameters to be given when instantiating the class.

The following script gives the structure of the function that defines the constrained expressions to be considered:

```
def constraints(x, u, w, p):

    # define the components of the constraints

    g = vertcat(...)
    return g
```

In the specific case of the PVTOL example, we consider the following constraints on the rate of change of the variable z and θ:

$$|\dot{z}| \leq \texttt{dzdt\_max} \quad ; \quad |\dot{\theta}| \leq \texttt{dthetadt\_max}, \tag{5.7}$$

which means that the function to be defined is the one that delivers:

$$\begin{bmatrix} \dot{z} \\ \dot{\theta} \end{bmatrix},$$

that corresponds to the following instantiation of the general function in the case of the PVTOL example:

```
def constraints_pvtol(x, u, w, p):

    # par = vertcat(yd, zd, qy, qz, qtheta, rho_f)

    g = vertcat(x[4], x[5])
    return g
```

5.4.6 The Bounds on the Control and the Constrained Expressions

The last problem-specific information the user needs to provide to the `mpc_solver` module concerns the saturation bounds on the actuators and the bounds on the components of constraints-related term g previously described.

These are real vectors which are arguments of the instantiation function of the class MPC (the main and unique class of the mpc_solver module).

In the case of the PVTOL example, these bounds are defined as follows:

```
umin = [-1.5, -0.5]
umax = [1.5, 0.5]
gmin = [-0.3, -0.2]
gmax = [0.3, 0.2]
```

since we have a two-dimensional control input vector as well as a two state-dependent constrained variables which are $\dot{z}$ and $\dot{\theta}$.

It is worth underlying that for computational efficiency reasons, it is important to separate the saturation-like constraints on the input vector which is used here as decision variable (Single-Shooting) while in theory, they can be added as components of the constraints vector g since the latter accepts u as one of its arguments.

At this stage, we have introduced all the elements that are required to declare an instance of the class MPC. After such an instance is created, it can be used to compute the feedback and to simulate the resulting closed-loop system.

This is explained in the next section.

5.5 mpc_solver: The Exported Methods

Recall that for the time being, only the use of the module mpc_solver is shown in a sort of black-box manner from the user point of view. The details of the implementation code are discussed later in Sect. 5.8.

In this section, the methods exported by the class (the ones that can be called using instances of the class) are discussed from the user point of view.

As usual, the first thing to do is to import the needed Python modules. This is the list of imports that are used in the following use-case illustration of the mpc_solver module[10]:

```
from mpc_solver import MPC
from casadi import vertcat
import numpy as np
import time
import matplotlib.pyplot as plt
```

[10] The Python module time is used in the computation of the CPU time needed here and there in the argumentation.

Here the class MPC is imported from the mpc_solver module. This class exports (makes callable) all the methods we shall need to get the MPC feedback and to simulate the resulting closed-loop behavior as shown below.

Once this class is imported, all the items and functions described in the previous Sect. 5.4 can be used as arguments to the function that creates an instance of the class MPC.

This is shown in the following script:

```
# create an instance of the class
mpc = MPC(
 nx=nx, nu=nu, nw=nw, npar=npar, N=N, dt=dt,
 ode=ode_pvtol,
 stage_cost=stage_cost_pvtol,
 terminal_cost=terminal_cost_pvtol,
 constraints=constraints_pvtol,
 umin=umin, umax=umax, gmin=gmin, gmax=gmax,
 max_iter=2000
 )
```

The MPC instance so-created exposes the following methods that can be used for various purposes:

5.5.1 The Method simulate_cl

The first method, named simulate_cl enables us to simulate the closed-loop system under the MPC feedback in which the OCP is solved via the **interior-point** solver Ipopt [3] which is made available by the CasaDi framework.

The call of this method takes the following syntax:

```
R  = mpc.simulate_cl(x0, U0, W_supposed, W_real, p, Nsim)
```

where the **input arguments are**:

- x0: the initial state.
- U0: the initial guess regarding the control profile which is the decision variable in the SS method. Note that U0 is flat Python vector[11] of dimension N·nu.
- W_supposed: the uncertainty profiles as supposed by the controller. This is a Python flat vector of dimension N·nw.

[11] See below for an example of creation of such initial guess.

This argument is obtained by flattening an originally defined **list of lists** of variables where the first element is the list of components of the uncertainty vector at the first instant in the prediction horizon and so on.
The deterministic OCP will be defined based on this value of the uncertainty profile over the prediction horizon of length `N`.

- `W_real` is the *real* uncertainty profile to be simulated under the MPC control algorithm that assumes the uncertainty profile to be equal to `W_supposed`.
  This enables to check the nominal robustness of the closed-loop without explicit handling of the statistics of the uncertainties. This argument has the same structure as the previously described `W_supposed` profile.
- `p`: the vector of known parameters as explained above.
- `Nsim` is the length of the closed-loop simulation in terms of the sampling period defined by the attribute `dt` of the class, namely `mpc.dt`.

Regarding the **output arguments**, they are collected in a Python object `R` that is an instance of a class having the following `attributes`:

`tcl, Xcl, Ucl, Jcl, cpu,`

where the first three outputs represent the time vector followed by the closed-loop profiles of the state and control respectively. `Jcl` gives the evolution in closed-loop of the *open-loop* achieved cost at the successive instants during the closed-loop simulation.

More precisely, `Jcl` is a list of values defined by

$$\left\{J(\mathbf{u}^\star(x_k^{cl}, \mathbf{w}_k))\right\}_{k=1}^{\texttt{Nsim-N}}. \tag{5.8}$$

Finally the field named `cpu` gives the sequence of CPU times needed to compute the optimal control at the successive updating instants during the closed-loop simulation.

! Open-Loop Costs Versus Closed-Loop Cost

It is important to keep in mind the difference between the resulting **closed-loop cost** under the receding-horizon MPC and the sequence of **open-loop costs** that are obtained after the optimization is stopped at each updating instants.

Indeed, there are as many open-loop cost values as there are updating instants during the closed-loop simulations while there is only a single closed-loop cost that can be obtained only one the whole closed-loop simulation is finished.

> Remember that the closed-loop cost involves the `Nsim` simulated instants while each open-loop cost involves N prediction instants where N is the prediction horizon used to formulate the open-loop OCP at each decision instant.

Notice that the method `simulate_cl` uses the interior-point solver `IPOPT` to compute the feedback that is applied at each decision instant. Obviously, the underlying optimization problem can be solved in many different ways, including using different available solvers inside `CasaDi`.

As a matter of fact, by the time this book has been written, only the `IPOPT` solver seemed to be easily callable.[12] An efficient solver called `ACADOS` [4] can be used inside `casadi` but this is conditioned by a specific instantiation of `ACADOS` dedicated class and the resulting implementation does not enable to manipulate all the tuning parameters that are considered in **Chap.** 6.

Nevertheless, it is also possible to use the `Jacobian` utility offered by `CasaDi` in order to get the gradient of the open-loop cost function with respect to the decision variable $\mathbf{u}$ to that a fast-gradient descent can be performed to solve the optimization problem.

The method described in the following section implements this fast-gradient option.

5.5.2 The Method `simulate_cl_fast_gradient`

The second method that is exposed by the `MPC` class in order to simulate the resulting closed-loop, named `simulate_cl_fast_gradient`, is similar to the previous one in the sense that it enables us to simulate the closed-loop behavior. The difference lies in the method used to solve the underlying optimization problem.

The call's syntax of this second method is the following:

```
R  = mpc.simulate_cl_fast_gradient(x0, U0, W_supposed,
    W_real, p, rho_soft, Nsim, gam, c, Niter)
```

The following arguments:

`x0, U0, W_supposed, W_real, p, Nsim,`

are exactly the same as for the previous method. The remaining arguments however, are specific to the so-called fast-gradient optimization algorithm. More precisely:

[12] As far as nonlinear optimization is concerned.

- `rho_soft` is the weight used in the formulation of the associated cost in which all the constraints **other than the saturation constraints on the inputs** are considered to be *soft constraints*.
  Recall that softening the constraints amounts at removing them from the list of *explicit* constraints and adding associated penalties on their violation to the cost function's definition.
  More precisely, the stage cost ℓ is replaced by a new one in which the constraints violation is penalized with a generally high penalty coefficient `rho_soft` and the resulting term is added to the original cost function:
  $$\ell \leftarrow \ell + \texttt{rho\_soft}\Big[\max\{0, g - g^{max}, g^{min} - g\}\Big]^2. \tag{5.9}$$
  The reason for this is that the fast-gradient method needs the only remaining constraints to be saturation-like constraints in the decision variables.
  Another reason is that all other constraints are typically soft constraints in nature which can be violated during a small amount of time. This also leads to formulations that always possess feasible solutions.
  The rationale in using soft constraints lies in the fact that, under some regularity assumptions, it can be shown that the constraints can be *theoretically* fulfilled up to any desired precision provided that the weighting coefficient `rho_soft` is taken sufficiently large.
  Notice however that using a large penalty on the soft constraints lead to stiff optimization problems that are harder to solve which might threaten the real-time implementability of the resulting MPC feedback. This is the reason why in **Chap.** 6, the penalty on the soft constraints is viewed as a part of the tuning parameters to be adequately chosen such that the resulting optimization problem can be conveniently solved within the allowable number of iterations.
  Notice that by the term: *conveniently solved* invoked above, it is meant that the quality of the sub-optimal solution is sufficient for the closed-loop to be *successful* in a sense that is more clearly precised in **Chap.** 6.

- In order to explain the two remaining arguments, namely `gam` and `c`, the definition of the fast-gradient iteration needs to be recalled.
  The details of this iteration is presented in more depth later on since one of the methods exhibited by the `MPC` class implements this iteration. For the time being, and in order to focus on understanding the arguments `gam` and `c` needed in the call of the method `simulate_cl_fast_gradient`, we simply need to look at the following updating rules that are used in the fast-gradient method:

```
z = U_previous - gam . Gradient(U_previous)
U = Project[z + c . (z-z_previous)]
```

  Roughly speaking the pair

  `(z_previous, U_previous)`

is a kind of *internal state* of the algorithm and the two above formulas define how this pair is updated. Namely:

- The first line implements a standard simple gradient updating rule with a step in the opposite direction of the gradient which starts at the previous update of the solution. The step is defined by the gain `gam`.
- The new update of the candidate solution, denoted by `U`, is then computed in the second line. More precisely, it is obtained from `z` by first adding the momentum `c(z-z_previous)` and then projecting the result on the admissible domain of the decision variable.[13]

To summarize, `gam` defines the step in the opposite direction of the gradient while `c` $\in (0, 1)$ determines the amount of the correction induced by *what happened* in the previous step (before projection) which makes the fast-gradient an order of magnitude more efficient than the simple gradient.[14]

Regarding the output `R` of this method, it is exactly the same as the one delivered by the method `simualte_cl` studied earlier.

The above mentioned two methods, namely

- `simulate_cl`, and
- `simulate_cl_fast_gradient`,

enable to simulate the closed-loop behavior under MPC feedback using two different solvers respectively in a rather *black-box* way.

Indeed, the user does not have to call for the computation of the best sequence of actions and then select the first one to feed a one-step ahead map in order to simulate the trajectory of the resulting closed-loop state evolution. Everything is done automatically including the use of different values of the uncertainty profile in the control step and in the simulation step.

This might be helpful when using the closed-loop trajectories in creating learning data or to quickly analyze the feasibility and the quality of the resulting MPC control. It can also be used to manually check the impact of different choices of the items used to define the cost function and the prediction horizon.

Nevertheless, calling these methods does not give access to the feedback law before the end of the closed-loop simulation. This renders impossible to use the delivered control in a state estimation algorithm that the user would be willing to run in parallel to the MPC controller to cite a single example.

More generally, it might be useful to dispose of the feedback law as a separate available function (not just inside a black-box simulator) and to be used in any user-

---

[13] This is precisely why we need the constraints to be saturation-like since in this case, the computation of the projection can be computed rapidly by simple component-wise clipping.

[14] More precisely, the fast-gradient method converges quadratically in the number of iterations while the simple gradient converges linearly.

specific workflow that needs this information and that can even impact the outcome of the simulation itself.

The following two methods, exported by the MPC class, address this need. They are precisely used inside the previous two closed-loop simulation methods at each sampling period to compute the associated (sub)-optimal control sequences.[15]

5.5.3 The Method compute_MPC

The third method, named compute_MPC, returns the CasaDi object containing the solution of the OCP defined by its arguments. The solution is obtained using the Interior-Point solver IPOPT [3].

More precisely, getting the solution object takes the following syntax:

```
sol = self.compute_MPC(U0_guess, x, W, p),
```

where the input arguments are quite self-explanatory:

- U0_guess is the initial guess regarding the sequence of controls over the prediction horizon.
- x is the current state vector at which the MPC feedback is to be computed.
- W is the presumed sequence of uncertainties while p is the vector of parameters as defined above.

The output of this method is a CasaDi object, named sol which can be used to get the optimal control sequence found by Ipopt through the call:

(Optimal sequence of actions) `sol['x']`,

while the corresponding value of the cost function can be obtained through

(Optimal value of the cost function) `sol['f']`.

Note that the maximum number of iterations of the Ipopt solver is defined by the argument max_iter used at the instantiation step of the class MPC (see the instantiation script at page 229).

5.5.4 The Method fast_gradient

The fourth method, named fast_gradient, implements the fast-gradient method to solve the underlying OCP. The call syntax is the following:

[15] The first action of which is implemented on the real system according to the receding-horizon principle that underlines the very definition of MPC.

```
Uopt = fast_gradient(U, x, W, p, rho_soft, gam, c, Niter),
```

were all the arguments have been explained in the previous discussion except `Niter` which stands for the number of iterations of the fast-gradient algorithm to be performed before the sub-optimal sequence is returned.

5.5.5 Other Methods Exported by the `mpc_solver` Module

In addition to the previously mentioned and detailed methods, there are some other methods that are available from the `mpc_solver` module which are used in the above four methods but can be of some interest for their own.

They are briefly listed below for the sake of completeness.

- `xnext = mpc.F(x,u,w,p,tau)`:
  This is the one-step-ahead predictor map with a specified sampling period `tau` that might be different from the one used in the MPC computation. The returned output is the next state vector.
- `X = mpc.sim(x,U,W,p,Nsim)`:
  This method simulates the system for a given sequence of control `U`, a given sequence of disturbance `W`, a given vector of parameters `p` and a given duration of the simulation defined by `Nsim` (defining the number of sampling periods to be simulated). The output result is a `vertcat`-concatenated state trajectory.
- `J = mpc.cost(U, x0, W, p)`:
  The cost over the prediction horizon in which the penalty on the constraints violation is not included. This enables to compare separately the level of the cost function and the level of satisfaction of the constraints The total cost including the soft constraints terms is provided via the `soft_cost` map.
- `Jsoft = mpc.soft_cost(x,u,w,p,rho_soft)`:
  This is the original cost augmented by the terms enabling to soften the constraints other than the saturation constraints on the control inputs as explained in Sect. 5.5.2.
- `G = mpc.grad(x,u,w,p,rho_soft)`:
  This function computes the gradient w.r.t. the decision variable `U` of the augmented soft cost (the previous one to be used in a gradient descent algorithm for instance).
- `mpc.plot(R)`: This is a method that plots the trajectories resulting from either closed-loop simulations (regardless of the solver being used). Notice that `R` is an object returned by the two simulation utilities mentioned above.

Before we dig into the concrete `Python` code that implements all the details of the `MPC_solver` described so far, let us have a look from the user's point of view and the kind of results that can be obtained using the module.

This is the aim of the next section.

5.6 Results Using simulate_cl (IPOPT)

Based on the above presentation of the available methods, let us start by using the simulate_cl method to simulate the closed-loop behavior of the PVTOL system under the MPC control based on the Ipopt solver through the CasaDi framework.

This can be done using the instance mpc of the MPC class of the mpc_solver package previously created (see page 229).

5.6.1 Nominal Design: No Model Mismatches

The following script defines the parameters of the scenario to be simulated:

- Initial state $x_0 = 0$.
- Desired values $y^d = 0.2$, $z^d = -0.4$.
- W_supposed=W_real=(0.4, 1). This simulates an ideal case where the controller gets the exact value of the model's parameters.
- Penalties on the state error: $q_z = 100$, $q_y = 10$, $q_\theta = 10$.

This leads to the following script:

```
x0 = np.zeros(nx)
U0 = np.zeros(mpc.N * nu)
Nsim = 201
W_supposed = [0.4, 1] * Nsim
W_real = [0.4, 1] * Nsim
p = [0.2, -0.4, 100, 10, 10, 1e2]

R  = mpc.simulate_cl(x0, U0, W_supposed, W_real, p, Nsim)
```

The result is shown in Fig. 5.1.

Notice that all the constraints on the state and the control are saturated at some instant during the closed-loop simulation scenario.

5.6.2 Check the Robustness Against Model Mismatches

In order to check the inherent robustness of the MPC feedback against model mismatches. Let us examine what happens if the supposed values of the model's parameters, namely W_supposed used by the control algorithm are different from the *true* values of the system's parameters used in the simulation, namely w_real.

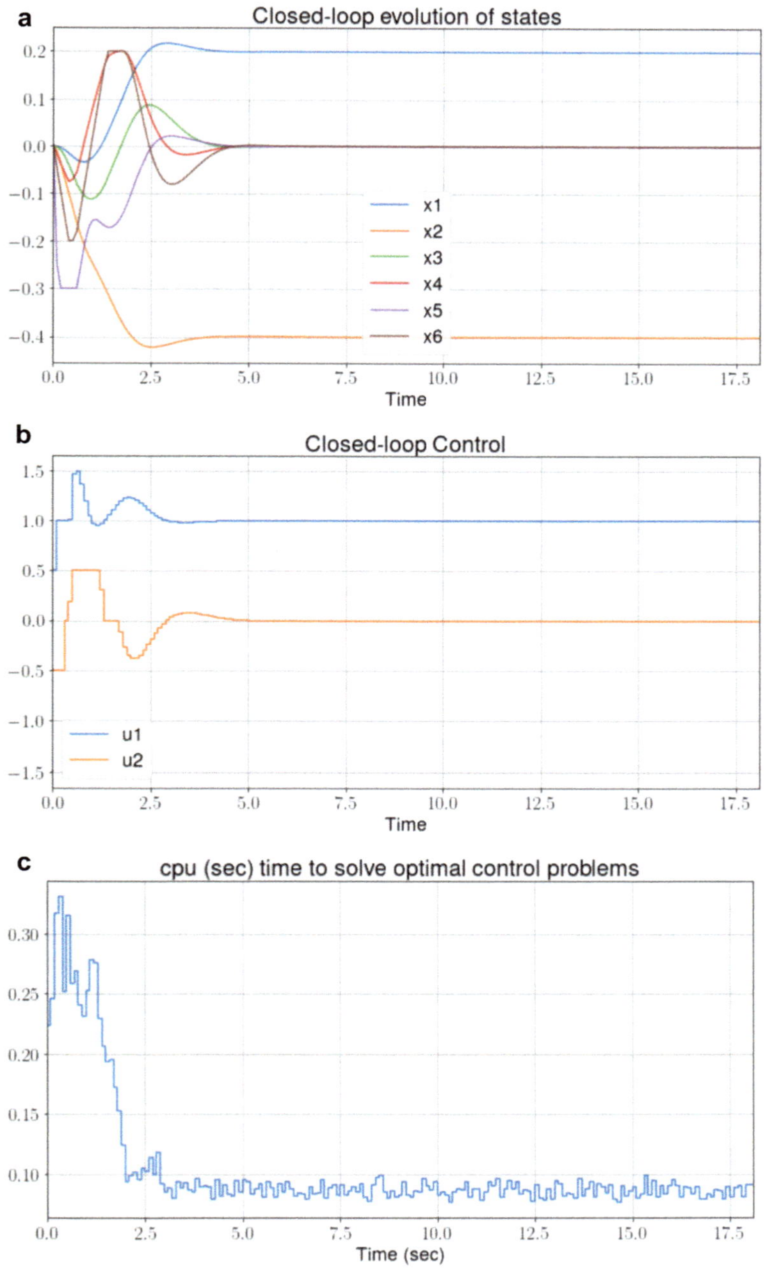

Fig. 5.1 PVTOL: Closed-loop simulation under perfect knowledge of the model's parameters and using the solver `Ipopt` to solve the underlying OCP. **a** Evolution of the state components. **b** closed-loop MPC control. **c** CPU time needed to solve the OCPs (`iter_max=2000`) during the closed-loop simulation. Recall that $|u_1| \leq 1.5$ and $|u_2| \leq 0.5$ while the derivatives of z and θ are respectively limited to 0.3 and 0.2

This can be done using the same previous script where the following values are used[16]:

`W_supposed=(0.25, 0.8)` and `W_real=(0.4, 1)`.

The results are shown in Fig. 5.2 where it can be observed that despite the strong error on the model parameters (-38%, -20%), the controller manages to stabilize the desired state while still almost systematically meeting the constraints on the state and the control values.

Now even in the absence of model mismatch, the observation of the CPU times needed to solve the successive open-loop OCPs shows that as far as the computational device used to perform the computation is concerned, the current setting of the MPC formulation is incompatible with real-time implementation requirements.

Indeed, recall that the sampling period used in the design of the MPC feedback is equal to `mpc.dt=0.1`. But some instances of the computation times shown in Figs. 5.1 and 5.2 lie largely above this threshold which simply means that:

! Real-Time Issue in the Simulation

The MPC feedback laws simulated on Figs. 5.1 and 5.2 are not real-time implementable since the computation time needed to get the feedback exceeds, at some instances, the control updating period which is fixed to `dt=100 msec`.

One way to try to address this issue is to reduce the maximum allowed number of iterations of the `Ipopt` method by changing the parameter `max_iter=2000` used in the instantiation of the `MPC` class to some lower values.

By doing so one might expect the computation time per updating period to be reduced at the price of deterioration of the optimality of the returned solution.

Nevertheless, since the computation during the next updating period starts from the previous sub-optimal solution while the problem would not have been changed dramatically, one can hope that this iterative process ends up by the solution being quite close to the optimal one at some later stage in the simulation.

This process is referred to as the process of **distributing the optimization over time**. Many works precisely addressed this issue but this is not the topics of this chapter.[17]

Here we are going to observe this process in action on the simple illustrative example by reducing the number of allowable iterations to `iter_max=5`. Recall that this is to be done in the class's instantiation script depicted in page 229 by changing the value of the last argument.

The results are shown in Fig. 5.3 where it can be observed that the quality of the closed-loop behavior is not drastically impacted (For a more clear examination of the impact on the quality of the cost function, see Fig. 5.5) while the statistics of the

[16] This parameters mismatch will be systematically used in the remainder of this section for the sake of comparison.

[17] See Chap. 6 for more details.

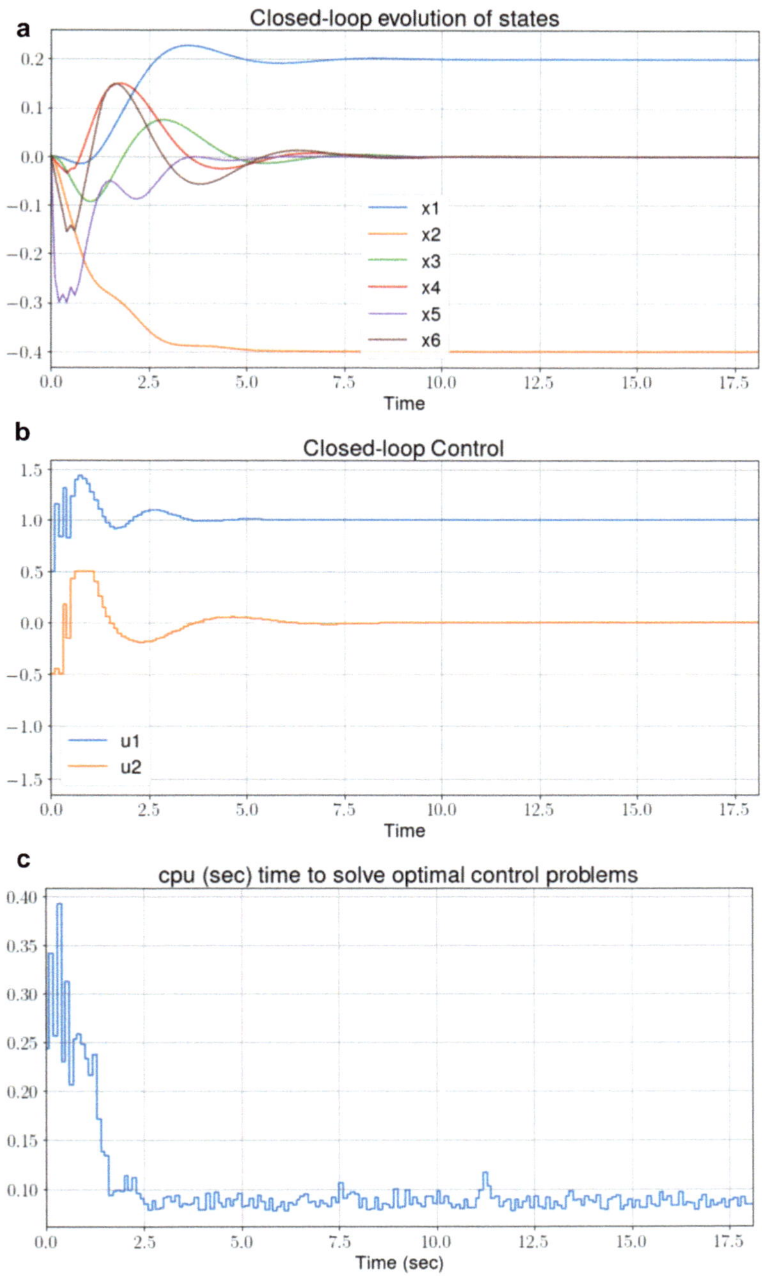

Fig. 5.2 PVTOL: Closed-loop simulation under strong model mismatches on the model's parameters using `Ipopt` to solve the underlying OCP. More precisely, the MPC controller uses `W_supposed=(0.25, 0.8)` while the true parameters are `W_real`=(0.4, 1). **a** Evolution of the state components. **b** closed-loop MPC control. **c** CPU time needed to solve the OCPs (`iter_max=2000`) during the closed-loop simulation. Recall that $|u_1| \leq 1.5$ and $|u_2| \leq 0.5$ while the derivatives of z and θ are respectively limited to 0.3 and 0.2

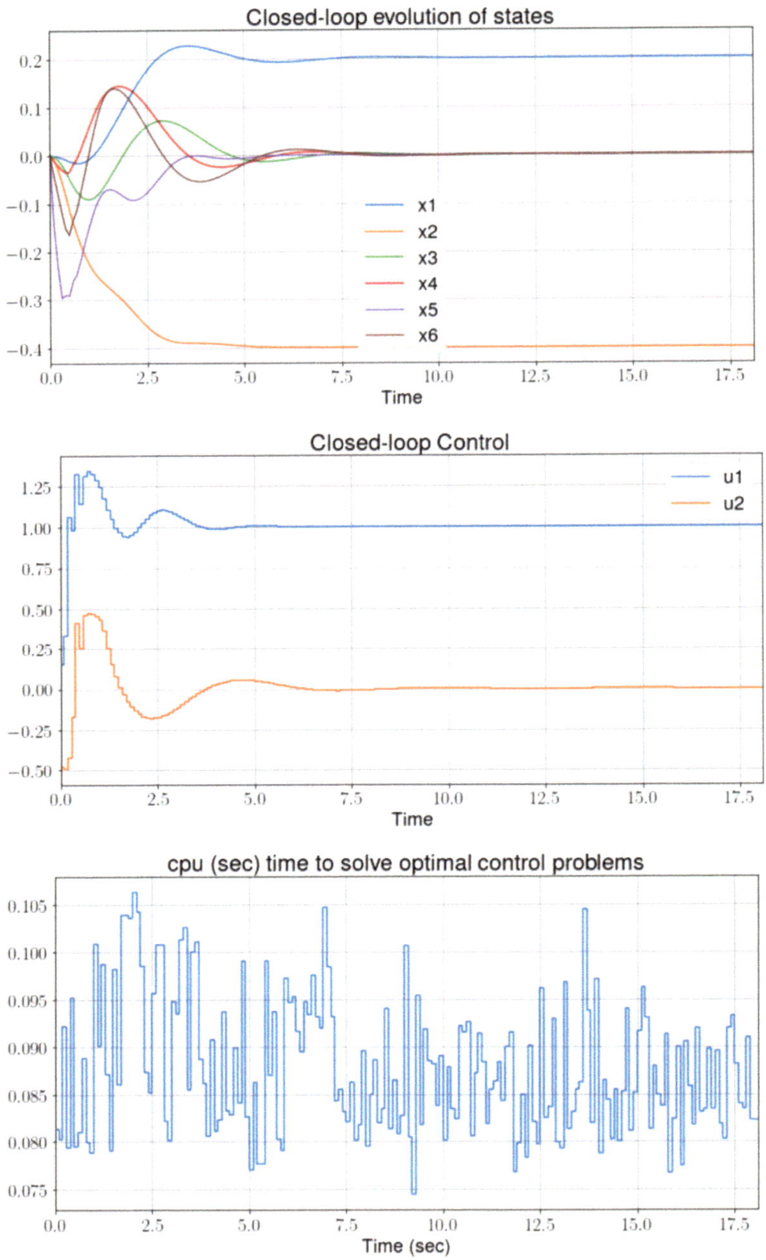

Fig. 5.3 Simulation of the PVTOL closed-loop system under the same model mismatches as the one used in Fig. 5.2 but with the maximum number of iterations set to `iter_max=5` instead of `2000`. Compare the statistics of computation time showing that the high values reaching previously 350 msec are no longer present

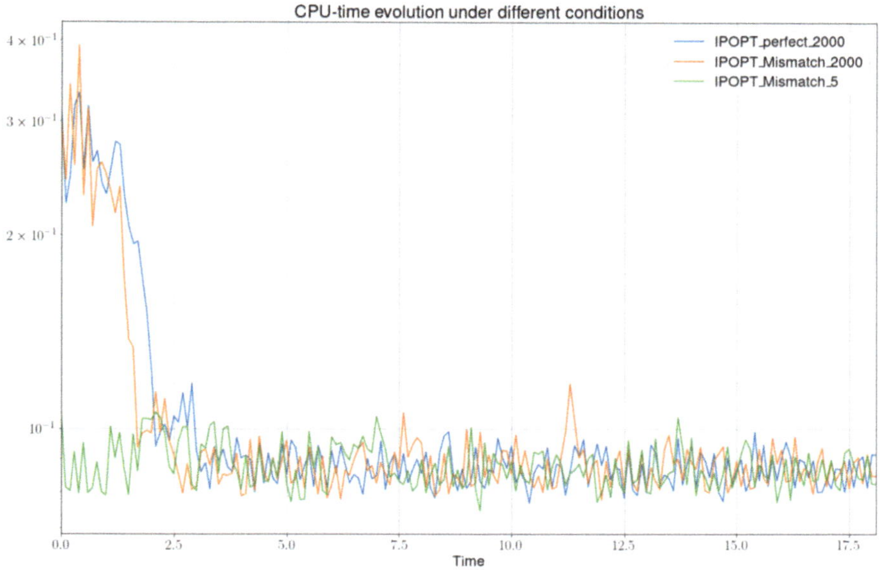

Fig. 5.4 PVTOL closed-loop simulation under `Ipopt`. Evolution of the CPU time needed during the closed-loop simulation under different conditions

computation time becomes almost compatible with real-time implementation since the maximum CPU time is not far from 100 msec.

Unfortunately, a further decrease in the allowed number of iterations leads to unstable closed-loop behavior!

Figures 5.4 and 5.5 summarize the computation times and the evolution of the cost function for the three previously discussed configurations.

Figure 5.4 shows that after an initial settling time period, the computation times of the three configurations converge to the same level as in this specific example, once the desired state is reached, the impact of the mismatch on the presumed and the true vector of parameters vanishes.

On the other hand Fig. 5.5 illustrates the impact of the parameters mismatch between reality and the values used by the controller which lead to a less decrease in the cost function value than in the case of perfect knowledge of the parameters but only the logarithmic scale makes this difference possible to grasp in the use-case under study.

Hopefully, this example helps illustrating that the issue of real-time implementation of MPC for general nonlinear systems might be an important issue and the *best* solution (if any) might almost certainly be problem-dependent.

The literature regarding the ways to handle real-time issues in the implementation of the MPC feedback is quite rich during the last two decade. Many works tried to address this issue in a more or less theoretical or heuristic way [5–7] but this is not the central topic of the current book.

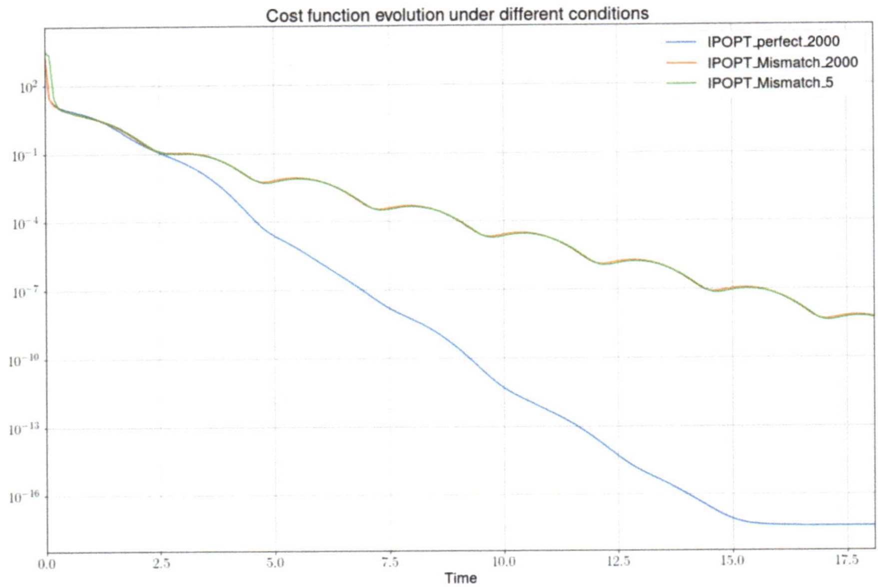

Fig. 5.5 PVTOL closed-loop simulation under `Ipopt`. Evolution of the predicted cost during the closed-loop simulation under different conditions

Nevertheless, in **Chap.** 6, it is shown that many items in the MPC formulation might impact the real-time feasibility. Some of these items are investigated in the `mpc_tuner Python` module proposed in **Chap.** 6.

In this chapter, we shall examine another easy-to-try option to address this issue by using the fast-gradient-based solver:

`simulate_cl_fast_gradient`,

mentioned above as shown in the next section.

But here again, there is no magic solution, this one comes also with some risks and drawbacks that we shall equally examine and discuss.

5.7 Results Using `simulate_cl_fast_gradient`

In this section, the method `simulate_cl_fast_gradient` is used to implement the MPC feedback while solving the underlying OCP using a prescribed number of iterations of the fast-gradient method described in page 232.

This is done using the following script:

```
x0 = np.zeros(nx)
U0 = np.zeros(mpc.N * nu)
Nsim = 201
W_supposed = [0.4, 1] * Nsim
W_real = [0.4, 0.8] * Nsim
p = [0.2, -0.4, 100, 10, 10, 1e2]
rho_soft = 1e2
gam, c, Niter = 1e-3, 0.9, 50

R  = mpc.simulate_cl_fast_gradient(x0, U0, W_supposed,
          W_real, p, rho_soft, Nsim, gam, c, Niter)
```

Note that the number of iterations here is chosen equal to `Niter=50`. Contrary to the `Ipopt` solver where the `iter_max` is an upper bound for the number of iterations meaning that the effective number of iterations might be less if the stopping condition are satisfied, here when `Niter=50` is chosen, there will be exactly that specific number of iterations.

The resulting closed-loop simulation is shown in Fig. 5.6 where it can be observed that globally, the behavior is quite similar to the optimal behavior provided by the `Ipopt` solver (compare to Fig. 5.1). The constraints seem to be also respected with obviously a less aggressive behavior (no activation over whole time intervals) which probably suggests a slight lack of optimality that will be examined later on.

However, the very good news here is the CPU time needed to perform the `Niter=50` iterations of the fast-gradient method. Indeed Compared to the 300 msec time required by the `Ipopt` solver, here **the maximum computation time needed does not exceed 22 msec** for roughly the same performances. This time is largely lower than the sampling period (100 msec). Using the same model mismatches as previously in order to check the robustness of the `Ipopt`-based MPC implementation, the resulting closed-loop behavior under the fast-gradient algorithm is shown in Fig. 5.7.

This simulation shows that the controller resists quite well to the disturbances induced by the model mismatches. **Notice however the slight transient violation of the constraints on the derivative of** z (that should be bounded by 0.3 in amplitude.)

Figure 5.8 shows the behavior of the closed-loop system when the fast-gradient is used with the number of iterations `Niter=15`. Again, the performance seems to be kept to a reasonable level while the maximum computation time goes down to less that **8 msec.** This is more than **35 times faster than the by-default setting of the `Ipopt` solver**.

> **Impact of the Solver on CPU Time**
> As far as this example is concerned, the choice of the solver together with the

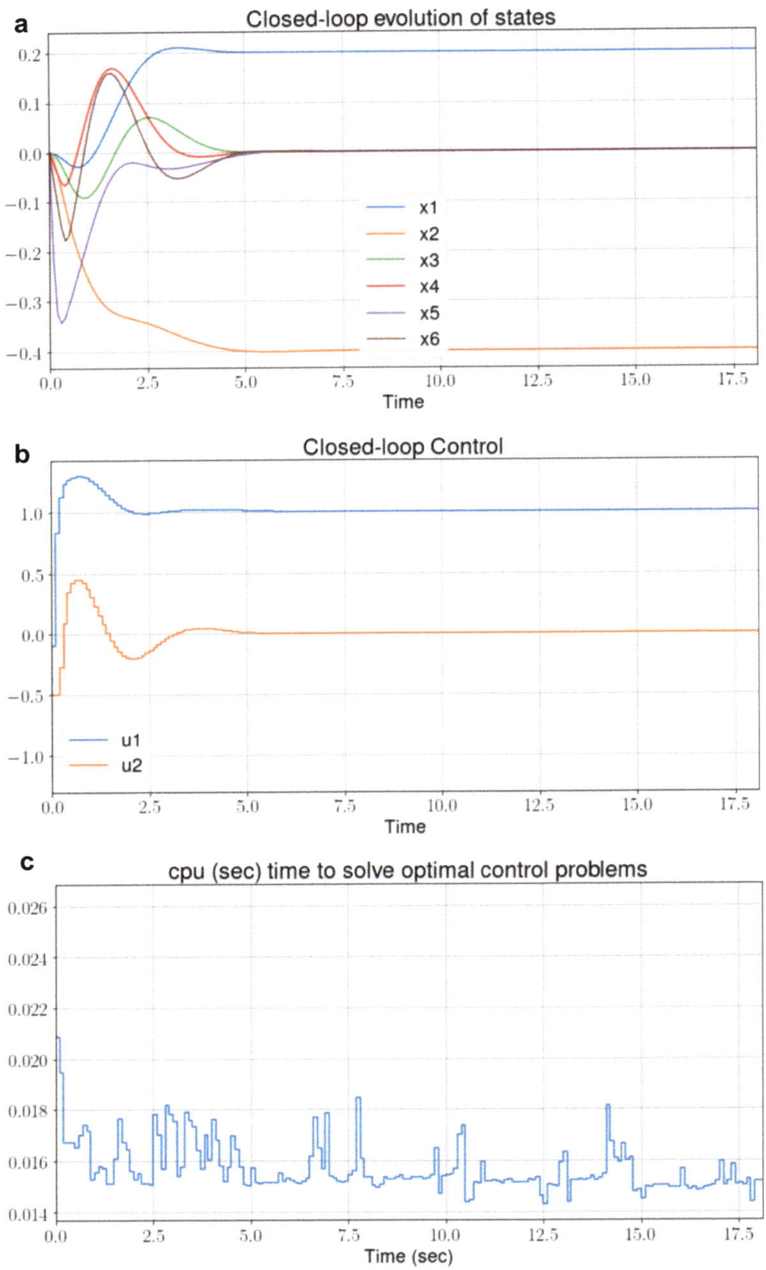

Fig. 5.6 Closed-loop simulation of the PVTOL under MPC feedback using fast-gradient algorithm (with `Niter=50`) to solve the underlying OCPs. This simulation is done under **perfect knowledge of the model's parameters**. **a** evolution of the state components. **b** Evolution of the control components. **c** CPU time needed to perform the computations

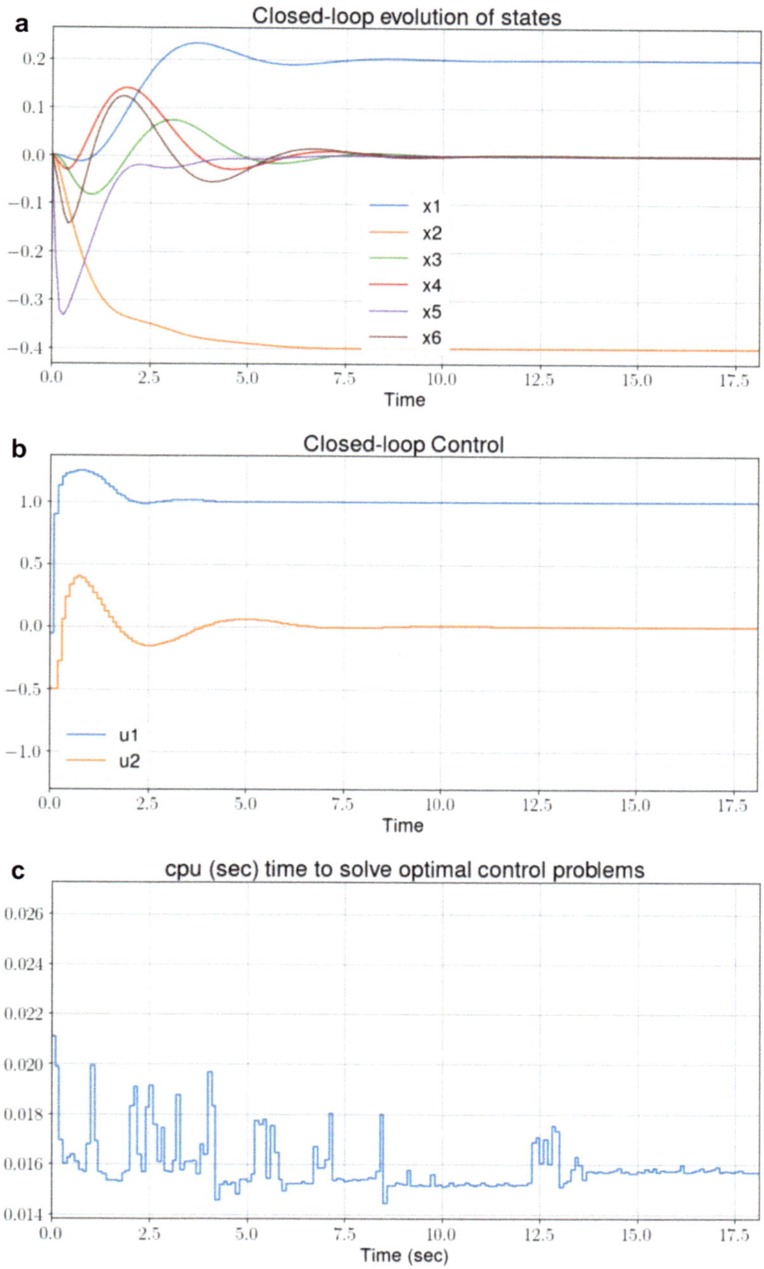

Fig. 5.7 Closed-loop simulation of the PVTOL under MPC feedback using fast-gradient algorithm (with `Niter=50`) to solve the underlying OCPs. This simulation is done under the same model mismatches used in the scenario of Figs. 5.2 and 5.3 of the model's parameters. **a** evolution of the state components. **b** Evolution of the control components. **c** CPU time needed to perform the 50 iterations of the fast-gradient

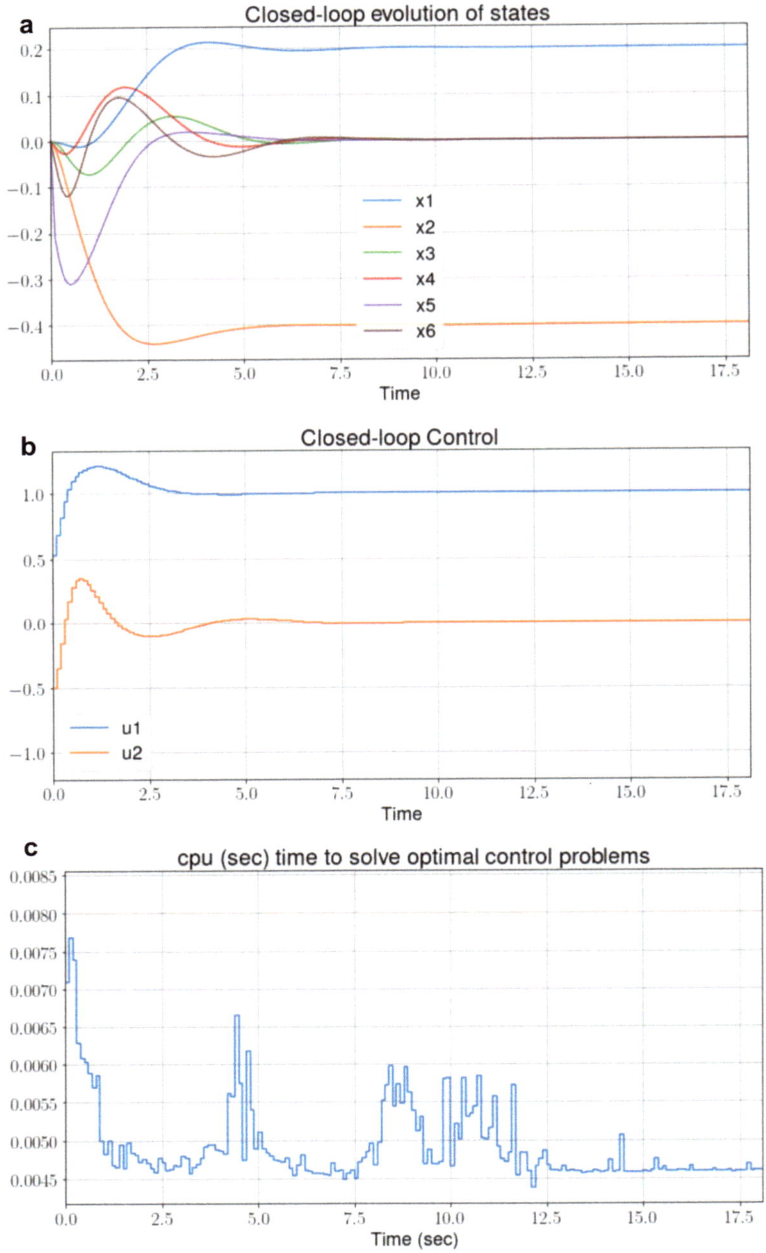

Fig. 5.8 Closed-loop simulation of the PVTOL under MPC feedback using fast-gradient algorithm (with `Niter=15`) to solve the underlying OCPs. This simulation is done under the same model mismatches used in the scenario of Figs. 5.2 and 5.3 of the model's parameters. **a** evolution of the state components. **b** Evolution of the control components. **c** CPU time needed to perform the 15 iterations of the fast-gradient

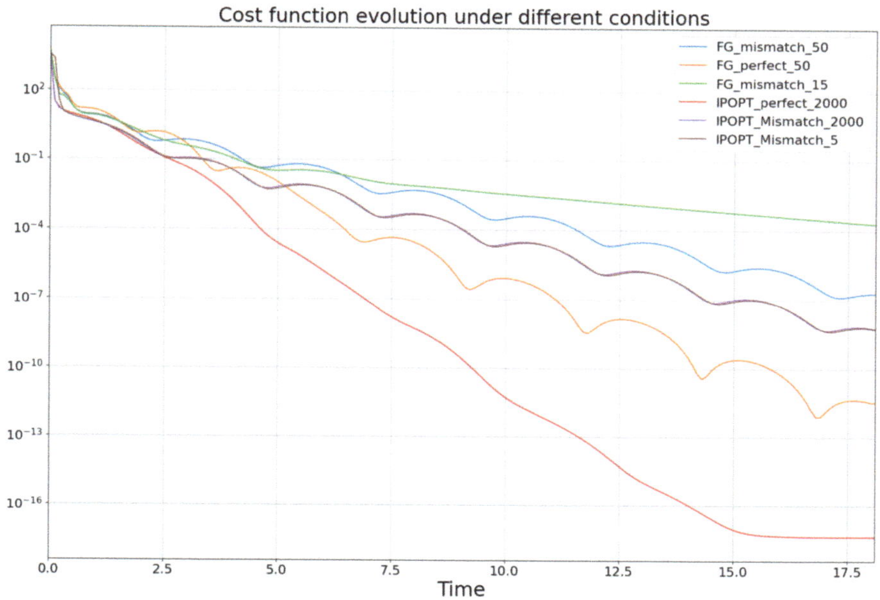

Fig. 5.9 PVTOL: Comparison of the evolution of the cost function during the closed-loop simulations under the different choices of solver and its parameters. The corresponding computation times are given in Fig. 5.10

maximum number of iterations **reduces the computation time up to 35 times** compared to the default setting of the `IPOPT` solver.

Let us end this use-case dedicated section by showing the CPU time and the cost function evolution for the six scenarios seen so far. This is given in Figs. 5.9 and 5.10 for the sake of completeness.

A careful examination of Figs. 5.9 and 5.10 clearly shows the trade-off between the computation time and the quality of the resulting behavior. This comparison should be done while keeping in mind that the computation time needs to be lower that the control updating period.

The *best* choice if any is difficult to find without a set of trial and errors process with an outcome that will probably remain problem-dependent. But Having in mind that the choice is not unique nor independent of the chosen parameters is a very good step towards a rationale and thoughtful use of MPC.

Figure 5.11 shows the evolution of the predicted cost function during the closed-loop simulations under the fast-gradient solver for different values of the moment parameter `c`. Note that high values of $\texttt{c} \in (0, 1)$ induce faster convergence rate at the price of some oscillations that might end by destabilizing the iterations.

Here again, the turning point above which instability might arise is problem-dependent and should be carefully checked.

Note that in the proof of convergence of the fast-gradient, the stability condition is expressed in terms of the strong convexity coefficient that enables, when it is

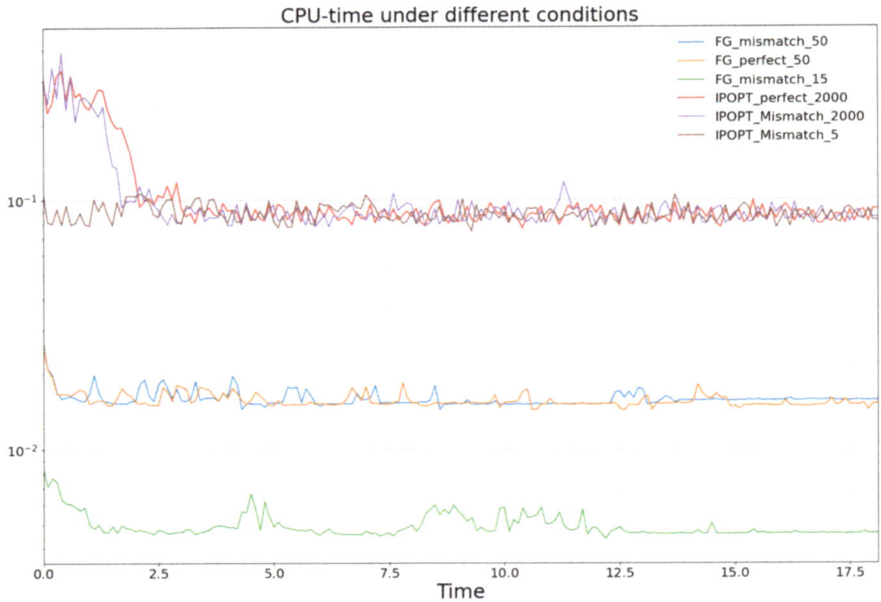

Fig. 5.10 PVTOL: Comparison of the evolution of the CPU-computation time used to solve the optimization problems during the closed-loop simulations under the different choices of solver and its parameters. The corresponding performance in terms of the evolution of the cost function are shown in Fig. 5.9

known, to fix the value of `gam`, used in the gradient descent, so that the convergence is guaranteed. More precisely, the convexity parameter is linked to the conditioning of the Hessian matrix which can be provided by taking the Jacobian of the Jacobian via the same `CasaDi` framework.

Nevertheless, this trick shows two drawbacks:

✓ The first is that in the general case, the hessian of the optimization problem in the MPC framework is generally dependent on the initial state and the control profile[19] This means that if one aims at guaranteeing the stability of the gradient-based iteration, it is necessary to update the maximum value of the step size on-line and at each updating instant. But the computation of the Hessian and its conditioning might itself take some time that makes the computation slower.

✓ More importantly, as it is shown in the remainder of this section dedicated to the presentation of the results under the fast-gradient approach, the result regarding the convergence when only cautious (small) steps are taken in the opposite direction of the gradient hold for a static optimization problem.
In the dynamic context where the number of iterations is limited and where the sub-optimal result of the truncated optimization is applied to a state-changing

[19] This is true in the single-shooting settings, otherwise it also depend on the state trajectory.

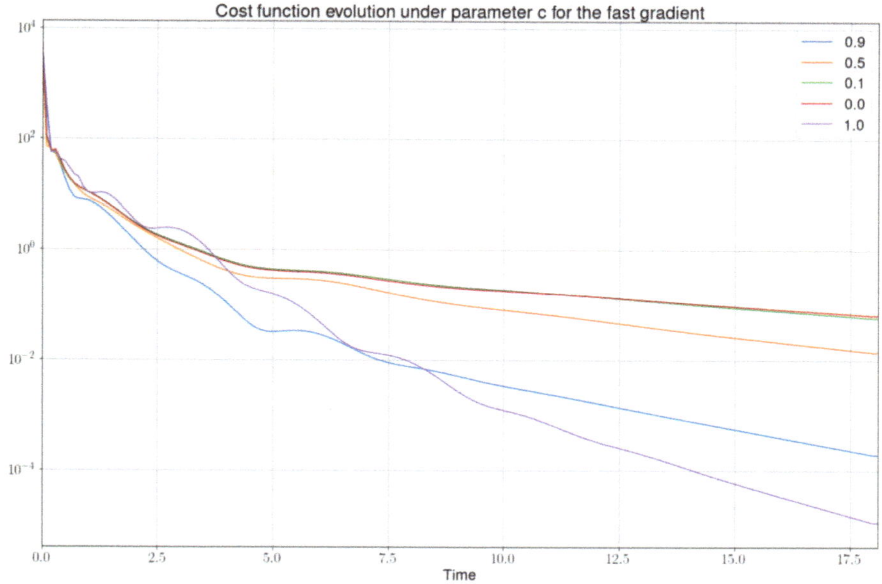

Fig. 5.11 PVTOL: Behavior of the cost function during the closed-loop simulation under the fast-gradient solver for different values of the parameter `c`

dynamical system, it might happen that a limited number of too small steps might induce a slow decrease in the open-loop cost that cannot compensate for the increase of the latter which is due to the change of the initial state that is inherent to the receding-horizon mechanism and hence, the overall convergence of the closed-loop dynamic would be threatened.

The previous discussion can be summarized as follows:

> **MPC Commissioning Is Necessary**
> There is no magical solution to this kind of context-dependent battles between an algorithm and real-life system. Instead, a first phase of MPC commissioning and outer tuning loop would anyway be necessary to perform for each specific combination of the model of the system, the level of uncertainty and a given computational device.
> A complete description of an example of such an MPC tuning framework is described in **Chap.** 6 together with the associated `GitHub` freely available `MPC_tuner` Python package.

The impact of the choice of the fast-gradient algorithm's parameter `gam` is shown on Figs. 5.12 and 5.13. Both figures show the evolution of the optimal predicted

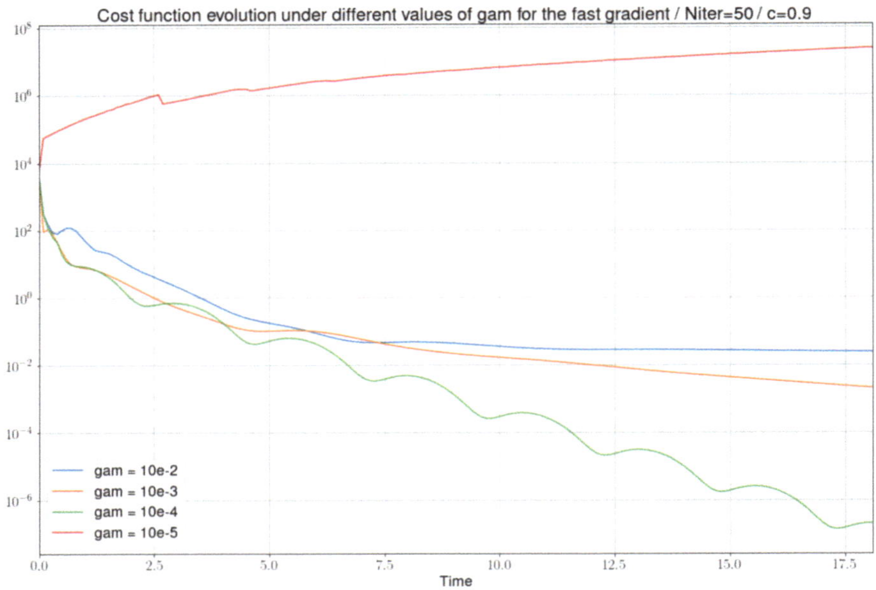

Fig. 5.12 PVTOL: Evolution of the predicted cost function in closed-loop under the fast-gradient algorithm with `Niter=50` for different values of the step parameter gain `gam`

cost in closed-loop when the fast-gradient algorithm is used with `c=0.9` for different values of `gam`. In the case of Fig. 5.12 the number of iterations used is `Niter=50` while in Fig. 5.13, `Niter=15` is used.

It can be observed that when `Niter=50` is used, the value `gamma=`10^{-2} leads to divergent cost because the induced step is too high given the convexity of the cost function that includes the constraints violation penalty terms.

On the other hand, when a too small number of iterations `Niter=15` is used, the choice `gam=` 10^{-5} also leads to divergent cost because the decrease rate of the cost function is not sufficient, given the low number of allowed iterations, to compensate for the change of the cost function that is induced by the change of the current state when the prediction horizon is shifted. Therefore, when `Niter=15` is used, the two extreme choices `gam=` 10^{-5} and `gam=` 10^{-2} becomes inappropriate for two different reasons!

We can summarize the observations made so far as follows:

> **! Fast-Gradient: A Delicate Setting!**
> The price to pay in order to get the small computation time associated to the fast-gradient method is a far more sensitive results to the choice of the algorithm's parameters c and gam.
> A randomized optimization approach can be used here to make a rational choice of these parameters following the lines of Sect. 2.11.

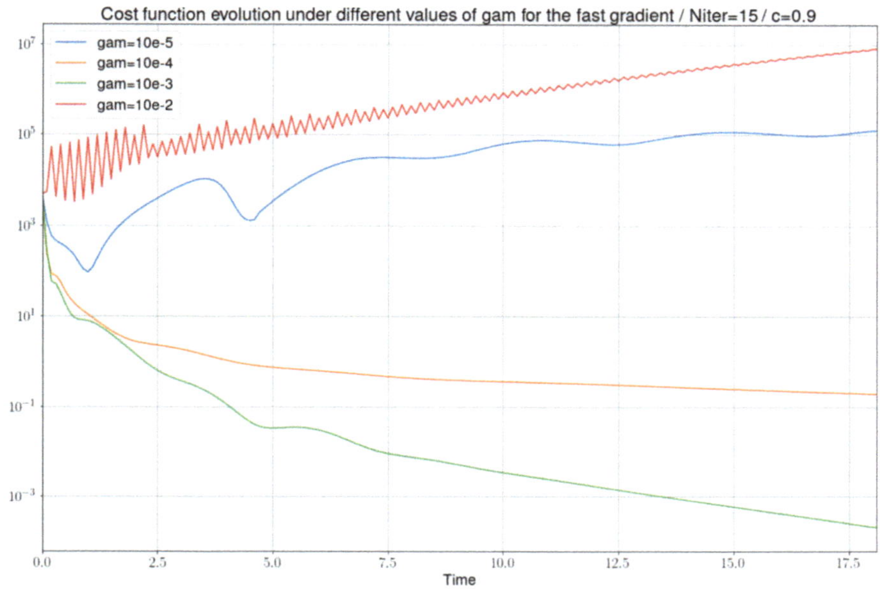

Fig. 5.13 PVTOL: Evolution of the predicted cost function in closed-loop under the fast-gradient algorithm with `Niter=15` for different values of the step parameter gain `gam`

Now that the use of the `mpc_solver` module has been sufficiently illustrated, it is time to see in some details how this module is coded. This is the aim of the next section.

5.8 Implementation of the `mpc_solver` Module

This section is dedicated to the implementation of the `mpc_solver` that has been presented in the previous section from the user's point of view.

The `Python` code can be freely downloaded at:

https://github.com/mazenalamir/mpc_solver,

but it might be interesting to give some comments regarding the choices that have been made here and there throughout the scripts in order to highlight the possibilities of improvement and/or extensions of the current version of the module.

5.8.1 The Needed Packages via import

The external modules that are used in the mpc_solver module are obtained via the following import instructions[20]:

```
import numpy as np
from casadi import MX, Function, nlpsol
from casadi import jacobian, if_else, power, vertcat
import time, copy
import matplotlib.pyplot as plt
```

5.8.2 The MPC Class Instantiation Function

As it is mentioned earlier, the module's main class is the MPC class and the instantiation __init__() method does a great deal of the job based on the list of user-defined input arguments explained earlier and exemplified using the PVTOL use-case.

The head of this method shows again the list of input arguments as shown in the following script:

The MPC Class (1)

```
class MPC:

    def __init__(self, nx, nu, nw, npar,
                 ode, stage_cost,
                 terminal_cost,
                 constraints, umin, umax,
                 gmin, gmax, dt, N, max_iter):

        self.nx = nx
        self.nu = nu
        self.nw = nw
        self.N = N
        self.dt = dt
        self.npar = npar
        self.ode = ode
        self.stage_cost = stage_cost
        self.terminal_cost = terminal_cost
        self.constraints = constraints
        self.max_iter = max_iter
```

Once these elements are provided to create the associated attributes ans methods of the class (for instant self.nx=nx, ..., etc.), the one-step-ahead predictor self.F

[20] Recall that MX is used to declare CasaDi variables, nlpsol is used to declare the solver, vertcat concatenates the variables, Jacobian computes exact sensitivities used in the gradient computation.

method is created via the following script which simply implements the 4-th order Runge–Kutta integration step as recalled in page 27.

The MPC Class (2)

```
# (inside __init__)
# Create discrete-time version of the dynamics
# this is the one-step-ahead-predictor

    x = MX.sym('x', nx)
    u = MX.sym('u', nu)
    w = MX.sym('w', nw)
    p = MX.sym('p', npar)
    tau = MX.sym('tau')
    k1 = ode(x, u, w, p)
    k2 = ode(x + 0.5 * k1 * tau, u, w, p)
    k3 = ode(x + 0.5 * k2 * tau, u, w, p)
    k4 = ode(x + k3 * tau, u, w, p)
    xplus = x +tau/6*(k1 + 2 * (k2+k3) + k4)
    self.F = Function('F', [x, u, w, p, tau], [xplus])
```

Then this one-step-ahead predictor method is used to create an open-loop simulator over `Nsim` steps via the following script[21] :

The MPC Class (3)

```
# (inside __init__)
# Create the multi-step-ahead predictor

    def sim(x0, U, W, p, Nsim):

        X = x0
        x = x0
        for i in range(Nsim):
            u = U[i*nu : (i+1) * nu]
            w = W[i*nw : (i+1) * nw]
            x = self.F(x, u, w, p, dt)
            X = vertcat(X, x)

        return X

    self.sim = sim
```

The next step is to create the optimization problem object for `CasaDi` so that it can be fed to the `Ipopt` solver.

[21] Notice how the slicing of the control and the uncertainty input is based on the one-dimensional flattened nature of the arguments `U` and `W`.

This is done by iterating over the instants of the prediction horizon while accumulating the stage cost into the cost function variable `J`.

In parallel, the constraint vector `g` is also created by using the `vertcat` method inside the iterations.

At the end of the iterations, the terminal cost is added by calling the `terminal_cost` method, given in the input argument. This leads to the creation of the following two `casadi` variables:

The value of the cost: `J`,
The value of the constraint vector: `g`.

Note that `CasaDi` keeps track of the symbolic chain of relations linking the *free* variables `x, U, W, p` and the above two *dependent* variables `J` and `g`.

This is done using the following script which also creates a new method called `self.cost` which computes the cost function in terms of the free variables `x, U, W, p` (Remember that we are still inside the `__init__` instantiation method of the class `MPC`):

The MPC Class (4)

```
# (inside __init__)
# Create the optimization problem and the bounds on the constraints

    x0 = MX.sym('x0', nx)
    U = MX.sym('U', N * nu)
    W = MX.sym('W', N * nw)
    par = MX.sym('par', npar)
    g = []
    J = 0
    x = x0
    for i in range(N):
        u = U[i * nu : (i+1) * nu]
        w = W[i * nw : (i+1) * nw]
        J += stage_cost(x, u, w, par) * dt
        gi = constraints(x, u, w, par)
        g = vertcat(g, gi)
        x = self.F(x, u, w, par, dt)
    J += terminal_cost(x, w, par)

    self.problem = {'f':J, 'x':U,
                    'g':vertcat(g), 'p':vertcat(x0, W, par)}

    self.cost = Function('cost', [U, x0, W, par], [J])
```

In order to solve the problem later on, we also need to define the bounds on the control profile (the decision variables in our single-shooting implementation) together with the bounds on the general nonlinear constraint components (defined by `g`).

This is done using the input arguments `umin`, `max`, `gmin` and `gmax` which are all defined point-wise and therefore they are concatenated to create the associated bounds on the profiles over the prediction horizon.

This is done using the following script inside the `__init__` method:

The MPC Class (5)

```
# (inside __init__)
    self.Umin = np.array(list(umin) * N).flatten()
    self.Umax = np.array(list(umax) * N).flatten()
    self.gmin = np.array(list(gmin) * N).flatten()
    self.gmax = np.array(list(gmax) * N).flatten()
```

Now remember that the use of the fast-gradient solution needs the cost function to be modified by adjoining a penalty on the constraints violation.

This is done in the instantiation method by defining the so-called `soft_cost` parameter. More precisely, during the iterations over the prediction horizon, the current value of `g` is compared to the two bounds on the constraints and if the bounds are exceeded, an associated penalty is added to the original cost as shown in the following script which builds on the previous computation of the vector of constraints `g`[22]:

The MPC Class: Creating the Cost Function

```
# Create the function that computes the gradient
# of the softened cost function

    rho_soft = MX.sym('rho_soft')
    Jsoft = J

    for i in range(len(self.gmin)):

        term = g[i]-self.gmax[i]
        Jsoft += if_else(term<0, 0, rho_soft * power(term, 2))
        term = self.gmin[i]-g[i]
        Jsoft += if_else(term<0, 0, rho_soft * power(term, 2))


    self.soft_cost = Function('soft_cost',
                        [U, x0, W, par, rho_soft], [Jsoft])
```

Based on the definition of the `soft_cost` parameter, we can use the `CasaDi` method `Jacobian` in order to compute the gradient of this cost w.r.t. the decision variable `U`. The resulting gradient `G` is again a `CasaDi` symbolic variables that can be used to define a new method which gives the gradient as a function of the free decision variable. This is done in the following script:

[22] Notice the use of the specific `CasaDi` instruction `if_else` since the standard `Python`'s `if` and `else` is not allowed here!

The MPC Class: Creating the Gradient of the Cost Function

```
# (inside __init__)
  G = jacobian(Jsoft, U)
  self.grad = Function('grad', [U, x0, W, par, rho_soft], [G])
```

This gradient function is to be used later on in order to define the fast-gradient-based solver.

As for the `CasaDi`'s `Ipopt` option, we need to define the solver and the feedback function through the following instructions:

The MPC Class: Defining the Solver

```
# Define the solver
self.solver = nlpsol('solver', 'ipopt', self.problem,
                    {'ipopt': {'max_iter':max_iter}})

# Define the function that runs the solver
self.compute_MPC = lambda U0, x0, W, par: self.solver(
                        x0=U0, lbx=self.Umin,
                        ubx=self.Umax, lbg=self.gmin,
                        ubg=self.gmax,
                        p=vertcat(x0, W, par))
```

Note that the first instruction is the `CasaDi` syntax to define the optimization problem which is a `Python` dictionary. Note that the second field is the `self.problem` previously defined.

The last instruction uses the `Python lambda`-based syntax for in-line functions' definition `lambda x:f(x)` meaning that from now on, if `mpc` is an instance of the class `MPC` then, the call:

```
mpc.compute_mpc(U0, x0, W, p),
```

solves the OCP with the initial guess `U0` when the current state is `x0`, the presumed uncertainty profile is `W` and the parameter vector is `p`.

This ends the definition of the instantiation method `__init__`.

In the following section, the methods enabling to compute the closed-loop simulation under the two possible options of the solver, namely:

(`Ipopt`, `fast-gradient`)

are introduced.

5.8.3 The Closed-Loop Simulation Method simulate_cl

The following script defines the method in the MPC class that simulates the closed loop using the IPOPT solver.

The MPC Class: Simulating the Closed-Loop Under the IPOPT Solver

```
def simulate_cl(self, x0, U0, W_supposed, W_real, p, Nsim):

    Xcl = x0
    Ucl = []
    Jcl = [self.cost(U0, x0, W_supposed[0:self.N * self.nw],
    p).full().flatten()[0]]
    cpu = []
    x = x0
    for k in range(Nsim-self.N):

        W = W_supposed[k * self.nw: (k+self.N) * self.nw]
        if k == 0:
            U0_guess = U0
        else:
            U0_guess = Uopt. # warm start

        t0 = time.time()
        sol = self.compute_MPC(U0_guess, x, W, p)
        cpu += [time.time()-t0]
        Uopt = sol['x'].full().flatten()
        Jcl = vertcat(Jcl, sol['f'])
        u = Uopt[0:self.nu]
        w = W_real[k * self.nw: (k+1) * self.nw]
        x = self.F(x, u, w, p, self.dt)
        Xcl = vertcat(Xcl, x)
        Ucl = vertcat(Ucl, u)

    Xcl = Xcl.full().flatten().reshape(-1, self.nx)
    Ucl = Ucl.full().flatten().reshape(-1, self.nu)
    Jcl = Jcl.full().flatten()
    tcl = np.array([i * self.dt for i in range(len(Xcl))])
    Ucl = np.array(list(Ucl)+[Ucl[-1]])
    cpu = np.array(cpu+[cpu[-1]])

    Result = Container()
    Result.tcl = tcl
    Result.Xcl = Xcl
    Result.Ucl = Ucl
    Result.Jcl = Jcl
    Result.cpu = cpu

    return Result
```

This method iterates over the `Nsim` sampling periods and uses the previously defined method `compute_MPC` in order to compute the optimal sequence of control inputs at each decision instant k.

The closed-loop evolution of the predicted optimal cost is also gathered in the variable `Jcl` as well as the associated computation times which are concatenated in the variable `cpu`. The resulting state trajectories `Xcl` and closed-loop control trajectories `Ucl` are also provided.

All the results are represented as attributes of the `object` class `Container` which is exported as the output of the method.

The .full() Method
Notice the use of the method `.full()` exported by the `CasaDi` object which delivers a `numpy` version of it.
Forgetting this transformation is at the origin of many of the coding errors when `Python` standard users start using `CasaDi`.

It is important to notice how the solution of the previous optimization problem, namely `Uopt`, is used to initialize the decision variable for the next round of optimization. This is referred to in the MPC literature as the *warm start* mechanism.

The warm start relies on the common sense idea according to which the slight change in the state between two successive updating instants is sufficiently small to consider that the optimal solution of the current OCP is not far from the solution of the previous instantiation of the problem making the initialization defined by

```
U0_guess=Uopt,
```

a good new initial guess.

5.8.4 The `fast_gradient` Method

Before we discuss the script of the method used to simulate the closed-loop system under the MPC using the fast-gradient to solve the OCP, we need to examine the method that implements (`Niter`) iterations of the fast-gradient method.

This method, named `fast_gradient`, will be then called by the method:

```
simulate_cl_fast_gradient,
```

that is shown later on.

The `fast_gradient` method implements the fast-gradient algorithm already recalled on page 232. It is implemented in the `mpc_solver` module using the following script:

The MPC Class: The Fast-Gradient Solver

```
def fast_gradient(self, U, x, W, p, rho_soft, gam, c, Niter):

        # The fast-gradient iteration
        # U is the current guess

        def project(U):

            U_projected = copy.copy(np.array(U.flatten()))
            for i in range(self.N * self.nu):
                    if U_projected[i] > self.Umax[i]:
                        U_projected[i] = self.Umax[i]
                    elif U_projected[i] < self.Umin[i]:
                        U_projected[i] = self.Umin[i]
            return U_projected

        Uopt_previous = copy.copy(U)
        z_previous = copy.copy(U)

        for _ in range(Niter):

            G = self.grad(Uopt_previous, x, W, p,
                    rho_soft).full().flatten()
                    z = Uopt_previous - gam * G
                    Uopt = project(z + c * (z-z_previous))
                    z_previous = copy.copy(z)
                    Uopt_previous = copy.copy(Uopt)

        return Uopt
```

In order to write the corresponding closed-loop simulator, all one has to do is to replace the call of `compute_MPC` method by the above defined fast-gradient iteration function.

This leads to the following script:

The MPC Class: The Closed-Loop Simulation Under the Fast Gradient Solver

```
def simulate_cl_fast_gradient(self, x0, U0, W_supposed, W_real,
        p, rho_soft, Nsim, gam, c, Niter):

        Xcl, Ucl, cpu, x, Jcl = x0, [], [], x0, []
        for k in range(Nsim-self.N):

            W = W_supposed[k * self.nw: (k+self.N) * self.nw]
            if k==0:
                J = self.soft_cost(U0, x, W, p,
                rho_soft).full().flatten()[0]
                Jcl = vertcat(Jcl, J)
            else:
```

```
                J = self.soft_cost(Uopt, x, W, p,
                rho_soft).full().flatten()[0]
                Jcl = vertcat(Jcl, J)
            t0 = time.time()
            if k==0:
                Uopt = copy.copy(U0)

            Uopt = self.fast_gradient(Uopt, x, W, p,
                        rho_soft, gam, c, Niter)
            u = Uopt[0:self.nu]
            cpu += [time.time()-t0]
            w = W_real[k * self.nw: (k+1) * self.nw]
            x = self.F(x, u, w, p, self.dt)
            Xcl = vertcat(Xcl, x)
            Ucl = vertcat(Ucl, u)

        Xcl = Xcl.full().flatten().reshape(-1, self.nx)
        Ucl = Ucl.full().flatten().reshape(-1, self.nu)
        Jcl = Jcl.full().flatten()
        tcl = np.array([i * self.dt for i in range(len(Xcl))])
        Ucl = np.array(list(Ucl)+[Ucl[-1]])
        Jcl = np.array(list(Jcl)+[Jcl[-1]])
        cpu = np.array(cpu+[cpu[-1]])

        Result = Container()
        Result.tcl = tcl
        Result.Xcl = Xcl
        Result.Ucl = Ucl
        Result.Jcl = Jcl
        Result.cpu = cpu

        return Result
```

where here again, the previous control profile is used as the initial value at which the fast-gradient iterations start hence implementing the warm start idea.

5.9 Conclusion

In this chapter, a `Python` module, named `solver_mpc`, is introduced that helps making easier the implementation of the MPC feedback design for a general nonlinear dynamical system governed by a set of ODEs in the **deterministic setting**, namely, an uncertainty vector involved in the mathematical model of the system is supposed to be given.

This module is introduced for later use in various contexts including the one where only nominal design is implemented leaving the uncertainty handling to the intrinsic robustness that is associated to any feedback action.

On the other hand, this framework can also be used in the case where the uncertainty vector is not known but where the solution of the stochastic design is done by solving deterministic problems as a step in generating the learning data from which an uncertainty-aware feedback design is derived. Different examples of such solutions are given in **Chaps.** 7 and 16.

Two algorithms are used in the `mpc_solver` that have each, its strong and weak points, namely the `Ipopt`-based solver and the fast-gradient-based solver. The former leads to more precise solutions in the ideal case where no real-time issue is involved with the potential counterpart of having higher computation times while the latter, when appropriately tuned and under certain circumstances, leads to faster-to-derive good sub-optimal solutions, the quality of which is improved by the repetitive character of the process.

In both cases, it has been shown that the number of iterations that might be allocated highly affects both the quality of the solution but also the real-time feasibility of the resulting feedback. These issues are active research areas in the control community that deserve an entire book for their own. Nevertheless, **Chap.** 6 addresses, at least, partially this critical tuning issue by means of the `Python` module `MPC_tuner`.

Let us end this chapter by recalling that `Python` packages that solve OCPs, or even that provide ready to use framework for the simulation of MPC-based closed-loop trajectories, pop up here and there, on an almost monthly basis.

This is the reason why a continuous study of the existing new offers is the best way to keep in touch with the best possible solutions to the MPC implementation.

Nevertheless, the aim of this chapter is to provide an overview of the key steps, items and keywords that appear unavoidably in any possible solution.

This should facilitate the comprehension and the use of present and future freely available packages.

References

1. Hart, W.E., Laird, C.D., Watson, J.P., Woodruff, D.L., Hackebeil, G.A., Nicholson, B.L., Siirola, J.D.: Pyomo—Optimization Modeling in Python, 2 edn. Springer Optimization and Its Applications. Springer (2017). https://doi.org/10.1007/978-3-319-58821-6, http://www.pyomo.org
2. Beal, L.D., Hill, D.J., Martin, R.S., Hedengren, J.D.: Gekko optimization suite. Processes **6**(8), 106 (2018). https://doi.org/10.3390/pr6080106, https://www.mdpi.com/2227-9717/6/8/106
3. W chter, A., Biegler, L.T.: On the implementation of an interior-point filter line-search algorithm for large-scale nonlinear programming. Math. Program. **106**(1), 25–57 (2006). https://doi.org/10.1007/s10107-004-0559-y
4. Houska, B., Ferreau, H.J., Diehl, M.: Acado toolkit: An open-source framework for automatic control and dynamic optimization. Optim Control Appl Methods **32**(3), 298–312 (2011)
5. Alamir, M.: A framework for real-time implementation of low-dimensional parameterized nmpc. Automatica **48**(1), 198–204 (2012)
6. Alamir, M.: A state-dependent updating period for certified real-time model predictive control. IEEE Trans Autom Control **62**(5), 2464–2469 (2016)
7. Pham, X., Alamir, M., Bonne, F., Bonnay, P.: Revisiting a fixed-point hierarchical control design for cryogenic refrigerators under constraints, nonlinearities and real-time considerations. Eur J Control **63**, 82–96 (2022)

A Framework and a Python-Package for Real-Time Nonlinear Model Predictive Control Parameters Settings

6

6.1 Chapter's Content at a Glance

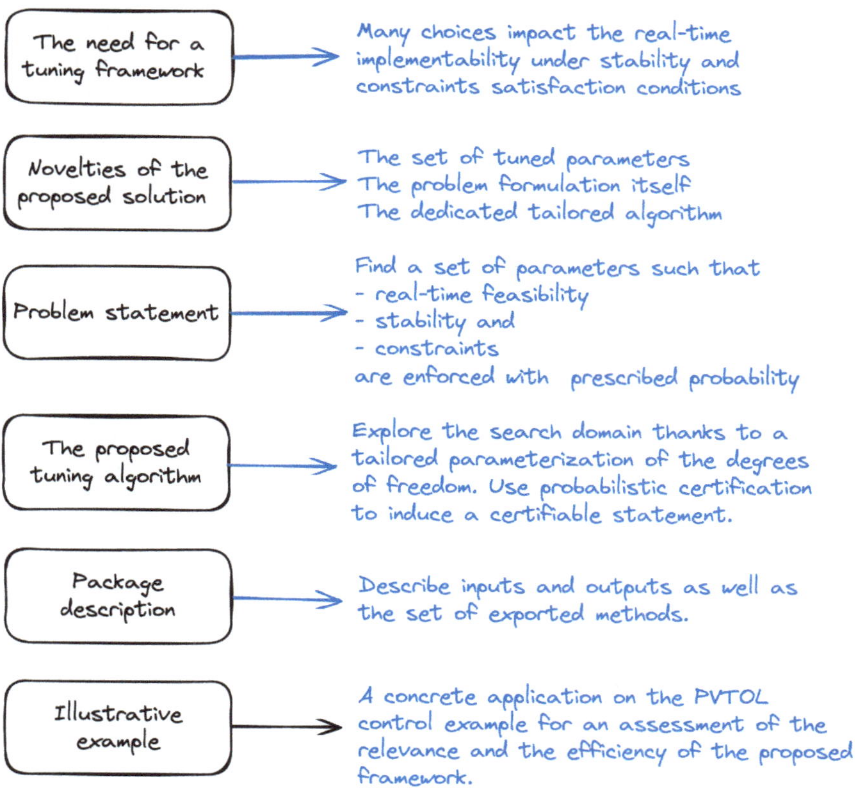

M. Alamir, *Nonlinear Control of Uncertain Systems*,
https://doi.org/10.1007/978-3-031-93287-8_6

6.2 Introduction

6.2.1 What Is This Chapter About?

In Chap. 5, a `Python` module is described that enables to formulate and solve open-loop optimal control problems (OCPs). It also provides the associated MPC feedback law as a `Python` function as well as methods to simulate the resulting closed-loop under two different solvers.

However, in the setting of Chap. 5, many of the details regarding the concrete choice of the different implementation parameters of the MPC control design are not discussed focusing on the main streamline of presentation.

This can give rise to the misleading idea that there are not many options except for the choice of the solver and the number of iterations of a given algorithm.

This chapter shows that there are many remaining choices that might heavily impact the performance of the resulting closed-loop behavior. Moreover, a dedicated `Python` module, the `mpc_tuner`, is discussed that enables to produce a list of successful settings of these parameters involved in the design.

More precisely, this chapter presents a framework that enables a systematic and rational choice of NMPC design components. While the principle of the framework applies to any possible choice of the set of parameters to be tuned, the present chapter and the current version of the `Python` module `MPC_tuner` addresses the tuning of the following set of implementation parameters[1]:

- the control updating period,
- the down-sampling period for prediction inside the MPC algorithm,
- the prediction horizon's length,
- the maximum number of iterations, and
- the penalties on the terminal cost and the soft constraints.

The chapter also shows typical results on an illustrative example using the previously studied PVTOL system.

But before digging into the details, let us first highlight the need for NMPC tuning when it comes to seek a concrete implementation on a specific computational device.

6.2.2 The Need for a Tuning Framework

As already discussed in Chap. 2, the theoretical foundations of Nonlinear Model Predictive Control (NMPC) [1] are now quite established. This came after years of development of various stability proofs for the resulting closed-loop systems. As mentioned earlier, this resulted in a long list of different formulations among which, several ones have been briefly described in Chap. 2.

[1] The precise meaning of these parameters is clearly explained in the following sections.

Moreover free and *easy-to-use* modeling frameworks such as the `CasaDi` framework used in the previous chapter offering the possibility to embedding multiple efficient and trustworthy dedicated solvers such as [2] and [3] and many others are now available. Other frameworks are also available such as `Pyomo` [4] and `Geckko` [5] are also briefly discussed in **Chap.** 3 any probably new frameworks are under preparation.

These reliable and easy-to-use frameworks render the problem of solving nonlinear optimization problems and consequently, the implementation of MPC control feedback quasi-transparent for the user in the nominal cases where the computation time and the real-time implementability are not the center of interest.

Handling the issue of real-time compatibility started in the early 2000s with the idea of *distributing the computation over time*. More precisely, the idea of performing a limited number of iterations of the solving algorithm as we did in Chap. 5 to show that almost the same level of performance can be achieved with much less computation effort per updating period.

The first time this idea was suggested and applied to a missile guidance control problem goes back to [6]. In this work, it was suggested to use a **single iteration of the gradient descent** at each sampling period.

In this early work, it has been then shown that the extremely short resulting computation time enables a high updating frequency which renders the probability of successful interception outcome much higher as the adaptation to target's unpredictable changes of trajectory becomes much more effective due to the high updating rate.

Later on, this idea of performing a single iteration per sampling period, has been put in a more formal and general setting [7] with the associated proof of stability (that required some ad hoc assumptions that might be difficult to guarantee in real-life applications).

Nevertheless, from a practical point of view, this conceptual advance enabled a huge extension of the number of successful applications of nonlinear MPC to *fast systems*. The concept is now widely and commonly referred to as the **real-time iteration** MPC.

Despite the impact of the real-iteration on the extension of the domain of application of MPC design, some legitimate questions remained unanswered, in particular:

1. The principle of real-time iteration does not involve the crucial information regarding the sampling period being used and how it relates to the characteristic time of the system.
2. This principle does not involve considerations regarding the level of uncertainties that affect the system. This seems conceptually strange since the raison d'être of the feedback is precisely the presence of such uncertainties and it seems obvious that the optimality of its concrete successful implementation might depend on the characteristics of these uncertainties.
3. Regardless of the previous lack of consideration of some key information, the successful convergence of the real-iteration scheme under a single iteration per period depends on some sufficient conditions showing a sort of tautological char-

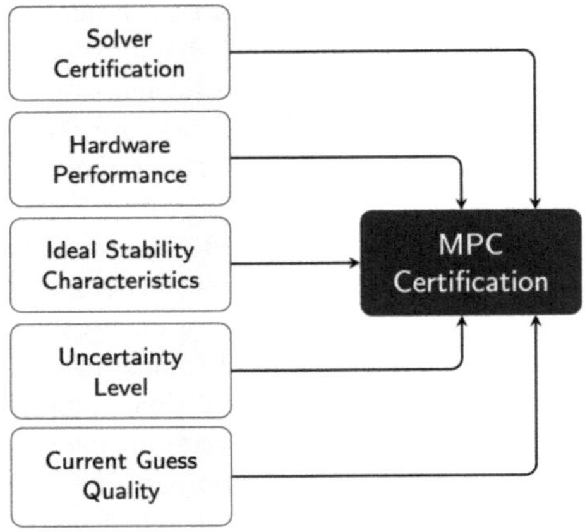

Fig. 6.1 List of items that impact the certification of a closed-loop MPC design

acter. This is the case of the condition regarding the distance of the initial guess to the truly optimal unknown solution that should be *small enough*!

Consequently,

Regarding the Number of Iterations

It seems that taking the number of iterations to perform per sampling period systematically equal to 1 is totally arbitrary and shows no reason to be the optimal choice or even to be an admissible one leaving aside its level of optimality.

In order to address these issues, later works such as [8] and [9] showed theoretically and then experimentally [10] that in general, the appropriate choice of the control updating period in NMPC must ideally be viewed as a **state-dependent** decision that is tightly linked to many factors such as

- the dynamics of the controlled system,
- the implementation hardware that executes the optimization algorithm,
- the solver being used and its certification[2] rules,
- the level of model discrepancies, and
- the length of the prediction horizon.

This can be summarized in the following Fig. 6.1.

Recall that in Sect. 2.7.4, an example of a heuristic inducing state-dependent control updating period has been mentioned which gives a single instantiation of this general principle in a very specific case.

[2] The certification of an optimization algorithm is the formulae that gives the number of iterations that leads to a given precision on the resulting solution in terms of distance to the optimal solution.

Notice however that despite the relevance of the above-cited references, they all exclusively attempt to rationalize the way the solution of a **given formulation** is *distributed on-line* over the real-life time of the process.

All the other ingredients of the formulation that are listed above, namely the prediction horizon, the down-sampling for prediction inside the controller and the terminal stability-related penalty, are considered to be given by the designer and put outside the set of possibly tunable parameters.

This short overview enables the objective of the present chapter to be stated as follows:

Problem Statement
The present chapter addresses the problem of *choosing off-line*, for a given optimization algorithm, the different major ingredients of NMPC formulation so that convergence, constraints satisfaction and real-time implementability are enhanced.

Beside the above rationale determination of the NMPC setting's parameters, the chapter briefly describes a freely available `Python`-based package that implements the proposed solution:

https://github.com/mazenalamir/MPC_tuner.

Notice that *MPC tuning* is not a new topic and the wording corresponds to different meanings that might induce some misleading interpretations.

In particular, it is common to include in this category of works, those involving Reinforcement Learning (RL) where the feedback parameter/model vectors θ are *tuned* to achieve good closed-loop performances of a repetitive task (see [11] and [12]).

Other works involve the use of the so-called Bayesian optimization in order to derive the set of NMPC parameters that minimizes the closed-loop costs [13–15]. Bayesian approaches amounts at using surrogate model of the closed-loop cost as function of the design parameters in order to accelerate the search for optimal design.

Some works rely on the use of genetic algorithms [16] to achieve the optimal choice of the parameter using again, the closed-loop cost as a the value to be minimized.

The approach presented in the present chapter differs from the ones mentioned above in various aspects as explained in the following section.

6.2.3 Highlights Regarding the Tuning Solution

Compared to the traditional solutions for NMPC tuning, the one discussed in the present chapter shows the following specific features:

- **The set of tuned parameters** includes relevant set of implementation settings (sub-sampling, control updating period, control parameterization, soft constraints penalty, soft/terminal constraints penalty, prediction horizon) that are rarely addressed together in the literature while being almost all commonly used by everyone on a trial-and-error basis.
- **The problem formulation** that mixes certified feasibility criteria (real-time, constraints satisfaction) and certified stability/performance criterion, both included in a certification framework leading to a non-conventional outcome which consists in **providing a list of feasible settings among which the designer can choose depending on its own criterion** such as:

  - short updating periods when high uncertainties are expected,
  - large updating periods when `cpu`-load is shared among multiple tasks,
  - closed-loop performance when this is the main goal, or
  - context-dependent time varying mix of the above.

  Providing a list of possible successful NMPC implementation options is a major difference of the solution proposed in this chapter compared to the traditional approaches. Indeed, all the approaches mentioned at the end of the preceding section, while using different algorithms, share the same feature, namely, the focus on the closed-loop cost as a guiding criterion. To say it differently: in the absence of cost function, the iterations inside all these algorithms can simply not be defined in the first place.
- **The Dedicated Algorithm.** The above-described specificity of the formulated problem involving mixed-integer decision variables that discontinuously connected to three complex targeted functions together with the need for a heavy computation associated to the certification requirement renders the problem intractable without a dedicated algorithm's design.
  In particular, even in the case where one focuses on the minimization of the closed-loop cost function, the Bayesian approach [15] that is commonly recommended for optimization problems involving heavy-to-compute cost functions would be inappropriate here.
  The reason is that this approach is based on incrementally building *fast-to-evaluate* surrogate models of the targeted functions (involved in the definition of the cost function and the constraints) with the hope that the approximation would be increasingly precise.
  In our case, this amounts at hoping that functions like the MPC closed-loop cost and constraints dependencies on the initial state, the sub-sampling, the control updating periods, the control parameterization, the maximum number of iterations, etc., can be captured by surrogate models and this with moderate number

of samples. This *smells* like a *wishful thinking* that does not resist (and probably not even worth) a serious check. This might probably explain why, in almost all instances of application of the method, the number of decision variable is less that three.
This being said, it remains true that improving the algorithm proposed in this chapter is a serious track of future research efforts and as such, the current proposal is worth to be viewed as an initial starting point. Some words about possible research directions are given in the conclusion of the chapter.

The chapter is organized as follows:
The problem under consideration is first clearly described at a more detailed and concrete level in Sect. 6.3. The implementation of the optimization algorithm that *tunes* the parameters of the NMPC setting is discussed in Sect. 6.4. A brief description of the `MPC_tuner` *Python* package that implements the algorithm is proposed in Sect. 6.5. A concrete example showing use-case involving the `mpc_tuner` module is depicted in Sect. 6.6. Finally the chapter ends with a conclusion and a brief discussion regarding possible extensions of the package scope and utilities.

6.3 Problem Statement

In order to clearly state the problem under consideration, we need to define the items to be fed as **inputs** to the algorithm and those NMPC parameters that need to be *tuned* by the algorithm and hence represent its delivered **output**.

6.3.1 The Input Items to the Algorithm

– **A set of Ordinary Differential Equations** (ODEs) governing the continuous-time dynamics of the system to be controlled which takes the form

$$\dot{x} = f(x, u, p),$$

where $x \in \mathbb{X} \subset \mathbb{R}^{n_x}$, $u \in \mathbb{U} \subset \mathbb{R}^{n_u}$ and $p \in \mathbb{P} \in \mathbb{R}^{n_p}$.

– **A basic sampling period** $\tau > 0$ which is tailored such that the associated Runge–Kutta (RK) integration scheme yields accurate forward predictions should the expression of f be perfectly known. This basic sampling period is to be distinguished from the *control updating period* that is a part of the set of implementation parameters to be tuned by the algorithm.

– **A stage cost and a terminal penalty** functions denoted respectively by

$$\texttt{(stage cost)} \qquad \ell(x, u, p, q), \tag{6.1}$$

and

$$\texttt{(Terminal penalty)} \quad \Psi(x, p, q), \tag{6.2}$$

in which $q \in \mathbb{R}^{n_q}$ is a vector of task-related parameters.[3]

– **A constraints-related map** of the form

$$\texttt{(Constraints)} \quad c(x, u, p, q) \leq 0 \in \mathbb{R}^{n_c}.$$

Notice that all the constraints other than the input saturation-related ones are treated as **soft constraints** through appropriate highly weighted exact penalties (to be tuned as shown later on).

– **Bounds on the search domain** that are detailed later on.

– **A random sampling function** that can be used to generate representative cloud of initial states x_0, vectors of parameters p and vectors of context parameters q.

6.3.2 The NMPC Parameters to be Tuned (Algorithm's Output)

– **The control updating period** $\tau_u = \kappa\tau$ expressed as an integer multiple κ of the basic sampling period τ invoked above.

This explicitly means that the computation rounds that update the control profiles and hence the implemented closed-loop control are separated by $\kappa\tau$ time units. The next figure illustrates this parameter for $\kappa = 3$.

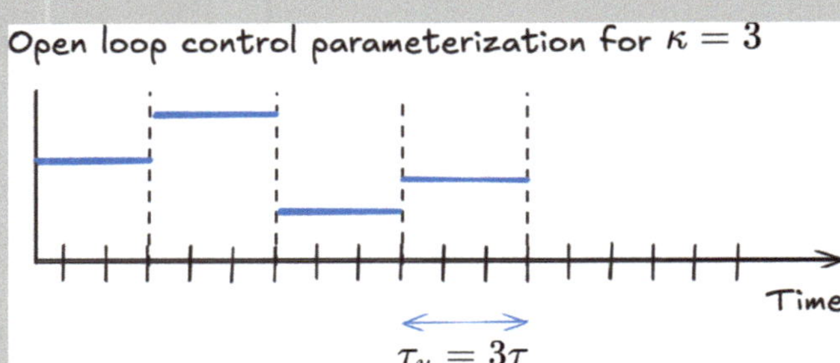

Illustration of the design parameter κ in the case where $\kappa = 3$.

Notice that this is not the only way to define a time structure for the control that is updated in a lower rate than the basic sampling rate. In particular, it would have

[3] This can be set-point vector, penalties involved in the stage cost definition, etc.

been possible to compute the optimal sequence that is defined using the same basic sampling period (τ) and then to apply, in open-loop, batches of κ successive control values. However, this corresponds to a higher dimensional optimization problem and its impact on the reduction of the computation time would be less significant.

The solution adopted here leads to the situation where, from the NMPC solver point of view, everything behaves as if the sampling period of the system is equal to τ_u so that the piece-wise constant trajectories of the control is defined by τ_u.

– **The prediction's precision parameter** $\mu_d \in [0, 1]$ that determines the precision of the integration used in the MPC-related computation. More precisely, if no precision drop is used, then κ steps of RK(τ) are performed with a sampling period of τ each in order to predict the τ_u-prediction step.

Now MPC practice suggests that the closed-loop might be successful while using `n_steps` ($\leq \kappa$) inner steps of RK($\frac{\tau_u}{\texttt{n\_steps}}$) to get a step prediction over τ_u, this is expressed by [4]

$$\texttt{n\_steps} := \lceil 1 + \mu_d(\kappa - 1) \rceil, \tag{6.3}$$

namely, $\mu_d = 0$ yields a single RK(τ_u) *large* step (low precision, faster computation) while $\mu_d = 1$ yields κ *small* steps of RK(τ) (high precision, longer computation).

This is illustrated in the figure below for the choices $\kappa = 3$ and $\mu_d = 0.5$ (hence leading to `n_steps=2`):

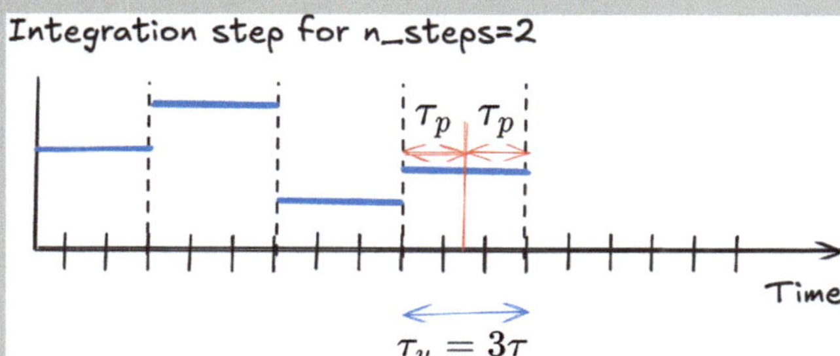

Illustration of the design parameters (κ, μ) when $\kappa = 3$ and $n\_steps = 2$.

– **The prediction horizon's length** defined as a multiple of the **control updating period** $\tau_u = \kappa\tau$, namely:

$$T = N_{\text{pred}} \times \kappa \times \tau = N_{\text{pred}} \times \tau_u.$$

[4] $\lceil r \rceil$ denotes the smallest integer greater than r.

Here again, this simply means that from the point of view of the NMPC controller, the discrete-time system that is used in the formulation is defined by the *one-step-ahead* simulator that corresponds to a sampling period of $\tau_u := \kappa\tau$.

– **The control horizon's length** n_{contr}. This is the number of updating instants over the prediction horizon before the control is frozen to the last value. This is a common practice for the definition of the control parameterization.

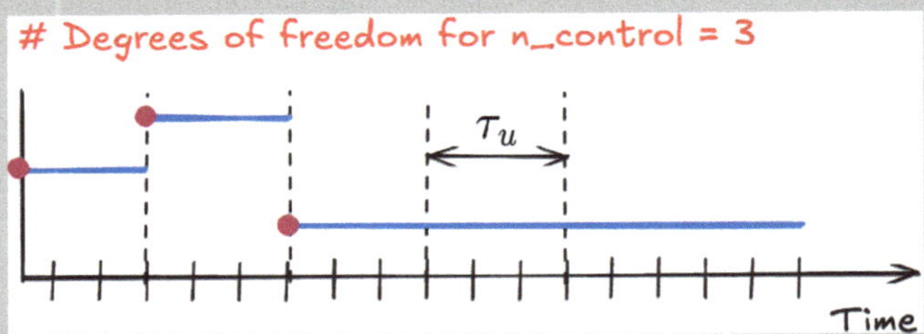

Illustration of the control horizon parameter `n_control` in the case where `n_control=3`. Three values are possible to choose before the open-loop trajectory is frozen.

Here again, this is not the only possible *control parameterization*. In particular, the affine parameterization already discussed in Chap. 2 (Sect. 2.6.3) could have been used also. Parameterization that are based on functional basis are also eligible.

Nevertheless, the one chosen here is simpler and easier to include in a research loop for possible instantiations since it is defined by a single integer called the control horizon, namely `n_control`. Moreover, it corresponds to the widely known concept of control horizon which is by definition the number of steps in the prediction horizon that correspond to free values before the control is frozen until the end of the prediction horizon.

– **The weighting penalties** $\rho_{\mathbf{f}}$ **and** $\rho_{\mathbf{constr}}$ used in the terminal penalty and the soft constraints respectively. Recall that the terminal penalty generally enhances the closed-loop stability while the soft constraint penalty enhances the fulfillment of the constraints. Theoretically, these penalties should be taken very high. it is shown through the illustrative example of this chapter that using high penalties induces stiffer optimization problem which implies smaller steps, being used in the optimization process, leading to bad performance under limited number of iterations.

This is precisely why a rational tuning framework is needed in order to choose the appropriate *cursor position* between the *sufficiently high* values and the *not-too-high* values.

– **The maximum number of iterations `max_iter`** used in the optimization process. Recall that in Chap. 5, it has been already shown that this parameter has a strong impact on the computation time on one hand, and that taking it too small might lead to quite bad performance and constraints satisfaction on the other hand. Here again, the need for fine tuning of this parameter is obvious.

To summarize:

The Vector of Design Parameter to Tune
The vector of NMPC design parameters to be tuned by the algorithm in order to satisfy the requirements regarding (1) Stability, real-time implementability and constraints satisfaction is defined by

$$\pi := \begin{bmatrix} \kappa \\ \mu_d \\ N_{\text{pred}} \\ n_{\text{contr}} \\ \rho_{\text{f}} \\ \rho_{\text{constr}} \\ \texttt{max\_iter} \end{bmatrix} \in \Pi := [\underline{\pi}, \bar{\pi}] \subset \mathbb{R}^{n_\pi}, \tag{6.4}$$

where, in the above statement, $\underline{\pi}$ and $\bar{\pi}$ are lower and upper bounds on the components of the design vector π that are to be given as inputs to the tuning algorithm.

This obviously corresponds to a quite **rich and non-convex** set of possibilities.

For each candidate setting π, the corresponding NMPC has to be evaluated **for the targeted implementation (device, algorithm)** over a high number of relevant scenarios needed for the certification.

Obviously, this cannot be done in an exhaustive way nor is it possible to consider an outer-loop optimizing π while an inner-loop performs a probabilistic certification using a high number of scenarios.

Some different heuristics should be derived which is explained in the remaining sections.

Consequently, the problem considered in the remainder of this chapter can be stated as follows:

Problem Statement
Given the above-defined input elements together with:

- an admissible domain

$$\Pi = [\underline{\pi}, \bar{\pi}],$$

of the design parameters,
- an implementation target.

Derive a *tractable heuristic* that:

–either yields a list of admissible rational choices of the vector of parameters π addressing **stability, constraints satisfaction and real-time implementability** concerns;

–or informs the user that these requirements cannot be met given the data of the problem.

Such a solution is proposed in the following section.

6.4 The Proposed Tuning Algorithm

As mentioned above, the set of parameters to be tuned is a non-convex set and the examination of the consequences of each choice inside this set on the stability, `cpu`-time and the constraints satisfaction is time consuming.

This is the reason why it is mandatory to define a parameterization of the elements of this set so that it can be spanned in an efficient way that is oriented towards the targeted objective.

Such a parameterization is proposed in the following section.

6.4.1 Parameterization of the Set of NMPC Design Parameters

In order to break the complexity of the search over the domain Π defined in (6.4), a first observation is worth making that can be stated as follows:

Two Types of Components
Each component π_i of π is of one of the two types shown in Fig. 6.2, namely, either the pairs (`cpu/quality`) are increasing functions of π_i (Type(+)) or they are decreasing functions of π_i.

Given the definition (6.4) of π, it can be easily checked that all the components are of type (+) except $\pi_1 = \kappa$ which is of type (–).

Nevertheless, this might not be the case if other implementation parameters are considered. This is why it is interesting to keep this categorization true in any possible extension of the framework (Table 6.1).

Notice however that for a given component, the shape of the monotonicity mentioned above for a given NMPC problem is difficult to know a priori.

Table 6.1 Types of the π vector's components

Components	Type
κ	(–)
ρ_f	(+)
μ_d	(+)
ρ_{constr}	(+)
N_{pred}	(+)
max_iter	(+)
n_{contr}	(+)

That is the reason why the shapes of the curves shown in Fig. 6.2 are determined hereafter via a set of random sampling of a **shaping parameter vector**[5] :

$$\sigma := [\sigma_1, \dots, \sigma_{n_\pi}] \in \mathcal{S}^{n_\pi},$$

where $n_\pi := \text{card}(\pi)(= 7$ in the current setting) while $\mathcal{S}$ is a set of allowed integers bounded by $\bar{\sigma}$ of the form:

$$\mathcal{S} := \{-\bar{\sigma}, \dots, -1, +1, \dots \bar{\sigma}\}.$$

For each sampled choice of the shaping parameter vector σ, the shape of the curve representing the monotonicity of π_i is defined via a function:

$$\phi_{\sigma_i}(\alpha) : [0, 1] \to [0, 1],$$

which depends on a scalar parameter $\alpha \in [0, 1]$:

$$\phi_{\sigma_i}(\alpha) := \begin{cases} \alpha^{\sigma_i} & \text{if } \sigma_i > 0 \quad \text{(positive curvature)}, \\ \alpha^{\frac{1}{|\sigma_i|}} & \text{if } \sigma_i < 0 \quad \text{(negative curvature)}, \end{cases} \tag{6.5}$$

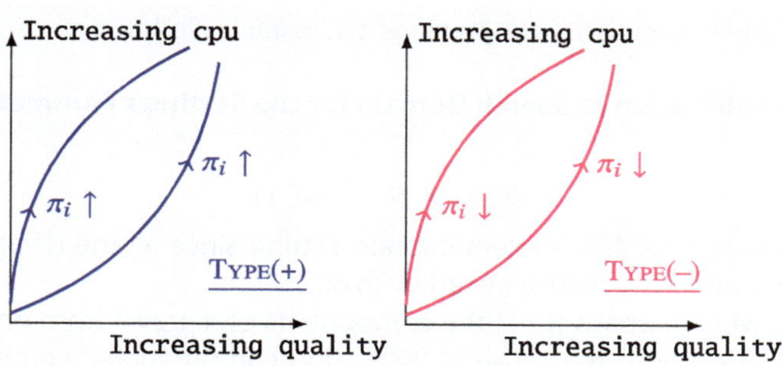

Fig. 6.2 Any component π_i of the design vector π is either of type(+) [left] or of type (–) [right]

[5] The appropriate shape may obviously not be the same for all the NMPC design parameters (components of π).

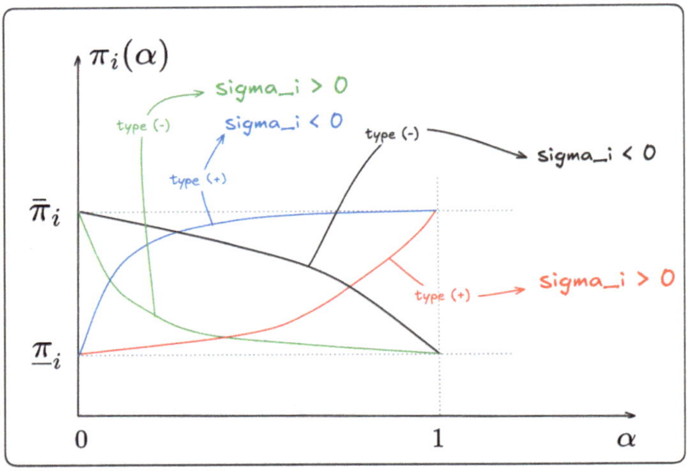

Fig. 6.3 Typical shapes of $\pi_i(\alpha)$ depending on the sign of σ_i and the type of the component π_i

and this curve is used to define the value of π_i according to:

$$\pi_i(\alpha) := \begin{cases} (1-\phi_{\sigma_i}(\alpha))\underline{\pi}_i + \phi_{\sigma_i}(\alpha)(\bar{\pi}_i - \underline{\pi}_i) \text{ Type (+)}, \\ (1-\phi_{\sigma_i}(\alpha))\bar{\pi}_i + \phi_{\sigma_i}(\alpha)(\underline{\pi}_i - \bar{\pi}_i) \text{ Type (–)}. \end{cases} \tag{6.6}$$

Consequently, the vector of shaping parameters σ is to be sampled in the set $\mathcal{S}^{n_\pi}$.

Figure 6.3 illustrates the above definitions for four different configurations of the pair (σ, type):

It is worth noticing that the parameterization defined by (6.6) is such that, regardless of the type of the component being considered, low values of $\alpha \in [0, 1]$ correspond to low computational complexity and low performance levels while high values induce high computation times and better performance levels **should the latter computations be possible within the updating period**.

It is possible to summarize the previous discussion as follows:

The Parametrization of Search Domain for the Setting's Parameter

Each pair

$$(\sigma, \alpha) \in \mathcal{S}^{n_\pi} \times [0, 1],$$

defines a specific NMPC implementation setting since all the design components of π defined by (6.4) are fixed by (6.6).

Moreover, when α spans $[0, 1]$ the components of π travel from one extreme value to the other in the sense of increasing computational complexity and ideally increasing quality should the concern regarding the real-time implementability be discarded.

Now that the domain of search is parameterized, the following section describes the heuristic of search for admissible solutions in which the parameters σ and α, previously discussed, play a key role.

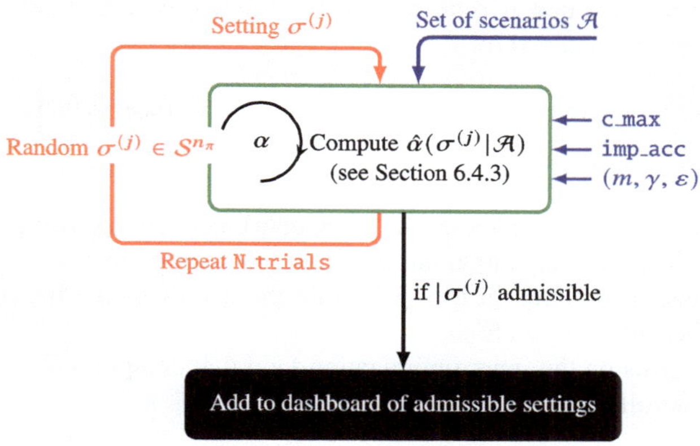

Fig. 6.4 Architecture of the two-layer algorithm for the determination of a set of admissible NMPC implementation setting pairs $(\sigma^{(j)}, \alpha^{(j)})$ for a given set of representative scenarios $\mathcal{A}$

6.4.2 The Ideal Computation Architecture at a Glance

The remaining task is to find a set of admissible pairs $(\sigma, \alpha) \in \mathcal{S}^{n_\pi} \times [0, 1]$ that are compatible with the real-time implementability, the constraints satisfaction and the closed-loop stability.

Obviously, the two components of the design, namely σ and α are to be handled differently. Indeed, it seems reasonably easy to select the scalar α for a given σ as the underlying *indicators* vary monotonically in α.

On the contrary, it is unrealistic to attempt a complete and rigorous optimization of $\sigma \in \mathcal{S}^{n_\pi}$ because of the highly combinatorics and the involved computational burden for each candidate value.

That is the reason why, the two-layer architecture shown in Fig. 6.4 is adopted in which `N_trials` values of the shaping parameter vectors:

$$\left\{\sigma^{(j)} \in \mathcal{S}^{n_\pi}\right\}_{j=1}^{\texttt{N\_trials}},$$

are sampled in $\mathcal{S}^{n_\pi}$ and for each sampled $\sigma^{(j)}$, a scalar constrained optimization on $\alpha \in [0, 1]$ is performed to check whether there is at least one value that meets the requirements (over a predefined set $\mathcal{A}$ of scenarios as explained later on) and if required, to select the *optimal* one,[6] denoted by $\hat{\alpha}(\sigma^{(j)}|\mathcal{A})$.

These requirements and the associated scalar optimization problem are explained in Sect. 6.4.3.

[5] Recall that this is not necessarily what one is looking for when the aim is to derive a list of admissible settings as discussed in the introduction to this chapter.

If a specific cost function $\mathcal{J}$ is the main focus, the *best* NMPC design setting pair $(\sigma^\star, \alpha^\star)$ is therefore obtained by

$$\sigma^\star \leftarrow \arg\min\Big\{\mathcal{J}(\sigma^{(j)}|\mathcal{A}) \quad j \in \text{ admissible dashboard}\Big\}, \tag{6.7}$$

$$\alpha^\star \leftarrow \hat{\alpha}(\sigma^\star|\mathcal{A}). \tag{6.8}$$

Otherwise, the designer can pick up the more convenient setting from the returned list of feasible settings using any multi-criterion selection rule.

This principle is depicted in Fig. 6.4 while the details of the boxes content is explained in the following sections.[7]

Let us first focus on the inner optimization loop that computes $\hat{\alpha}(\sigma^{(j)}|\mathcal{A})$. This is the aim of the following section.

6.4.3 The Admissibility Criteria: Computing $\hat{\alpha}(\sigma\,|\,\mathcal{A})$

For a candidate value of the shaping vector σ, admissible values of α are those that make the NMPC setting defined by (σ, α) compatible with the requirements regarding the real-time implementation, the constraints satisfaction and the stability.

More precisely, **Considering** m **updating steps** over a closed-loop simulation scenario:

$$\texttt{sc} \in \mathcal{A},$$

corresponding to a given initial state, a given model's vector of parameters p and an exogenous task-related parameter vector q (see Sect. 6.3), there are three concerns when it comes to evaluate the successful (or not) use of the NMPC setting on this specific scenario, namely:

(1) The Real-Time Feasibility Constraint

> **Real-Time Feasibility Constraint**
> This is represented by the following constraint:
>
> $$\text{RT-feasibility} \tag{6.9}$$
>
> $$C_{RT}(\alpha, \sigma|\texttt{sc}) := \max_{k=1,\dots,m}\left\lfloor \frac{\tau_{\text{solver}}}{\tau_u}(k) - 1\right\rfloor_+ = 0,$$
>
> where $\lfloor \xi \rfloor_+ := \max(0, \xi)$,

where $\tau_{\text{solver}}(k)$ denotes the time needed to solve the underlying k-th optimization problem encountered in the scenario and given the chosen maximum number of iterations $\pi_7(\alpha) = \phi_{\sigma_7}(\alpha)$ corresponding to the currently evaluated σ.

[7] In particular, the significance of m, γ, ε, `imp_acc` and `c_max` is given in Sect. 6.4.3.

Indeed, if the above expression is equal to 0, this means that for all k, one has $\tau_{\text{solver}}(k) \leq \tau_u$.

> **Using a Different Computation Device**
> Notice that be replacing τ_u in (6.9) by
>
> $$\texttt{imp\_acc} \times \tau_u,$$
>
> real-time implementatbility can be checked for a targeted implementation that is (certifiably) `imp_acc` faster than the one on which the evaluation is done (or `imp_acc` times slower in the case where `imp_acc` < 1).

That is the reason why the parameter `imp_acc` is provided to the box that determines the admissibility in Fig. 6.4 in addition to other information that are described in the next items.

(2) The Contraction Constraints
This constraint implicitly expresses that the MPC formulation is such that a decrease of the cost function is expected over the closed-loop trajectory although not necessarily at each step.

This contraction property is common to almost all *provably* stable NMPC formulations as it simply means that there is a multi-step Lyapunov function that expresses the stability of the closed-loop behavior.

Therefore, using m-step contraction horizon, the associated requirement writes:

The Contraction Constraints (Stability)
This is represented by the following constraint:

$$\gamma\text{-Contraction} \qquad C_\gamma(\alpha, \sigma\,|\,\texttt{sc}) := \lfloor J_{ol}(m) - \gamma J_{ol}(1) \rfloor_+ = 0, \tag{6.10}$$

where $J_{ol}(k)$ is the *best open-loop cost value* returned by the solver at the k-th updating computation[8] while $\gamma \in (0, 1)$ is some predefined contraction rate.

Notice that the contraction rate γ is also added to the list of parameters that are delivered to the box in Fig. 6.4 in order to decide whether the corresponding implementation setting is admissible or not.

(3) The constraints satisfaction condition which applies mainly to soft exact penalty constraints since it is reasonably assumed that the input hard constraints are structurally enforced by the optimization algorithm, for instance, via projection on the admissible hypercube.

This can be stated as follows:

[8] Given the allowed maximum number of iterations corresponding to the design parameter $\pi_7 =$ `max_iter` [see (6.4)].

Constraints Satisfaction Condition
This is represented by the following constraint:

$$\text{Constraints satisfaction} \qquad C_{\mathsf{cstr}}(\alpha, \sigma | \mathtt{sc}) := \max_{k=1,\dots,m} \lfloor \max_{i \le n_c} c_i(k) \rfloor_+ = 0. \tag{6.11}$$

Notice that the above formulation requires that the constraints are perfectly satisfied over the m steps. As these are soft constraints, it is also possible to accept some level of constraints violation. This can be expressed by replacing the last condition by the following less stringent one

$$C_{\mathrm{cstr}}(\alpha, \sigma | \mathtt{sc}) := \max_{k=1,\dots,m} \lfloor \max_{i \le n_c} c_i(k) \rfloor_+ \le \mathtt{c\_max},$$

for some threshold c_max≥ 0. This is the reason why the parameter c_max is added to the list of exogenous parameters to be delivered to the box of Fig. 6.4.

Notice that the above-mentioned *success conditions*, namely (6.9), (6.10) and (6.11), concern a specific scenario sc. But obviously, the overall assessment of a candidate NMPC setting pair:

$$(\sigma, \alpha) \in \mathcal{S}^{n_\pi} \times [0, 1],$$

has to hold *whatever* is the scenario or at least, success has to be certified with a sufficiently high probability of a population of possible scenarios.

This is the reason why a set $\mathcal{A}$ of *representative scenarios* is considered. As it is discussed later, the cardinality of this set can be determined following the probabilistic certification formulas [17].

For the remainder of the presentation, it is hence assumed that one disposes of a generator ofthe form:

$$\mathcal{A} \leftarrow \mathtt{Generate\_A}(n_{\mathrm{sc}}),$$

that provides n_{sc} representative scenarios.

The set of scenarios $\mathcal{A}$ is used to compute the *best* value $\hat{\alpha}(\sigma|\mathcal{A})$ **for a given** σ by solving the following constrained scalar optimization problem:

The α-Optimization Problem
For a given σ, the optimal α is the maximum one that meets the three above-defined constraints, namely:

$$\hat{\alpha}(\sigma|\mathcal{A}) \leftarrow \max_{\alpha \in [0,1]} [\alpha], \tag{6.12a}$$

$$\text{under} \left| \begin{array}{l} \max_{\mathtt{sc} \in \mathcal{A}} C_{RT}(\alpha, \sigma | \mathtt{sc}) = 0, \\ \max_{\mathtt{sc} \in \mathcal{A}} C_{\gamma}(\alpha, \sigma | \mathtt{sc}) = 0, \\ \max_{\mathtt{sc} \in \mathcal{A}} C_{\mathsf{cstr}}(\alpha, \sigma | \mathtt{sc}) \le \mathtt{c\_max}, \end{array} \right. \tag{6.12b}$$

where again, `c_max` is a threshold on the level of possible violation of the soft constraints.

Although one can view (6.12) as a general constrained optimization problem, the specificity of the problem enables a simple method to derive quite good sub-optimal solutions with less computational effort than by using general purpose mixed-integer nonlinear programming methods.

This is even mandatory given the computation effort needed to evaluate the terms involved in (6.12) for any candidate value of α (since all the scenarios included in $\mathcal{A}$ are involved).

The specific process which is mainly based on a binary search is described in details by the algorithm detailed in Fig. 6.5.

This algorithm delivers for each candidate shaping parameter vector σ and for a given set of scenarios $\mathcal{A}$, a success-related boolean and in the case of success an associated $\hat{\alpha}(\sigma|\mathcal{A})$ together with the associated optimal cost (if required) denoted by $\mathcal{J}(\sigma|\mathcal{A})$.

As it can be noticed by examining the steps of the algorithm, the algorithm starts by examining the real-time implementability by analyzing the associated condition for the value $\alpha = 0$. If feasibility condition is not satisfied for this value, then there is obviously no need to perform binary search for the value of α. In case the real-time feasibility is validated, Step 2 is executed to deliver the highest value of α for which feasibility is granted. Finally, the last step checks for those values of α that lead to the satisfaction of the two remaining constraints.

Unfortunately, this process is still computationally very expensive as explained in the next section where an alternative sub-optimal but far more tractable alternative is proposed.

6.4.4 A Sub-optimal Tractable Algorithm

The previous formulation requires a quite large amount of computation should the certification level be high.[9] Indeed for each σ, a binary search have to be performed in order to compute the optimal $\hat{\alpha}(\sigma|\mathcal{A})$ (if any) in which the expressions involved in the constraints (6.12b) has to be evaluated through m-steps closed-loop simulations and this, **for all the scenarios included in the high cardinality set** $\mathcal{A}$ (see Table 2.1).

Assuming a precision of ε on α, this induces a *worst case* number of OCPs that compares to:

$$\texttt{N\_trials} \times \text{card}(\mathcal{A}) \times m \times \log(1/\varepsilon). \tag{6.13}$$

Note that card($\mathcal{A}$) is determined by the precision ($\eta \in (0, 1)$) and the confidence ($\delta \in (0, 1)$) parameters required for the certification of the success when using the corresponding setting [17] as it is already recalled in the section dedicated to the probabilistic certification of Chap. 2 [see Sect. 2.11].

[9] Leading to large number of computed scenarios represented by the cardinality of the set $\mathcal{A}$.

Computing $\hat{\alpha}(\sigma|\mathcal{A})$

1) Start with $\alpha = 0$.
   If (6.12) is not feasible because of the RT constraint, namely if

$$\max_{\texttt{sc}} C_{RT}(0, \sigma|\texttt{sc}) > 0,$$

   return a non feasibility flag for σ, namely:

$$\hat{\alpha}(\sigma|\mathcal{A}) \leftarrow \texttt{None},$$

   and stop.
   Otherwise proceed into Step 2)

2) Evaluate RT-feasibility for $\alpha = 1$.
   If the problem is feasible then return

$$\hat{\alpha}(\sigma|\mathcal{A}) = 1,$$

   Otherwise, perform a **binary** search, starting with 0 and 1 as initial extreme values, in order to determine the largest value $\alpha_{\max}(\sigma)$ for which RT-feasibility holds.

3) **If** at least one of the remaining constraints is violated at $\alpha = \alpha_{\max}(\sigma)$, namely

$$\max_{\texttt{sc}} C_{RT}(0, \sigma|\texttt{sc}) > 0,$$

   or

$$C_{\text{cstr}}(\alpha, \sigma|\texttt{sc}) > \texttt{c\_max},$$

   return a non feasibility, namely:

$$\hat{\alpha}(\sigma|\mathcal{A}) \leftarrow \texttt{None},$$

   and stop.

   Otherwise, returns

$$\hat{\alpha}(\sigma) = \alpha_{\max}(\sigma) \quad \text{and} \quad \mathcal{J}(\sigma|\mathcal{A}).$$

   The sum of closed-loop costs corresponding to the use of $\alpha_{\max}(\sigma)$ in the scenarios contained in $\mathcal{A}$.

Fig. 6.5 The algorithm exploiting the specificity of the constrained scalar optimization problem (6.12) to compute the optimal $\hat{\alpha}(\sigma|\mathcal{A})$ for a given candidate value of the shaping vector σ and a set of scenarios $\mathcal{A}$

For the sake of easiness of reading the corresponding table showing the number of samples to be drawn in order to get the probabilistic certification is recalled on Table 6.2 for the confidence parameter $\delta = 10^{-3}$.

That is the reason why a sub-optimal version of the formulation is adopted in which, the process is split into the following two sub-processes:

1. First, a randomly generated set $\mathcal{A}_0$ containing a reduced number of scenarios, namely:

$$n_0 := \text{card}(\mathcal{A}_0) \ll \text{card}(\mathcal{A}),$$

Table 6.2 The required card($\mathcal{A}$) as a function of the precision parameter η for a confidence parameter of $\delta = 10^{-3}$ and a number of admitted failure = 1

N_trials	$\eta = 0.1$	$\eta = 0.05$	$\eta = 0.01$	$\eta = 0.001$
1	132	264	1317	13164
5	154	308	1536	15354
10	163	326	1628	16280
100	193	386	1930	19299
1000	223	445	2225	22249

is considered for which the previously described optimal formulation is applied in order to determine $\hat{\alpha}(\sigma^{(j)}|\mathcal{A}_0)$ for $j = 1, \dots, \texttt{N\_trials}$.

2. **The so-obtained** $\hat{\alpha}(\sigma^{(j)}|\mathcal{A}_0)$ **are now frozen** and the $\sigma^{(j)}$ are individually checked (which means that we can use the first line of Table 2.1) over a sequence of subsets $\mathcal{A}^{[\ell]}$ that forms a partition of the original set $\mathcal{A}$, namely:

$$\mathcal{A} := \bigcup_{\ell=1}^{n_s} \mathcal{A}^{[\ell]},$$

and in the case where cost-oriented choice is needed, the associated optimal costs are summed up over the subsets $\mathcal{A}^{[\ell]}$ in order to ultimately form the corresponding cost $\mathcal{J}(\sigma^{[j]}\,|\mathcal{A})$.

During this process, as soon as a failure is detected for some $\sigma^{[j]}$ for one of the scenarios of a subset $\mathcal{A}^{[\ell]}$, this $\sigma^{[j]}$ is removed from the set of candidate values and is no more considered for the remaining subsets of scenarios $\mathcal{A}^{[\ell+\cdot]}$.

This solution strongly reduces the number of evaluations since binary search is restricted to $\mathcal{A}_0$ on one hand and the number uselessly visited evaluations is drastically reduced by progressively removing unsuccessful $\sigma^{[j]}$ after failure on intermediate low cardinality subset of scenarios $\mathcal{A}^{[\ell]}$ on the other hand.

Obviously the counter-part of this simplification is that the frozen values of $\hat{\alpha}(\sigma^{(j)})$ do not take into account all the scenarios of $\mathcal{A}$ and might hence be wrongly biased by the small number n_0 of initial scenarios contained in $\mathcal{A}_0$.

While the above discussion totally describes the methodology, the effective implementation is slightly different in the sense that the complete set of scenarios $\mathcal{A}$ is decomposed into n_batch batches of subsets:

$$\{\mathcal{A}_\ell\}_{\ell=1}^{\texttt{n\_batch}},$$

each of which showing a cardinality n_sc_batch= card($\mathcal{A}_\ell$).

For each current set $\mathcal{A}_\ell$, the problem $\mathcal{P}(\sigma, \mathcal{A}_\ell)$ is solved and the feasibility indicator enables to remove those parameters $\sigma^{(j)}$ that leads to unfeasible problems avoiding thus to examine them for the remaining set of scenarios to come.

By so doing, the set of indices $\mathcal{I}_{\texttt{sig}}$ For the remaining feasible settings, initially set to $\{1, \dots, \texttt{N\_trials}\}$ is updated by progressively removing the unfeasible indices.

For the feasible ones and in the case where a cost-oriented selection is to be done, the sum of the corresponding closed-loop costs are kept so that once, the total subsets of scenarios are examined, the best configuration $\sigma^\star$ is determined and the associated $\alpha^\star$ is recomputed for the total set of scenario $\mathcal{A}$ as a last step.

This achieves the presentation of the design algorithm.

In the next section, the proposed `Python`-based package is briefly described.

6.5 Brief Description of the `Python`-Package

In this section, a brief description of the `Python` package `mpc_tuner` is provided while a complete description can be found at the following `gitHub` repository:

https://github.com/mazenalamir/MPC_tuner.

Recall that, in order to use the package, the user needs to provide an instance `pb` of the `Container` class with the attributes and the methods shown in Fig. 6.6.

This object `pb` gathers the problem-dependent items and is used in the package's utilities.

The package contains the following main classes and their main *exposed* methods (Fig. 6.6):

<u>`Class: Sigma`:</u>

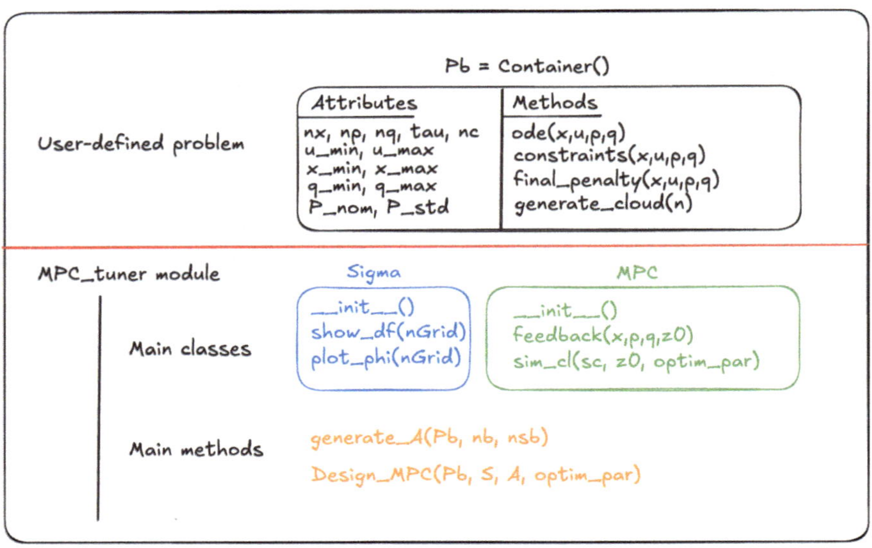

Fig. 6.6 Main classes and methods exported by the `MPC_tuner` package and the user-defined object to prepare for the a specific control problem

An instance of this class corresponds to a specific value of the σ vector. This class exports the maps $\kappa(\alpha)$, N_pred(α), ..., etc., which are the maps $\pi_i(\alpha)$ described in (6.6).

<u>Class: MPC</u>:

An instance of this class is a specific MPC setting defined by the triplet (pb, σ, α). The main methods of this class are

```
MPC.feedback(x,p,q,z0),
MPC.sim_cl(sc, z0, optim_par),
```

where the feedback function delivers the feedback input vector while sim_cl simulates the closed-loop associated to the scenario with initial guess z_0 at the initial instant followed by warm start initialization.

The computation of the success condition uses the object:

optim_par:=(γ, ε, imp_acc, c_max),

as shown in Fig. 6.4.

Notice that in the current version of the package, the optimization uses the framework CasADi [18] with the solver option IPOPT [19] in single-shooting mode.

Beside the above classes, the package contains the following main global methods:

<u>Function: Generate_A(pb,nb,nsb)</u>:

This function generates a list of nb subsets $\mathcal{A}^{[\ell]}$, $\ell = 1, \dots,$ nb of cardinality nsb each, leading to a total set of scenarios $\mathcal{A} := \bigcup_{\ell=1}^{\mathtt{nb}} \mathcal{A}^{[\ell]}$ of cardinality card($\mathcal{A}$) = nb $\times$ nsb.

The first of these sets plays the role of the initial set $\mathcal{A}_0 := \mathcal{A}^{[1]}$ invoked above and used to determine the values of $\hat{\alpha}(\sigma^{(j)}|\mathcal{A}_0)$ to be certified using the remaining subsets $\mathcal{A}^{[\ell]}$, for $\ell = 2, \dots,$ nb.

The function Generate_A(pb,nb,nsb) calls the method pb.generate_cloud of the user-defined object pb.

<u>Function: Design_MPC(...)</u>:

This is the main function that enables to look for a list of admissible NMPC settings following the algorithm described in Sect. 6.4.4.

The call of this main function admits the following input arguments:

1. the user-defined object pb;
2. a list S of possible values of the shaping parameter σ generated by successive calls (N_trials) of instance generation of the class Sigma described above;
3. the list of sets $\mathcal{A} := \bigcup_{\ell=1}^{n_s} \mathcal{A}^{[\ell]}$; and
4. the object optim_par described above.

This function returns a `pandas` data frame representing the set of admissible NMPC implementation settings defined by (σ, α) with their associated cumulated closed-loop costs.

6.6 An Illustrative Example

In order to illustrate the framework and the use of the associated `Python`-based package `MPC_tuner`, let us consider the control of the previously described PVTOL system that obeys the following normalized dynamics:

$$\ddot{y} = -u_1 \sin\theta + p_1 u_2 \cos\theta, \tag{6.14a}$$

$$\ddot{z} = u_1 \cos\theta + p_1 u_2 \sin\theta - 1, \tag{6.14b}$$

$$\ddot{\theta} = p_2 u_2, \tag{6.14c}$$

which involves a six-dimensional state vector given by

$$x := (y, z, \theta, \dot{y}, \dot{z}, \dot{\theta}),$$

and a control vector of dimension $n_u = 2$. The model also depends on two-dimensional vector of parameters p ($n_p = 2$).

The control objective is to regulate the aircraft around reference values of y and z that are denoted hereafter by q_1 and q_2 respectively. These are the first two components of the context vector q mentioned in the previous sections as in the expressions of the stage cost (6.1) as well as the terminal penalty term (6.2).

This leads to the desired targeted state:

$$x_d := (q_1, q_2, 0, \ldots, 0),$$

corresponding to the steady control:

$$u_d := (1, 0).$$

Therefore the economic stage cost is given by

$$\ell(x, u, p, q) := \|x - x_d\|_Q^2 + \|u - u_d\|_R^2,$$

with predefined Q and R weighting matrices which are not to be tuned. In the numerical investigation, the following weighting matrices are used:

$$Q = \text{diag}(10^3, 10^3, 10^3, 1, 1, 1),$$

and

$$R = \text{diag}(0.1, 0.1).$$

On the contrary, the terminal penalty function defined by

$$\Psi(x, p, q) := \rho_{\text{f}}\|x - x_d\|_Q,$$

does involve the tunable parameter $\pi_5 := \rho_{\text{f}}$ [see (6.4)]. All the scenarios considered hereafter last a duration of 0.5 normalized time units which induces different number of updating periods and hence different number of repeated optimization instances depending on the values of the parameters that induce different values of τ_u. The contraction rate $\gamma = 0.98$ is used. The precision on α in the binary search is fixed to:

`optim_par.eps`$= 0.15.$

As for the constraints, beside the input saturation defined by

$$u \in [-50, +50]^2,$$

the following two constraints are imposed:

$$|\dot{\theta}| \le q_3,$$

and

$$|\theta| \le q_4,$$

which defines two other components of the context vector q (hence $n_q = 4$).

The following parameter determining the admissible level of constraints violation is used:

$$c_{max} = 0.1,$$

is used on the **normalized** constraints.

This defines the constraints map $c(x, u, p, q)$ that encodes $n_c = 4$ constraints.

Figure 6.7 shows the `Python` script that defines the necessary elements to be used by the `MPC_tuner` in order to perform the search for a set of admissible NMPC implementation settings.

Regarding the script of Fig. 6.7, the following comments are worth giving in order to highlight some important features, namely:

✓ The `Container` class is simply a void class that accepts new attributes to be added on the flow. Notice that the object `pvtol` plays the role of the object `pb` invoked in the previous discussion.

✓ Notice the systematic use of the `CasaDi`'s `vertcat` command that is useful to define symbolic vectors. This is mandatory since the `CasaDi` framework is used here to solve the optimization problem.

✓ The definition of the bounds on q enables different set-points to be explored inside $[-1, +1]^2$. However, q_3 and q_4 are taken constant here meaning that the bounds on θ and $\dot{\theta}$ show no reason to be modified among different instantiations of this problem.

```
import numpy as np
from casadi import vertcat

class Container:
    def __init__(self):
        pass

pvtol = Container()
pvtol.nx, pvtol.nu = 6, 2
pvtol.n_p, pvtol.n_q = 2, 4
pvtol.u_min = np.array([-5e1, -5e1])
pvtol.u_max = np.array([+5e1, +5e1])
pvtol.x_min = np.array([-2, -2, -0.8 * np.pi, \
                        -0.1, -0.1, -0.1])
pvtol.x_max = -pvtol.x_min
pvtol.q_min = [-1, -1, 1, np.pi]
pvtol.q_max = [+1, +1, 1, np.pi]
pvtol.p_nom = pvtol.p
pvtol.p_std, pvtol.tau, pvtol.nc = 0.1, 0.02, 4
#------
def p_vtol(x, u, p):
    xdot = vertcat(...)
    return xdot
#------
def constraints(x, u, p, q):
    d_theta_dt_max, theta_max = q[2], q[3]
    c = vertcat(x[5]/d_theta_dt_max-1,...)
    return c
#------
def stage_cost(x, u, p, q):
    ...
    return ell
#------
def final_penalty(x, p, q):
    ...
    ef = 0
    for i in range(6):
        ef += q_xf[i] * (x[i] - xd[i]) ** 2
    return ef
#------
def generate_cloud(nSamples=None):
    x_min, x_max = pvtol.x_min, pvtol.x_max
    ...
    A_sc = Container()
    A_sc.x0, A_sc.p, A_sc.q = X, P, Q
    return A_sc
#------
pvtol.ode = p_vtol
pvtol.constraints = constraints
pvtol.final_penalty = final_penalty
pvtol.stage_cost = stage_cost
pvtol.generate_cloud = generate_cloud
```

Fig. 6.7 The user-defined Python file `user_defined_pvtol.py` that creates the instance `pvtol` of the `Container` class. This instance is imported by the main script file that performs the search for the admissible NMPC implementation settings

✓ Notice that this script defines the nominal value p_nom of the vector of parameters together with its vector of standard deviations p_std. This information, together with the bounds on x and q defined in the beginning of the script is used in the last function (generate_cloud) in order to generate a representative set of scenarios. For instance A_sc.x0[i] and A_sc.p[i] provides the initial state x_0 and the vector of parameters p for the scenario number i.

The problem-dependent object defined in the user-defined script, namely pvtol is imported in the main script shown below:

```
import numpy as np
from MPC_tuner import Design_MPC, Sigma, generate_A, OptimPar
from user_defined_pvtol import pvtol

# Generate the set of scenarios
#---------------------------------------------------

nb, nsb = 30, 10
A = generate_A(pvtol, nb, nsb)

# Generate the set of candidate sigma's
#---------------------------------------------------
# replace by N_trial=100 to conform to the paper

N_trials = 100
S = [Sigma() for _ in range(N_trials)]

# Set the Design meta-parameters and run the Design
#---------------------------------------------------
optim_par = OptimPar(gam=0.98, c_max=0.1,
                     imp_acc=1.0, T=0.5)

R_design_log = Design_MPC(pvtol, S, A, optim_par)
```

This script starts by using the function generate_A in order to create a set $\mathcal{A}$ of scenarios with cardinality 300 that is decomposed into nb = 30 batches of nsb = 10 scenarios each. Then the script creates a set of N_trials = 100 candidate shaping parameter vectors $\sigma^{[j]}$, $j = 1, \dots,$ N_trials.

Notice that the chosen values of nb and nsb induce an initial set of scenarios $\mathcal{A}_0$ with cardinality $n_0 = 10$ and a number of certification scenarios equal to 290 which, according to the first line of Table 2.1, is sufficient to certify the successful design for a precision level of $\eta = 0.05$ and a confidence level of $\delta = 10^{-3}$.

6.6.1 Results and Discussion

Table 6.3 shows the default bounds used in the random sampling of the shaping parameter σ when using the instantiation call `Sigma()`.

Figure 6.8 shows the tuning results for two different targeted hardware devices for the implementations corresponding to `imp_acc=1` (top) and `imp_acc=2` (bottom) for the same randomly generated set of scenarios $\mathcal{A}$ and candidate set of parameters $\sigma^{(j)}$, $j = 1, \dots,$ `N_trials`.

For each case, two results are shown, namely, the data frame showing the list of admissible settings together with their individual parameters as well as the cumulative closed-loop cost over the scenarios contained in the set $\mathcal{A}$ of 300 scenarios.

Regarding the results, the following observations are worth making:

Only a small portion of configurations is eligible:

Despite the quite reasonable bounds given in Table 6.3, it is remarkable that only a small portion of the randomly sampled settings are admissible, namely:

- 5% for a computation target given by `imp_acc=1`;
- 13% for a computation target given by `imp_acc=2`.

This simple fact suggests that the problem addressed in this chapter is quite relevant.

High values of ρ_f generally lead to non-feasibility:

The results suggest that high values of ρ_f lead to inadmissible settings and/or high values of the cost. Recall that the reported closed-loop costs do not incorporate the terminal penalty as the latter is generally used for stability reasons.

However, this cost includes the possible non-vanishing terms coming from the constraint being violated by less than the authorized threshold $c_{max} \neq 0$.

The results suggest that too high terminal penalty might lead to bad performances when the number of iterations is limited. This is intuitively sound because in this case, the problem is stiffer and the resulting step size, used in the optimization process, is consequently small.

Table 6.3 Bounds used on the design parameters

Design parameter	Min-value	Max-value
`N_pred`	5	25
κ	1	10
ρ_f	1	10^3
`rho_cstr`	10^3	10^7
`max_iter`	5	20

imp_acc = 1

κ	N_pred	n_steps	n_ctr	ρ_{ctr}	ρ_{f}	τ	τ_{u}	max_iter	cost	α
9	7	3	8	9.6×10^6	178.9	0.02	0.18	11	78.6×10^2	0.750
4	6	2	4	9.1×10^6	24.2	0.02	0.08	16	44.0×10^2	0.625
2	5	2	2	7.9×10^6	1.24	0.02	0.04	8	24.3×10^2	0.250
8	6	5	2	1.2×10^6	2.95	0.02	0.16	18	34.0×10^2	0.500
9	6	7	7	9.1×10^6	8.80	0.02	0.18	6	37.6×10^2	0.500

imp_acc = 2

κ	N_pred	n_steps	n_ctr	ρ_{ctr}	ρ_{f}	τ	τ_{u}	max_iter	cost	α
7	17	4	12	8.4×10^6	16.6	0.02	0.14	5	32.1×10^2	0.250
9	7	3	8	9.6×10^6	178.8	0.02	0.18	11	78.6×10^2	0.750
9	20	2	19	8.4×10^6	820.5	0.02	0.18	8	37.7×10^2	0.250
6	17	3	16	6.1×10^6	3.8	0.02	0.12	5	32.9×10^2	0.375
4	6	2	4	9.1×10^6	24.2	0.02	0.08	16	44.0×10^2	0.625
9	5	8	4	6.3×10^5	3.0	0.02	0.18	18	31.3×10^2	0.500
8	12	2	11	7.6×10^4	1.1	0.02	0.16	17	28.9×10^2	0.375
7	9	5	2	2.4×10^6	15.5	0.02	0.14	19	35.9×10^2	0.625
9	5	8	6	9.3×10^6	38.2	0.02	0.18	18	46.1×10^2	0.625
2	12	2	2	8.8×10^6	1.1	0.02	0.04	7	34.1×10^2	0.375
6	6	6	2	5.2×10^5	1.1	0.02	0.12	18	33.5×10^2	0.375
9	18	2	13	3.5×10^6	125.9	0.02	0.18	12	32.5×10^2	0.125
6	19	2	2	7.4×10^6	3.0	0.02	0.12	14	33.6×10^2	0.125

Fig. 6.8 Two instances of results given by the function Design_MPC for different values of the imp_acc input for the algorithm, namely, imp_acc = 1 (upper items) and imp_acc=2 (lower items). In both cases, the configurations are sampled inside the domains defined by the bounds shown on Table 6.3

High values of rho_cstr are needed:

On the contrary, the high majority of admissible settings correspond to high values of soft constraints penalty rho_cstr. More precisely, these values seem to lie exclusively in a region that is close to the upper bound of the admissible interval $[10^3, 10^7]$.

This is intuitively quite tricky to guess and enforces, if still needed, the relevance of the problem addressed in this chapter.

This also suggests that it might be interesting to re-run the algorithm with the bounds of rho_cstr shifted towards higher values, for instance $[10^5, 10^9]$ following the conjecture according to which, given the random sampling, too many sampled settings fail in meeting the constraints with low values of this parameter which penalizes a rather important subset of the randomly generated set of configurations.

Nevertheless, without the first round of search for admissible configuration, the region of appropriate values for rho_cstr would have been not easy to guess.

The scenarios are constraints-challenging:

The previous fact also suggests that the set of scenarios used in the certification does involve constraint-violation-risky initial conditions that have been managed using high penalty on the soft constraints, otherwise, the random sampling would have provided successful settings with lower values of ρ_{constr} since the latter would have had no effect on the success/failure status.

Large eligible possibilities for the prediction horizon:

Notice that among the set of admissible implementation settings, the set of successful prediction horizon lengths spans values from 0.2 up to 1.26 when `imp_acc`= 1 and from 0.48 to 3.24 when `imp_acc`=2.

This can be explained by the definition of $\hat{\alpha}(\sigma\,|\mathcal{A})$ being the maximum allowable value since this definition enhance longer prediction horizons `N_pred` × `tau_u`.

The final choice of the NMPC design might favor not too short prediction horizons for obvious reasons even if this corresponds to slightly higher closed-loop cost since the connection between the truly obtained closed-loop performances and the ones predicted on the short term simulation used in the algorithm is not so strong as it is widely accepted by NMPC practitioners. Longer prediction horizon enable a long sight regarding the choices.

On the other hand, if the uncertainties in the model expressions are high, longer prediction horizon might lead to irrelevant *planning* with consequences that might not be as good as expected.

Obviously, the higher the cardinality of the set of candidates σ is, the more exhaustive is the resulting list of admissible configurations. This is a matter of computation time.

The computation times for the design examples lie around 1 h each on a

```
MacBook Pro, 2.4 GHz Intel Core i9.
```

Keep in mind however that the computation time depends on the randomly sampled design settings.

6.7 Conclusion

In this chapter, a systematic approach and an associated freely available `Python`-package (`MPC_tuner`) are discussed for the generation of a set of admissible NMPC implementation settings for a specific computational target.

Despite encouraging preliminary results, it might be conjectured that additional/different tricks might be used in order to accelerate the computation and or to reduce the level of sub-optimality.

One option would be to use machine learning tools to derive preliminary *feasibility predictors* that can be trained over a cloud of (σ, α) without the binary search used in a first step which greatly impacts the computation time, the resulting model can then be used to make better guesses reducing hence the number of useless randomly generated samples.

On the other hand, future inclusion of solvers in the `CasaDi` framework can immediately be included in the search instead of the only `IPOPT` that is currently used by default.

By the time this book is written, this was not possible to do using the `ACADOS` solver because the later does not *give the control* on some of the design parameters such as the control horizon parameter `n_contr` and, especially, the under-sampling parameter `n_steps` (or equivalently μ_d) which seem to play an important role managing the feasibility of the implementation settings as it can be observed on the results of Fig. 6.8 where the successful settings never corresponds to `n_steps=1` or `n_steps` equals to the prediction horizon's length.

For the time being, recall that any other implementation conditions (device, algorithm) can be examined through the current `mpc_tuner` though the parameter `imp_acc`.

References

1. Rawlings, J.B., Mayne, D.Q., Diehl, M.: Model predictive control: Theory, computation, and design, vol. 2. Nob Hill Publishing Madison, WI (2017)
2. Biegler, L.T., Zavala, V.M.: Large-scale nonlinear programming using ipopt: An integrating framework for enterprise-wide dynamic optimization. Comput. Chem. Eng. **33**(3), 575–582 (2009)
3. Houska, B., Ferreau, H.J., Diehl, M.: Acado toolkit: An open-source framework for automatic control and dynamic optimization. Optim. Control Appl. Methods **32**(3), 298–312 (2011)
4. Hart, W.E., Laird, C.D., Watson, J.P., Woodruff, D.L., Hackebeil, G.A., Nicholson, B.L., Siirola, J.D.: Pyomo—Optimization Modeling in Python, 2 edn. Springer Optimization and Its Applications. Springer (2017). https://doi.org/10.1007/978-3-319-58821-6, http://www.pyomo.org
5. Beal, L.D., Hill, D.J., Martin, R.S., Hedengren, J.D.: Gekko optimization suite. Processes **6**(8), 106 (2018). https://doi.org/10.3390/pr6080106, https://www.mdpi.com/2227-9717/6/8/106
6. Alamir, M.: Nonlinear receding horizon sub-optimal guidance law for the minimum interception time problem. Control Eng. Pract. **9**(1), 107–116 (2001). https://doi.org/10.1016/S0967-0661(00)00085-X, https://www.sciencedirect.com/science/article/pii/S096706610000085X
7. Diehl, M., Bock, H.G., Schlöder, J.P., Findeisen, R., Nagy, Z., Allgöwer, F.: Real-time optimization and nonlinear model predictive control of processes governed by differential-algebraic equations. J. Process Control **12**(4), 577–585 (2002)
8. Alamir, M.: A framework for monitoring control updating period in real-time NMPC schemes. Nonlinear Model Predictive Control: Towards New Challenging Applications, pp. 433–445 (2009)
9. Alamir, M.: A state-dependent updating period for certified real-time model predictive control. IEEE Trans. Autom. Control **62**(5), 2464–2469 (2016)
10. Bonne, F., Alamir, M., Bonnay, P.: Experimental investigation of control updating period monitoring in industrial PLC-based fast MPC: Application to the constrained control of a cryogenic refrigerator. Control Theory Technol. **15**(2), 92–108 (2017)
11. Bøhn, E., Gros, S., Moe, S., Johansen, T.A.: Optimization of the model predictive control meta-parameters through reinforcement learning. Eng. Appl. Artif. Intell. **123**, 106211 (2023)
12. Sorourifar, F., Makrygirgos, G., Mesbah, A., Paulson, J.A.: A data-driven automatic tuning method for mpc under uncertainty using constrained bayesian optimization. IFAC-

PapersOnLine **54**(3), 243–250 (2021). https://doi.org/10.1016/j.ifacol.2021.08.249, https://www.sciencedirect.com/science/article/pii/S2405896321010223. 16th IFAC Symposium on Advanced Control of Chemical Processes ADCHEM 2021
13. Stenger, D., Ay, M., Abel, D.: Robust parametrization of a model predictive controller for a cnc machining center using Bayesian optimization. IFAC-PapersOnLine **53**(2), 10388–10394 (2020). https://doi.org/10.1016/j.ifacol.2020.12.2778, https://www.sciencedirect.com/science/article/pii/S2405896320335412. 21st IFAC World Congress
14. Neumann-Brosig, M., Marco, A., Schwarzmann, D., Trimpe, S.: Data-efficient autotuning with Bayesian optimization: An industrial control study. IEEE Trans. Control Syst. Technol. **28**(3), 730–740 (2019)
15. Lu, Q., Kumar, R., Zavala, V.M.: Mpc controller tuning using bayesian optimization techniques. ArXiv preprint arXiv:2009.14175 (2020)
16. Mohammadi, A., Asadi, H., Mohamed, S., Nelson, K., Nahavandi, S.: Multiobjective and interactive genetic algorithms for weight tuning of a model predictive control-based motion cueing algorithm. IEEE Trans. Cybern. **49**(9), 3471–3481 (2019). https://doi.org/10.1109/TCYB.2018.2845661
17. Alamo, T., Tempo, R., Camacho, E.F.: Randomized strategies for probabilistic solutions of uncertain feasibility and optimization problems. Autom. Control IEEE Trans. **54**(11), 2545–2559 (2009)
18. Andersson, J.A.E., Gillis, J., Horn, G., Rawlings, J.B., Diehl, M.: CasADi—A software framework for nonlinear optimization and optimal control. Mathematical Programming Computation (2018)
19. Biegler, L., Zavala, V.: Large-scale nonlinear programming using ipopt: An integrating framework for enterprise-wide dynamic optimization. Comput. Chem. Eng. **33**(3), 575–582 (2009). https://doi.org/10.1016/j.compchemeng.2008.08.006, https://www.sciencedirect.com/science/article/pii/S0098135408001646. Selected Papers from the 17th European Symposium on Computer Aided Process Engineering held in Bucharest, Romania, May 2007

7 Designing Uncertainty-Aware Output Feedback via Deterministic Optimal Solutions

7.1 Chapter's Content at a Glance

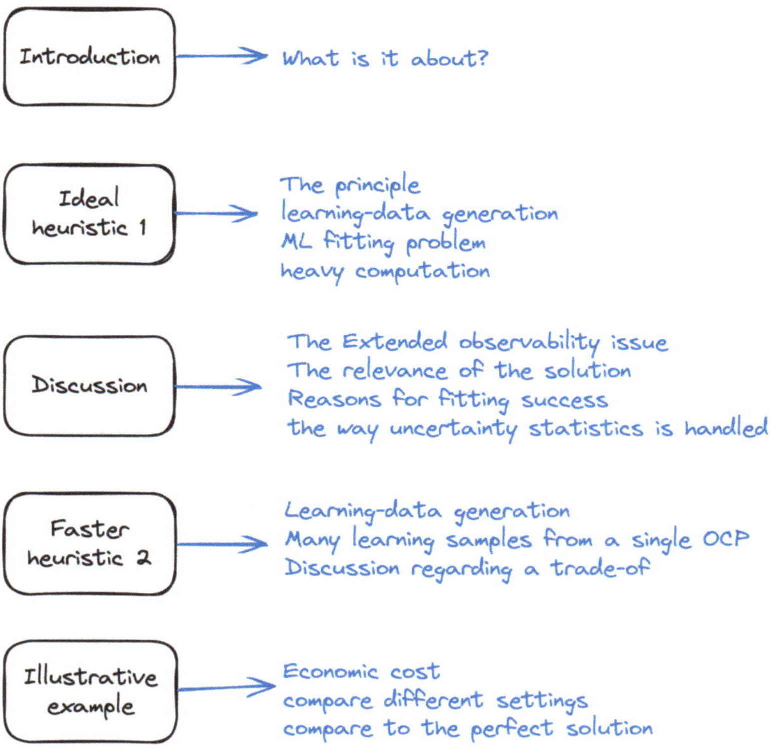

M. Alamir, *Nonlinear Control of Uncertain Systems*,
https://doi.org/10.1007/978-3-031-93287-8_7

7.2 Introduction

In this chapter, it is shown how stochastic OCPs can be approximately solved by solving a cloud of deterministic OCPs. Each of these problems can be solved using deterministic OCPs' solvers such as the ones depicted in **Chap.** 5 or any of the many other freely available solvers therein mentioned.

Indeed, it is shown that the results of the cloud of solutions of a set of deterministic OCPs can be used to build a learning dataset. The latter can be used by an appropriate choice of a Machine Learning (ML) pair of features and labels in order to derive a dynamic output feedback control design that can be qualified as an *Uncertainty-aware* effective feedback control design.

An example is used in the present chapter in order to illustrate the proposed methodology, while in **Chap.** 16, the latter is used to solve the real-life problem of designing a rational power management in a hybrid vehicle along a specific track in the suburb of Tokyo, Japan.

The presentation of the heuristic proposed in this chapter is done in two steps. In the first, an ideal scheme is proposed that is rather computationally involved. Then the heuristic is modified in order to make it more computationally tractable by using arguments from the Bellman's Minimum Principle of optimality.

So let us start by the ideal version that enables to understand the rationale behind the design methodology. This is the aim of the next section.

7.3 The Main Idea: Ideal Heuristic 1

In this chapter, we are interested in controlling nonlinear dynamical systems that are governed by the following discrete-time state evolution equation of the form[1]:

$$x^+ = F(x, u, w, p), \tag{7.1}$$

where

- $x \in \mathbb{R}^{n_x}$ is the state vector.
- $u \in \mathbb{R}^{n_u}$ is the vector of manipulated variables (input).
- $w \in \mathbb{R}^{n_w}$ represents the vector of uncertainties.
- $p \in \mathbb{R}^{n_p}$ gathers all the **known** constant quantities that are needed to completely define the problem, including those that might be needed to express the cost functions, the constraints as discussed in earlier chapters.

[1] Recall that this discrete-time form can be obtained from continuous-time Ordinary Differential Equations (ODEs)-based representation after computing the one-step ahead prediction map for a given sampling period τ. Continuous-time Partial Differential Equations (PDEs) can also be initially used to derive the discrete-time Eq. (7.1).

Contrary to the general context, it is assumed here that the uncertainty vector is constant (although unknown) or slowly varying:

Constant Uncertainties with Known Bounds

The uncertainty vector $w \in \mathbb{R}^{n_w}$ is supposed to be constant or slowly varying.

In other words, the problem in this chapter is the one consisting in controlling dynamics involving badly known constant parameters.

It is also assumed that there is a **known** subset $\mathbb{W} \subset \mathbb{R}^{n_w}$ to which the unknown vector of parameter w belongs:

$$w_k \in \mathbb{W} := [w^{min}, w^{max}] \subset \mathbb{R}^{n_w} \quad \forall k. \tag{7.2}$$

We also assume that we dispose of tools such as the one presented in **Chap.** 5 (the `mpc_solver` module) that enables to simulate the closed-loop behavior under the **MPC feedback that disposes of some given value of the vector of uncertainties**.

Using such tools enables to perform a cloud of simulations in which, the initial state x_0 as well as the instantiation of constant uncertainty vector (w) as well as the constant parameters (p) defining the problem are randomly sampled over their respective sets denoted hereafter by $\mathbb{X}_0$, $\mathbb{W}$ and $\mathbb{P}$ according to their presumably known statistics.

! About Knowing the Statistics

It is a fact that knowing these statistics of sampling of x_0 and w inside their respective sets $\mathbb{X}_0$ and $\mathbb{W}$ is difficult if not generically impossible.

The need for this assumption means that the results any heuristic can provide are valid as long as the relevance of the sampling that helps building the learning set is valid or close to reality.

Each time x_0, w and p are sampled, one disposes of a so-called closed-loop scenario. In the sequel, the i-th scenario is denoted by

$$s^{(i)} := (x_0^{(i)}, w^{(i)}, p^{(i)}). \tag{7.3}$$

The associated closed-loop simulations provide the following set of closed-loop trajectories:

$$\mathcal{D} := \left\{\mathbf{x}^{cl}(s^{(i)}), \mathbf{u}^{cl}(s^{(i)})\right\}_{i=1}^{n_{sc}}, \tag{7.4}$$

where

- n_{sc} is an integer representing the number of scenarios.
- $\mathbf{x}^{cl}(s^{(i)})$ and $\mathbf{u}^{cl}(s^{(i)})$ are the state and control closed-loop trajectories[2] starting from the initial state $x_0^{(i)}$ under the uncertainty realization $w^{(i)}$ and the problem's parameter vector $p^{(i)}$ corresponding to the scenario $s^{(i)}$. More precisely:

$$\mathbf{x}^{cl}(s^{(i)}) := \begin{bmatrix} x_0^{cl}(s^{(i)}) \\ x_1^{cl}(s^{(i)}) \\ \vdots \\ x_{n_i}^{cl}(s^{(i)})) \end{bmatrix} \in [\mathbb{R}^{n_x}]^{n_i+1}, \tag{7.5}$$

$$\mathbf{u}^{cl}(s^{(i)}) := \begin{bmatrix} u_0^{cl}(s^{(i)}) \\ u_1^{cl}(s^{(i)}) \\ \vdots \\ u_{n_i-1}^{cl}(s^{(i)})) \end{bmatrix} \in [\mathbb{R}^{n_u}]^{n_i}. \tag{7.6}$$

- n_i is the number of periods in the closed-loop simulation corresponding to the scenario $s^{(i)}$.

Now if we assume that at each decision instant the solver reached the (unique) optimal solution of the optimization problem[3] then the previous elements are related through the following very definition of the optimal solution:

$$u_k^{cl}(s^{(i)}) := u_0^{\star}\Big(\underbrace{x_k^{cl}(s^{(i)}),\quad w^{(i)},\quad p^{(i)}}_{\text{Ingredients of the optimal control problem over } [k,k+N]}\Big), \tag{7.7}$$

$$:= \text{MPC}\Big(x_k^{cl}(s^{(i)}),\, w^{(i)},\, p^{(i)}\Big). \tag{7.8}$$

Now using the notation of Sect. 2.10 regarding the observability and the extended observability paradigm, assume that there exists some measured vector y given by[4]

$$y = h(x, u, p), \tag{7.9}$$

such that the extended observability condition holds for some map $\mathcal{O}$ and some observation horizon N_O, namely:

$$\begin{bmatrix} x_k^{cl}(s^{(i)}) \\ w^{(i)} \end{bmatrix} = \mathcal{O}(\mathbf{y}_k^{cl(-)}(s^{(i)})), \tag{7.10}$$

[2] Under the deterministic MPC that perfectly knows the value of the uncertainty vector $w^{(i)}$.

[3] The uniqueness of optimal solutions is not guaranteed without dedicated assumptions. The argumentation used here is rather theoretical but it enables to convey the main idea behind the heuristic proposed in the present chapter.

[4] Notice that y does not depend on the uncertainty vector w!

where here again, the notation of Sect. 2.10 are used, namely [see (2.93)]:

$$\mathbf{y}_k^{cl(-)}(s^{(i)}) := \begin{bmatrix} y_k^{cl} \\ y_{k-1}^{cl} \\ \vdots \\ y_{k-N_O}^{cl} \end{bmatrix}. \tag{7.11}$$

Now injecting this in (7.8) leads to:

$$u_k^{cl}(s^{(i)}) = \text{MPC}\Big(\underbrace{x_k^{cl}(s^{(i)}), w^{(i)}}_{\mathcal{O}(\mathbf{y}_k^{cl(-)}(s^{(i)}))}, p^{(i)}\Big), \tag{7.12}$$

and consequently:

$$u_k^{cl}(s^{(i)}) = \text{MPC}\Big(\mathcal{O}(\mathbf{y}_k^{cl(-)}(s^{(i)})), p^{(i)}\Big), \tag{7.13}$$

and since the scenarios $s^{(i)}, i = 1, \dots, n_{sc}$ are only implicitly used to generate rich instances of the above relationship which would be satisfied for all of them,[5] the reference to them can be removed from the last relationship to yield:

$$u_k^{cl} = \text{MPC}\Big(\mathcal{O}(\mathbf{y}_k^{cl(-)}), p\Big). \tag{7.14}$$

This simply means that, under the used assumptions, there exists a map, say ML such that the following holds true[6]

$$u_k^{cl} = \texttt{ML}\Big(\mathbf{y}_k^{cl(-)}, p\Big), \tag{7.15}$$

which can be viewed as a ML fitting problem:

$$\underbrace{u_k^{cl}}_{\text{label}_k} = \texttt{ML}\Big(\underbrace{\mathbf{y}_k^{cl(-)}, p}_{\text{feature}_k}\Big). \tag{7.16}$$

> **Stochastic MPC Design as a ML Problem**
> If the extended observability assumption holds for the measurement vector y, then by concatenating the data collected from the closed-loop simulations for all the scenarios that are sampled according to relevant probabilities of initial states and uncertainty vectors, the resulting ML problem constructed according to (7.16) should admit a rather precise solution.

[5] Thanks to the extended observability assumption!

[6] Readers may have guessed that ML is used here by anticipation of the use of ML identification of the underlying map.

This can be summarized in the following heuristic:

> **The Ideal Heuristic 1**
> Consider the dynamic system given in discrete-time by
>
> $$x^+ = F(x, u, w, p),$$
>
> where the uncertainty vector w is supposed to be constant but unknown. Draw n_{sc} scenarios $s^{(i)}$ for $i = 1, \dots, n_{sc}$ such that:
>
> $$s^{(i)} = (x_0^{(i)}, w^{(i)}, p^{(i)}).$$
>
> Simulate the n_{sc} closed-loop under the MPC feedback that perfectly knows the components of the scenarios. Consider
>
> - a candidate measurement vector map $y = h(x, u, p)$, and
> - an observation horizon N_O
>
> and build the features matrix using the features collected from all the closed-loop simulations:
>
> $$\text{feature}_k^{(i)} = \left[\mathbf{y}_k^{cl(-)}(s^{(i)})\ p^{(i)}\right], \tag{7.17}$$
>
> together with the label vector:
>
> $$\text{label}_k^{(i)} := u_k^{cl}(s^{(i)}), \tag{7.18}$$
>
> where (k, i) is taken such that:
>
> $$i \in \{1, \dots, n_{sc}\} \quad \textit{and} \quad k \in \{N_O, \dots, n_i\}. \tag{7.19}$$
>
> This provides a number of samples in the learning data which is equal to:
>
> $$\texttt{nSamples} = \sum_{i=1}^{n_{sc}} (n_i - N_O). \tag{7.20}$$
>
> Use this learning data to fit a $\texttt{ML}$ map s.t.:
>
> $$\texttt{label} \approx \texttt{ML}(\texttt{feature}).$$

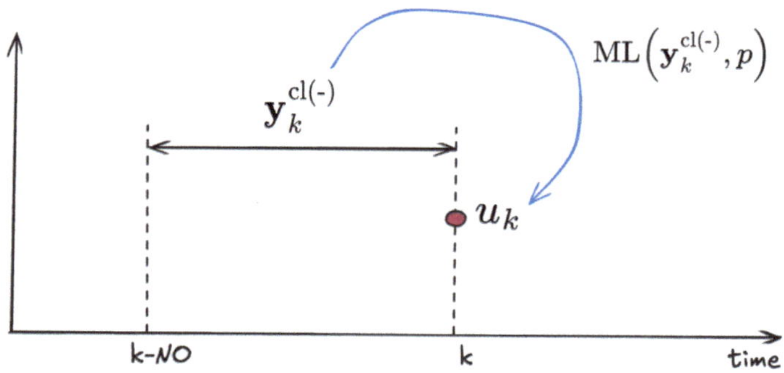

Fig. 7.1 On-line use of the ML map: Use the stored previous measurements over the observation horizon of length N_O and the known problem's parameter p in order to compute the action to be applied over the next sampling period.

Once this road map is applied and a fitted ML map is derived, one disposes of the dynamic output feedback given by

$$u_k = \mathtt{ML}(\mathbf{y}_k^{(-)}, p). \tag{7.21}$$

Regarding this map it is important to underline a rather attractive feature that can be stated as follows:

No Need for Observer Design!
By opposition to state feedback which would need an observer to reconstruct the state before the feedback can be applied. Here, the feedback needs only the measurements over some time window of size N_O.

As a matter of fact, the design of the partial inversion map that extracts from the measurement the needed information that *determines* the control value is done by the fitting process in a hidden way. Moreover, the approximation is weighted by the *topology* of the sampled scenarios that should reflect the randomness of dispersion of the parameters in a relevant way.[7]

Figure 7.1 shows the principle of use of the so-obtained feedback ML-Based on-line.

The following section discusses some issues regarding the assumptions that underline the framework. On the other hand, since the computational burden of the above-described *ideal* framework is rather heavy, a faster approximate version of the heuristic is proposed in Sect. 7.5 and then applied to an illustrative example.

[7] By the very assumption regarding the relevance of the sampling law used in the creation of the set of scenarios.

7.4 Discussion

Regarding the above-described heuristic, the following comments are worth keeping in mind.

1. For general nonlinear systems, the assumption regarding the extended observability is **extremely difficult to check** as any global property related to general nonlinear systems.
   Fortunately enough, the satisfaction of this assumption is not *constructive* in the sense that it is not needed to proceed into the execution of the computational road map proposed above.
   The unique reason for which this implicit assumption is used is to give an understanding of the ideal conditions that support the heuristic formulation.
   More precisely, it helps understanding that in an ideal world, the process of looking for a fitting map `ML` should be perfectly successful if the observability condition holds. If not, one can still hope that an approximate matching can be derived with a narrow dispersion.
2. While checking the satisfaction of the assumption is not an option, what does really matter is the quality of the fitted function, denoted above by `ML`.
   This can obviously be appreciated once a choice of the output map h and the observation horizon length N_O have been made and used in the road map and a fitted feedback map is obtained.
   If the fitting is obtained with a reasonable precision, the so-fitted feedback map should be successful in tackling the presence of uncertain parameters.
   The veracity of the assumption is not needed if good fitting can be obtained using representative dataset and a rigorous train/split procedure as explained in the section dedicated to ML recalls (see Sect. 3.4.3).
3. Note that even in the case where successful fitting is obtained, this does not mean that the extended observability assumption does hold. This is because the fitting success can be due to two possible reasons:

   ✓ Either the sampled elements, namely the initial states and the realization of the uncertainty vector are not representative enough meaning that the quality of the fit is only partial and no extrapolation can be done regarding this quality on the regions of possibilities that were not visited. This *danger* can be reduced by using more simulations so that asymptotically, this risk goes vanishing.

   ✓ Or it might be possible that not all the components of the uncertainty vector do really impact the value of the best control over the closed-loop simulation. In this case, the fit might be successful because extended observability holds on the components that really matter. The remaining components might not be reconstructible from the measurement profile but this has no harmful impact on the quality of the fit because these last components do not impact the label anyway!

4. One might ask:

Taking into Account the Uncertainties?

What is the mechanism through which the fitted feedback `ML`, resulting from the above-described procedure, takes into account the statistics of dispersion of the uncertain vector w?

A first *lazy* answer to this question consists in recalling that the statistics of the scenarios are used in the sampling process!

This answer simply suggests that the results do depend on the statistics but it does not really provide a comprehension regarding the link between the statistics and the resulting feedback.

In order to give a more satisfactory answer, one should remember that `ML` regression algorithms naturally produce results that are more precise over the regions of the features space where there are more samples.[8]

As a consequence, if a sample lies in a region where the probability of occurrence is small, it is likely that the learning will not give it much attention and its precision might be sacrificed in order to better fit the results over those regions containing much more samples according to the associated statistics.

On the contrary, if the distribution of probability is rather uniform, then all the regions will be given almost equal importance.

In other words:

Interpretation of the Fitted Feedback Map
Since the fitting process induces somehow a least-squares minimization, then the weight of the highly probable regions would naturally stem from the associated number of samples making the optimal result close to the optimal solution of the optimization problem aiming at minimizing the statistical mean of the cost function.

This is precisely what one tries to minimize when solving stochastic optimal control leading to stochastic MPC feedback.

Since[9] the fitting process induces somehow a least-squares minimization, then the weight of the highly probable regions would naturally stem from the associated number of samples making the optimal result close to the optimal solution of the optimization problem aiming at minimizing the statistical mean of the cost function.

[8] Obviously, the amplitudes of the label's values also shape the regions where the precision migrates, but for the sake of the intuitive explanation we are seeking, the formulation is not totally wrong!
[9] In which no penalties are applied to the variance term.

5. Obviously the previous discussion does not prove that the resulting feedback is the one that **minimizes** the expectation of the cost function as it is rigorously defined in Sect. 2.6.2 and this would be anyway probably rigorously wrong!
   Nevertheless, the discussion above simply gives some rationale to which one should add the fact that the proposed heuristic, contrary to the exact solution of the stochastic MPC problem, is tractable in terms of computational burden.

While the above discussion explains how the proposed heuristic 1 can be understood as a tractable approximation of some stochastically relevant solution, it remains true that the number of closed-loop simulations it might require before the quality of this sub-optimality becomes acceptable might be extremely large making the ideal heuristic intractable.

Indeed, it is worth recalling that each closed-loop scenario needs a high number of optimization problems to be solved because of the very principle of the MPC feedback. More clearly, each sample in the learning data is obtained at the price of solving a deterministic OCP.

Nevertheless, one might argue that the *warm starting* technique, invoked in Chap. 5 consisting in appropriately set the initial guess close to the previous solution, might enable to reduce the computation time for a class of problems and scenarios by reducing the computation time for successive instants where no set-point changes nor parameter changes intervene. This is only partially true because too close successive problems do not give enough new information for the training dataset to be sufficiently rich and relevant. In other words, We need different and *distant* instances of the features and label to build a consistent learning dataset anyway.

This suggests the need for a `CPU`-aware faster version of the above-proposed heuristic.

This is the aim of the following section.

7.5 The Faster Version: Heuristic 2

Let us first of all summarize the features and the labels used in the formulation of the ML problem included in the `heuristic 1`.

Recall that according to the proposed design, one should perform n_{sc} closed-loop simulations that we consider here, for the sake of simplicity, to be of equal length

$$n_i = N_{\text{sim}},$$

each. Recall that n_{sc} is the number of scenarios included in the learning dataset.

Recall that **each of the collected (features/label) pairs**, namely:

$$\text{features: } \mathbf{y}_k^{cl(-)}(s^{(i)}) \quad / \quad \text{label}^{10}: \text{MPC}(x_k^{cl}(s^{(i)}), w^{(i)}, p^{(i)}),$$

is obtained at the price of a call for the solver that is used to compute the MPC feedback.

This means that applying the `heuristic 1` provides

$$n_{\text{samples}} = N_{\text{sim}} \cdot n_{\text{sc}},$$

samples at the cost of n_{samples} solutions of optimization problems.

Given that the success and the statistical relevance of the fitted map might need tens of thousands of samples, this might become cumbersome quite rapidly for some problems where the solution of a single problem can be quite expensive in terms of computation time (see the previous discussion).

When considering the fact that the heuristic itself is defined up to some choices (some of the Nonlinear Model Predictive Control (NMPC) implementation setting's parameters are studied in **Chap.** 6 that is dedicated to the NMPC tuning algorithms).[11] This induces the need to apply the whole heuristic several times.

The conclusion is that the CPU time needed to use heuristic 1 can become rapidly dissuasive.

This is the reason why, in the present section, a *heuristic* is used to generate a learning dataset in a cheaper way which provides a high number of samples in the learning dataset at the price of a lower number of solutions of Nonlinear Programming Problems (NLP).

The main idea is to extract several learning samples from each single solution of an open-loop OCP associated to a single scenario s.

The idea is graphically shown in Fig. 7.2 that is explained below.

The top plot of Fig. 7.2 shows the open-loop OCP's items associated to the scenario s that is defined by some prediction horizon N.

The definition of the past output related feature is given by the choice of an observation horizon $N_O = M$.

The feature vector is then given by the associated past measurement vector $y_M^{\star(-)}(s)$ at instant M. The superscript * refers to the fact that we are dealing with open-loop optimal trajectories.

Now the top plot of Fig. 7.2 shows that considering instant M one can get a first sample of the learning data consisting in the feature vector $\boldsymbol{y}_M^{\star(-)}(s)$ and the control

[10] $\text{MPC}(x, w, p)$ is the first action of the optimal sequence solution to the open-loop OCP defined by the scenario $s^{(i)} = (x_0^{(i)}, w^{(i)}, p^{(i)})$.

[11] This can be the statistics of the uncertainties, the solver options (type of solver, the maximum number of iterations, γ and c for a fast gradient solvers, the prediction horizon, the penalty being used in the formulation of the stage and the terminal costs to cite but few possible options.

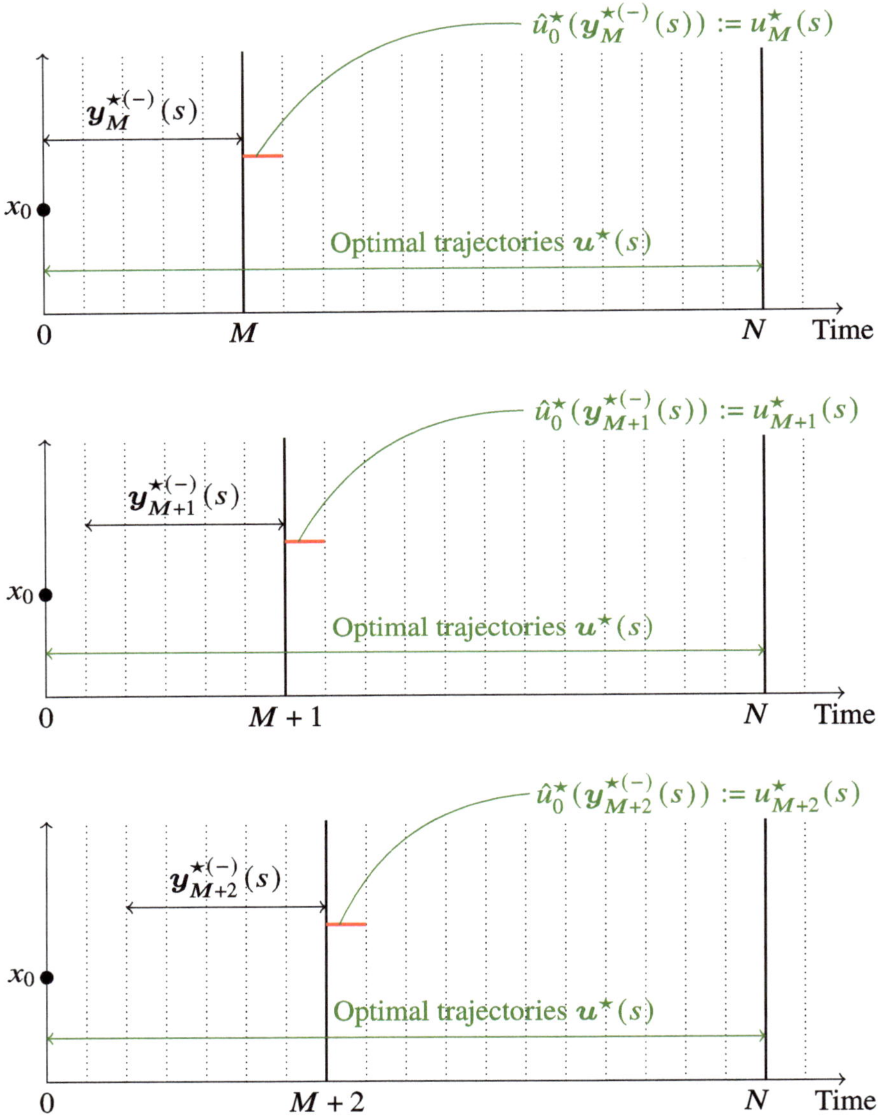

Fig. 7.2 Construction of the learning data $\mathcal{D}(s)$ for a given scenario $s = (x_0, w, p)$: by going forward in a moving window along the open-loop optimal trajectory computed for the scenario s, it is possible to generate a high number of different samples of pairs $(\boldsymbol{y}^{(-)}, u)$ that can be used in the construction of the learning data. Note here that the integer M plays the role of the so-called observation horizon N_O in the presentation of Sect. 7.3

input $u^\star_M(s)$, namely: the M-th sample extracted from the optimal sequence $\mathbf{u}^\star(s)$ for scenario s is given by the pair

$$\left(y_M^{\star(-)}(s), u^\star_M(s)\right).$$

Note that this is not exactly what `heuristic 1` was suggesting. Indeed, according to heuristic 1, we should work only on closed-loop simulations in which the optimization problem is solved at each sampling instant. Consequently, only labels of the form:

$$\mathbf{u}_0^\star(\texttt{some state on some closed-loop trajectory})$$

would have been eligible should heuristic 1 be applied.

Instead, here, on one hand, we use the feature vector:

$$\mathbf{y}_M^{\star(-)}(s),$$

that lies on the **open-loop** optimal trajectory that might have been never visited should closed-loop simulation be used as suggested in Heuristic 1. On the top of that, we associate to this feature vector the control:

$$u^\star_M(s),$$

which is simply the M-th control in the open-loop optimal sequence of actions associated to the scenario $s = (x_0, w, p)$.

In the spirit of heuristic 1, the associated label should be the one given by

$$u_0^\star(x_M^*(s), w, p),$$

which would be the first action of the optimal sequence associated to a scenario in which the initial state is the state reached at instant M and the associated w and p defined by the scenario s.

The modification to heuristic 1 can be summarized as follows:

Changes to Heuristic 1

There are two sources of approximation in the new rule of selection of learning samples compared to heuristic 1:

1. First, the closed-loop trajectory that would have been visited should heuristic 1 be used may not contain the feature vector $\mathbf{y}_M^{\star(-)}(s)$ on one hand.
2. On the other hand, the control $u^\star_M(s)$ is not rigorously the first action of the optimal sequence associated to the scenario $(x^*_M(s), w, p)$, rather, it is the M-th action on the open-loop optimal sequence defined by the scenario (x_0, w, p).

Nevertheless, the rationale behind this *trick* lies in the already discussed Bellman's principle of optimality (see page 45) which states that *the remaining part of an optimal trajectory is optimal.*

One of the consequences of this simple principle is that the above two sources of approximation would lead to negligible effects if the prediction horizon N is sufficiently large[12] since for an infinite prediction horizon, there would be no approximation at all!

Indeed, in this case, the feature vector $\boldsymbol{y}_M^{\star(-)}(s)$ would have been truly visited even in closed-loop and $u_M^\star(s)$ would have been the exact first action of the exact optimal infinite sequence of actions that would have been computed at instant M along the closed-loop trajectory.[13]

Building on this trick, the middle and the top plots of Fig. 7.2 simply suggests that the operation might be repeated at the next instants $M+1$, $M+2$, ...so that the following pairs are also added to the learning dataset, namely:

$$\left(\boldsymbol{y}_{M+1}^{\star(-)}(s), u_{M+1}^\star(s)\right), \left(\boldsymbol{y}_{M+2}^{\star(-)}(s), u_{M+2}^\star(s)\right), \dots.$$

More generally, by considering m samples extracted using the process described above at the sampling instants:

$$M, M+1, \dots, M+m-1,$$

on the open-loop optimal trajectory, one can extract from the single solution of the OCP associated to the scenario s, the following m learning samples:

$$\mathcal{D}(s) := \left\{\left(\boldsymbol{y}_{M+j}^{\star(-)}(s), u_{M+j}^\star(s)\right)\right\}_{j=0}^{m-1}, \tag{7.22}$$

and by using n_{sc} scenarios $s^{(i)}$, $i = 1, \dots, n_{\text{sc}}$, one gets the learning dataset given by

$$\mathcal{D} := \left\{\mathcal{D}(s^{(i)})\right\}_{i=1}^{n_{\text{sc}}}. \tag{7.23}$$

This dataset contains $m \times n_{\text{sc}}$ samples at the price of n_{sc} solution of NLPs.

It appears consequently that the integer m representing the number of samples (in ML language) to be extracted from a single solution of an OCP is a parameter of the new version of the heuristic.

The next section discusses the consequences and the qualitative arguments that should govern the choice of the integer m introduced above.

[12] Recall that the Bellman principle holds only for OCPs with an infinite prediction horizon.

[13] Which would be the same as the open-loop optimal trajectory!

7.5.1 Regarding the Number m of Samples Per Scenario

As discussed earlier, the relevance of the *trick* that underlines the heuristic 2 depends on the extent to which the situation is close to the one where the Bellman principle of optimality holds. Namely, when the prediction horizon is infinite.

Now obviously, if we take m very close to N, we would be very far from satisfying this assumption. That is the reason why moderate values of m should be preferred intuitively.

On the other hand, recall that, for a given computational burden (mainly related to the number of NLPs solutions) the number of samples is proportional to m. This means that for the same computation load, the quality of the fitted model

$$u = \texttt{ML}(\boldsymbol{y}^{(-)}, p),$$

and its ability to generalize to unseen data increases with m.

The above two opposite *effects* lead to the presence, **for a given computational effort** used in building the learning dataset, of an optimal choice of m that obviously depends on the problem at hand, namely:

- ✓ the dynamics;
- ✓ the cost function;
- ✓ the constraints;
- ✓ the prediction horizon's length N;
- ✓ the statistics of uncertainties;
- ✓ the number of scenarios n_{sc};
- ✓ the observation horizon M;
- ✓ etc.

In other words and to say it very shortly, the optimal choice of m is problem-dependent and can hardly be optimally chosen without trying out several values and look at the quality of the resulting model on unseen datasets.

This trade-off is schematically illustrated in Fig. 7.3.

In the remainder of this chapter, the use of heuristic 2 is illustrated through a complete example.

7.6 Illustrative Example: Maximizing the Expected Production of a Parallel Reactor

In this section, we examine the application of the proposed heuristic 2 to an economic feedback control problem in order to induce a better understanding of the heuristic as well as the implication of the different choices of its parameters on the quality of the result.

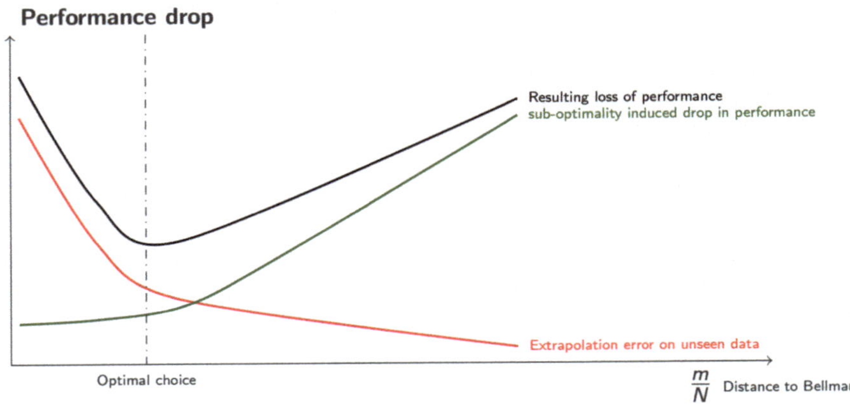

Fig. 7.3 For a given number of NLP's solutions defined by the number of randomly drawn scenarios n_{sc}, there might exist an optimal choice of m that cannot be a priori guessed since it is strongly problem dependent

7.6.1 The System, Statistics of Uncertainties and Objective

Let us consider the parallel reactor system [1] commonly used in the study of deterministic Economic NMPC:

$$\dot{x}_1 = 1 - \mathbf{w}_1 x_1^2 e^{-1/x_3} - \mathbf{w}_2 x_1 e^{-\mathbf{w}_3/x_3} - x_1, \tag{7.24}$$

$$\dot{x}_2 = \mathbf{w}_1 x_1^2 e^{-1/x_3} - x_2, \tag{7.25}$$

$$\dot{x}_3 = u - x_3, \tag{7.26}$$

where x_1 and x_2 stand for the concentrations of reactant and product respectively, while x_3 represents the temperature of the mixture in the reactor.[14]

The control variable is given by the heat flow u which is constrained to remain in the following interval:

$$u \in \mathbb{U} := [0.049, 0.449].$$

The objective of the control is to maximize the amount of product x_2 which is precisely the only measured variable other than the control input u.

This leads to the following cost function[15]:

$$J(\boldsymbol{u} \mid x_0, w) = -\sum_{i=1}^{N} C x_i^{(\boldsymbol{u})}(x_0, w) \quad \text{with} \quad C = (0, 1, 0),$$

[14] Note that in this example, there is no need to a parameter vector p. That is the reason why p is absent in the notation of this section.

[15] Notice that the – sign is used in order to be able to use minimizing paradigm, while the objective is to maximize the production.

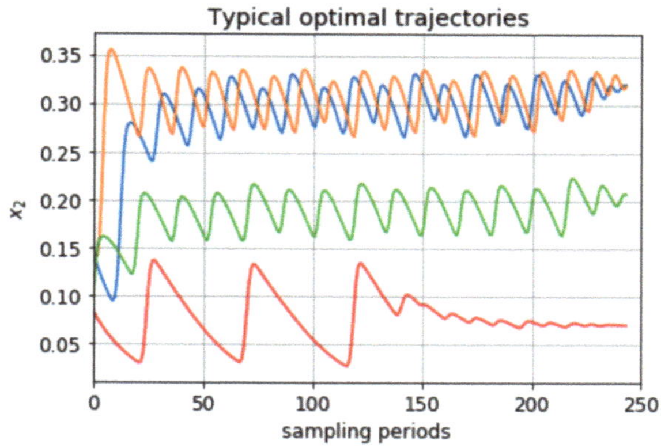

Fig. 7.4 Typical optimal open-loop trajectories of the system. Observing this kind of figure enables to choose the prediction horizon $N = 250$ as the one used in the implementation of the heuristic 2

where $x_i^{(\boldsymbol{u})}(x_0, w)$ denotes the state, at instant i on the open-loop trajectory that starts at the initial state x_0 under the control profile $\boldsymbol{u}$ when the uncertainty vector value is given by w.

In the forthcoming computations, a sampling period of:

$$\tau = 0.02,$$

time units is used (Fig. 7.4).

Based on this value and observing typical optimal trajectories, the prediction horizon length:

$$N = 250,$$

is considered as it corresponds to a prediction horizon that is slightly greater than the settling time.

The past measurement horizon length used to define $\boldsymbol{y}_k^{(-)}$ is taken equal to:

$$M = 10.$$

This choice is motivated by the fact that if one has in mind the paradigm of the extended observation, then the number of variables to be recovered which represent the pair (x_0, w) is equal to $3 + 3 = 6 < 10$.

Recall however that no explicit construction of such an extended estimation algorithm is used in heuristic 2. The previous argument might be rather used to exclude values of M that are lower than 6.

As stated in the beginning of this chapter, the methodology assumes that the statistics of the dispersion of uncertainties is known. In the following development, it is assumed that the uncertainty vector w is normally distributed around a nominal value of:

$$w^0 := (10^4, 400, 0.55),$$

with a **relative** standard deviation $\sigma = 0.33$.

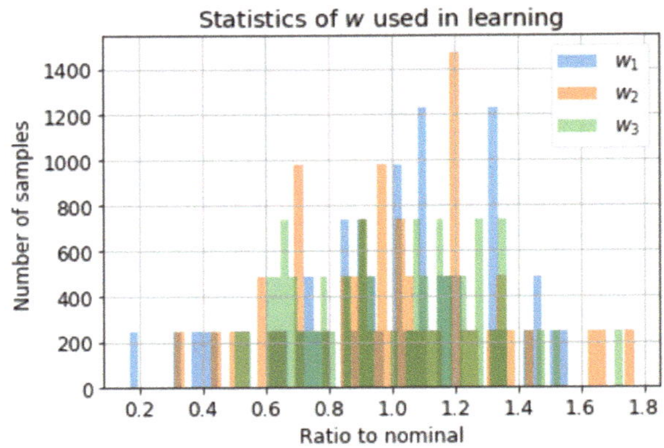

Fig. 7.5 Typical dispersion of the parameters in the learning data in terms of ratio to the nominal components. (Ratios to nominal values are shown)

More precisely, it is assumed that the following rule is relevant for the random sampling of the uncertainty components:

$$w_i = (1 + \nu_i)w_i^0 \quad \text{with} \quad \nu_i \in \mathcal{N}(0, \sigma_i) \quad \sigma_i = 0.33. \tag{7.27}$$

Figure 7.5 shows the dispersion of the values of the parameters that results from the random sampling law (7.27).

Note that the variations of the parameters span the interval going from 20% to 180% of the nominal values.

Regarding the initial state random sampling, it is assumed that a relevant approach consists in sampling the initial state x_0 uniformly inside the set defined by

$$\mathcal{X}_0 := [10^{-4}, 0.5] \times [10^{-4}, 0.2] \times [10^{-4}, 0.25], \tag{7.28}$$

which is known to include the values of the state that are encountered along all relevant profiles of the system under various control designs observed in the literature.

Regarding Traditional Approaches

This is a typical problem that can hardly be addressed using the standard approaches such as the one described in Chap. 4. This is due to the lack of reference trajectory to be tracked around which some high-gain approach can be used to compensate for the presence of parametric uncertainties. Moreover, the level of uncertainties and the long term consequences that should be considered here makes a dedicated uncertainty-aware approach mandatory.

In the next section, the details regarding the generation of the learning data following the heuristic described previously are detailed for the specific example.

7.6.2 Generating the Learning and Validation Data

Based on the previously given definition of a scenario given by

$$s := (x_0, w),$$

a number of

$$n_{sc} = 500,$$

scenarios is generated using the statistics of x_0 and w described in the previous section. This leads to a learning set of cardinality $(500 \times m)$ for each choice of the integer parameter m described in this chapter.

Recall that m represents the number of learning samples extracted from a single solution of an OCP associated to a single scenario s.

This value m is to be compared to the prediction horizon $N = 250$ previously chosen for the definition of the open-loop OCP.[16]

Reminder
Recall that the aim of heuristic 2 is to *fit* a ML map $\texttt{ML}$ such that, when at some current instant, the previous measurement vector

$$\boldsymbol{y}^{(-)} \in \mathbb{R}^M,$$

is available together with the vector of known parameters[17] p, then the value of

$$\texttt{ML}(\boldsymbol{y}^{(-)}, p),$$

can be used as dynamic output feedback that is oriented towards the maximization of the expectation of the production of x_2.

In what follows, several applications of the heuristic 2 are tested (and hence different identified feedback maps $\texttt{ML}$ are obtained) for different values of m, namely[18]:

$$m \in \{10, 50, 100, 150, 200\},$$

leading to learning datasets of different sizes, namely:

$$\{5000, 25000, 50000, 75000, 100000\} \quad \text{samples},$$

having all in common the same computation time, namely the time needed to solve $n_q = 500$ deterministic NLP problems, while the distance to optimality (closeness

---

[16] Recall that since N is chosen so that it can be considered to be sufficiently high, when $m \ll 500$, then the situation is close to the set of settings where the Bellman principle is approximately satisfied.
[17] In the specific example under investigation, there is no such a vector p that intervene in the formulation of the problem.
[18] Recall that m is the number of samples extracted from a single solution of an open-loop OCP.

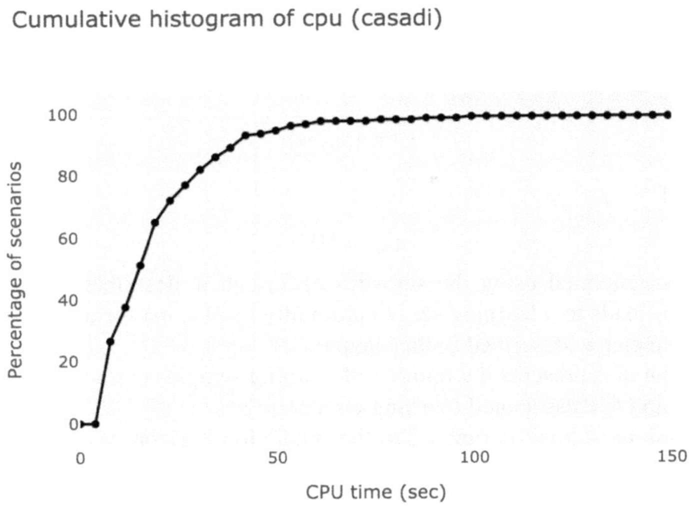

Fig. 7.6 Cumulative histogram of computation time (sec) needed by `Casadi/IPOPT` (**multiple-shooting**) to solve a single OCP when creating the learning data for the illustrative example. The histogram is constructed using the computation times needed to solve the $n_{\text{sc}} = 500$ randomly sampled scenarios

to the condition of the Bellman optimality condition) decreases as the cardinality increases.

It is important to underline that all the above-mentioned datasets can be built using the same set of open-loop optimal trajectories corresponding to the $n_{\text{sc}} = 500$ generated scenarios. Each choice of m inside the above-mentioned list of values amounts at including, in the learning dataset, more or less instants of each optimal solution over the prediction horizon.

There is no need to re-run open-loop optimization for each of the above-mentioned learning sets of different sizes.

Recall that each of the $n_{\text{sc}} = 500$ OCPs invoked above can be solved using the `mpc_solver` package described in **Chap.** 5 since for each value of the vector w, one has to solve deterministic OCP with a given vector of uncertainty.

As it was mentioned in **Chap.** 5, instead of the single shooting formulation used ***mpc***`_solver` package, a multiple-shooting version can be developed in which the decision variables is defined to be the control and the state profiles over the prediction horizon.

The results shown hereafter are obtained using a multiple-shooting version of the solver.

The cumulative histogram of computation time (in sec) needed by a multiple-shooting implementation of the optimal control solver `IPOPT` through `CasaDi` is shown in Fig. 7.6.

The computation time shown in Fig. 7.6 might seem too long for readers who commonly use `CasaDi/IPOPT` to solve standard regulation problems. Indeed, the median time is close to 15 sec!

This is due to the oscillating character (see Fig. 7.4) of typical optimal trajectories. This feature is very specific to the underlying *economic* NMPC formulation and to the relatively long prediction horizon $N = 250$ being used which lead to a decision variable of dimension:

$$N(n_x + n_u) = 1000.$$

Indeed, the economic formulation does not correspond to a steady terminal regime as the oscillatory behavior enables a better economic outcome, in terms of produced quantity of x_2.

This histogram of computation time justifies the need for the heuristic 2 compared to heuristic 1.

Having at hand each of the above learning datasets, the next step is to identify the relationship that links the previous output measurements $\boldsymbol{y}^{(-)}$ to the associated control u as explained earlier.

This is examined in the nest section.

7.6.3 Identifying the ML Feedback Map

Let us first have a look on the histogram of the encountered values of the label, namely the histogram of the instances of the form:

$$u^{\star}_{M+j}(s^{(i)}) \quad where \quad (i, j) \in \{1, \ldots, n_{sc}\} \times \{0, \ldots, m-1\},$$

where the index i refers to the scenario's number, while the index j stands for the time index over the prediction horizon at which the sample is extracted according to the heuristic 2.

Figure 7.7 shows the histogram of the $125000 = 500 \times 250$ values of control input u contained in the set of all the first $N = 250$ instants of the 500 optimal control profiles that are obtained when solving the open-loop OCPs associated to the set of n_{sc} randomly sampled scenarios.

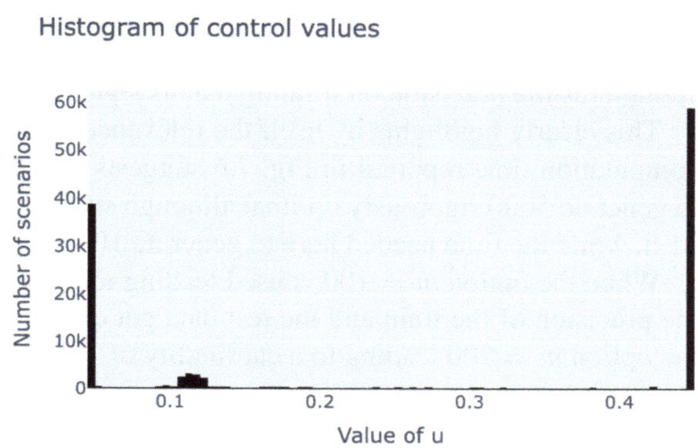

Fig. 7.7 Histogram of the values of the control u present in the 500 scenarios (of length $N = 250$ each) contained in the learning data

This histogram suggests that classification tools might be more suitable for the learning of the control using ML tools compared to regression tools.[19]

Indeed, the following three-valued set of label values can be used:

$$\text{label} = \begin{cases} 1 \text{ if} u \le 0.07, \\ 2 \text{ if} u \in]0.07, 0.14], \\ 3 \text{ otherwise}, \end{cases} \tag{7.29}$$

with this definition of the label, it is possible to identify a **classifier** that associates to any time series of measurement $\boldsymbol{y}^{(-)}$ an element of the set $\{1, 2, 3\}$ and hence its corresponding control value in $\{0.049, 0.11, 0.449\}$.

More precisely, denoting by `cl` the fitted classifier, namely:

$$\text{label} = \texttt{cl}(\boldsymbol{y}^{(-)}),$$

the feedback function one is looking for becomes:

$$u = \texttt{ML}(\boldsymbol{y}^{(-)}) := \begin{cases} 0.049 & \text{if } \texttt{cl}(\boldsymbol{y}^{(-)}) = 1, \\ 0.11 & \text{if } \texttt{cl}(\boldsymbol{y}^{(-)}) = 2, \\ 0.449 & \text{if } \texttt{cl}(\boldsymbol{y}^{(-)}) = 3. \end{cases} \tag{7.30}$$

Notice that if the values of the control input were more *continuously distributed* between $u^{min} = 0.049$ and $u^{max} = 0.449$, a regressor would have been more appropriate to address the identification problem.

The feature vector of past measurements is constructed using the measured quantity x_2 as output.

The Identification of the map $\texttt{ML}(\boldsymbol{y}^{(-)})$ has been obtained using a `RandomForest Classifier` from `scikit-learn` package [2] which has been already described in Sect. 3.4.3.

In order to avoid over-fitting and enhance the quality of the extrapolation on unseen data, the `max_leaf_nodes` parameter has been set to 500.

The confusion matrices corresponding to the training and the test data (using a test-size ratio of 33% of the learning data) are shown in Fig. 7.8 for $m \in \{10, 100, 200\}$.

The examination of these matrices clearly suggests that when $m = 10$ is used (cardinality of the learning set = 5000) the learning data is still not representative enough (as the precision on training data is significantly better than on test data).

This clearly highlights by itself the relevance of the proposed heuristic since the computation time reported in Fig. 7.6 suggests that the time that would be needed to generate 5000 rigorously optimal although still insufficient data would be around 21 h, while the time needed here to generate 100000 samples is around 2h.

When the option $m = 100$ is used leading to a learning set of cardinality 50000, the precision of the train and the test data get closer and get even more closer with the option $m = 200$ leading to a cardinality of 100000.

[19] For this specific example.

Results for m = 10

Train	Predicted 1	Predicted 2	Predicted 3	Test	Predicted 1	Predicted 2	Predicted 3
True 1	**0.991888**	0.000000	0.008112	True 1	**0.519095**	0.000000	0.480905
True 2	0.015625	**0.937500**	0.046875	True 2	0.307692	**0.153846**	0.538462
True 3	0.001040	0.000000	**0.998960**	True 3	0.163020	0.001094	**0.835886**

Results for m = 100

Train	Predicted 1	Predicted 2	Predicted 3	Test	Predicted 1	Predicted 2	Predicted 3
True 1	**0.827709**	0.001611	0.170680	True 1	**0.770533**	0.002540	0.226926
True 2	0.035165	**0.896389**	0.068446	True 2	0.040978	**0.883675**	0.075347
True 3	0.063519	0.059739	**0.876742**	True 3	0.108823	0.061350	**0.829827**

Results for m = 200

Train	Predicted 1	Predicted 2	Predicted 3	Test	Predicted 1	Predicted 2	Predicted 3
True 1	**0.865979**	0.000699	0.133322	True 1	**0.829027**	0.000796	0.170177
True 2	0.036988	**0.908058**	0.054954	True 2	0.041939	**0.907135**	0.050926
True 3	0.060827	0.070979	**0.868194**	True 3	0.081609	0.074119	**0.844272**

Fig. 7.8 Fitting results on the training and test data (split ratio (33%) for the different values of receding horizon moves m used in the construction of the learning data. Recall that the label the classifier tries to guess lies in $\{1, 2, 3\}$ as explained in (7.29)

Notice however that this does not necessarily mean that one should choose the model learned with the dataset built using the option $m = 200$. Indeed, the quality of the extrapolation that improves when the results of training and test are close concerns only the quality of the fitted map with the provided features and labels.

It is the relevance of the labels as the near optimal values to fit that might be affected by higher values of m as explained earlier (see Fig. 7.3).

The relevance of the possible values of the parameter m can only be inspected by using the different controllers on a new set of randomly sampled scenarios and by comparing the statistics of the corresponding closed-loop results regarding the expectation of the produced x_2.

It is only then that the trade-off between a good extrapolation quality of the so-obtained solution (enhanced by high values of m) and the distance to optimality (reduced by small values of m) can be definitively resolved (Fig. 7.9).

This is the object of the next section.

7.6.4 Performance Evaluation

In order to evaluate the quality of the different identified models, each corresponding to a specific value of learning dataset building parameter m, a new set of 100 pairs of (x_0, w) is generated to evaluate the performance of the different controllers on this set of unseen initial states and parameters values w.

Optimal value (Casadi)

Perfect NMPC that computes the optimal solution profiles for each scenario **knowing perfectly** the value of the uncertainty w of that specific scenario. This is the **ground** truth serving as an ideal controller that is impossible to design.

Identified on Nominal

The learned controller from the 100 new scenarios used in the validation step. Since the controller is learned using the same data that is used in the validation, the result of this controller is an **optimistic evaluation** of an identified nominal controller

Heuristic 2 (5 options)

These are 5 different settings of Heuristic 2 obtained by using different values of m, namely:

$$m \in \{10, 50, 100, 150, 200\}$$

leading to five different stochastic MPC controllers.

Fig. 7.9 Description of the different settings whose closed-loop performance are compared

Table 7.1 Comparison of the closed-loop performances on the validation dataset of the different setting described in Fig. 7.9

Identification setting	Mean value	Relative to nominal
Optimal value (Casadi)	0.145	1.313
Identified_on_nominal	**0.110**	**1.000**
Identified with m = 200	0.137	1.246
Identified with m = 150	0.137	1.238
Identified with m = 100	0.136	1.234
Identified with m = 50	0.134	1.207
Identified with m = 10	0.115	1.040

Beside this, the 100 initial states contained in these scenarios are used to generate learning data using the nominal value w^0 of the parameter vector. Using this dataset, one can fit a map that represent the nominal controller that does not take into account **explicitly** the dispersion of the parameters of the model.

This is done in order to evaluate the benefit from explicitly learning the feedback from the randomly sampled vectors of parameters compared to a nominal design that only use the nominal expected value.

To summarize, we are going to compare 7 different controllers that are described as follows:

Then for each value of m, the corresponding identified feedback law is used to generate 100 sets of closed-loop trajectories in which the newly drawn samples of the parameter vector (which are unknown to the previously identified feedback) are used in the simulated model.

The results are shown in Table 7.1 where one can find

1. The ideal optimal closed-loop cost given by the NLP solver `IPOPT` in the **unrealistic case where the disturbance is perfectly known for each sampled scenario**.
2. The closed-loop performance obtained using the nominal feedback described above.
3. The closed-loop performance of the identified feedback using the different values of m.

From Table 7.1, the following observations can be made:

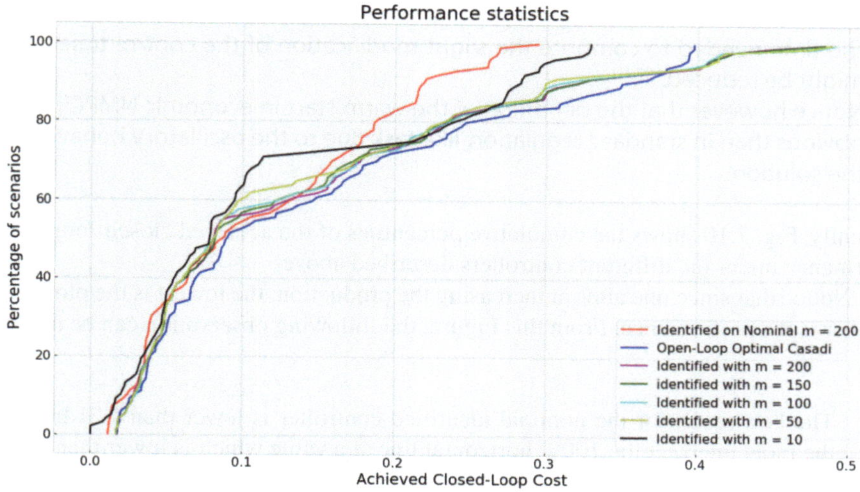

Fig. 7.10 Cumulative percentile of the achieved closed-loop performance for the different controllers. This figure shows that in 10% of the scenarios, the previously identified stochastic feedback laws designed following heuristic 2 **perform better** than the necessarily numerically imperfect NMPC solver that is aware of the true values of the uncertain parameter vector!

- The optimal achievable values of the cost **when the parameter vector is perfectly known** is 31% better (in average) than what the nominal identified controller enables to achieve.
- The proposed methodology enables to **recover 78% of the advantages**[20] **of knowing perfectly the value of the parameter vector** when $m = 200$ is used in the construction of the learning data. This corresponds to 24% better average than the nominally optimal identified controller.
  Very similar results are obtained when $m \in \{150, 100\}$ is used.
- The average performance drops to 20% (66% of recovered advantage) when $m = 50$ is used, while a drastic decrease corresponding to 3.9% (and only 12% of recovered advantage) when short $m = 10$ is used in the building of the learning data.

It is important to underline that if strict optimality is used in the dataset building step (only one sample per NLP solution), the time that would be needed to solve the 100000 NLP problems [21] would have been equal to approximately 17 days.

! The Role of Warm Start!

As a matter of fact, the use of warm start (that does not help in the case of use of an interior point optimization solver such as `IPOPT`) might drastically reduce the computation time associated to the first ideally defined heuristic.

[20] This is the ratio 0.245/0.312.

[21] Cardinality of the learning set for $m = 200$.

This is because when sliding the prediction horizon, while using the warm start, the time needed to compute the slight modification of the control trajectory might be reduced.
Notice however that the definition of the warm start in economic NMPC is less obvious than in standard regulation-like task due to the oscillatory behavior of the solution.

Finally, Fig. 7.10 shows the cumulative percentiles of the achieved closed-loop performance under the different controllers described above.

Notice that since one aims at increasing the production, the lowest is the plots the better is the performance. From this Figure, the following observation can be made:

- ✓ The best result for the nominal identified controller is lower than 0.3! Indeed the plots intersect the 100% horizontal line at a value which is lower than 0.3.
- ✓ In the case of the perfect NMPC controller, this maximal values is close to 0.4, while for the group of stochastic MPC controllers corresponding to $m \in \{50, 100, 150, 200\}$, the maximum achieved performance is close to 0.48. **That is greater than the ideal perfect NMPC in roughly 10% of the scenarios**! This seems rigorously contradictory!
  As a matter fact, this simply comes from the fact that the `IPOPT` solver inside `CasaDi` is used with a limited number of iterations and even without this limitation, an NLP solver does not always reach optimal solution.
  This is the malediction of non-convex optimization.

It remains interesting to underline the fact that:

A Good Surprise!
For 10% of the scenarios, the identified stochastic feedback laws designed following heuristic 2 perform better than the necessarily numerically imperfect NMPC solver that is aware of the true values of the uncertain parameter vector!

7.7 Conclusion

In this chapter, a complete framework is presented in order to design an uncertainty-aware dynamic output feedback controller that is oriented towards minimizing the expectation of some cost function.

The framework applied to general dynamical systems that are described by discrete-time state equations that depend on some uncertain vector of parameters for which it is assumed that a statistical description of the dispersion is given.

It is shown that the execution of the heuristic 2 needed to build the feedback function is based on a tool such as the one described and fully provided in Chap. 5 that solves general deterministic nonlinear open-loop OCPs.

The framework of the present chapter is also used in **Chap.** 16 to address the driving cycle use-case aiming at optimizing the energy consumption and the battery usage under driving profile's uncertainties.

Despite the promising results of this chapter, it is worth keeping in mind the following facts:

✓ The type of problems that might be addressed using the methodology described in this chapter is the one where the control objective is *economic* such as the one studied in this chapter where one aims at maximizing the expectation of some cost function. For instance, the problem of cancer treatment invoked in Sect. 2.11.1 should be possible to handle using heuristic 2 of the present chapter.
However, **this approach might not work in stability-critical problems** where the parameters of the model might need to be rigorously estimated on-line so that precise regulation around some highly unstable targeted state can be achieved.

✓ The heuristic 2 remains a guideline procedure that should be almost always coupled with specific system-dependent handling. In the current chapter's use-case for instance, the use of a classifier in order to better parameterize the `ML`-fitted map is very specific to the histogram of control input describing the values encountered in the learning data.
Moreover, the use of the past values of the only x_2 measurement (and not adding the past values of u to those of x_2) is also a specific choice that fits the specificity of the current problem.
This is because the extreme values appearing in the histograms of the control inputs values as shown in Fig. 7.7 induce highly unbalanced learning data that might lead to a classifier that systematically copies the last previous values of the control input as the one that should hold after since this choice corresponds to highly precise metrics.
The use of only past values of x_2 in this specific case, helped overcoming this unbalance issue.

References

1. Müller, M.A., Angeli, D., Allgöwer, F., Amrit, R., Rawlings, J.B.: Convergence in economic model predictive control with average constraints. Automatica **50**(12), 3100–3111 (2014)
2. Pedregosa, F., Varoquaux, G., Gramfort, A., Michel, V., Thirion, B., Grisel, O., Blondel, M., Prettenhofer, P., Weiss, R., Dubourg, V., Vanderplas, J., Passos, A., Cournapeau, D., Brucher, M., Perrot, M., Duchesnay, E.: Scikit-learn: Machine learning in Python. J. Mach. Learn. Res. **12**, 2825–2830 (2011)

8 Design of Parameterized NMPC via Data-Driven Cost Function Identification

8.1 Chapter's Content at a Glance

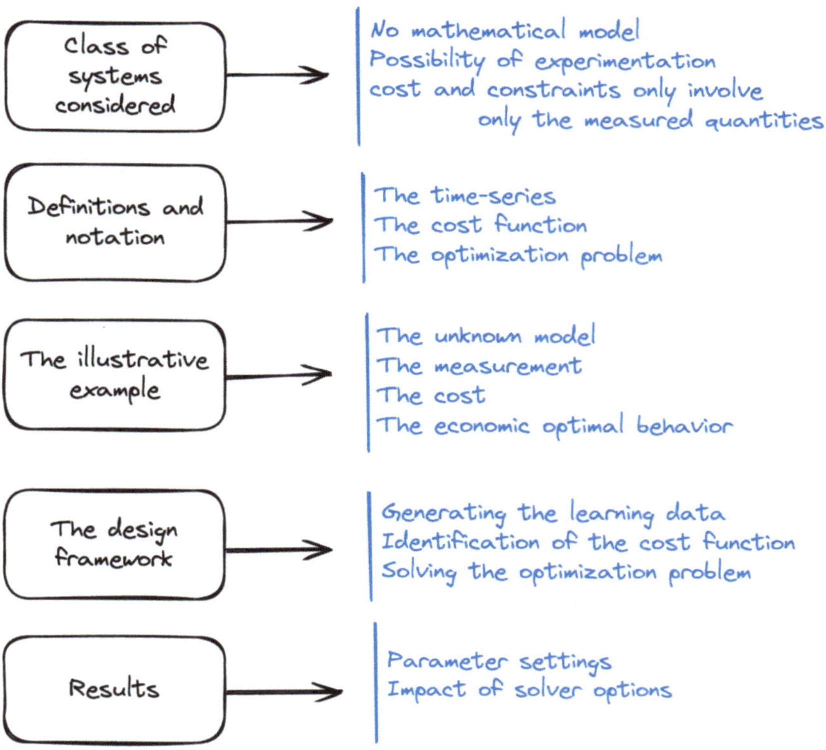

M. Alamir, *Nonlinear Control of Uncertain Systems*,
https://doi.org/10.1007/978-3-031-93287-8_8

8.2 Introduction

In **Chap.** 7, a control design framework is proposed using extensive computation of deterministic model based open-loop optimal control problems (OCPs). This assumes that the dynamic model of the system is available, possibly with badly known vector of parameters.

In this chapter, we consider another class of problems that is characterized by the following features:

1. The dynamic model of the system is not available. However,
2. it is possible to perform a large number of experiments (under open-loop control or *loosely defined* closed-loop control) during each of which,
3. it is possible to collect the time-series of a vector of output measurements and control inputs, denoted by[1] $y \in \mathbb{R}^{n_y}$ and $u \in \mathbb{R}^{n_u}$. Moreover,
4. the information contained in the measurement profiles (this include the output measurements and the control input measurements) is sufficient to express the cost function of interest, including the exact penalty term representing the contribution of the soft constraints. To say it differently, the cost function is expressed in terms of the sensors's output over some window.

Notice that items 2. and 4. enumerated above represent a set of rather **strong** assumptions that are not necessarily satisfied for a wide class of problems. Indeed:

- **Item 2** assumes that one can operate the system over many experiments without having a dynamical model based on which a control feedback can be designed. This can be the case under two possible circumstances:
  - The first corresponds to systems that are open-loop stable which show bounded responses to bounded excitations. The issue is then to calibrate the input excitation such that the system's response remain in a *reasonable* domain of variations. Notice that even in this case, operating the system might correspond to some economic cost or a time duration that can be impossible to accept.
  - The second situation corresponds to systems for which some *loosely defined* feedback is available that is potentially far from optimality but which keeps the system's evolution within an admissible bounded domain in order to gather a sufficiently large data-set for the learning step to come. This might be for instance the case in building management where a rules-based strategies can be imagined in order to explore a reasonably large cloud of operating points in the space of the different variables of interest.

[1] Notice that while the input is a measured quantities, we distinguish in this chapter between the control input and the remaining measured quantities because they play different roles in the formulation from a technical point of view.

- **Item 4** assumes that the control objective and constraints depends on the only measurable quantities gathered in $y \in \mathbb{R}^{n_y}$ and $u \in \mathbb{R}^{n_u}$. This is a **restrictive property** since in many situations, the measured quantities do not cover all the key variables and that is the reason why state estimators are necessary as mentioned in Sect. 2.10 (see also **Chap.** 9).
  Notice however that observer design requires the knowledge of the system's equations which is assumed to be unavailable by assumption in this chapter.
  This being said, it is crucial to keep in mind the following fact:

! A Restriction Shared by all Data-Driven Solutions
The restrictions introduced via **Items 2 and 4** are common to all **model-free data-driven solutions** to the control design problem[2] although it is not sufficiently underlined and brought to the attention of readers.

Indeed, as far as a solution framework pretends to be dependent only on the measured data, all the elements of the problem's formulation including the cost function and the constraints should be exclusively expressed in terms of the latter and for the data to be collected, one should be able to *operate the system* over data-collection experiments. These conditions are precisely the content of items 2. and 4. discussed above.

The previous discussion can be summarized as follows:

Class of Systems Considered
It is assumed that although the **dynamic model is not available**, it is possible to perform experiments (in open or loosely closed-loop) on the system, during which, all the quantities that impact the cost function and the constraints are collected.

Obviously, this is the ultimate example of the presence of uncertainties since nothing is a priori known about the equations governing the system and this is why we need the above discussed strong assumptions to be able to do something.

The remainder of this chapter is organized as follows: First of all, Sect. 8.3 introduces some definitions and notation used in the presentation of the methodology. The illustrative example is introduced in Sect. 8.4. The control design framework is discussed in Sect. 8.5 and its results are discussed in Sect. 8.6.

[2] In particular for Reinforcement Learning (RL).

8.3 Definitions and Notation

Let us consider a nonlinear system that is governed by the following **unknown dynamics** given by:

$$x^+ = f(x, u) \qquad y = h(x, u), \tag{8.1}$$

where $x \in \mathbb{R}^{n_x}$, $u \in \mathbb{R}^{n_u}$ and $y \in \mathbb{R}^{n_y}$ stand for the state vector, the control input vector and the vector of measured variables respectively.

According to the discussion of the previous section, the maps f, h and the very definition of the state vector x are supposed to be unknown to the control designer.

On the other hand, it is assumed that a cost function that expresses the quality of the system's trajectories over some prediction horizon of length N can be expressed, at each decision instant k, by a function of the following form:

$$\texttt{cost}(\boldsymbol{y}_k, \boldsymbol{u}_k),$$

that involves the measured time-series over the past interval of the only measured quantities, namely the output $\boldsymbol{y}_k$ and the control input $\boldsymbol{u}_k$ where:

$$\boldsymbol{y}_k := \begin{bmatrix} y_{k+1} & \dots & y_{k+N} \end{bmatrix} \in [\mathbb{R}^{n_y}]^N,$$

$$\boldsymbol{u}_k := \begin{bmatrix} u_k & \dots & u_{k+N-1} \end{bmatrix} \in [\mathbb{R}^{n_u}]^N.$$

It is important to underline that the possibility to design a successful output feedback is implicitly based on the assumption according to which, the knowledge of a sufficient number of instances of the past measurements enables to determine how the future key variables depend on the future inputs.

More rigorously, this assumption implies that for a sufficiently high N, there exists some static map O such that the following relationship holds true:

$$\boldsymbol{y}_k = O(\boldsymbol{y}_{k-N}, \boldsymbol{u}_{k-N-1}, \boldsymbol{u}_k). \tag{8.2}$$

Indeed this simply stems from the fact that the observable sub-state[3] x_k is inferred from $(\boldsymbol{y}_{k-N}, \boldsymbol{u}_{k-N-1})$ which together with $\boldsymbol{u}_k$ determine the future $\boldsymbol{y}_k$ (see Fig. 8.1).

A direct consequence of the observability property is that the cost function $\texttt{cost}(\boldsymbol{y}_k, \boldsymbol{u}_k)$ can be written in the following form

$$\texttt{cost}(\boldsymbol{y}_k, \boldsymbol{u}_k) = \texttt{cost}(O(\boldsymbol{y}_{k-N}, \boldsymbol{u}_{k-N-1}), \boldsymbol{u}_k),$$

[3] The part of the state that impacts the future evolution of the output.

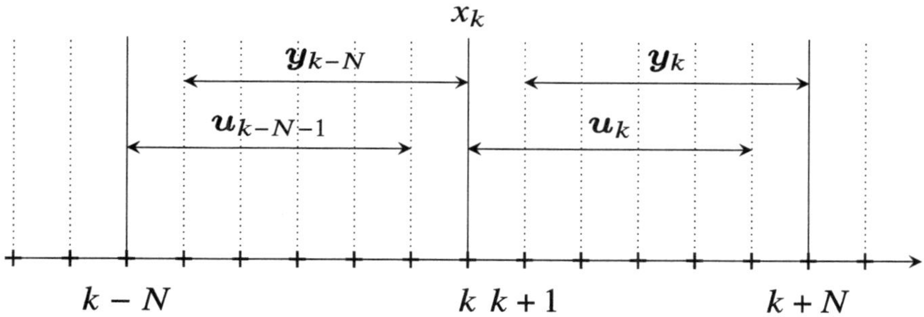

Fig. 8.1 Illustration of the observability property: (y_{k-N}, u_{k-N-1}) uniquely determines x_k which, together with u_k uniquely determines y_k

or equivalently, using a straightforward definition of J:

$$\text{cost}(\, O(\overbrace{y_{k-N}, u_{k-N-1}}^{\text{Past measurement}}), \quad \overbrace{u_k}^{\text{Future action}} \,), \tag{8.3}$$

$$=: J(u_k \mid y_{k-N}, u_{k-N-1}), \tag{8.4}$$

where (8.4) is derived from (8.3) by reordering the terms so that J can be interpreted as follows:

Reading the Cost
The expression:

$$J(u_k \mid y_{k-N}, u_{k-N-1}),$$

refers to the cost induced by a future control sequence u_k given the past measurements (y_{k-N}, u_{k-N-1}).

This cost can then obviously be used to design NMPC output feedback by solving at each instant k the following optimization problem:

$$u_k^* \leftarrow \arg\min_{u \in \mathbb{U}^N} J(u \mid y_{k-N}, u_{k-N-1}). \tag{8.5}$$

Moreover the solution can be used in a receding horizon manner, namely by applying the first control u_k^* in the so-computed optimal sequence u_k^* following the principle of Model Predictive Control (MPC) discussed in Sect. 2.7.

Note that in (8.5), the set $\mathbb{U} \subset \mathbb{R}^{n_u}$ stands for the admissible domain for the input vector. This implements box-like hard constraints on the control.

In the case where output constraints need be incorporated, an associated exact penalty term has to be incorporated in the the definition of the cost function as described in Chap. 2.

In the next section and before we dig into the details of the technicalities associated to the design framework, an illustrative example is introduced so that the concepts can be associated to a concrete use-case. This example is also used in the numerical investigation-related section.

8.4 The Illustrative Example

In order to illustrate the design methodology explained in the sequel, we shall consider the same example, already used in Chap. 7 (see Sect. 7.6) which is governed by the following set of ODEs:

$$\dot{x}_1 = 1 - 10^4 x_1^2 e^{-1/x_3} - 400 x_1 e^{-0.55/x_3} - x_1, \tag{8.6a}$$

$$\dot{x}_2 = 10^4 x_1^2 e^{-1/x_3} - x_2, \tag{8.6b}$$

$$\dot{x}_3 = u - x_3, \tag{8.6c}$$

that describes the reaction:

$$R \to P_1 \qquad R \to P_2,$$

where x_1 and x_2 stand for the concentrations of the reactant R and the product P_1 respectively while x_3 represents the temperature of the mixture in the reactor. P_2 represents the waste product.

The control variable represents the heat flow given by u that belongs to an interval $\mathbb{U}$ of admissible values:

$$u \in \mathbb{U} := [0.049, 0.449].$$

The control objective is to maximize the amount of produced $x_2 = P_1$ which is precisely the only measured variable, other than u. It can be easily checked that with these two measured quantities, the whole state is reconstructible (since the system is fully observable). But unfortunately:

! State Reconstruction Is Not Possible
Notice that this state reconstruction is not possible since the mathematical model is not available. Nevertheless, it is mentioned here to support the existence of the maps invoked earlier in the discussion.

This dynamic system is quite famous within the economic MPC community as it is commonly used as a supporting example. Recall that economic MPC **EMPC** [1] refers to the MPC formulations where the cost function is not based on a priori known desired steady state. Instead, only an economic stage cost can be used which, for our supporting example, can be defined by:

$$\ell(x) = -x_2 = -y,$$

in order to enhance the maximization of the production.

As a result, the measurement-dependent definition of the cost function might be given by:

$$J(\boldsymbol{u}_k|\, \boldsymbol{y}_{k-N}, \boldsymbol{u}_{k-N-1}) := -\big(\frac{1}{N}\mathbf{1}^T \cdot \boldsymbol{y}_k + \alpha y_{k+N}\big), \tag{8.7}$$

where

- $\boldsymbol{y}_k$ depends on the arguments of J through (8.2).
- The first term is noting but the average value over the prediction horizon while.
- The second terms implements a terminal penalty at the end of the prediction horizon following standard stability-related recommendations [2] already discussed in Chap. 2.
- The minus sign recalls that maximization is required.

For a better understanding of the forthcoming results, the following known facts regarding the system under study are worth to be stated:

1. Under the Economic MPC formulation defined above and when the model of the system is perfectly known and used in the design of the EMPC, the resulting optimal trajectories are not steady but rather oscillating around the optimal steady state as this oscillating behavior induces larger produced yield.
2. As it is already noticed in **Chap.** 7, because of the absence of steady behavior, the computation time of the OCP increases when the prediction horizon increases in a continuous way and this even when sophisticated solvers [3] are used. This means that the underlying optimization problems are quite challenging. This fact is worth to keep in mind when appreciating the quality of the results proposed in this chapter.

In the forthcoming developments, a sampling period of $\tau = 0.02$ time units is used for the integration of the ODE for the validation purpose.[4]

Finally, it is worth mentioning that for the parameters used in the simulation, among all possible **steady pairs**, the optimal one corresponds to:

$$x_{st} = (0.0832, 0.0846, 0.149) \quad ; \quad u_{st} = 0.146 \quad \text{and} \quad y_{st} = 0.0846.$$

Notice however that all the above values are supposed to be **unknown** since the model itself is supposed to be so.

[4] Notice that the equations above are normalized so that the absolute time is meaningless.

8.5 The Proposed Framework

In order to understand the key steps in the model-free, data-driven control design, it is worth recalling that in standard model-based MPC, the dynamic model is used to compute the components of the optimization problem to be solved. This includes:

- The cost function that links, for the current value of the state, say x_k, the future control sequence $\boldsymbol{u}_k$ to the associated value of the cost function. This relationship implicitly incorporates the computation of the state trajectory by integrating the model's equations of evolution.
- The constraints violation function which is also based on the computation of the state trajectory and the evaluation of the degree of violation of the constraints.

Once these functions are available, in the standard MPC formulation, the optimization problem is ready to be formulated and solved leading to the optimal control sequence $\boldsymbol{u}_k^\star$ from which, the first control is applied following the principle of MPC feedback. This is the only *used* outcome of the whole computation process.

Therefore, the key point in the above process is to define an optimization problem that is parameterized by the current situation which is *summarized* by the current value of the state.

This reminder serves in establishing the following fact:

One Only Needs a Parameterized Optimization Problem
MPC is all about being able, for a given current situation, to compute the best sequence of actions ($\boldsymbol{u}_k^\star$) that minimizes the cost[5] given the current situation.

Now in the absence of the model's equation and the absence of knowledge regarding the very definition of the state of the system under study, the above mentioned paradigm is handled through the following two *tricks*:

1. The *current situation* is represented by the past time-series of the output and the control input, namely:
$$\boldsymbol{y}_{k-N} \quad \text{and} \quad \boldsymbol{u}_{k-N-1}.$$
2. The algorithmic evaluation of the cost function is replaced by a *fitted* function that is *learned* based on the availability of experimentally acquired measurements of the system whose very possibility is guaranteed by the working assumption discussed above.

Based on the above discussion, the three main tasks to be achieved that are discussed in the remainder of this section are the following:

[5] The constraints are supposed to be incorporated in the cost function via appropriately weighted penalty.

1. Make a decision about the experiment to be conducted so that a sufficiently rich learning data is acquired for the learning process. This might be an open-loop sufficiently long excitation or a *loosely defined feedback law*. This depends on the system to be controlled. The above discussed assumptions guarantee that at least one of the two options is available.
   This aspect is discussed in Sect. 8.5.1 below.
2. Use the so collected data to identify the map (8.4) which incorporates the possible constraints (other than the control saturation constraints) through exact penalty soft constraints associated term. This is done using `Machine Learning` (ML) like identification structures.
   This aspect is discussed in Sect. 8.5.2 hereafter.
3. While the above steps are done once for all and off-line, the last task is the one that is to be repeatedly done at each control updating period, namely, solving the current instance of the optimization problem that depends on the past measurements. Note however that, depending on the structure of the map used in the second step, some specific solvers might be needed for a successful resolution of the optimization problem. This is in particular the case when non differentiable functions are used in the identification of the cost function.
   This aspect is discussed in Sect. 8.5.3 below.

8.5.1 Generate the Learning Data

As mentioned above, this task is problem-dependent. For some systems, some non necessarily optimal feedback might be possible to design that maintains the system in a bounded excitation regimen that enable to grasp information about the hidden unknown relationships. For other systems, the open-loop stability renders possible to excite the system with a sufficiently rich signal in order to collect the measurement needed to construct the learning data.

For the example we are interested in, it is the second alternative that is assumed to hold. Namely, a rather long sequence of inputs is sampled uniformly inside the admissible set given by:

$$\mathbb{U} := [0.049, 0.449].$$

This sequence of inputs is applied to the system and the resulting measured output sequence is collected to build the learning dataset.

More precisely, the following procedure is applied to intentionally induce non uniform piece-wise time segment length where constant input is used:

- A set of 1000 random instances of interval duration is generated. These values are sampled randomly in the set of integer multiples of τ ranging from τ to $100\,\tau$. This results in a mean duration of roughly 50τ and a randomly distributed duration ranging between 0.02 and 2 time units.[6]

[6] Remember that the time is normalized.

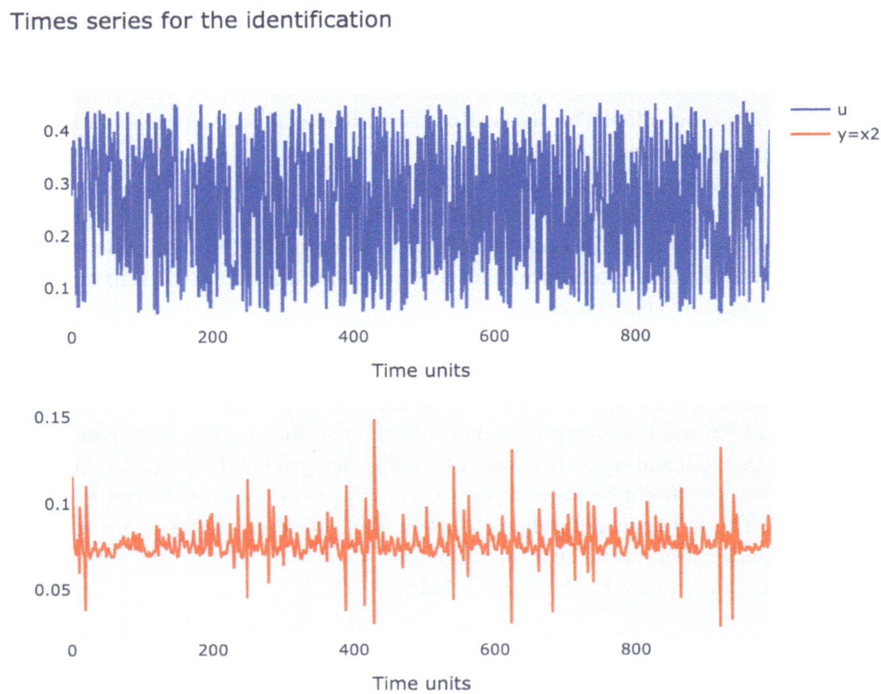

Fig. 8.2 Typical input and output time-series for randomly generated input sequence used to build the learning data for the cost function

- A corresponding set of 1000 values of the input is uniformly sampled inside the admissible set of control values $\mathbb{U}$.

Typical instances of the inputs and outputs measurements are shown in Fig. 8.2. Regarding the sensitivity of the overall results to the specific instance of generated excitation input, the experience showed that the quality of the results is quite robust provided that the sequence is sufficiently long.

Nevertheless, this assessment holds for the example used in this chapter. The sensitivity of any data-driven solution to the quality of the training data is a general fact.

8.5.2 Identification of the Cost Function

Recall that the objective here is to identify a map that fits the relationship (8.3) [or equivalently (8.4)] that holds between the past measured time-series and the future control sequence on one hand and the associated value of the cost function on the other hand.

More precisely, recall that the value of the cost function is built from the collected measurements using the definition (8.7) where it appears clearly that the cost function only depends on the collected measurements (here precisely the output measurement y).

This problem is a standard **regression** ML problem where one looks for a function F that maps a features vector ξ to some continuous label η, namely[7]:

$$\text{(Regression problem)}\ \ \text{Find } F \text{ s.t } \ F(\xi) \approx \eta, \tag{8.8}$$

where the $\approx$ is to be understood in terms of some measure of the statistics of the errors:

$$\eta_i - F(\xi_i),$$

over a set of realizations:

$$\mathcal{D} := \{(\xi_i, \eta_i)\}_{i=1}^{n_r},$$

that is referred to as the learning data set.

Notice that in the case of the example under study, the features vector and the label are defined by (see Fig. 8.3):

$$\xi = (\boldsymbol{y}_{k-N}, \boldsymbol{u}_{k-N-1}, \boldsymbol{u}_k), \tag{8.9}$$

$$\eta = J(\boldsymbol{u}_k \mid \boldsymbol{y}_{k-N}, \boldsymbol{u}_{k-N-1}). \tag{8.10}$$

From the long experimentally induced measurements and using a rolling window of length N over the time-series depicted in Fig. 8.2 that contains 50324 lines, a number of $n_r = 50324 - 2N$ of samples can be extracted in which, each line corresponds to the features and label contained in the rolling window. This end up with the the training data set $\mathcal{D}$ mentioned above (see Fig. 8.3). An illustration of the *rolling window process* is schematically shown in Fig. 8.4.

As it is recalled in the section dedicated to ML of Chap. 2, the quality of the identified map F should be evaluated using a test set containing new *unseen realizations* in order to check the generalization power of the obtained map.[8]

As it is shown later (see the validation Sect. 8.6), the size N of the rolling window, used in the above process, can be as large as 100 when using cost functions that are defined on a prediction horizon of 2 time units.[9] Therefore, when considering the features vector ξ defined by (8.9), the process induces a features vector of dimension 300 which is obviously prone to over-fitting (too many parameters). Consequently, something needs to be done before stepping to the ML fitting step.

[7] Refer to **Chap.** 2 for a reminder of Machine learning terminology and concepts.

[8] Or to say it in other words, in order to evaluate the degree of over-fitting that might occur when using a too complex model compared to the available data.

[9] Recall that the sampling period is $\tau = 0.02$ which is needed to ensure stability of the numerical integration as the system dynamics is stiff.

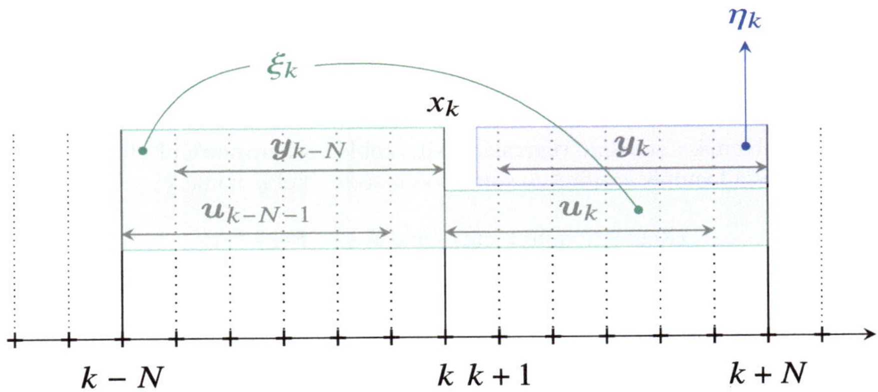

Fig. 8.3 Definition of ξ_k and η_k: The learning data is built by a moving window over the generated open-loop trajectory. Using a rolling window of length N over the time series depicted in Fig. 8.2 that contains 50324 lines, it is possible to extract a number of $n_r = 50324 - 2N$ samples to build the data set $\mathcal{D}$

A typical way to reduce the dimension of the features vector is to use Principal Component Analysis.[10] While this a straightforward and easy-to-implement solution, it is not necessarily the appropriate one when handling physical time series. In this chapter, a different *moment-like* defined features are used which amounts at considering **the average of the successive derivatives of a smoothed version of the signals over the time intervals**.

More precisely, given a signal s defined on some interval I, the following features are defined:

$$\phi_j(s) := \frac{1}{|I|}\sum_{i=1}^{|I|} s^{(j)}(i) \quad j \in \{0, 1, \dots, m-1\}, \tag{8.11}$$

where the notation $s^{(j)}(i)$ is used to refer to the value at sampling instant i of the j-th **filtered** derivative of the signal s with the convention $s^{(0)} \equiv s$. These filtered versions can be obtained by feeding the set of filters of the form:

$$D_j(p) := \Big[\frac{p}{1+\tau_f p}\Big]^j \tag{8.12}$$

with the measured time-series.

This features extraction process is implemented in the `feature_gen` method of the `siso_predictor` module that is available on the GitHub repository [4]:

https://github.com/mazenalamir/siso_predictor,

.

[10] Using for instance the `sklearn.decomposition.PCA` module.

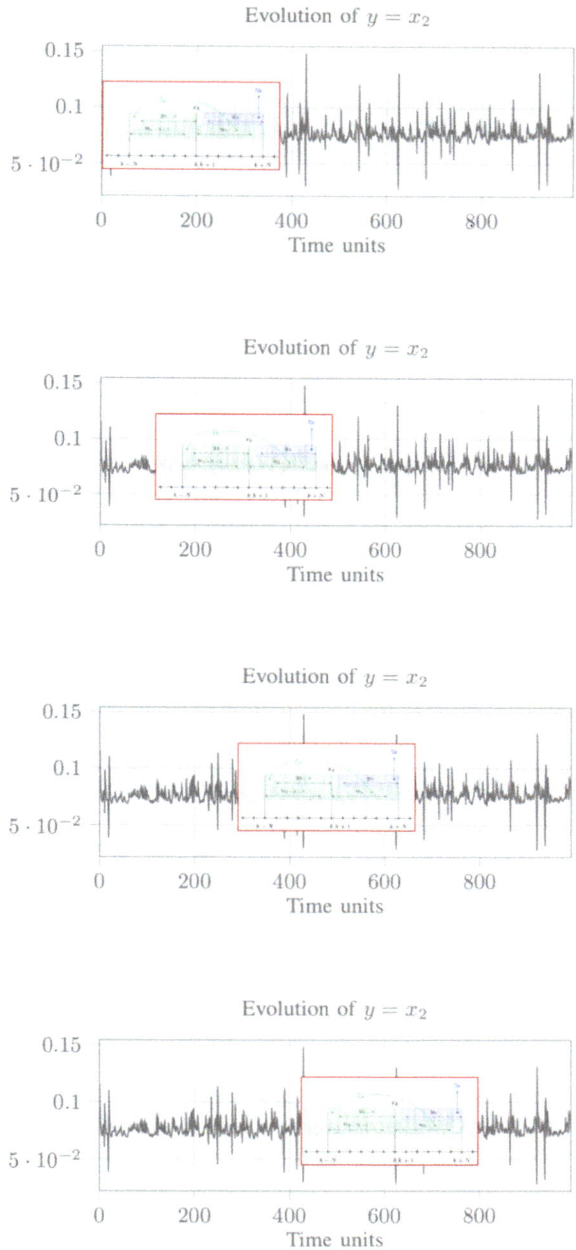

Fig. 8.4 Illustration of the rolling window process enabling to create the matrix of features from a long experiment's measurement collection

which is more detailed below but before, let us answer the following legitimate question:

About the Use of High Derivatives?

Is it wise to use high derivatives of physical signals?

The answer is that the degree of filtering in the above features generation option (tailored by the parameter τ_f involved in (8.12)) might be quite strong while still providing reasonably discrimninant features. This is because the feature extraction process is not expected to provide the *true*, nor quite close values to the true derivatives. This is because the resulting features are not used as *representative* of the true values of the derivatives. Rather, they are expected to give enough information regarding the current configuration of the hidden state of the process.

In other words, the feature extraction process is just a way to give a characterization of the current situation (state) of the system by using a past window of measurements as such characterization is more likely to produce meaningful and trustworthy results than standard PCA-related features that would have been extracted using measurement time-series.

The characterization that is represented by the resulting derivatives-related features can be qualified as follows: *The state of the system is the one that leads to these averages of the filtered time-series of the three first derivatives!*

Notice however that *ultimately*, it is the effectiveness of this choice in terms of the reconstruction of the cost function that it to be used in order to evaluate the success of this features extraction choice. The previous arguments are given in order to convey the intuition behind this choice that showed efficient for the specific example at hand.[11]

Now by extracting the m features of $\boldsymbol{y}_{k-N}$, $\boldsymbol{u}_{k-N}$ and $\boldsymbol{u}_k$ that are involved in the definition (8.9), one defines the following features extraction map:

$$\begin{aligned}\Phi(\xi) := \Big[&\phi_0(\boldsymbol{y}_{k-N}), \dots, \phi_{m-1}(\boldsymbol{y}_{k-N}),\\ &\phi_0(\boldsymbol{u}_{k-N}), \dots, \phi_{m-1}(\boldsymbol{u}_{k-N}),\\ &\phi_0(\boldsymbol{u}_k), \dots, \phi_{m-1}(\boldsymbol{u}_k)\Big] \in \mathbb{R}^{3m}. \qquad (8.13)\end{aligned}$$

Dimension of the Features Vector

As far as the example is concerned, $m = 3$ is used in the implementation leading to a features vector of dimension 9.

Having this reduced vector of features $\Phi(\xi)$ at hand, it is possible to choose one of the various available *ML* nonlinear structures (SVM, Random Forest (RF), Neural

[11] Nevertheless, it is the conviction of its wide applicability that justified making it available in the corresponding `GitHub` repository and very likely soon as a `pip-installable` python module.

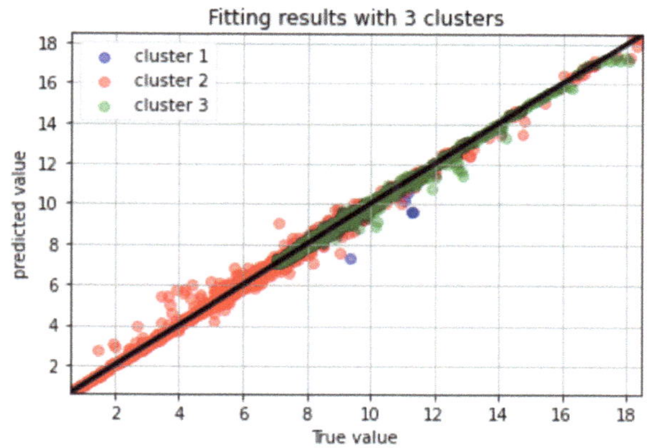


```
-----
Mean abs(Error) on all cluster= 0.022830
-----
abs(Error) - cluster 1 = 0.008946
abs(Error) - cluster 2 = 0.037074
abs(Error) - cluster 3 = 0.022471
```

Fig. 8.5 Typical fitting results, obtained by the `siso_predictor` module, with statistics on the prediction error. $N = 50$ and $\alpha = 100$ is used in the terminal penalty (8.7)

Network (NN), Ridge, Lasso, to cite but a few of the available structures) in order to solve the regression problem that can be shortly denoted as follows:

$$\eta \approx \text{ML}(\Phi(\xi)) =: \hat{J}(\xi). \tag{8.14}$$

As far as the example is concerned, a `RandomForest` structure with a controlled maximum number of leaves nodes has been used.[12]

As a matter of fact, the method `learn_model` of the module `siso_predictor` enables to build a piece-wise `RandomForestRegressor` model with a desired number of clusters in order to increase the precision of the model while controlling the complexity of each sub-model).

An example of the fitting results that can be achieved using a 66-33% of training/testing split ratios of the training data set is shown in Fig. 8.5.

As for the clustering (if any), it is obtained using the $3m$ features contained in $\Phi(\xi)$ using a standard `K-Nearest-Neighbors` (KNN) clustering algorithm of the `scikit-learn` library [5].

8.5.2.1 More About the `siso_predictor` Module

The `python` module `siso_predictor` that is used in Sect. 8.5.2 is available at

https://github.com/mazenalamir/siso_predictor,

[12] `sklearn.ensemble.RandomForestRegressor`.

and the useful methods can be imported via the following import script:

```
from siso_predictor import learn_model, construct_regressor
from siso_predictor import predict, GenericResults
```

where all the methods and the classes used hereafter are imported.

Once the time-series collected through the experiments are available, the `learn_model` method that *fits* the model of the cost function is called using the following script:

```
sol = learn_model(
    y=y,
    U=U,
    ydef=ydef,
    N=100,
    n_clusters=3,
    nJump=1,
    max_leaf_nodes=1200,
    test_size=0.33,
    validation_mode='all',
    plot=True
    )
```

where the **input arguments** of the call can be described as follows:

- `y` the time-series representing the output measurements during the experiment;
- `U` the time-series representing the input measurements during the experiment; and
- `ydef` The function that defines the measurement-dependent value defining the label for the `ML` regression problem, namely, the cost function.

The calling syntax of the `ydef` function is the following (see Fig. 8.3 for the significance of the arguments):

```
def ydef(ypast=None, upast=None, yplus=None, uplus=None):
    ...
    J = ....
return J
```

For instance, in the case of the example we are working on, this function is defined by the following script which simply indicates that the targeted value (which will be minimized by an appropriate choice of the future sequence of control) is the difference between the future mean value of the output $y = x_2$ on the future interval, which is the result of the choice of the future control sequence, and the mean value of the output on the previous time interval which is independent of the decision variable and can hence be viewed as constant from the point of view of the optimizer,[13] namely:

```
def ydef(ypast=None, upast=None, yplus=None, uplus=None):
    J = (yplus.mean()-ypast.mean()) + alpha * yplus[-1]
return J
```

which obviously implements the definition given by (8.7) (without the minus sign) including the terminal penalty-related term.

`N`	The width of the rolling window used to separate the *past* and the *future* time-series.
`n_clusters`	The number of clusters used to define the piece-wise computed fitted maps.
`nJump`	The under-sampling parameters used in the extraction of the features/label lines from the learning data. This might be use-

[13] The reason for which it is the increment that is identified and not the future average is that this leads to more robust identification since for small value of the window size N, the regression problem might be solved using a trivial map that simply predicts the future value to be equal to the previous one.

	ful for huge datasets since by taking `nJump>1` the training time might be reduced.[15]
`max_leaf_nodes`	The maximum number of leaves to be allowed in the RF sub-models when fitting the map on the different clusters.
`test_size`	The ratio used when splitting the data into train and test sub-sets. For instance, the use of `test_size=0.3` leads to training the model using 70% of the data.
`validation_mode`	This option ∈ {`all`, `learning`, `test`} enables to choose the set used to plot the quality of the prediction in the (`true`, `predicted`) 2D plan. More precisely, the (`all`) choice lead to plotting all the data, the (`learning`) choice shows the matching results only on the training data while the (`test`) choice restricts the plot to the unseen samples in the test data. Notice that this concerns not only the plots that show the cloud of points but also the statistics regarding the dispersion of the absolute values of the prediction error as shown in Fig. 8.5.
`plot`	Boolean option indicating whether the plot is required or not.

Once the model is satisfactory fitted as it can be witnessed by the statistics of the errors on the test set, it is possible to use the resulting object, namely `sol` in the prediction of the targeted variable (here the cost). This prediction takes the following syntax that uses the `predict` method imported earlier:

```
J = predict(ypast, upast, uplus, sol)
```

where the first three arguments are self-explanatory while the returned solution `sol` by the `learn_model` method is used as the last argument of the `predict` method.

This is just a quick overview of the content of the `siso_predictor` module. A full description might be obtained by examining the scripts available at the previously mentioned `GitHub` repository. Moreover, the module can be strongly enriched by adding an option that enables the use of `XGBoost` regressor which shows quite often better performance than the Random-Forest one. Moreover, it is important to underline that that the **current version is limited to scalar output measurements**.

In the next, section, the `predict` method is used to define the cost function that is to be solved in order to implement the data-driven model-free NMPC which is the main objective of the present chapter.

[15] Notice that the computation time for the example studied in this chapter is negligible and does not justify the use of large value of `nJump`.

8.5.3 Solving the Optimization Problem

This task uses the previously identified map in order to find the *best* sequence of future control inputs given the previous measurements of output and control input over some past rolling measurement window of length N.

The optimization problem can therefore be expressed as follows:

$$P(\boldsymbol{y}_{k-N}, \boldsymbol{u}_{k-N}): \quad \boldsymbol{u}_k^* \leftarrow \arg\min_{\boldsymbol{u}\in\mathbb{U}^N} \hat{J}(\boldsymbol{y}_{k-N}, \boldsymbol{u}_{k-N}, \boldsymbol{u}), \tag{8.15}$$

in which

- $(\boldsymbol{y}_{k-N}, \boldsymbol{u}_{k-N})$ are the time-series gathering the previous measurements (obtained from the past rolling window $[k-N, k]$). These are viewed as the parameters of the problem,
- $\boldsymbol{u}$ is the future sequence of inputs which serves as the decision variable for the optimization problem.

Notice however that the dimension of the decision variable $\mathbf{u}$ is (Nn_u) which can be quite large. Fortunately, as recalled in Sect. 2.6.3, the number of decision variables can be drastically reduced using the control parameterization trick without necessarily a large impact on the quality of the closed-loop cost.

More precisely, the control parameterization enables to keep as free variables only the values of the control input at some few chosen specific future instants on the prediction horizon. The remaining values are inferred from the latter by linear interpolation.

Figure 8.6 shows an example of such piece-wise linear control profile over a prediction horizon of length $N = 50$ and using three degrees of freedom which are the control applied during the sampling periods starting at instants $k = 0$, 10 and 30.

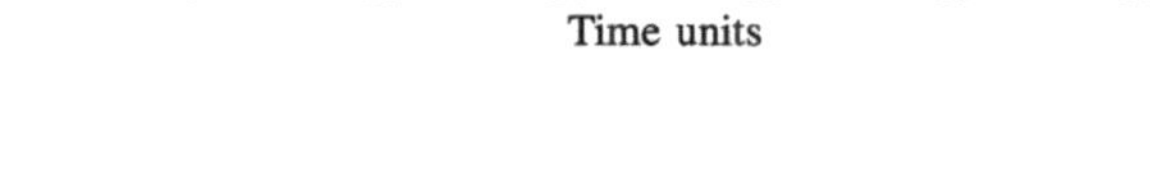

Fig. 8.6 Example of piece-wise linear control profile resulting from a parameterization that involves three-degrees of freedom which are the values of the control at instants $k = 0$, 10 and 30. $N = 50$ in this example

Using this control parameterization, one can denote the future control profile by the following notation that explicitly indicates that it depends on the reduced dimensional vector of decision variables p:

$$\boldsymbol{u} := \mathcal{U}(p).$$

By so doing, the vector p becomes the new vector of degrees of freedom and the optimization problem (8.15) to be solved becomes:

$$P_r(\boldsymbol{y}_{k-N}, \boldsymbol{u}_{k-N}) : \quad \boldsymbol{p}_k^* \leftarrow \arg \min_{\boldsymbol{p} \in \mathbb{U}^{n_p}} \hat{J}(\boldsymbol{y}_{k-N}, \boldsymbol{u}_{k-N}, \mathcal{U}(\boldsymbol{p})). \tag{8.16}$$

Now, at this stage, one might might consider that the problem is the one consisting in solving an optimization problem which can be done using many available solvers.

This is only partially true because many of the `ML` structures that can be invoked from the available libraries in order to solve similar regression problems and hence to fit a nonlinear function as expressed in (8.14) are **discontinuous in their arguments by nature**.

This is in particular true for the RF regressor which is mainly expressed via `if-then-else` rules. This means that the optimization method should be selected with some care since it has to accommodate for cost functions that are not only non differentiable but even discontinuous.

To address this issue, it might be useful to use a modified version of the derivative-free Torczon algorithm [6,7] that is made available [8] through a `python` module that can be installed using the simple script:

```
pip install torczon
```

This algorithm is particularly adapted to discontinuous cost functions since it implements a derivative-free algorithm that only needs to call some *black-box* defined function. It also incorporates a multiple starting point mechanism in order to reduce as far as possible the risk of the iterations being *trapped* by local minima.

Before we proceed in using this module and to simulate the resulting closed-loop behavior of the controlled system, this `python` module is briefly described in the next section. It is also possible to consult the project `Pypi` home page at:

https://pypi.org/project/torczon/,

for a full description of the module and its use together with some use-case examples.

Notice however that this tool is to be used only for moderate size problems in terms of the number of decision variables. This was another reason that justifies the use of the control parameterization step described above.

8.5.3.1 Brief Description of the torczon Module

In a nutshell, the Torczon algorithm starts with a regular polygon in $\mathbb{R}^n$ where n is the dimension of the decision variable and improves it iteratively based on the values of the cost functions at its nodes.

The updating process that aims at improving the quality of the polygon, in the sense invoked above, involves three types of transforming operations, namely (see Fig. 8.7):

- contraction,
- reflection, and
- extension.

The one used at some stage depends on the success/failure of the last transformation. More precisely, success lead to extension while failure might lead to either contraction or reflection. Notice that during this process, only function evaluation is required and discontinuity of the cost function induces no specific issue.

Figure 8.7 illustrates the three transformations starting from an initial polygonal to which a reflection, extension and contraction are successively shown. In order to use the optimization solver implemented in the Torczon module, the cost function should be provided through a python function showing the following syntax:

```
def f(x,par):
    c =
    return c
```

where par is any regular python object that might be needed in order to compute the output returned by the function.

The above function and the value of the argument par are used as the value of the input arguments f_user and par respectively as shown in the script below:

```
solve(f_user=f,
      par=par,
      x0=x0,
      xmin=xmin,
      xmax=xmax,
      Nguess=Nguess,
      Niter=Niter,
      initial_box_width=0.1)
```

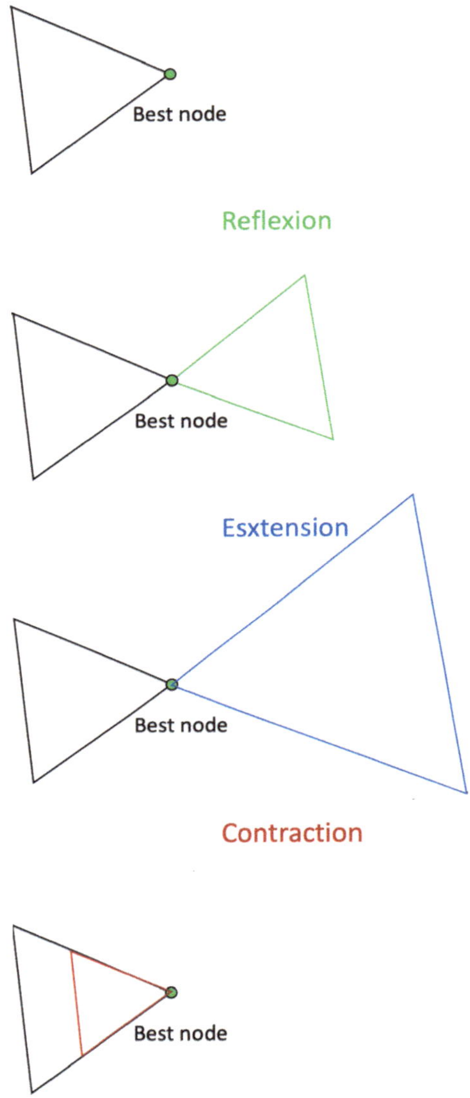

Fig. 8.7 Illustration of the three polygon's transformations used in the derivative-free Torczon algorithm in the case where the optimization problem shows two decision variables

In addition to the previous two arguments, the following input arguments need to be provided:

`x0`	The initial guess for the decision variable.
`xmin`	The vector of lower bounds on the components of the decision variable.

`xmax`	The vector of upper bounds on the components of the decision variable.
`Nguess`	The number of multiple starting points that helps avoiding local minima.
`Niter`	The maximum number of iterations per starting point.
`initial_box_width`	The size of the initial polygon.

8.6 Results and Discussion

The following parameters are used in the closed-loop simulations:

- Prediction horizon $N = 100$, sampling period $\tau = 0.02$.
- In order to examine the impact of the control parameterization on the quality of the closed-loop, two control parameterization options are used and their results are compared. The first option is defined so that the computation time is reduced while the second option is computationally more involved a priori.
  More precisely the two compared options are defined as follows:

  1. In the **first option**, a control parameterization involving two degrees of freedom is considered in which, the values of the control at the sampling instants $\{0, 10\}$ are taken as free decision variables to define the piece-wise affine control profile (see Fig. 8.6). As for the number of starting points, one single initial guess ($N_{\text{guess}} = 1$) is used and the maximum number of iterations of the Torczon algorithm is fixed to $N_{\text{iter}} = 30$.
  2. In the **second option**, four degrees of freedom for the control parameterization are used, namely the values of the control during the sampling periods that start at the sampling instants $\{0, 5, 10, 20\}$. The number of starting points is taken equal to $N_{\text{guess}} = 3$ and the number of allowed iterations per initial guess is taken equal to $N_{\text{iter}} = 50$.

  These two options are summarized in Fig. 8.8.
  Four each of these two options, **nine different initial states** have been used to check the robustness of the results to context variations. For each of these initial states, the closed-loop system under each of the above settings is simulated.
  The results are shown in Figs. 8.9 and 8.10.
- The identification of the model has been done using a random forest regressor with a single cluster and a maximum number of leaves nodes set to 1200.
- Note that since the cost function's prediction model needs the measurement over the past window of length N, the first N control actions are taken randomly waiting for the buffer to be filled. This corresponds to the first 2 time units of the simulations.

Setting 1

- $p := (u(0), u(10)) \in \mathbb{U}^2$
- $N_{guess} = 1$, $n_{iter}^{max} = 30$ (Torczon)
- $\rightarrow 1 \times 30 \times (2+4) = 180$ functions evals

Setting 2

- $p := (u(0), u(5), u(10), u(20)) \in \mathbb{U}^4$
- $N_{guess} = 3$, $n_{iter}^{max} = 50$ (Torczon)
- $\rightarrow 3 \times 50 \times (4+4) = 1200$ functions evals

Fig. 8.8 Summary of the parameters used in the two options defining the data-driven MPC feedback

It can be noticed that, despite the absence of knowledge regarding the mathematical equations governing the unknown system, the results are globally quite good in terms of the convergence towards the vicinity of the optimal a priori **unknown value** 0.0846.

Notice however that the oscillations discussed above that are the signature of the rigorously optimal regimen are not present. This is **most probably due to the low dimensional control parameterization** that prevents the cycles from taking place since its open-loop realization is not inside the possible options offered by the parameterized control profile.[16]

It can also be observed that the quality of closed-loop is slightly better under the second settings which enables more iterations, more initial guesses and a more rich control parameterization. However, it should be kept in mind that the maximum number of iterations is 5 times higher in this setting compared to the first (3×50 compared to 1×30).

As a matter of fact, what does really matter here is to assess the fact that the identified model seems appropriate and the remaining issue concerns the tuning of the hyper-parameters which is a typical problem/hardware dependent trade-off.

8.6.1 Computation Time

The computation leading to the results shown in Figs. 8.9 and 8.10 is performed using `Python 3 jupyter-notebook` on a MacBook-Pro 2017, Mojave 10.14.5 with a 2.9 GHz Core i7 processor.

The computation of a single evaluation of the identified Random-Forest cost function (8.14) takes less than 1.7 ms. Since the Torczon algorithm performs $(n+4)$ evaluations per iteration (where n is the number of decision variables), this leads to a maximum MPC computation time when $n = 2$, $N_{\text{iter}} = 30$ and $N_{\text{guess}} = 1$ (the scenario of Fig. 8.9) which is about 0.3 s against 1.53 s for the settings of Fig. 8.10.

[16] On the other hand, the resulting quasi-steady behavior might after all be the desired one since the slight increase in production induced by the fully optimal limit cycles comes at the price of undesirable fatigue on the actuator.

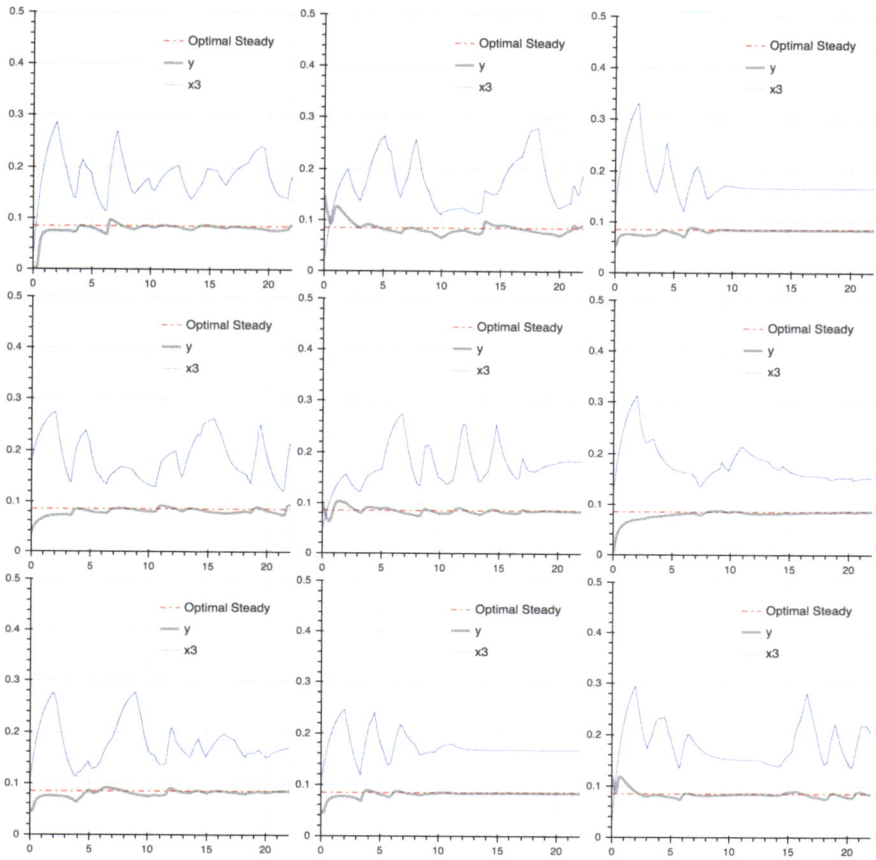

Fig. 8.9 Evolution of the closed-loop system under the data-driven-based MPC for **nine different initial states**. Width of the rolling window $N = 100$, maximum number of iterations $N_{\text{iter}} = 30$ and number of initial guesses $N_{\text{guess}} = 1$. The control parameterization is defined by the values of the control during the sampling period starting at instants $\{0, 10\}$

Now the ultimately used options depend on the characteristic time of the system compared to the required computation times. For slow processes, there might be room for a quite large number of iterations, starting points and degrees of freedom used in the control parametrization.

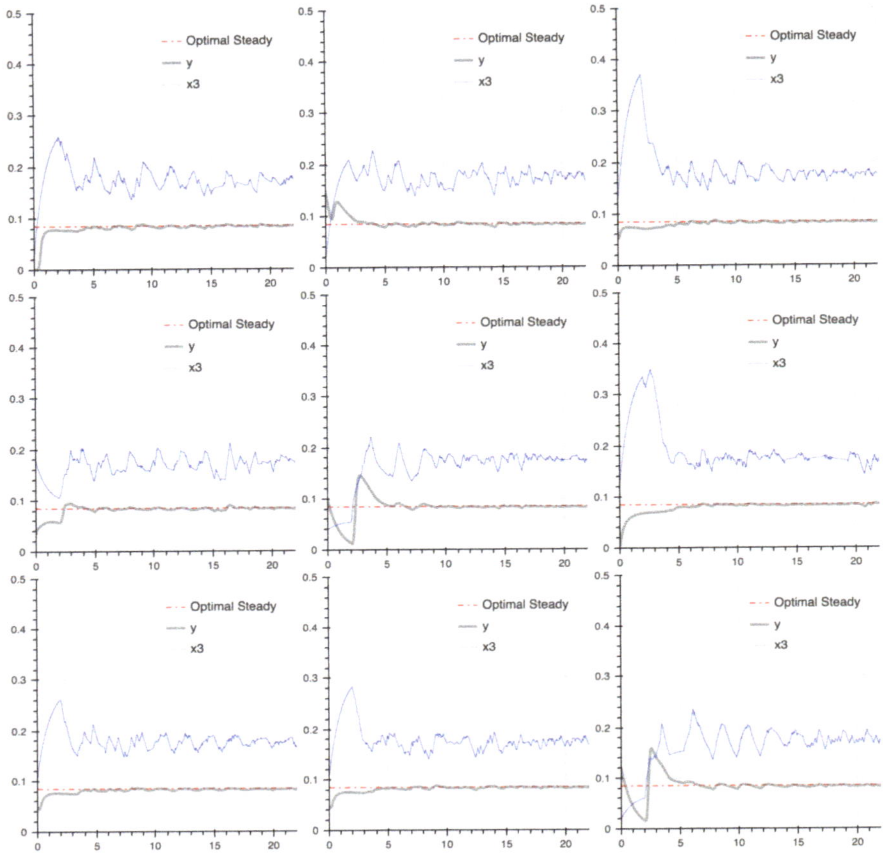

Fig. 8.10 Evolution of the closed-loop system under the data-driven-based model predictive controller for **nine different initial states**. Width of the rolling window $N = 100$, maximum number of iterations $N_{\text{iter}} = 30$ and number of initial guesses $N_{\text{guess}} = 1$, three decision variables at instants $\{0, 5, 10, 20\}$

8.7 Conclusion and Future Work

In this chapter, a complete framework for data-driven control for systems with unknown mathematical prediction model is presented and tested through a rather challenging nonlinear control problem.

As it is clearly stated in the beginning of this chapter, the framework applies to the class of dynamical systems for which it is possible to apply an open-loop or a priori simple output feedback control law which can be far from being optimal in order to create the learning data. Moreover, it is required that the control objective as well as the constraints to be enfored can be expressed in terms of the measured variables.

A specific, tailored choice of the features extraction is proposed and validated on the illustrative example. Moreover, the associated freely available `python` modules are explained and their use illustrated.

To summarize, the solution proposed in this chapter is not universal but it should be added to the panorama of tools that might thought of when the above conditions are fulfilled. This is the case of all the solutions proposed in this book. It is only when all of them are understood and their advantages and drawbacks are fully mastered that one can be fully armed to address the different real-life problems that require, as always, problem-specific customization of the finally adopted solution.

References

1. Müller, M.A., Angeli, D., Allgöwer, F., Amrit, R., Rawlings, J.B.: Convergence in economic MPC with average constraints. Automatica **50**(12), 3100–3111 (2014)
2. Mayne, D.Q., Rawlings, J., Rao, C.V., Scokaert, P.O.M.: Constrained model predictive control: Stability and optimality. Automatica **36**, 789–814 (2000)
3. Andersson, J.A.E., Gillis, J., Horn, G., Rawlings, J.B., Diehl, M.: CasADi—A software framework for nonlinear optimization and optimal control. Mathematical Programming Computation (2018)
4. Alamir, M.: siso_predictor: A python source for the implementation of the identification framework. GitHub repository: https://github.com/mazenalamir/siso_predictor (2020)
5. Pedregosa, F., Varoquaux, G., Gramfort, A., Michel, V., Thirion, B., Grisel, O., Blondel, M., Prettenhofer, P., Weiss, R., Dubourg, V., Vanderplas, J., Passos, A., Cournapeau, D., Brucher, M., Perrot, M., Duchesnay, E.: Scikit-learn: Machine learning in Python. J. Mach. Learn. Res. **12**, 2825–2830 (2011)
6. Dennis, J.E., Jr., Torczon, V.: Direct search methods on parallel machines. SIAM J. Optim. **1**(4), 448–474 (1991)
7. Torczon, V.: On the convergence of pattern search algorithms. SIAM J. Optim. **7**(1), 1–25 (1997)
8. Alamir, M.: Torczon: A simple python implementation of torczon algorithm. GitHub repository: https://github.com/mazenalamir/torczon (2020)

9 Nonlinear Moving-Horizon Extended Observers

9.1 Chapter's Content at a Glance

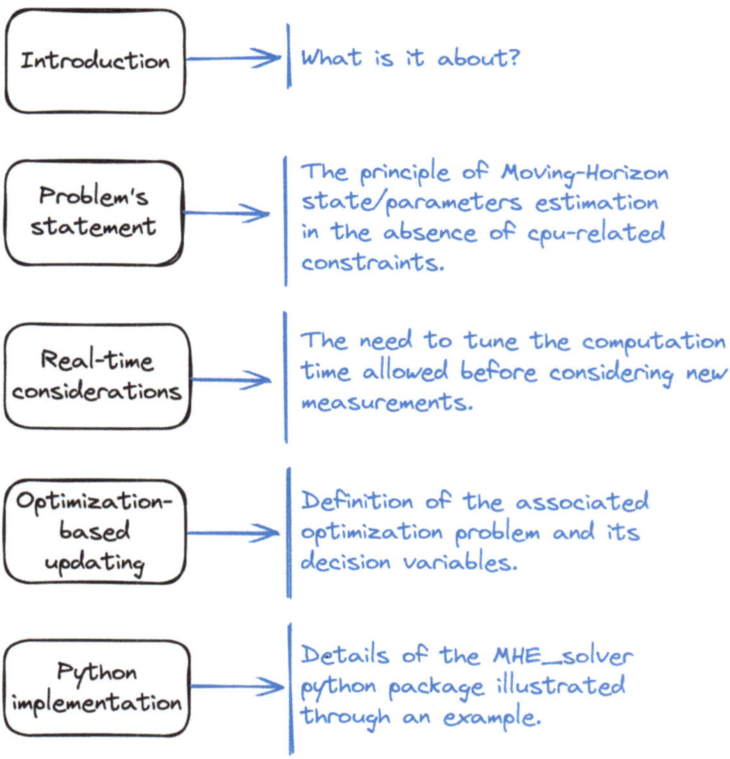

M. Alamir, *Nonlinear Control of Uncertain Systems*,
https://doi.org/10.1007/978-3-031-93287-8_9

9.2 Introduction

In this chapter, we shall address one of the major techniques that enables to address the presence of uncertain parameters in the model's equations, namely, the on-line parameters estimation via the design of dynamic observers.

Indeed, in many situations, one does not need to view the uncertainties in the model as totally unknown quantities that might play against the control objective and, as such, should be addressed on a worst-case basis or even on a statistical or certification-oriented basis.

Rather, they can be considered as unknown quantities that can be *reconstructed* on-line and updated during the life-time of the system thanks to the on-line measurements coming from the available sensors.

These measurements, coupled with the knowledge of the mathematical model of the system provide *precious* pieces of information regarding the true values of the model's parameters.

This idea has already been discussed in **Chap.** 2 (see Sect. 2.10) from a high level conceptual point of view. Since this chapter is dedicated to the concrete implementation of this idea, some minor differences might exist because we need here to be fully rigorous in order to come out with the associated algorithms that leave no room for a comprehensive *fuzzy* presentation.

But do not worry, these are really minor differences that many readers would even not notice. The other readers should simply consider that the notation of this chapter are the ones to adopt.

One important aspect to be underlined is that this chapter addresses the problem of full simultaneous state and parameters estimation as described in Sect. 2.10.1 by opposition to the partial reconstruction of an observation target z that is mentioned in Sect. 2.10.3.

The reason for this specific treatment associated to the full estimation problem is that the partial estimation problem can *simply* be viewed as a regression problem that can be addressed using standard `Machine Learning` (ML) tools where the concatenation of the previous instances of the measurements are used as features while the observation target z is viewed as the label to be matched. More precisely the lines that are added to the features and the label matrix and vector respectively to represent the relationship at instant k translate the following relationship:

$$z_k \approx \texttt{ML}\left(\begin{bmatrix} y_k \\ \vdots \\ y_{k-N_O} \end{bmatrix}\right) := \mathcal{O}\left(\mathbf{y}_k^{(-)}\right). \tag{9.1}$$

So the principle is quite simple although the generation of the learning data might be a delicate question in terms of representativity of the chosen scenarios, namely: *How to choose the set of initial states, the vectors of parameters and the input trajectories (if any) so that the resulting set of time-series from which the ingredients involved in* (9.1) *can be collected is sufficiently representative for a solid identification of the map* $\mathcal{O}$ *invoked in* (9.1).

One of the advantages of the full state/parameter reconstruction algorithms studied in the present chapter lies in the fact that they are used on-line on any instance of the initial state, the vector of parameters and the scenario being used. The quality of the reconstruction does not rely on the quality of some *learning datasets that might not be representative enough.*.

However this is obtained at the cost of an on-line optimization that might be time consuming and hence might require a special care regarding the choice of the implementation parameters as it is already shown in the case of Nonlinear Model Predictive Control studied in **Chaps.** 5 and 6.

Recall however that the success of the on-line full-state and parameters estimation requires that the quantities one aims at reconstructing are fully observable for the chosen set of measurements which is not always granted.

The resulting conclusion is that each type of reconstruction, namely, partial or full, comes with its advantages, drawbacks and its appropriate instances of application and therefore, as it is always the case, it is worth to known both approaches and choose which one to use depending on the context.

As in all the chapters of this book, the first step is to precisely give the class of systems we are interested in and some of the underlying assumptions behind the philosophy of the solution and algorithms.

9.3 Problem Statement

In this chapter, we consider dynamical systems that are governed by the following set of discrete-time state and measurement equations of the form:

$$x_{k+1} = f(x_k, u_k, p), \tag{9.2a}$$

$$y_k = h(x_k, u_k, p) + \nu, \tag{9.2b}$$

where

✓ $x \in \mathbb{R}^{n_x}$ is the state vector.

✓ $u \in \mathbb{R}^{n_u}$ is the exogenous input vector that gathers control input but also any measured exogenous inputs such as set-points, or other measured non controlled signals, etc. Indeed since we are interested in the problem of state/parameters estimation, there is no fundamental difference between the control inputs and the other known quantities since from the point of view of the estimator, these represent known information whatever is the reasons that *give* them such or such behavior.

✓ $p \in \mathbb{R}^{n_p}$ is the vector of parameters of the model that might in some circumstances need to be estimated from on-line measurements. This vector of parameters is supposed to be constant or very slowly varying in principle even if the proposed estimation scheme might be successful in some situations where the vector of parameters varies quite rapidly. Nevertheless, the spirit that lies behind the formulation proposed in this chapter is based on the above *slow-variation* assumption.

✓ y is the vector of measured quantities **including the input vector u.**

✓ $\nu \in \mathbb{R}^{n_y}$ is the vector of measurement noise that affect the sensors outputs.

✓ $k \in \mathbb{N}$ is the sampling time index under the assumption that the discrete-time version of the dynamics is obtained using the sampling acquisition period of the measurement which is therefore assumed to be the same for all sensors. It is important to keep in mind that this last assumption is not always satisfied *to say the least* since the characteristic time of the different phenomena are generally quite different. Nevertheless:

! About the Uniform Sampling Assumption

In the case where the sensors show different acquisition periods, the uniform-sampling assumption means that some *preprocessing* is applied, such as interpolation or down-sampling, making the assumption satisfied.
This preprocessing is possible because, as it is shown later on, the computation performed by the estimation algorithm is done over a window of past measurements which enables to apply advanced signal preprocessing prior to the application of the reconstruction algorithm described hereafter.

It is worth underlying that the notation used in this book differ from the one used in standard state estimation textbooks where a distinction between the control input u and measurement vector y is used. This is generally due to the fact that the input vector u plays a specific role in the control algorithms to which the estimation results are generally fed.

As mentioned earlier, from the estimation point of view, this distinction is artificial since in all these books, the vector of inputs u is assumed to be measured and since y stands for the vector of measurements, u should be a part of y which is what is done in the remainder of this chapter. More precisely[1]:

! u Is a part of y

In this chapter, the notation y denotes all the measured quantities **including the vector of exogenous inputs u.** More clearly:

$$h(x, u, p) := \begin{bmatrix} \vdots \\ u \end{bmatrix},$$

[1] As a matter of fact this distinction might become relevant at least when some paradigms are studied, namely:

- The case where one studies the impact of the control on the observability leading to the concepts of uniform/(non uniform) observability.
- The case where the control applies intentionally excitation signals in order to enhance the observability of the system.

where the elements referred to by the dots $\vdots$ contain measurements of quantities that **depend on the value of the state** and hence cannot be considered exogenous!

The problem we are about to address can be schematically shown through the **preliminary version** shown in Fig. 9.1.

More precisely, the problem can be stated as follows in its **preliminary version**:

> **Problem Statement: Preliminary Version**
> Given the past measurements Y_{k-1} and the last estimation of the pair $(\hat{x}_{k-1}, \hat{p}_{k-1})$ of the previous state and the vector of parameters, use the available time τ_u to compute the estimation updates $\hat{x}_k$ and $\hat{p}_k$.

In this statement, some details should be emphasized that are not commonly considered in some textbooks regarding the simultaneous estimation of states and parameters, namely:

✓ The past measurement that can be processed in order to deliver an estimation of the quantities at instant k **stops**[2] **at instant** $k - 1$. This is because the processing cannot be instantaneous as it involves a non-convex optimization process and hence, requires some incompressible amount of time, denoted by τ_u in Fig. 9.1. Consequently, taking into account the information that is available at instant k to deliver the estimation of the quantities as instant k is simply impossible.

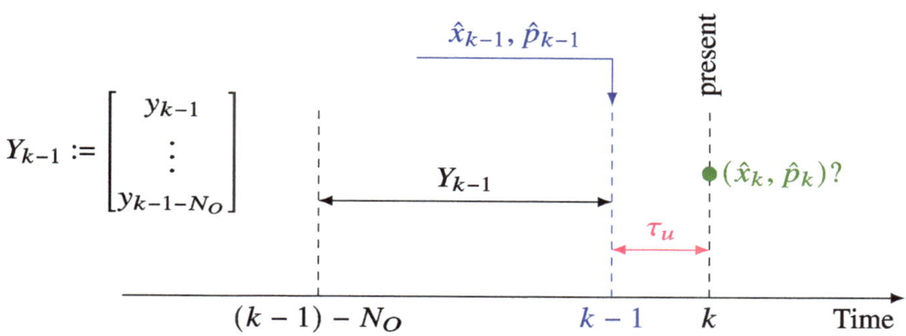

Fig. 9.1 Schematic of the preliminary version of the problem of state/parameter estimation, given the past measurements Y_{k-1} and the last estimation of the pair $(\hat{x}_{k-1}, \hat{p}_{k-1})$, use the available time τ_u to provide the estimation updates $\hat{x}_k$ and $\hat{p}_k$

[2] There is a delay that expresses the fact that the estimation of the quantities at instant k is based on the information up to instant $k - 1$.

✓ As it is shown in detail in the present chapter, the processing of the information contained in Y_{k-1} consists in finding the best value of the *Unknown* vector z defined by:

$$z = \begin{bmatrix} \tilde{X} \\ \tilde{p} \end{bmatrix} \in \left[\mathbb{R}^{n_x}\right]^{N_O+1} \times \mathbb{R}^{n_p}, \tag{9.3}$$

where $\tilde{X}$ is a candidate value for the state trajectory:

$$\tilde{X} := \begin{bmatrix} \tilde{x}_k \\ \vdots \\ \tilde{x}_{k-N_O} \end{bmatrix}, \tag{9.4}$$

and $\tilde{p}$ is a candidate value for the vector of parameters such that the corresponding states and parameter vector **are compatible** with the following available information:

- the equations of the system.
- the past measurements contained in Y_{k-1}, and
- the last estimates $\hat{x}_{k-1}$ and $\hat{p}_{k-1}$ of the state and parameter vectors at the sampling instant $k-1$.

Once this task[3] of finding $z^\star$ that corresponds to the best candidate, one gets the updated estimation $x^\star_{k-1}$ of the state at instant $k-1$ but remember that the target updating instant for the estimation is k and one needs an estimation of the state at instant k.

As a mater of fact, having $(x^\star_{k-1}, p^\star_{k-1})$ together with u_{k-1} (which is a part of Y_{k-1}), enables to compute the model-based estimation of the next state, namely

$$x^\star_k = f(x^\star_{k-1}, u_{k-1}, p^\star_{k-1}), \tag{9.5}$$

which provides an estimation of the current state at instant k.

Notice that the previous discussion assumes that the computation needed to get $z^\star$ can be done within a computation time of duration τ_u (see Fig. 9.1). This assumption has no reason to hold for any specific choice of the algorithm and the available computation time τ_u. Something has to be specifically done in order to take into account this real-time related aspect of the matter.

This is the aim of the following section that provides a second formulation of the state/parameter reconstruction problem.

[3] It seems obvious that this task cannot be done in a single and unique way. In other words, the implementation of this task involves some parameters that can be tuned and which can be viewed as parameters of the task. The relative weight of the confidence one should have in the model on one side and in the measurement on the other side is an example of such tuning parameters.

9.4 Real-Time Considerations

There is a priori no reason to assign to the computation time τ_u the same value as the measurement acquisition period which is assumed here to be the sampling time which is used to get the discrete-time dynamic model of the system.

The reason for which the problem stated so far is referred to as a preliminary version is that the above setting forces the computation time to be exactly equal to the measurement acquisition period.

Indeed, τ_u in Fig. 9.1 is the time that separates sampling instants $k-1$ and k. Obviously, there is no reason for this choice to be optimal, or even to be simply feasible. In general, the acquisition time can be very small and limiting the computation time to this period is generically inappropriate.

In order to define a more general setting from which the previous choice can be obtained as a particular case, the schematic of Fig. 9.2 is used in the development presented in the remainder of this chapter.

In this new version, the past measurements that are used in the search for a best updating stops at instant $k-q$ rather than $k-1$ **leaving a computation time for the optimization process that lasts q acquisition periods rather than a single one** in the preliminary setting of Fig. 9.1.

Notice that since during the computation time taking place over the time interval $[k-q, k]$, no updating of the estimation is possible, the estimated states $\hat{x}_{k-q+j}$ for $j = 1, \dots, q-1$ are obtained through open-loop simulation of the model, starting from the previously available estimations[4] $(\hat{x}_{k-q}, \hat{p}_{k-q})$, namely:

$$\hat{p}_{k-q+j} = \hat{p}_{k-q} \quad j \in \{1, \dots, q-1\}, \tag{9.6a}$$

$$\hat{x}_{k-q+j} = f(\hat{x}_{k-q+j-1}, u_{k+q+j-1}, \hat{p}_{k-q}). \tag{9.6b}$$

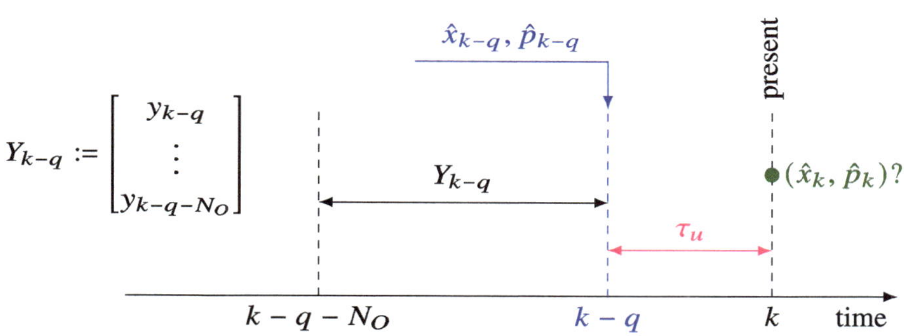

Fig. 9.2 Schematic of the definitive version of the problem of state/parameter estimation, given the past measurements Y_{k-q} and the last estimation $\hat{x}_{k-q}$ and $\hat{p}_{k-q}$, use the available time τ_u (which lasts q acquisition periods) to provide the estimation updates $\hat{x}_k$ and $\hat{p}_k$

[4] Which is not yet updated based on the outcome of the in process optimization.

Indeed, Eq. (9.6a) expresses the fact that the parameters are supposed to be constant while (9.6b) is nothing but the dynamic equation representing the model of the system defined by (9.2a).

Notice that the previous discussion involves two different pairs of state and parameters, namely $(\hat{x}_{k-q}, \hat{p}_{k-q})$ and $(x^\star_{k-q}, p^\star_{k-q})$. Understanding the forthcoming development requires to fully understand the difference between these two pairs. This difference is clearly recalled hereafter:

! **Difference Between** $(\hat{x}_{k-q}, \hat{p}_{k-q})$ **and** $(x^\star_{k-q}, p^\star_{k-q})$

These are respectively the estimations of the values of the state and parameter vectors at instant $(k-q)$ before and after the updating based on the outcome of the optimization process that uses the information Y_{k-q} and which takes place during the time interval $(k-q, k)$.

Indeed, since during the optimization process, $x^\star_{k-q}$ is not yet available, the real-time estimations at instants

$$\left\{k - q + j\right\}_{j\in\{1,\dots,q-1}},$$

can only be based on $(\hat{x}_{k-q}, \hat{p}_{k-q})$ and not on $(x^\star_{k-q}, p^\star_{k-q})$.

Now the optimization process (that is not yet explained) leading to the computation of $(x^\star_{k-q}, p^\star_{k-q})$ remains identical to the one that would have been undertaken should the preliminary setting be used, except that the used information is the measurement acquired **over the past observation window** which now spans over the time interval

$$[k - q - N_O, k - q],$$

instead of

$$[k - N_O, k - 1].$$

This information is used to produce the updating estimation $\hat{x}_k$ through the same simulation process as in (9.6b) but starting from the updated estimation $x^\star_{k-q}$ rather than the previously estimated $\hat{x}_{k-q}$, namely:

$$\xi_0 \leftarrow x^\star_{k-q}, \tag{9.7a}$$

$$\xi_j \leftarrow f(\xi_{j-1}, u_{k-q+j-1}, p^\star_{k-q}) \quad j \in \{1, q\}, \tag{9.7b}$$

$$\hat{x}_k \leftarrow \xi_q := P(x^\star_{k-q}, U^{k-1}_{k-q}, p^\star_{k-q}), \tag{9.7c}$$

$$\hat{p}_k = p^\star_{k-q}, \tag{9.7d}$$

which is supposed to be an almost zero-time operation since it consists in calling some explicit functions few times.

Notice that the outcome of the computation involved in (9.7) is denoted using the following short notation:

$$\hat{p}_k = p^\star_{k-q}, \tag{9.8a}$$
$$\hat{x}_k = P(x^\star_{k-q}, U^{k-1}_{k-q}, p^\star_{k-q}), \tag{9.8b}$$

where the r.h.s of (9.8b) simply denotes the state obtained by simulating the model starting from the initial state $x^\star_{k-q}$ and using the successive control $u_{k-q}, \dots, u_{k-1}$ contained in the sequence of controls U^{k-1}_{k-q} while using $p^\star_{k-q}$ as the value of the vector of parameters in the simulation.

It is important to notice that in the previous iterations, the value of p is taken constant in conformity with the assumption stated above regarding the constant or slow varying character of the parameters. On the other hand, the time-varying profile of the exogenous measured signal u, accumulated on the time interval $[k-q, k-1]$, is fully injected in the current state prediction.

It is also important to keep in mind that once a multiple, q, of the basic acquisition period is adopted, the previous development and the notation in Fig. 9.2 implicitly assume that k is a multiple of q.

In other words, the estimation rules previously established can be now stated as follows:

> If k is a multiple of q then use the optimization-based updating rule (9.8), otherwise, use the simulation-based updating rule (9.6).

Now instead of distinguishing those indices k that are multiple of q and those which are not, it might be more convenient to acknowledge that there are two different counters with different periods, namely, the original one that is defined by the acquisition period and a new one that shows q-times slower (updating) rate.

More precisely, let us define the following under-sampling index:

$$\mu_k = \left\lfloor \frac{k}{q} \right\rfloor \times q. \tag{9.9}$$

This leads to the evolution of μ_k as depicted in Fig. 9.3 which shows the updating instants indices as a function of the basic acquisition sampling index k.

Let us update Fig. 9.2 using this new notation. This leads to Fig. 9.4 in which all the key instants are expressed in terms of the new time counter μ_k. This enables to clearly underline the under-sampling character of the new scheme.

This new notation enables to write the updating rules previously expressed through (9.8) and (9.6) as follows:

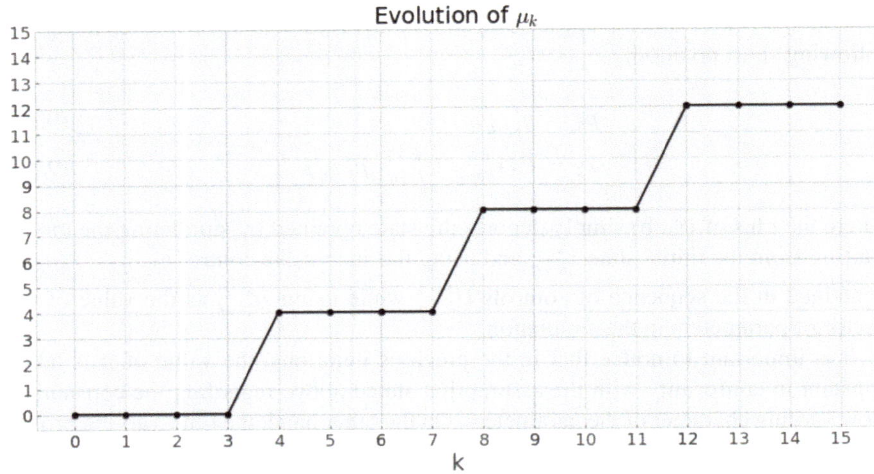

Fig. 9.3 Evolution of the under-sampling index μ_k defined by (9.9)

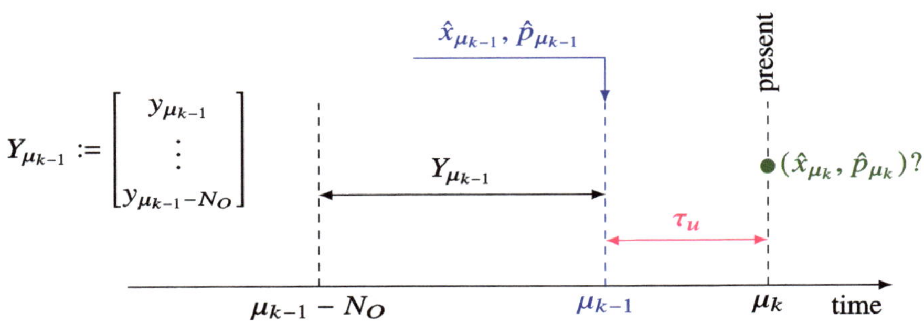

Fig. 9.4 Schematic of the definitive version of the problem of state/parameter estimation while using the under-sampling index μ_k, given the past measurements $Y_{\mu_{k-1}}$ and the last estimation $\hat{x}_{\mu_{k-1}}$ and $\hat{p}_{\mu_{k-1}}$, use the available time τ_u to provide the estimation updates $\hat{x}_{\mu_k}$ and $\hat{p}_{\mu_k}$

This leads to the final updating rule that is defined as follows:

Updating Rule for $k \in [\mu_{k-1}, \mu_k]$

$$(x^\star_{\mu_k}, p^\star_{\mu_k}) \leftarrow \text{Process}(Y_{\mu_k}, \hat{x}_{\mu_k}, \hat{p}_{\mu_k}) \quad \text{not available before } \mu_{k+1} \tag{9.10a}$$

$$(\hat{x}_k, \hat{p}_k) = \begin{cases} \left(P(x^\star_{\mu_{k-1}}, U^{k-1}_{k-q}, p^\star_{\mu_{k-1}}), p^\star_{\mu_{k-1}}\right) & \text{if } k = \mu_k, \\ \left(f(\hat{x}_{k-1}, u_{k-1}, \hat{p}_{\mu_{k-1}}), \hat{p}_{\mu_{k-1}}\right) & \text{otherwise,} \end{cases} \tag{9.10b}$$

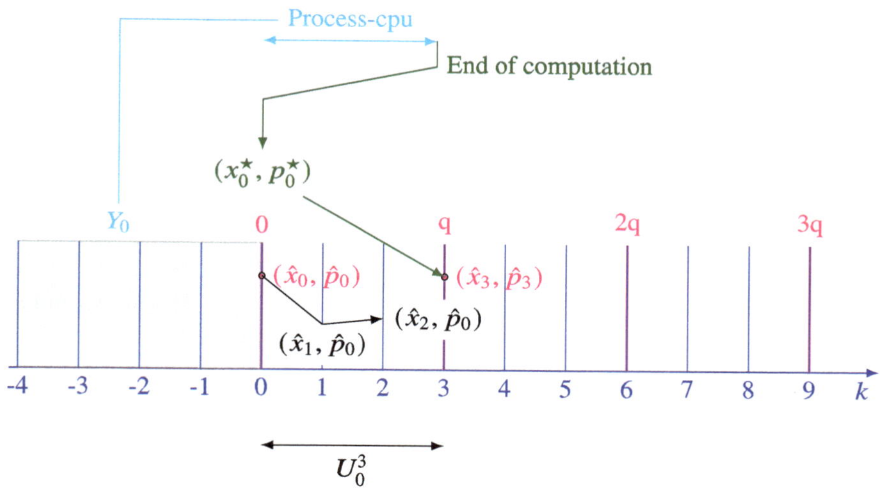

Fig. 9.5 Schematic of the updating rule and its scheduling over time for the special case $q = 3$ and $N_O = 4$: View of **a** the sampling (acquisition) instants (unit step) and **b** the optimization based updating instants (q-step). Note that $k = 0$ corresponds to the instant where a buffer of size N_O is available

which might be better understood through the following comments:

- ✓ Notice first of all that the second branch of the updating law is the one adopted at all instants except when k is a multiple of q (which is equivalent to $k = \mu_k$).
- ✓ This rule simply uses the model's simulation for the updating of $\hat{x}$ and leaves $\hat{p}$ unchanged.
- ✓ When $k = \mu_k$, the result of the computation $(x^\star_{\mu_{k-1}}, p^\star_{\mu_{k-1}})$ is available and can therefore be used in the updating through the map $P(\cdot)$ defined by (9.7).

These updating rules are depicted graphically over time as shown in Fig. 9.5.

Regarding Fig. 9.5, the following comments might be helpful to be understand some features:

- ✓ Figure 9.5 instantiates the situation where $q = 3$ and $N_O = 4$ are used in the estimation scheme.
- ✓ Notice that the origin of time ($k = 0$) starts as soon as a buffer of measurements of length $N_O = 4$ is available. At that time the past measurements are gathered in Y_0 in conformity with the notation used so far.
- ✓ Moreover, at instant $k = 0$ some initial estimation of the state and parameters is supposed to be available which is denoted by $(\hat{x}_0, \hat{p}_0)$. This might be chosen arbitrarily since a strong correction is shortly expected, nevertheless, choosing a relevant initialization, if available, is always recommended.

✓ During the computation time[5] $\tau_u = q\tau = 3\tau$ the information is processed (an associated optimization problem is solved or more precisely, a number of iterations of some solver are performed as shown later).

✓ While the computation process is in progress, the estimation of the state is propagated by simply simulating the system's model using the initial estimation $(\hat{x}_0, \hat{p}_0)$ and the measured exogenous input values gathered in

$$U_0^3 = (u_0, u_1, u_2).$$

As a mater of fact, in the case illustrated in the figure ($q = 3$), only u_0 and u_1 are used[6] to produce the following updating of the state:

$$\hat{x}_1 = f(\hat{x}_0, u_0, \hat{p}_0) \quad \text{and} \quad \hat{x}_2 = f(\hat{x}_1, u_1, \hat{p}_0),$$

while $\hat{p}_1 = \hat{p}_2 = \hat{p}_0$ in accordance with the working assumptions.

✓ When the results of the optimization are available (just before instant $k = 3$), the pair $(x_0^\star, p_0^\star)$ that explains the measurement Y_0 gathered on the previous observation window while taking into account the available estimation $(\hat{x}_0, \hat{p}_0)$ is by now available and the map P can be invoked to compute the estimation of the state and the parameters at instant $k = q = 3$, namely:

$$\hat{x}_3 = P(x_0^\star, U_0^2, p_0^\star) \quad \text{and} \quad \hat{p}_3 = p_0^\star. \tag{9.11}$$

✓ At the end of this process, the whole sequence can be repeated since Y_3, $\hat{x}_3$ and $\hat{p}_3$ are now available for a new round where they can play the role of Y_0, $\hat{x}_0$ and $\hat{p}_0$ respectively.
This is shown in Fig. 9.6.

Indeed, the two phases are again repeated, in the first, the updated estimation based on the simulation of the model with the new initial values $(\hat{x}_3, \hat{p}_3)$ leading to:

$$\hat{x}_4 = f(\hat{x}_3, u_3, \hat{p}_3) \quad \text{and} \quad \hat{x}_5 = f(\hat{x}_4, u_4, \hat{p}_3),$$

and as soon as the optimization process is stopped just before instant $k = 2q = 6$, the estimation of $(\hat{x}_6, \hat{p}_6)$ based on the resulting sub-optimal values $(x_3^\star, p_3^\star)$ is computed using:

$$\hat{x}_6 = P(x_3^\star, U_3^5, p_3^\star) \quad \text{and} \quad \hat{p}_3 = p_0^\star, \tag{9.12}$$

and the process is again ready to be repeated indefinitely.

---

[5] τ refers to the basic sampling period for the discrete-time dynamics and the measurement which are supposed to be identical.

[6] Before a strong correction is used after the end of the optimization process. u_2 will then be used in a new open-loop simulation that starts at the computed sub-optimal pair $(x_0^\star, p_0^\star)$ in order to compute the estimated $(\hat{x}_3, \hat{p}_3)$.

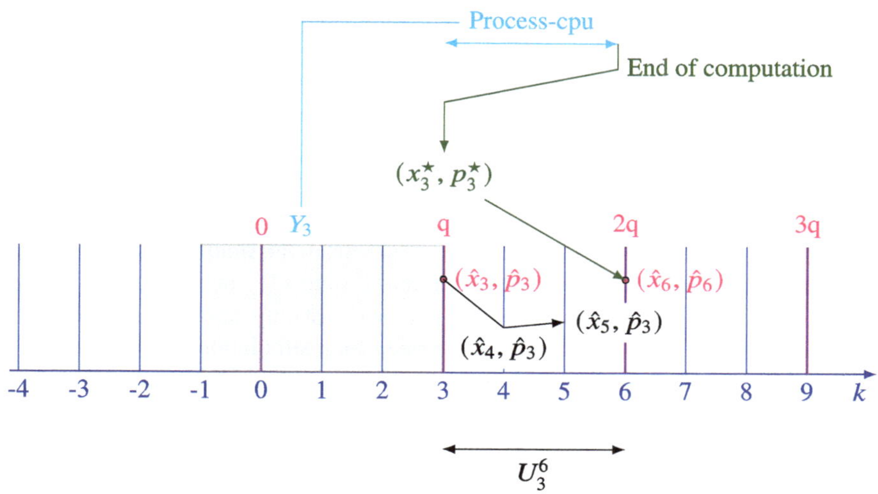

Fig. 9.6 Schematic of the updating rule and its scheduling over time for the special case $q = 3$ and $N_O = 4$: Starting of the second round of optimization after a time shift of q sampling periods

A careful reader might already *grasped* the connection between the real-time implementation problem described above and the one we already encountered in the implementation trade-off encountered in the previous chapters regarding the real-time implementation of Model Predictive Control (MPC). Indeed, both problems need the computation time $\tau_u = q\tau$ to be carefully chosen and there is no obvious optimal value as this depends on both the dynamics of the system, the level of uncertainties and the available computation power.

The following section discusses this issue in more details.

9.4.1 Real-Time Trade-Off in Moving-Horizon Estimator

Let us start the discussion starts with the following question:

> **Regarding the Choice of** q Since high values of $q > 1$ leave more time for the optimization to provide more accurate, why don't we use very large computation time corresponding to very large values of the parameter q?

The answer to this question is linked to the possible presence of non expected incidents / disturbances that might induce important errors on the estimated $x_k^\star$ even if the estimated $(x_{k-q}^\star, p_{k-q}^\star)$ is perfect, namely: even if $x_{k-q}^\star \equiv x_{k-q}$ and $p_{k-q}^\star = p$.

To formulate it differently, even if the estimation is perfect at instant $k - q$, estimating the value of the state q sampling periods later using open-loop simulations exposes the outcome to the risk of non predictable disturbances that are not modeled and that might occur during the *long interval of time that last q periods.*

This can be expressed through the following inequality[7]:

$$\|\hat{x}_k - x_k\| \leq E_e(q, \|x^\star_{k-q} - x_{k-q}\|, \|p^\star_{k-q} - p\|) + E_D(q), \tag{9.13}$$

which expresses the fact that the error on the estimation of the state at instant k when using the above process with the parameter q stems from **two sources of errors**:

- ✓ The first one, expressed by the E_e term, comes from the simulation of the system starting from an erroneously estimated initial state $x^\star_{k-q}$ and/or an erroneously estimated vector of parameters $p^\star_{k-q}$. This error not only depends on these initial errors, but also on the duration of the open-loop simulation starting from this erroneous initialization, this is why q is an argument of E_e.
- ✓ The second source of errors represented by the term $E_D(q)$ comes from potential non modeled and non predictable disturbances/incidents that might have occurred during the time interval $[k-q, k]$ over which open-loop simulation of the model are performed in order to provide the estimated quantities.

In order to push further the qualitative discussion, let us acknowledge that the larger q is the smaller is the error $(x^\star_{k-q}, p^\star_{k-q})$ thanks to the quality of optimization outcome. On the other hand if this error is not zero, the larger q is the higher the term E_e will be.

These two antagonistic effects can be qualitatively modeled by a function of the form:

$$\|\hat{x}_k - x_k\| \leq \varphi_1(q^{-1}) \cdot \varphi_2(q) + \varphi_3(q), \tag{9.14}$$

where all the φ_i are monotonically increasing functions of their argument q.

More precisely,

- φ_1 expresses the fact that longer optimization lead to smaller errors in the second and third arguments of E_e used in (9.13) which explains the use of q^{-1} instead of q.
- φ_2 expresses the fact that longer open-loop simulations propagate the initial error.
- φ_3 is a increasing function of its argument expressing the fact that the longer we leave the estimation in open-loop, the higher is the risk of being blind regarding unpredictable events that might render the resulting estimation erroneous.

[7] The argumentation here is qualitative and aims at conveying the intuition that underlines the trade-off that might be thought off when it comes to implement real-time state/parameters. The functions E_e and E_D are not mathematically tractable nor are they expected to be known or even estimated. The needed trade-off resolution, if any, should be done through extended simulations and or data-driven learning process as it is shown later.

For the sake of illustration, let us consider the following instantiation of the maps φ_i invoked in (9.14):

$$\varphi_1(q) = aq \quad , \quad \varphi_2(q) = 1 + q \quad , \quad \varphi_3(q) = cq^2. \tag{9.15}$$

Notice that in the above expressions and given the roles played by the different maps, the two parameters a and c play the following roles:

✓ high values of a means that the value of the optimization error regarding $x^\star$ and $p^\star$ is high if no sufficient iterations are allowed (because of a high values of q^{-1}),

✓ on the other hand, when c is high, this means that the level of uncertain and unpredictable incidents is high.

Figure 9.7 shows the impact of different instantiations of the pair (a, c) used in the model (9.15) on the estimation error bound (9.14).

More precisely, Fig. 9.7a shows the situation for a given algorithm efficiency $a = 5$ for four different values of the unpredictable uncertainty level $c \in \{0, 0.1, 0.2, 0.5\}$.

This figure clearly shows than when there are no uncertainties (*ideal unrealistic situation*), one should spend a longer time for the computation, on the other hand, as soon as uncertainties are present, there always be an optimal choice of q that minimizes the estimation error (in the case of the figure this value lies in $\{2, 3\}$).

Figure 9.7b, c show the situation for a given level of unpredictable uncertainty $c = 0.2$ or $c + 0.05$ and different values of the algorithm's efficiency $a \in \{1, 3, 5\}$.

From the previous illustrative examples, it appears clearly that:

A Delicate Trade-Off!

When the solver's *local* efficiency is not sufficiently high compared to the level of uncertainties, small values of q should be used while in the presence of low level of uncertainties it is sometimes better to take longer time in the optimization process.

It is difficult to overestimate the difficulty in fitting the model (9.15) so that the real situation at each instant during the system real-life time can be captured.

This is because the fitting map depends on the current state and the configuration of the exogenous inputs. The only aim of the current discussion is to underline that the choice $q = 1$ which is used in the majority of textbooks has no reason to be optimal in terms of the resulting estimation error.

As a matter of fact, an attempt to derive an on-line adaptive scheme of the measurement inclusion rate (which is somehow related to adaptive values of q being used based on the decrease of the cost function) has been undertaken in [1]. The principle

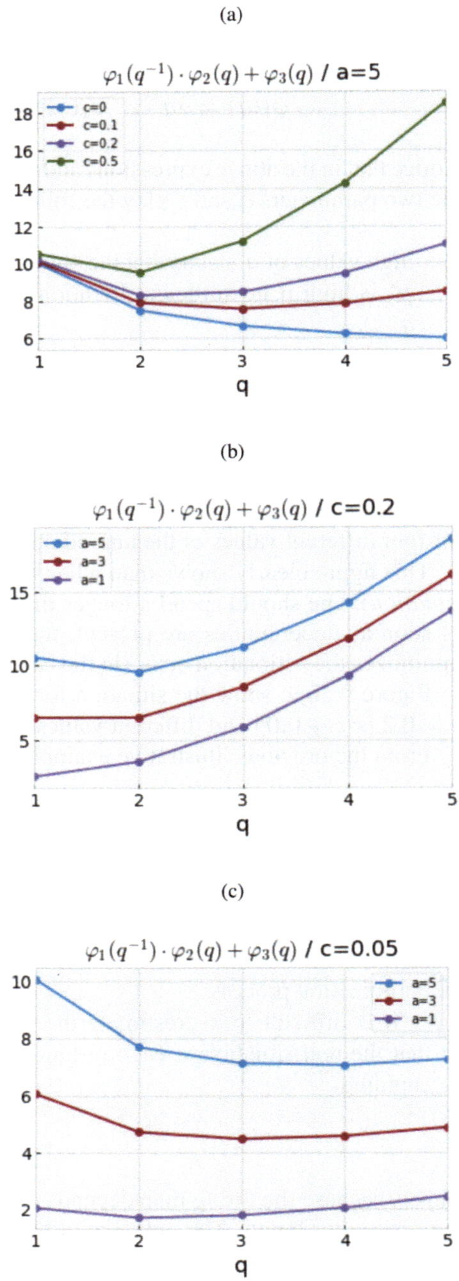

Fig. 9.7 Examples of possible instantiations of the state/parameters estimation errors corresponding to different values of the parameters **a** and **c** used in (9.15) to model the possible impact of the context on the estimation errors

of the algorithm is based on an adaptation of what has been done in [2] regarding the on-line adaptation of the control updating period used in NMPC design. The study showed promising results on a specific example.

The least one can do is to undertake an extensive set of simulations of the estimator for different values of q and different configurations of initial states and parameters in order to rationalize the choice of this key parameter.

Now let us go back to our flow of presentation of the full state/parameters estimation algorithm. Recall that in the above discussion, we referred to an expensive optimization-based computation that yields the updated quantities $(x_k^\star, p_k^\star)$. We also mentioned that this optimization process takes place during successive computation time slots that last q sampling (acquisition) periods each.

It is now time to describe this process in detail. This is the aim of the next section.

9.5 Optimization-Based Updating/Computing $(x_k^\star, p_k^\star)$

Recall that we are about describing the computation process that is executed during the time interval[8] $[k, k+q)$ at the beginning of which, one disposes of the following information (see Fig. 9.6):

✓ The previous vector of measurement profile, namely:

$$Y_k := \begin{bmatrix} y_k \\ \vdots \\ y_{k-N_O} \end{bmatrix} \in [\mathbb{R}^{n_y}]^{N_O+1}. \tag{9.16}$$

✓ The previous vector of estimation $(\hat{x}_k, \hat{p}_k)$ which comes from the previous round of computation.[9]

Notice that the available estimation $(\hat{x}_k, \hat{p}_k)$ is obtained by the open-loop simulation procedure denoted previously by $P(\cdot)$ [see (9.8b)] which involves the model in which the last updated value of its parameters based on the previous measurement information Y_{k-q}.

The task we are about to explain is the one aiming at modifying this previous estimation $(\hat{x}_k, \hat{p}_k)$ based on the measurement collected since it has been made. The new estimation is denoted by $(x_k^\star, p_k^\star)$.

It comes out that this new estimation has to be a kind of consensus between two requirements, namely: first, it should accommodate (explains) the new series of measurements included in Y_k on one hand, but it should also take into account the last estimation $(\hat{x}_k, \hat{p}_k)$ that summarizes the previous computation.

---

[8] Figure 9.6 shows the case where $k = 3$, and $q = 3$ and $N_O = 4$.

[9] In the specific case where $k = 3$, this estimation comes from the round of computation that is represented in Fig. 9.5.

The trade-off between these two potentially incompatible requirements lies in the relative *confidence* one has in the last series of measurements included in Y_k and the confidence granted to the previous estimation.

To understand this dilemma, let us imagine that there is a single value of the unknown (x_k, p_k) that is perfectly compatible with the last measurement information Y_k.

In this case, if our confidence in the measurement is perfect, then we should take this unique value as the new estimation, namely $(x_k^\star, p_k^\star)$, of (x_k, p_k) **no matter how far this information is from the previous estimation** $(\hat{x}_k, \hat{p}_k)$.

On the other hand, if our measurement is affected by a high level of measurement noise and/or possible bias, then we should show some moderation in the adoption of this presumably unique value because it might fit imperfectly collected information that is not totally reliable.

In order to concretely handle this trade-off, weighting matrices are used in the formulation of an optimization problem in which the decision variable is the one that has been already introduced [see (9.3)] which is rewritten hereafter in a slightly different manner for the sake of clarify of exposition of the optimization problem:

$$z = \begin{bmatrix} z.x_0 \\ \\ \vdots \\ z.x_{N_O} \\ z.p \end{bmatrix} \in \left[\mathbb{R}^{n_x}\right]^{N_O+1} \times \mathbb{R}^{n_p}, \tag{9.17}$$

where

- $z.x_0$ stands for candidate value of the state vector at the beginning of the observation window involved in the optimization problem.
- $z.x_1$ is the next candidate value for the state vector over this window and so on.

More precisely, when the measurement profile is Y_k, one gets the following correspondences (see Fig. 9.8):

$$\begin{aligned} &z.x_0 \quad \text{candidate value for} \quad x_{k-N_O}, \\ &z.x_1 \quad \text{candidate value for} \quad x_{k-N_O+1}, \\ &\dots \\ &z.x_{N_O} \quad \text{candidate value for} \quad x_k. \end{aligned}$$

This leads to the following optimization problem in the decision variable z that is only a preliminary version since we need to add the soft constraints-related penalty term:

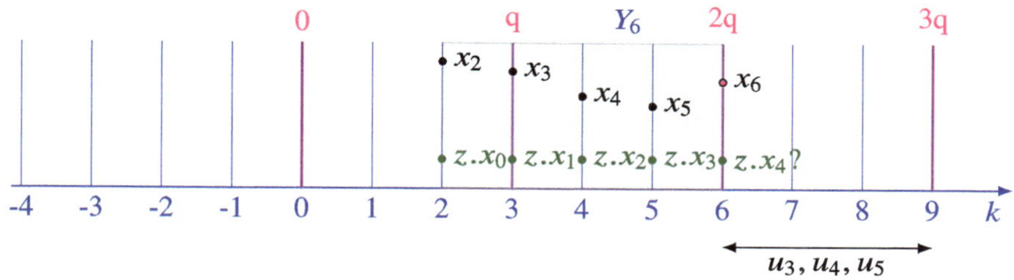

Fig. 9.8 Illustration of the definition of the decision variable z given in (9.18) for the optimization problem defined at instant $k = 6$, $N_O = 4$ and $q = 3$

> **The MHE's Optimization Problem: Version 1**

$$z^\star := \arg\min_{z\in\mathbb{Z}} J(z\,|Y_k, \hat{x}_k, \hat{p}_k, \lambda) :=$$

$$\frac{\lambda}{N_O}\sum_{i=0}^{N_O-1} \|z.x_{i+1} - f(z.x_i, u_{k-N_O+i}, z.p)\|_Q^2 + \tag{9.19a}$$

$$+\frac{\lambda}{N_O}\sum_{i=0}^{N_O} \|y_{k-N_O+i} - h(z.x_i, u_{k-N_O+i}, z.p))\|_R^2 + \tag{9.19b}$$

$$+(1-\lambda)\Big[\|z.x_{N_O} - \hat{x}_k\|_{Q_f}^2 + \|z.p - \hat{p}_k\|_{P_f}^2\Big], \tag{9.19c}$$

in which

✓ $\lambda \in (0, 1]$ is a filtering parameter in the sense that when λ is close to 0, the new values $(x_k^\star, p_k^\star)$ would be very close to the previously estimated one $(\hat{x}_k, \hat{p}_k)$ as almost only the last term (9.19c) expressing the distance to that estimation would be considered.
When $\lambda = 1$ the measurement windows totally determines the outcome of the computation without any consideration for the previously estimated state at the same instant k.
Having this tuning parameter is especially interesting in the case where the estimated values are to be used in feedback control as the level of chattering of the estimated quantities might induce a harmful fatigue in the actuator devices. Notice how this filtering effect can also be monitored in a differentiated manner among the components of x and p through the weighting matrices Q_f and P_f in order to selectively filter the evolution of the different components of the decision variable z.

✓ The first term defined by (9.19a), namely:

$$\frac{1}{N_O}\sum_{i=0}^{N_O-1}\|z.x_{i+1}-f(z.x_i,u_{k-N_O+i},z.p)\|_Q^2,$$

represents Q-weighted[10] penalty on the violation of the discrete-time state equation that would be induced by a candidate value of z in which the components of $z.x$ (which represent successive values of the state vector) and $z.p$ are not compatible with the dynamical model of the system and the previously applied control instances that are part of the measurement vector Y_k.

✓ The second term defined by (9.19b), namely:

$$\frac{\lambda}{N_O}\sum_{i=0}^{N_O-1}\|y_{k-N_O+i}-h(z.x_i,u_{k-N_O+i},z.p))\|_R^2,$$

represents the R-penalty[11] on the output prediction error which is based on the measurement vector Y_k and the output map h that is a part of the system's model (9.2b).

✓ Notice finally that the optimization over candidate values of z is constrained by the the expression:

$$z\in\mathbb{Z}\subseteq[R^{n_x}]^{N_O+1}\times\mathbb{R}^{n_p},$$

where $\mathbb{Z}$ is an admissible set that is to be defined for each specific problem's setting.

In the case of use of a fast-gradient-like descent algorithm to solve the optimization problem, it might be helpful to restrict the admissible set $\mathbb{Z}$ to be a hypercube (saturation-like constrained set) while all other sorts of constraints are enforced through soft constraints inclusion in the definition of the cost function.

More precisely, assume that one knows that there are some non-saturation like constraints that have to be satisfied by the state and the parameter, namely:

$$g(x,p)\le 0\in\mathbb{R}^{n_c}. \tag{9.20}$$

In this case, the following soft constraint-related term has to be added to the definition of the cost function previously defined by[12] (9.19):

$$\frac{\rho_{\text{constr}}}{n_c\cdot N_O}\sum_{j=1}^{n_c}\sum_{i=0}^{N_O}\lfloor g_j(z.x_i,z.p)\rfloor_+ . \tag{9.21}$$

---

[10] We'll come back to the definition of Q later in the sequel.

[11] See later.

[12] $\lfloor r\rfloor := \max\{0,r\}$.

This leads to the following modified formulation of the MHE's optimization problem in which the $\mathbb{Z}$-related constraint has been replaced by a saturation constraint associated to some minimum and maximum vector values $\underline{z}$ and $\bar{z}$ while the above soft ρ_{constr}-penalty based term is added to the previously defined cost function:

> **The MHE's Optimization Problem (Version 2)**

$$z^\star := \arg\min_{z\in[\underline{z},\bar{z}]} J(z\,|Y_k, \hat{x}_k, \hat{p}_k, \lambda) :=$$

$$\frac{\lambda}{N_O}\sum_{i=0}^{N_O-1} \|z.x_{i+1} - f(z.x_i, u_{k-N_O+i}, z.p)\|_Q^2 + \tag{9.22a}$$

$$+\frac{\lambda}{N_O}\sum_{i=0}^{N_O-1} \|y_{k-N_O+i} - h(z.x_i, u_{k-N_O+i}, z.p))\|_R^2 + \tag{9.22b}$$

$$+(1-\lambda)\Big[\|z.x_{N_O} - \hat{x}_k\|_{Q_f}^2 + \|z.p - \hat{p}_k\|_{P_f}^2\Big] + \tag{9.22c}$$

$$+\frac{\rho_{\text{constr}}}{n_c\cdot N_O}\sum_{j=1}^{n_c}\sum_{i=0}^{N_O} \lfloor g_j(z.x_i, z.p)\rfloor_+^2 . \tag{9.22d}$$

This ends the presentation of all the details of the algorithm. Nevertheless, the best way to fully understand the above formulation is to follow the `python` implementation shown in the next section.

9.6 Python Implementation: The `MHE_solver`

In this section, an implementation of the Moving-Horizon Estimator (MHE) whose principle has been described in the previous section is detailed.

Many of the `Python` functions and commands used in the implementation are the same as the equivalent ones used in the previously studied `MPC_solver` presented in **Chap.** 5 which is dedicated to the design of deterministic MPC feedback.

Before we dig into the implementation details, let us first summarize the features we want to implement.

These features have already been explained in the previous section, but it is worth recalling them so that the different parameters and options developed in the code can be fully understood.

9.6.1 Features and Options of the MHE_solver

Our aim is to derive a Python-based implementation of a MHE that shows the following features, options and degrees of freedom:

- ✓ We need the Python package to perform an extended state/parameter estimation for a dynamic system that is governed by a model that is defined through uncertain parameters that are constant or slowly varying.
- ✓ The implementation should make it possible to tune the computation time by defining the number of sampling delays q as explained in the previous section.
- ✓ It should be made possible to incorporate and handle constraints on the state components and on the parameters vector. While a part of these constraints might be saturation-like, general nonlinear constraints should also be possible to incorporate via dedicated user-defined functions.
- ✓ It should be possible to use either Interior-Point optimization algorithm via CasaDi or a fast-gradient algorithm with tunable parameters in a similar way as for the MPC_solver presented in **Chap.** 5.
- ✓ The solution should accept user-defined selective filtering parameters on the different components of the estimated vector. This enables to test the observability of the different components of the unknown vector under different assumptions.

Now before digging into the details of the Python code, let us describe a simple illustrative example so that we can check progressively the different functions as we build them. The same example is also used at the end of the chapter in order to check the performance and usability of the resulting package.

9.6.2 The Illustrative Example: The Forced Lorentz Oscillator

In order to illustrate the MHE_solver code and the behavior of the corresponding extended state/parameters estimator under different configurations, we consider the forced Lorentz systems which is governed by the following set of ODEs:

$$\dot{x}_1 = p_1(x_2 - x_1) + u_1, \tag{9.23a}$$
$$\dot{x}_2 = x_1(p_2 - x_3) - x_2, \tag{9.23b}$$
$$\dot{x}_3 = x_1 x_2 - p_3 x_3 + p_4 u_2, \tag{9.23c}$$

that depend on four uncertain parameters denoted by $p_1, \dots p_4$ that belong to the following compact set:

$$\mathbb{P} := [4, 6] \times [0, 4] \times [15, 30] \times [0, 0.2]. \tag{9.24}$$

In order to test the implementation of the non saturation-like constraints such as the ones expressed through (9.20), the following two constraints are considered:

$$c(x, u, p) := \begin{bmatrix} p_4 - p_2 \\ p_2 - p_1 \end{bmatrix} \le 0. \tag{9.25}$$

In order to collect time-series representing the sensors measurement stream, let us go through the following protocol:

1. define some exogenous input profiles $\mathbf{u}$,
2. simulate the system for a set of parameter vector p,
3. collect the corresponding measurements profile y, and
4. check whether the estimator, when initialized with some set of erroneous values of the state and the parameters, is able to *asymptotically* recover the correct values of the state and parameter.

Moreover, let us follow this protocol under different assumptions regarding the definition of the measurement vector, namely:

$$h_1(x, u) := \begin{bmatrix} x_1 \\ x_3 \\ u \end{bmatrix} \quad or \quad h_2(x, u) := \begin{bmatrix} x_1 \\ u \end{bmatrix}. \tag{9.26}$$

Regarding the exogenous input u, a sinusoidal basis involving n_f modes is used to generated quite a rich set of behaviors.

More precisely, let us consider the following expression:

$$u(t) := \frac{1}{n_f}\sum_{i=1}^{n_f} \alpha_i \sin(\frac{2i\pi t}{T} + \phi_i) \in \mathbb{R}^2, \tag{9.27}$$

where $\alpha_i \in [-1, 1]^2$ and $\phi_1 \in [0, 2\pi]$ are randomly generated numbers and T, an arbitrarily chosen basic period.

We first create a `python` file names `lorentz.py` that contains all the non generic (user-defined) problem-dependent data, namely:

- the different dimensions and bounds on the state and parameter vectors' components as as well as,
- the functions that define the ODEs, the measurement and the constraints (other than saturation) related maps.

The following script inside the `lorentz.py` file defines the ODEs describing the Forced Lorentz System defined by (9.23). The first configuration regarding the measurement map is also defined.

Notice, in particular, the function `u_lorentz` which extracts the exogenous input vector from the whole measurement vector y.

```
# Inside lorentz.py: Definition of dimensions,
#                    bounds parameters and functions.

import numpy as np
from casadi import vertcat

nx, n_par, nu, nc, ny = 3, 4, 2, 2, 4
p_min = np.array([4, 0, 15, 0.0])
p_max = np.array([6, 4, 30, 0.2])
x_min = np.array([-10000]*nx)
x_max = np.array([+10000]*nx)

def ode_lorentz(x,u,p):

    xdot = vertcat(
        p[0]*(x[1]-x[0])+u[0],
        x[0]*(p[1]-x[2])-x[1],
        x[0]*x[1]-p[2]*x[2]+p[3]*u[1]
    )
    return xdot

def h_lorentz(x,u,p):

    y = vertcat(x[0], x[2], u)
    return y

def u_lorentz(y):
    return y[-2:]

def c_lorentz(x, u, p):

    c = vertcat(
        p[3]-p[1],
        p[1]-p[0]
    )
    return c
```

The main file `MHE_solver.py` which is precisely the `Python` module, contains all the generic processes that use the user-provided information as described above in order to build the different functions and utilities that are needed to design the MHE and simulate the estimation scenarios.

The main class of this package is the MHE class that is initialized using the user-defined information as well the set of options that impact the definition of the estimator as explained in the previous sections.

Notice also that the file MHE_solver.py starts by importing the needed packages, functions and utilities including the ones provided in the package CasaDi that are extensively used in the building and the solution of the associated optimization problem as well as the computation of the gradient of the cost function augmented by the soft constraints-related terms.

This leads to the head of the file MHE_solver.py that is shown in Fig. 9.9.

Note that the role of the max_iter is to set the maximum number of iterations the optimizer is allowed to perform during the q sampling periods that are allowed to the computation of the optimal solution of the underlying optimization problem.

The next step is to create the CasaDi version of the discrete-time dynamics, called, the one_step_ahead map as well as the corresponding output and the constraints functions.

This is done in the following part of the __init__ function that is depicted in Fig. 9.10.

Notice that the one-step-ahead discrete-time map F is defined using the Runge–Kutta maximal sampling period dt that is used in the creation of the instances of the class MHE.

The next step is to create the cost function (9.22) which incorporates four contributions:

1. The error on the state equations that describes the compatibility of current candidate value z of the decision variable with the system equations. This is denoted in the program using J_eq.
2. The prediction error on the measurements. This is denoted in the program using J_meas.
3. The distance to the previous estimations of the state and parameter vectors. This is referred to in the program using J_previous.
4. The violation of the constraints induced by the value of z. Denoted hereafter by J_constr in the code.

For an easier diagnosis capabilities, these four contributions are computed separately through dedicated functions and then summed up to deliver the total cost to be optimized.

Before to get into this involved task, let us recall that for the sake of compatibility with the CasaDi framework, we need the decision variable z to be a single flat vector.

This is done by adopting the order already used in (9.17) when z has been introduced. This leads to a vector of dimension:

$$(N_O + 1)n_x + n_p,$$

as it is defined in the __init__ function shown in Fig. 9.10.

```
# From MHE_solver.py

import numpy as np
import pandas as pd
import time
from casadi import vertcat, Function, MX, mtimes
from casadi import nlpsol, reshape
from casadi import jacobian, if_else, power

class Container:
    pass

class MHE:

    def __init__(self,
                 nx, n_par, nu, nc, ny,
                 x_min, x_max,
                 p_min, p_max,
                 ode,
                 meas_func,
                 constraints,
                 u_from_y, # map extracting u from y
                 dt,      # maximum dt for runge kutta
                 N_O,     # observation horizon
                 q,       # observation delay
                 Q, R, Qf, Pf, rho_c, # Penalty matrices
                 max_iter=500
                 ):

        self.nx = nx
        self.nu = nu
        self.np = n_par
        self.nc = nc
        self.ny = ny
        self.x_min = x_min
        self.x_max = x_max
        self.p_min = p_min
        self.p_max = p_max
        self.dt = dt
        self.N_O = N_O
        self.q = q
        self.Q = Q
        self.R = R
        self.Qf = Qf,
        self.Pf = Pf
        self.rho_c = rho_c
        self.nz = (N_O+1)*nx+n_par
        self.max_iter=max_iter
        self.u_from_y = u_from_y
```

Fig. 9.9 Head of the file `MHE_solver.py` that defines the main class MHE. This part of the script simply creates attributes of the instance of the class being created that corresponds to the input argument of the instance creation call

```
# Create the casadi maps
x = MX.sym('x',nx)
u = MX.sym('u',nu)
p = MX.sym('p',n_par)
xdot = ode(x,u,p)
# r.h.s of the ode (continuous time)
self.f = Function('f',[x,u,p],[xdot])
# One-step-ahead map (discrete-time)
k1 = self.f(x, u, p)
k2 = self.f(x+0.5*k1*dt,u,p)
k3 = self.f(x+0.5*k2*dt,u,p)
k4 = self.f(x+k3*dt,u,p)
x_next = x+dt/6*(k1+2*(k2+k3)+k4)
self.F = Function('F',[x,u,p],[x_next])
# Measurement map
y = meas_func(x, u, p)
self.h = Function('h',[x,u,p],[y])
# Constraints map
c = constraints(x, u, p)
self.c = Function('c',[x,u,p],[c])
```

Fig. 9.10 Second part of the `__init__` function of the MHE class that creates the functions attributes `f, F, h` and `c`

Notice that for the sake of easiness of coding, the measurement profile is also arranged such that the oldest instances corresponds to the smallest indices which might contradict the presentation of the conceptual framework which adopts the principle of a buffer in which the last items appears at the top.

More precisely, the Y_k in the program is indexed as shown below:

$$\texttt{Program notation:} \quad Y_k := \begin{bmatrix} y_{k-N_0} \\ \vdots \\ y_k \end{bmatrix}, \tag{9.28}$$

in conformity with the notation of z.

Figure 9.11 shows the script inside the `__init__` function that is dedicated to the definition of the different parts of the cost function.

Moreover, notice that two main cost functions are defined, namely:

✓ The cost function `J_ipopt` that will be used when the optimization framework is chosen to be the one based on the `Ipopt` interior point algorithm through the `CasaDi` framework. In this case, the constraints will be defined through the list `g` that is constructed upon iteration over the observation horizon.

✓ The cost function `J_grad` that will be used when the fast-gradient is chosen in order to solve the soft-constrained version of the underlying optimization problem. Even if this option is chosen, the gradient of the cost function `J_grad` is obtained thanks to the utility `jacobian` provided by the `CasaDi` framework.

```
# Inside the init map of the MHE class

# The cost functions
x_hat = MX.sym('xhat', nx)
p_hat = MX.sym('phat', n_par)
lam = MX.sym('lam', 1)
z = MX.sym('z', self.nz)
Y = MX.sym('Y', N_O*ny)
J_eq, J_meas  = 0, 0
J_constr, J_previous = 0, 0
g, lbg, ubg = [], [], [] # for Casadi-Ipopt formulation
p = z[-n_par:]

#------ loop over the observation horizon

for i in range(N_O):
    xi = z[i*nx:(i+1)*nx]
    xip1 = z[(i+1)*nx:(i+2)*nx]
    yi = Y[i*ny:(i+1)*ny]
    ui = u_from_y(yi)
    e_x = xip1-self.F(xi, ui, p)
    e_y = yi-meas_func(xi, ui, p)
    ci = self.c(xi, ui, p)
    g += [ci[j] for j in range(self.nc)]
    lbg += [-1e12] * self.nc
    ubg += [0] * self.nc
    J_eq += mtimes(e_x.T, mtimes(Q, e_x))
    J_meas += mtimes(e_y.T, mtimes(R, e_y))
    for j in range(nc):
        J_constr += if_else(ci[j]<0, 0, ci[j]*ci[j])

e_xf = xip1 - x_hat
e_p = p - p_hat
J_previous += mtimes(e_xf.T, mtimes(Qf, e_xf))
J_previous += mtimes(e_p.T, mtimes(Pf, e_p))
#-- Filter and Normalise wrt to time
J_eq *= lam / (N_O * nx * self.dt)
J_meas *= lam / (N_O * ny)
J_previous *= (1-lam)
J_constr *= rho_c/(nc * N_O)
J_ipopt = J_eq+J_meas+J_previous
J_grad = J_ipopt + J_constr

#---  Define the exported cost functions

self.J_eq = Function('J_eq', [z, Y, x_hat, p_hat, lam], [J_eq])
self.J_meas = Function('J_meas', [z, Y, x_hat, p_hat, lam], [J_meas])
self.J_previous = Function('J_previous', [z, Y, x_hat, p_hat, lam],[J_previous])
self.J_constr = Function('J_constr', [z, Y, x_hat, p_hat, lam], [J_constr])
self.J_ipopt = Function('J_ipopt', [z, Y, x_hat, p_hat, lam], [J_ipopt])
self.J_grad = Function('J_grad', [z, Y, x_hat, p_hat, lam], [J_grad])
```

Fig. 9.11 The third part of the script of the `__init__` function which defines the different cost function provided by the class MHE

Notice the use in the script of Fig. 9.11 of the matrix multiplication facility `mtimes` of `Casadi`.[13]

The structure of this code is rather standard: First the `CasaDi` symbolic vectors are defined, then the cost function variable is initialized to zero before an iteration on the observation horizon is performed in order to add the penalties of different kinds, namely:

[13] Recall that the matrix multiplication operator @ can also be used instead of `mtimes`.

- the term expressing the incompatibility of z with the system's model resulting in J_eq,
- the term that expresses the difference between the expected measurement and the acquired ones that lead to J_meas,
- the contribution of the soft constraints term J_constr as well as,
- the contribution expressing the difference between the optimization based pair $(x^\star, p^\star)$ and the expected pair $(\hat{x}, \hat{p})$ which induces the cost J_previous.

Notice that the following two main costs are defined as described above[14]:

$$\texttt{J\_ipopt} := \texttt{J\_eq} + \texttt{J\_meas} + \texttt{J\_previous}, \tag{9.29}$$

$$\texttt{J\_grad} := \texttt{J\_ipopt} + \texttt{J\_constr}. \tag{9.30}$$

Now before we define the optimization problem and the two mentioned solvers to check the MHE performance, we need to produce the validation scenarios.

In order to do so, we shall need two utilities:

✓ The first one generates random trajectories of the exogenous vector u. This is done by the function generate_random_U which is defined as a method of the instances of the class MHE.
The script of this function is given in Fig. 9.12.

✓ The second utility uses the so-generated list of values representing the profile of u to simulate the trajectories of the model for a given initial state x_0 and a given vector of parameters p. This is done by the function simulate which is also a method of the MHE class.

The script of this function is shown in Fig. 9.13. Notice that this function returns the output arguments t, X and Y where X, Y are matrices in which, each line is the value of the state [resp. the measurement vector] at an instant in the simulation.

Consequently, if the simulation contains N_sim sampling instants, then the shape of the matrices X and Y would be (N_sim×nx) and (N_sim × ny) respectively.

Before going further, let us examine the way these functions can be used in the generation of the simulated trajectories of states and measurements. This enables us for instance to check that when the decision variable z is initialized at the correct values of the initial state and the parameter vector, the associated costs vanish Provided that the previous estimation $\hat{x}$, $\hat{p}$ are also conveniently set.

Figure 9.14 shows a script that uses the two files described above, namely, Lorentz.py and MHE_solver.py in order to generate an 8 time unit simulation of the Lorentz system.

[14] The Ipopt solver does not need the soft penalty term since it handles the constraints as hard constraints through the list of constraint values gathered in the list g that is constructed during the iteration loop on the observation horizon.

```
def generate_random_U(self, nf=10, T=5, t_sim=10):

    N_sim = int(t_sim/self.dt)
    t = np.linspace(0,t_sim,N_sim)
    alpha = [np.random.randn(nf) for j in range(self.nu)]
    phi=2*np.pi*np.random.rand(nf)

    U = np.array([np.array([
        alpha[j][i]*np.sin(2*i*np.pi*t+phi[i]) for i in range(nf)
    ]).sum(axis=0)
       for j in range(self.nu)]).T

    U = list(U)
    return U
```

Fig. 9.12 The utility that generates random sinusoidal exogenous profiles according to (9.27). This script is a part of the class MHE

```
# Inside the script of the MHE class
def simulate(self, x0, U, p):
    # utilities that simulates the system for
    # a given sequence of exogenous inputs
    X = [list(x0)]
    Y = []
    for i in range(len(U)):
        u = U[i]
        Y.append(list(self.h(X[-1],u,p).full().flatten()))
        X.append(list(self.F(X[-1],u, p).full().flatten()))
    X = np.array(X).reshape(-1, self.nx)
    Y = np.array(Y).reshape(-1, self.ny)
    t = np.array([i*self.dt for i in range(X.shape[0])])
    return t, X, Y
```

Fig. 9.13 The function that simulates the model of the system for a given initial state x_0, a list of exogenous input values (that might be delivered by the function `generate_random_U` of Fig. 9.12). This function is a method of the class MHE

The figure does not include the plotting instructions that are now hopefully quite well known to the reader. The kind of time profiles of the state and the exogenous signals that result from the execution of the script of Fig. 9.14 are shown in Fig. 9.15.

These trajectories can be now used to check the consistency of the cost functions we already created. This can be done using the following steps that can be repeated for different values of the *running* instant k:

1. Randomly choose an index i to play the role of $k - N_O$, namely the initial time in the observation horizon.
2. The vector `[X[i:i+N_O+1, :].ravel(),p]` is used as the correct value of the decision variable z.
3. The measurement vector is therefore given by:

```
Y[i:i+N_O].ravel().
```

```
# Inside the script of the MHE class
from mhe_solver import MHE
from lorentz import ode_lorentz, h_lorentz, nx, nu, n_par, nc, ny,
                    x_min, x_max, p_min, p_max, c_lorentz, u_lorentz

Q, R = np.eye(nx), np.eye(ny)
Qf, Pf = np.eye(nx), np.eye(n_par)
dt, rho_c, q, N_O =0.01, 1e2, 3, 10

mhe = MHE(nx=nx, nu=nu,
          n_par=n_par,
          nc=nc, ny=ny,
          x_min= x_min,
          x_max=x_max,
          p_min=p_min,
          p_max=p_max,
          ode=ode_lorentz,
          meas_func=h_lorentz,
          constraints=c_lorentz,
          u_from_y=u_lorentz,
          dt=dt, N_O=N_O, q=q,
          Q=Q, R=R, Qf=Qf, Pf=Pf,
          rho_c=rho_c)

p = np.array([5, 10/3, 18, 0.1])
U = mhe.generate_random_U(nf=10, T=5, t_sim=8)
x0 = np.random.randn(nx)
t, X, Y = mhe.simulate(x0, U, p)
```

Fig. 9.14 Script that uses `generate_random_U` and `simulate` in order to generate the state and measurement trajectories of the model

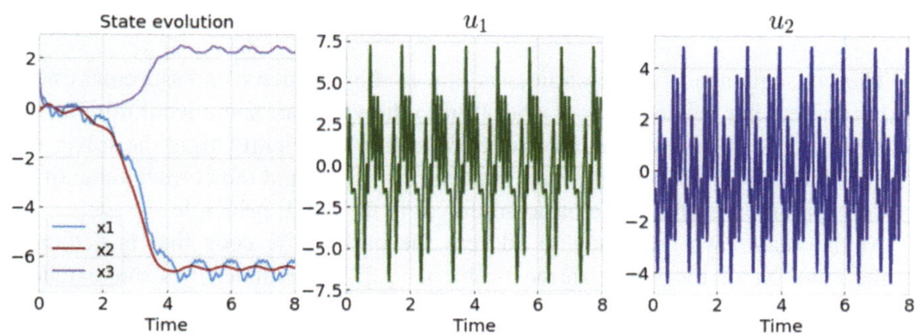

Fig. 9.15 State and exogenous signals profiles used in the validation

```
i = 15
z = np.array(list(X[i:i+mhe.N_0+1,:].ravel())+list(p))
Y_minus = Y[i:i+mhe.N_0,:].ravel()
x_hat = X[i+mhe.N_0,:]
p_hat = p
lam = 0.9

J_eq = mhe.J_eq(z, Y_minus, x_hat, p_hat, lam)
J_meas = mhe.J_meas(z, Y_minus, x_hat, p_hat, lam)
J_previous = mhe.J_previous(z, Y_minus, x_hat, p_hat, lam)
J_constr = mhe.J_constr(z, Y_minus, x_hat, p_hat, lam)
J_ipopt = mhe.J_ipopt(z, Y_minus, x_hat, p_hat, lam)
J_grad = mhe.J_grad(z, Y_minus, x_hat, p_hat, lam)

print(f'J_eq       = {J_eq}')
print(f'J_meas     = {J_meas}')
print(f'J_previous = {J_previous}')
print(f'J_constr   = {J_constr}')
print(f'J_ipopt    = {J_ipopt}')
print(f'J_grad     = {J_grad}')
```

```
J_eq       = 0
J_meas     = 0
J_previous = 0
J_constr   = 0
J_ipopt    = 0
J_grad     = 0
```

Fig. 9.16 The script that checks the consistency of the definition of the cost functions. The decision variable is instantiated to the correct values given the state trajectories and the value of the parameter p leading to zero costs

4. In order for z and Y defined above to lead to 0 costs, we need to take as previous estimation of the state, namely $\hat{x}$, the values of the state at the end of the observation horizon, namely `X[i+N_0]`[15] while the correct estimation $\hat{p}$ of the parameter vector should be taken equal to the value p that has been used to generate the trajectories `X` and `Y` in the first place.

The consequence of these choices is that in the absence of measurement noise and provided that the solver does not fall into the valley of some local minima and provided that[16] the allowed number of iterations is sufficiently high, the solver will provide the correct guess of the state trajectories values and the correct value of the model's parameter. This is the basic argument in the MHE principle.

At this stage, we are ready to address the part in the code that is dedicated to the solution of the optimization problem (9.22) that underlines our extended state/parameter estimation.

A large part of the associated development are quite similar to the ones that are used in **Chap.** 5 dedicated to the design of deterministic MPC control design.

The continuation of the script of the instantiation map `__init__` of the MHE class is shown in Fig. 9.17. In this script, the `Ipopt` solver is declared in the case where the associated option (interior point solver) is chosen to solve the optimization problem.

[15] Remember that in python, the last line in `X[i:i+N_0+1,:]` is precisely `X[i+N_0,:]`.

[16] This makes a long list of working assumptions!

```
# Inside the init function of the MHE class

# Define the Casadi-Ipopt solver
self.problem = {'f':J_ipopt,
                'x':z,
                'g':vertcat(*g),
                'p':vertcat(Y, x_hat, p_hat, lam)}

self.solver = nlpsol('solver', 'ipopt',
                self.problem, {'ipopt': {'max_iter': self.max_iter}})

lb_state = np.array(list(x_min) * (self.N_0 + 1)).flatten()
ub_state = np.array(list(x_max) * (self.N_0 + 1)).flatten()
lb_p = np.array(p_min).flatten()
ub_p = np.array(p_max).flatten()
self.lbz = np.array(list(lb_state) + list(lb_p))
self.ubz = np.array(list(ub_state) + list(ub_p))

self.compute_mhe_ipopt = lambda z0, Y, x_hat, p_hat, lam: \
                    self.solver(x0=z0, lbx=self.lbz, ubx=self.ubz,
                                lbg=lbg,ubg=ubg,
                                p=vertcat(Y, x_hat, p_hat, lam))

# Compute the gradient for the fast-gradient algorithm

G = jacobian(J_grad, z)
self.grad = Function('grad', [z, Y, x_hat, p_hat, lam], [G])
```

Fig. 9.17 The script inside the `__init__` function of the MHE class that declares the `Ipopt` solver as well as the gradient of the soft constraints cost for the fast-gradient algorithm's option

Notice that this declaration requires the vector of lower and upper bounds on the decision variable, namely `lbz` and `ubz` to be computed. On the other hand, the lower and upper bound of the constraints, namely `lbg` and `ubg` are also needed.

Notice that the definition of those last bounds on the constraints is based on the assumption that the definition of the `constraints` map is such that the returned values are negative if the constraints are satisfied. That is the reason why `lbg` is set to 10^{12} while `ubg` is set to 0 in the script of Fig. 9.17.

The script also uses the `CasaDi` function `jacobian` in order to compute the gradient of the previously computed soft constraints-version of the cost function, namely `J_grad` with respect to the decision variable z.

Now computing the gradient is only a step in the fast-gradient-based solver version. Indeed, this gradient needs to be called iteratively following the fast-gradient algorithm.

This is performed by the function that is depicted in Fig. 9.18.

Note that the algorithm uses a projection function that projects the current value of the decision variable into the hypercube of admissible values.

Notice also the additional parameters `gam` and `c` that are used in the fast-gradient algorithm which represent respectively the step size parameter and the moment coefficient that accelerates the convergence compared to the simple gradient algorithm as it has been discussed previously in **Chap.** 5.

The fact that these two parameters are not defined as attributes of the instance of the class MHE but rather as parameters of the call of the fast-gradient algorithm

```
# Inside the MHE class

def compute_mhe_fg(self, z, Y, x_hat, p_hat, lam, gam, c, Niter):

def project(z):

    # projection on the hypercube of admissible value
    z_proj = copy.copy(np.array(z).flatten())
    for i in range(self.nz):
        if z_proj[i] > self.ubz[i]:
            z_proj[i] = self.ubz[i]
        elif z_proj[i] < self.lbz[i]:
            z_proj[i] = self.lbz[i]

    return z_proj

zstar_previous = copy.copy(z)
eta_previous = copy.copy(z)

for _ in range(Niter):

    G = self.grad(zstar_previous, Y, x_hat, p_hat, lam).full().flatten()
    eta = zstar_previous - gam * G
    zstar = project(eta + c * (eta - eta_previous))
    eta_previous = copy.copy(eta)
    zstar_previous = copy.copy(zstar)

return zstar
```

Fig. 9.18 Script of the fast-gradient algorithm for the extended moving-horizon observer

enables them to be modified on-line depending on the measured decrease of the cost function in the previous iterations.

Now that we have the two options (`ipopt/fast-gradient`) for the solver to be used in order to implement the MHE optimization problem, let us experiment them on some randomly chosen samples of the trajectories of the state and measurements that we have generated previously.

The idea is to check whether the correct values of the decision variables can be retrieved despite some incorrect initial guesses provided to the solver.

On the other hand, we seek an estimation of the computation times that might be involved in order to grasp an idea about the delay parameter q that should be considered.

Check The Ipopt-Based Solver

Here we are not checking the overall behavior of the MHE, but only whether **the single call** of the solver for a given list of arguments including:

- an initial guess z_0,
- a sequence Y of previous measurements,
- a pair of current estimation $(\hat{x}, \hat{p})$, and
- a filtering parameter `lam`,

does provide the *the correct* values of the state and parameter gathered in the solution $z^\star$ even when provided with a wrong erroneous initial guess z_0.

In order to check this fact, we use different values of the index i used in the script of Fig. 9.16 to collect the sequence of measurements, denoted by Y_minus and associated to corresponding value of the z as well as the corresponding values of the correct estimation x_hat and p_hat at the end of the observation horizon. Having y_minus, x_hat, p_hat we first call the method compute_mhe_ipopt with $z_0 = 0$ and lam=1 using the following script:

```
from time import time
t0= time()
R = mhe.compute_mhe_ipopt(0*z, Y_minus, x_hat, p_hat, lam=1)
cpu = time()-t0
```

where the time Python package is used to estimate the computation time needed by the solver to get the *optimal* solution.

Notice that by using lam=1 the solver does not take into account the information contained in the previous and correct estimation of the state and parameters, namely x_hat and p_hat. Only the output and the model's equations are considered.

Before we examine the optimization results and the computation time let us first examine how bad is the initial guess $z_0 = 0$ by looking at the values of the different cost functions at this initial value.

The results are shown below:

```
J_eq = mhe.J_eq(0*z, Y_minus, x_hat, p_hat, lam)
J_meas = mhe.J_meas(0*z, Y_minus, x_hat, p_hat, lam)
J_previous = mhe.J_previous(0*z, Y_minus, x_hat, p_hat, lam)
J_constr = mhe.J_constr(0*z, Y_minus, x_hat, p_hat, lam)
J_ipopt = mhe.J_ipopt(0*z, Y_minus, x_hat, p_hat, lam)
J_grad = mhe.J_grad(0*z, Y_minus, x_hat, p_hat, lam)

print(f'J_eq       = {J_eq}')
print(f'J_meas     = {J_meas}')
print(f'J_previous = {J_previous}')
print(f'J_constr   = {J_constr}')
print(f'J_ipopt    = {J_ipopt}')
print(f'J_grad     = {J_grad}')
```

```
J_eq       = 0.0182178
J_meas     = 11.9543
J_previous = 0
J_constr   = 0
J_ipopt    = 11.9725
J_grad     = 11.9725
```

```
zstar = R['x'].full().flatten()
e = zstar-z
print(f'cpu = {cpu}')
print('----')
print(f'abs(X)={abs(z[:-mhe.np]).max()}')
print(f'abs(e_x)={abs(e[:-mhe.np]).max()}')
print('------')
print(f'abs(p)={abs(z[-mhe.np:]).max()}')
print(f'abs(e_p)={abs(e[-mhe.np:]).max()}')
```

```
cpu = 0.10766196250915527
----
abs(X)=6.534928046871588
abs(e_x)=0.0012355607564158433
------
abs(p)=18.0
abs(e_p)=0.05406175655924361
```

Fig. 9.19 Example of computation time and some indicators of resulting estimation errors when using the `Ipopt` option through the `compute_mhe_ipopt` method of the MHE class

The computation time and the resulting maximum error on the estimated quantities are shown in Fig. 9.19. This figure shows that a typical computation time is around 0.1 sec (recall that `mhe.dt=0.01`) meaning that $q \geq 10$ might be needed if this options is to be used without further considerations (see later).

On the other hand the figure shows that the error on the state and the parameters are not fully 0 but they are reasonably small compared to the order of magnitude of the estimated quantities.

Now comparing the max norm of the true and the estimated quantity is not truly sufficient to assess that the solver is doing the expected job. This is the reason why in Fig. 9.20 the resulting cost functions at the solution are compared to the ones corresponding to the initial guess shown previously. The results of Fig. 9.20 clearly suggest that the solver does greatly the assigned job by minimizing the values of the cost functions while meeting the constraints.

The fact that the order of magnitude of the error on the components of the decision variable z might indicate that there is probably an observability issue in the sense that the cost function might be explained by slightly biased combination of state and parameters' values.

Now recall that in the previous attempt, we used `lam=1` which means that the previous (and here correct) estimation `x_hat` and `p_hat` of the state and parameters values, while provided to the solver, are not used since the corresponding terms in the cost function are multiplied by (`lam-1`)=0.

So let us try to use the value lam=0.9 through the following script:

```
from time import time
t0= time()
R = mhe.compute_mhe_ipopt(0*z, Y_minus,
                    x_hat, p_hat, lam=0.9)
cpu = time()-t0
```

which leads to the following obviously better results

```
zstar = R['x'].full().flatten()
e = zstar-z
print(f'cpu = {cpu}')
print('----')
print(f'abs(X)={abs(z[:-mhe.np]).max()}')
print(f'abs(e_x)={abs(e[:-mhe.np]).max()}')
print('------')
print(f'abs(p)={abs(z[-mhe.np:]).max()}')
print(f'abs(e_p)={abs(e[-mhe.np:]).max()}')
```

```
cpu = 0.09730815887451172
----
abs(X)=6.534928046871588
abs(e_x)=1.572248508807661e-05
------
abs(p)=18.0
abs(e_p)=0.00015267998989622333
```

```
J_eq = mhe.J_eq(zstar, Y_minus, x_hat, p_hat, lam)
J_meas = mhe.J_meas(zstar, Y_minus, x_hat, p_hat, lam)
J_previous = mhe.J_previous(zstar, Y_minus, x_hat, p_hat, lam)
J_constr = mhe.J_constr(zstar, Y_minus, x_hat, p_hat, lam)
J_ipopt = mhe.J_ipopt(zstar, Y_minus, x_hat, p_hat, lam)
J_grad = mhe.J_grad(zstar, Y_minus, x_hat, p_hat, lam)

print(f'J_eq       = {J_eq}')
print(f'J_meas     = {J_meas}')
print(f'J_previous = {J_previous}')
print(f'J_constr   = {J_constr}')
print(f'J_ipopt    = {J_ipopt}')
print(f'J_grad     = {J_grad}')
```

```
J_eq       = 1.14761e-10
J_meas     = 2.86378e-09
J_previous = 0
J_constr   = 0
J_ipopt    = 2.97854e-09
J_grad     = 2.97854e-09
```

Fig. 9.20 Values of the different costs at the solution provided by the method compute_mhe_ipopt of the MHE class

showing almost perfect estimation (despite the quite slight modification of `lam`) and almost the same computation time.

Notice that the maximum number of iterations of `Ipopt` is by default set to 500 in the definition of the class. In general, it is always a good practice to try the performance of the solver when this number is taken smaller.

In this specific case, when using the following setting:

`iter_max=25,`

the computation time remains almost identical as well as the result.[17] If smaller values are taken, however, the cpu-time might be reduced but the quality of the resulting solution is significantly worse.

Obviously the only effective assessment of the MHE can be done over an entire set of simulations and not on a single trials. This will be done later on but the success of such individual tests is a necessary condition for a first simple validation of the code before going further in the implementation of entire simulation scenarios.

Before doing that, let us first examine, using the same principle, the behavior of the fast-gradient solver implemented in the `compute_mhe_fg` method of the MHE class.

Checking the Fast-Gradient-Based Solver

To check the fast-gradient-based solver, we use the `compute_mhe_fg` method of the MHE class:

```
from time import time
t0= time()
zstar = mhe.compute_mhe_fg(0*z, Y_minus, x_hat, p_hat,
          lam=0.9, gam=0.1, c=0.9, Niter=2000)
cpu = time()-t0
```

which produces the following results:

[17] This means that the default number of maximum number of iterations is not fully used since the solver converged with much less number of iterations.

```
e = zstar-z
print(f'cpu = {cpu}')
print('----')
print(f'abs(X)={abs(z[:-mhe.np]).max()}')
print(f'abs(e_x)={abs(e[:-mhe.np]).max()}')
print('------')
print(f'abs(X)={abs(z[-mhe.np:]).max()}')
print(f'abs(e_x)={abs(e[-mhe.np:]).max()}')
```

```
cpu = 0.658301830291748
----
abs(X)=6.541806721526534
abs(e_x)=5.795746316206163e-05
------
abs(X)=18.0
abs(e_x)=1.2212063774441617e-05
```

These results obviously show that the solver does the job but a high number of iterations is needed resulting in a significantly higher computation time than the Ipopt solver *for this instance of the problem.*

Indeed **when using only 200 iterations**, we get significantly worse results as it is shown below:

```
cpu = 0.08867597579956055
----
abs(X)=6.541806721526534
abs(e_x)=2.215806664686288
------
abs(X)=18.0
abs(e_x)=0.46149101456109287
```

Finally in order to appreciate the contribution of the fast-gradient compared to classical gradient ($c = 0$), the results with 2000 iterations and $c = 0$ are shown below:

```
cpu = 0.7255580425262451
----
abs(X)=6.541806721526534
abs(e_x)=2.66106725710123
------
abs(X)=18.0
abs(e_x)=0.5489825223462286
```

Now despite the fact that when studying a single shot of the problem's solution, it might appear that the Ipopt solver perform better, it is important to keep in mind that in a real-time context, it might be interesting to *distribute the optimization over time* in order to be able to use more frequent updating of the estimated values (by lower values of q) and still get asymptotic convergence toward the best solution.

To reformulate it differently, the fast-gradient might converge slowly per iteration but the elementary operation of a fast-gradient is much less expensive than the interior point one.

In order to show this, one can observe the results of the fast-gradient method when only 100 iterations are allowed and using $c = 0.9$.

```
cpu = 0.045034885406494 14
----
abs(X)=6.541806721526534
abs(e_x)=3.94930495951132
------
abs(X)=18.0
abs(e_x)=0.6666666666666665
```

where we can still observer a significant decrease in the error while the computation time is drastically reduced so that $q = 5$ can be used rather than $q = 10$ for the interior point option.

Even if the decrease is not sufficient for an immediate recovery of the good state/parameter vectors, an additional decrease would be obtained in the next round and asymptotically the estimation would be as good as under the interior point option while enabling a twice more frequent use of the updated measurements.

Simulating the Observer over Complete Scenarios

Now that we have assessed the basic elementary behavior of our code, let us add the attributes and the methods that can be used to simulate the whole operation involving the MHE.

The corresponding method is `simulate_mhe` which is shown in Fig. 9.21. Note that this function admits the following arguments that define an estimation scenario:

✓ The *real* system trajectories of state `X` and measurements `Y` and the associated time vector `t` as generated by the previously described `simulate` method (see Fig. 9.13).
✓ The initial estimated values of the state and the parameter vectors, namely `x_hat_0` and `p_hat_0`.
✓ The filtering parameter `lam` described above and used in the definition of the cost function (9.22).
✓ The parameter `option` ∈ {`ipopt, fg` } that determines the solver to be used among the two available ones.
✓ The fast-gradient parameters `gam, c` and `Niter` in case `option=fg` is used.

Note that the method `simulate_mhe` contains mainly two parts. In the first, the initial estimated values `x_hat_0` and `p_hat_0` are used to produce estimations that are purely based on open-loop simulations since in real-life the measurement contained in `Y` are not accessible before `N_O` sampling periods have been elapsed.

Additional `q` sampling periods are still necessary to simulate in order to emulate the computation time that is allowed for the solution of the optimization problem.

At the end of this open-loop simulation, the problem is solved and the first measurement-based update of the estimation is computed by calling the internal method `compute_zstar` which delivers the estimation of the state and parameters that hold at the previous instant $k - q$.

Having these estimated values, the estimation of the state between instant $k - q$ and k can then be produced by calling the internal method `update_x_hat`.

```
def simulate_mhe(self, t, X, Y, x_hat_0, p_hat_0, lam,
                 option='ipopt', gam=0.05, c=0.8, Niter=20):

# t, X and Y are outputs of self.sim
# they are matrices.

#---------- some reusable functions

def compute_zstar(k, Y, X_hat, P_hat):
 #---- detailed later
    return x_star, p_star, cpu_

def update_x_hat(k, Y, x_star, p_star):
 # ---- detailed later
   return x.full()
#-------------------------------------------------
# Fill up the measurement buffer for the first
# optimization to be possible to perform

X_hat, P_hat = [x_hat_0], [p_hat_0]
cpu, t_cpu = [], []
k = 0 # the real time of the system
# fill the Y_minus for the first time
for i in range(self.N_O+self.q):
    x = X_hat[-1]
    u = self.u_from_y(Y[k,:])
    p = P_hat[-1]
    X_hat.append(self.F(x, u, p).full().flatten())
    P_hat.append(p)
    k += 1

# Run the optimizer for the first time
# and update x_hat accordingly
x_star, p_star, cpu_ = compute_zstar(k, Y, X_hat, P_hat)
cpu.append(cpu_)
t_cpu.append(k * self.dt)
X_hat[k] = update_x_hat(k, Y, x_star, p_star)
P_hat[k] = p_star

# Ready to start the  main loop
# until the end of the simulated data
while k < len(X) - self.q:

    for i in range(self.q):
        x = X_hat[-1]
        u = self.u_from_y(Y[k,:])
        p = P_hat[-1]
        X_hat.append(self.F(x, u, p).full().flatten())
        P_hat.append(p)
        k += 1

    x_star, p_star, cpu_ = compute_zstar(k, Y, X_hat, P_hat)
    cpu.append(cpu_)
    t_cpu.append(k * self.dt)

    # Update the value at instant k
    X_hat[k] = update_x_hat(k, Y, x_star, p_star)
    P_hat[k] = p_star

X_hat.append(X_hat[-1])
X_hat = np.array([ell.flatten() for ell in X_hat])
P_hat.append(P_hat[-1])
P_hat = np.array([ell.flatten() for ell in P_hat])
t_cpu, cpu = np.array(t_cpu), np.array(cpu)

return X_hat, P_hat, t_cpu, cpu
```

Fig. 9.21 Script of the method `simulate_mhe` of the MHE class. This method uses two internal functions, namely `compute_zstar` and `update_x_hat` that are given later on

The second part of the algorithm reproduces repetitively the previous computation by shifting the observation window by an amount of `q` sampling periods, solving the optimization problem to gather the sub-optimal solution before the associated estimation is propagated via the open-loop simulation over `q` sampling periods and so on.

Notice that each time the optimization is performed, the associated computation time is recorded for the real-time implementability assessment of the scenario.

Two internal methods are used in `simulate_mhe`, namely:

- ✓ `compute_zstar` which solves the optimization problem that is defined at some instant k given the previous measurement and the current estimations. Notice that the later are extracted from the on-going built sequences `Y`, `X_hat` and `P_hat` that are internal variables of the main method `simulate_mhe`. This method delivers the estimated values `x_star` and `p_star` as well as the computation time needed in the optimization process.
- ✓ `update_x_hat` which uses the results of the previous method in order to perform an open-loop model-based simulation so that the new estimated values of the state and the parameter vectors can be updated for the next round.

Figure 9.22 shows the script of the internal method `compute_zstar`. This is the method that runs the optimization problem solver depending on the `option` arguments used in the call of `simulate_mhe`. At each computation round, the cpu-time is also collected and exported together with the solution of the optimization problem.

Finally Fig. 9.23 shows the script of the method `update_xhat` that uses the optimal solution in order to compute updated estimation of $\hat{x}_k$ given the optimal solution $x^\star$ and $p^\star$ that stems from the solution of the optimization problem that involves the measurement on the observation interval $[k - N_O - q, k - q]$.

We are now ready to observe the *closed-loop* behavior of the MHE. This is done using the following steps:

1. A random profile of the exogenous input u is first generated.
2. An initial state is chosen as well as a value of the parameter vector p.
3. The system is simulated starting from the above chosen initial state, parameters and the exogenous profile in order to produce the measurement profile.
4. Different settings of the MHE estimator and/or different values of the initial guesses on the state and the vector of parameters can be simulated using the above described `simulate_mhe` method.

This road map is used in the remainder of this chapter in order to grasp a better understanding of the role played by the different ingredients of the estimation algorithm.

The scenario is generated using variants of the script depicted in Fig. 9.24. This script produces the time, state and measurement vectors that are the first arguments of the call of `simulate_mhe` (see Fig. 9.21).

The script of Fig. 9.25 shows typical call of the observer's method `simulate_mhe` that produces the state/parameter estimation profiles for a specific choice of

```
def compute_zstar(k, Y, X_hat, P_hat):

    # Determine the instants delimiting the
    # observation window.
    i1 = k - self.N_O - self.q
    i2 = i1 + self.N_O
    # collect the corresponding measurement and
    # previous estimation
    Y_minus = Y[i1:i2, :].ravel()
    x_hat = X_hat[i2]
    p_hat = P_hat[i2]
    # warm start for the initial guess
    lex = np.array([ell.flatten() for ell
                    in X_hat[i1:i2+1]]).flatten()
    z = np.array(list(lex) + list(p_hat)).flatten()
    # solve the optimization problem
    if option == 'ipopt':
        t1 = time()
        R = self.compute_mhe_ipopt(z, Y_minus,
                                   x_hat, p_hat, lam)
        cpu_ = time()-t1
        z_star = R['x'].full().flatten()
    elif option == 'fg':
        t1 = time()
        z_star = self.compute_mhe_fg(z, Y_minus,
                                     x_hat, p_hat,
                                     lam, gam, c, Niter)
        cpu_ = time()-t1
    # extract x_star and p_star from the computed solution
    x_star = z_star[0:-self.np].reshape(-1, self.nx)[-1, :]
    p_star = z_star[-self.np:]

    return x_star, p_star, cpu_
```

Fig. 9.22 The script of the internal method `compute_zstar` used in `simulate_mhe`

```
def update_x_hat(k, Y, x_star, p_star):

    x = x_star
    for i in range(self.q):
        u = self.u_from_y(Y[k - self.q + i, :])
        x = self.F(x, u, p_star)

    return x.full()
```

Fig. 9.23 Script of the internal method `update_x_hat` that update the last estimation of the state and parameter vectors based on the last solution of the optimization problem

```
Nsim = 501
x0 = 0.2 * np.ones(mhe.nx)
t_sim = (Nsim+1) * mhe.dt
U_sim = mhe.generate_random_U(nf=10, T=5,t_sim=t_sim)
p_sim = np.array([5,10/3,18,0.1])
t, X, Y = mhe.simulate(x0,U_sim,p_sim)
```

Fig. 9.24 Typical script serving in the generation of the scenario on which the observer behavior is evaluated

```
x_hat_0 = 1.5*np.ones(mhe.nx)
p_hat_0 = np.diag(0.3*np.random.randn(mhe.np)) @ p_sim
lam = 1

# In case fast-gradient is used
gam, c, Niter = 0.1, 0.9, 300

Y_noisy = Y + 0.02 * np.random.randn(*Y.shape)
R = mhe.simulate_mhe(t, X, Y, x_hat_0, p_hat_0,
                     lam, gam=gam, c=c,
                     Niter=Niter, option='ipopt')

X_hat, P_hat, t_cpu, cpu = *R
```

Fig. 9.25 Script that calls the method `simulate_mhe` of the class MHE's instance in order to examine the behavior of the MHE

the call together with the choices already specified in the creation of the class MHE instance shown before.

Notice that it is possible to introduce a measurement noise leading to the noisy measurement vector `Y_noisy` which can be used in the call of `simualte_mhe` instead of `Y`.

The first example of setting uses the ideal situation where noise-free measurements are used and a perfect confidence in the measurement is considered by setting $\lambda = 1$. The results are shown in Fig. 9.26.

In this case, it appears clearly that when using the `Ipopt` option, the perfectly correct values of the components of the state and parameters are recovered as soon as the first optimization are performed and from that instant, the estimations remain correct since the model is perfectly known and the measurements are noise-free.

This figure also clearly shows that given the delay $q = 10$ that is used in the creation of the MHE instance, the resulting estimator is real-time implementable[18] as the computation time is largely lower that the available computation time given by:

$$q\tau = 10 \times 0.02 = 0.2,$$

time units.

[18] As far as the computational device used to produce the results is concerned.

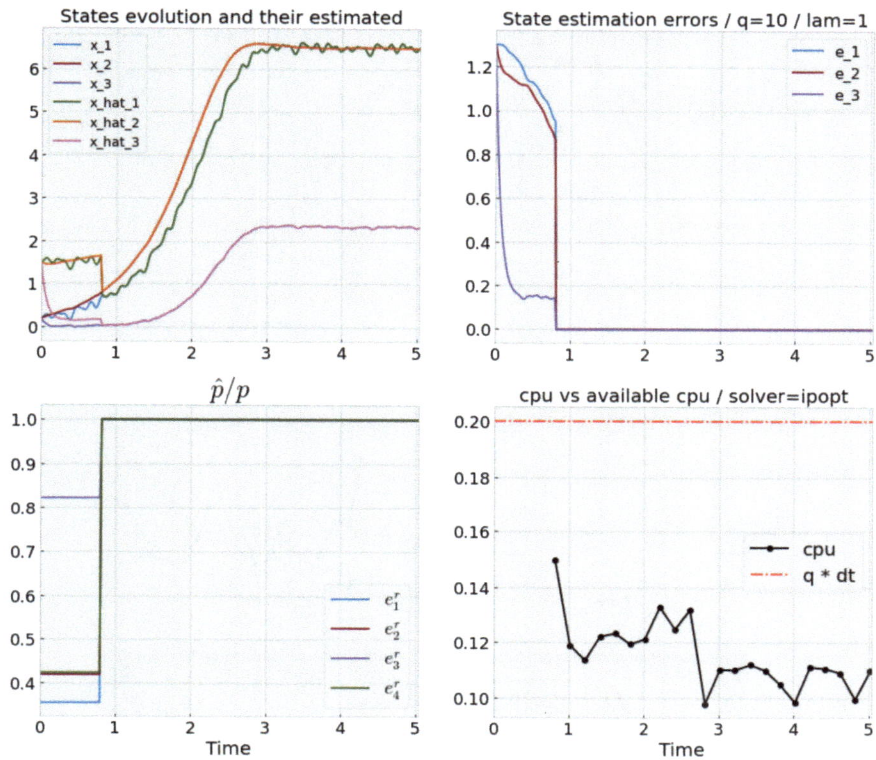

Fig. 9.26 MHE validation. Noise-free simulation with the parameters $q = 10,\ \lambda = 1,$ `dt=0.02` and the `Ipopt` solver

Notice that given the different orders of magnitude of the parameters' values, the convergence is shown in terms of the ratio $p/\hat{p}$ which should converge to 1 in case of perfect estimation.

Figure 9.27 shows the behavior of the MHE when the fast-gradient algorithm is used with two different maximum number of iterations, namely, 300 and 1000.

Obviously, for the example at hand, the sampling period and the computation hardware, the interior point option is to be preferred. This is why this option is maintained in the following investigations.

The next feature we want to investigate is related to the filtering parameter λ. Recall that this parameter tunes the extent to which one is ready to change the estimation based on the new coming measurement compared to the estimation that would result from the previous estimation using the model of the system.

Recall that in the previous simulation, $\lambda = 1$ is used expressing a total confidence in the measurements compared to the previous estimation. This is obviously a reasonable choice given the very large initial estimation error. Indeed, when using $\lambda = 0.99$, the estimation results are those shown in Fig. 9.28.

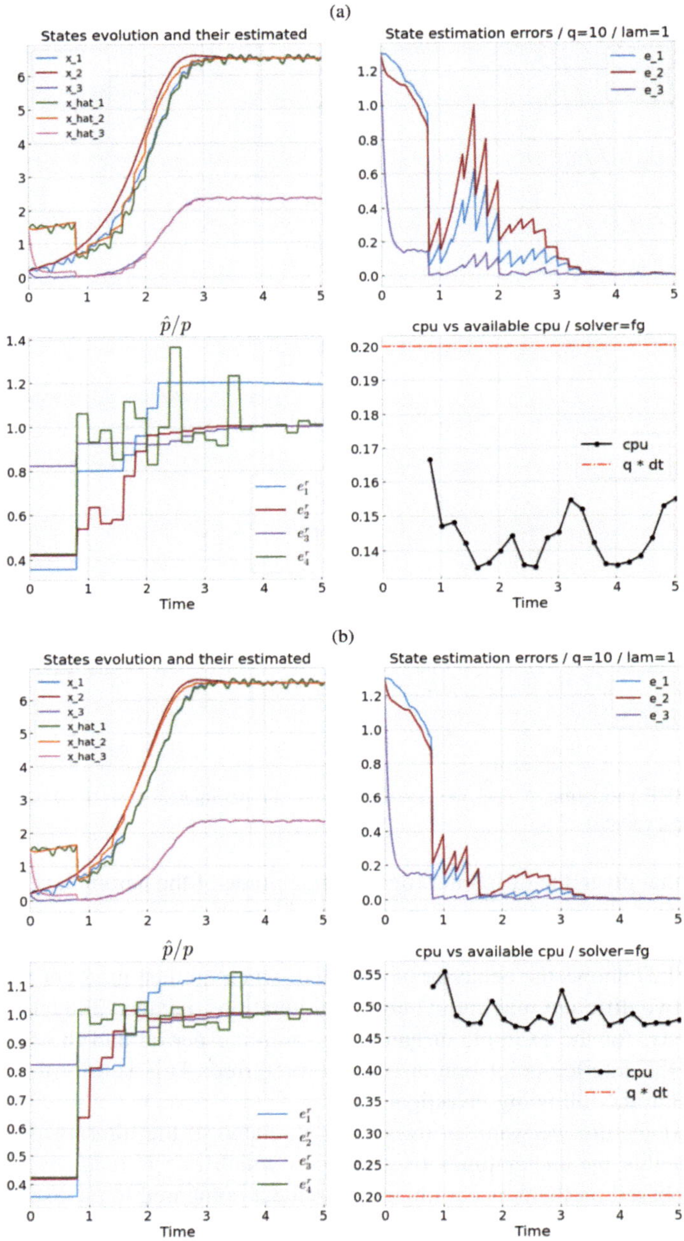

Fig. 9.27 Behavior of the MHE under the fast-gradient algorithm with maximum number of iteration = 300 **a** or 1000 **b**. The scenario is the one used in Fig. 9.26 where option= `ipopt` is used

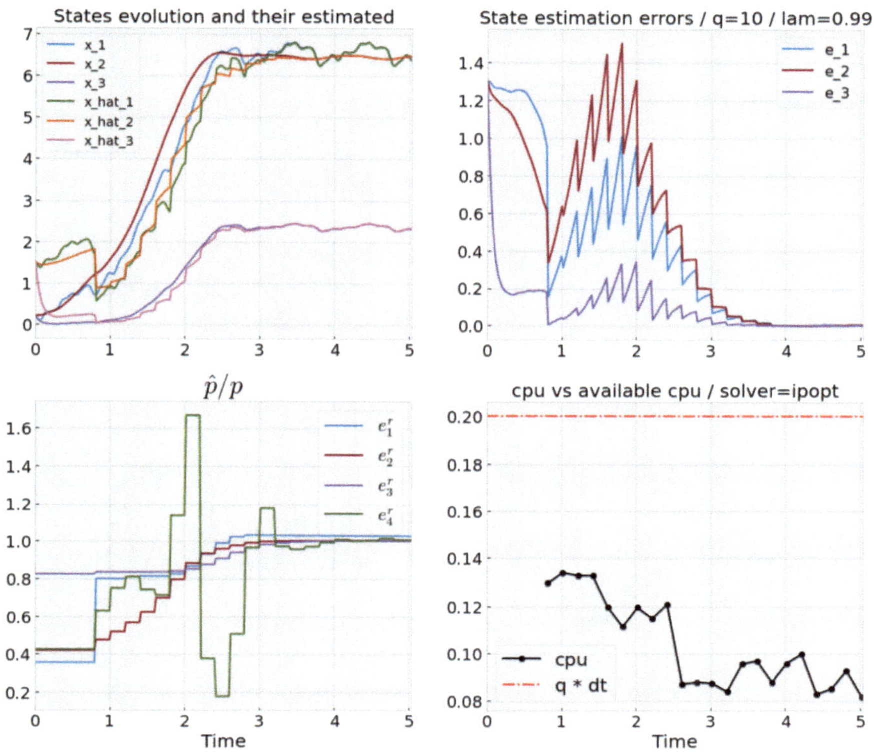

Fig. 9.28 MHE validation. Noise-free simulation with $q = 10$, $\lambda = 0.99$ and dt=0.02

The seemingly high impact of a slight variation of the parameter $\lambda = 0.99$ instead of 1 comes from the artificially high error in the initial guess which, when introduced in the definition of the cost function induces high change in the optimal solution. But this affects the result only in a transient way.

Note also that for the same value of λ, the impact on the estimation lasts even more when q is high since the number of updating is lower for the same duration of the observed scenario.

As a matter of fact, the benefit from the parameter λ appears more evidently asymptotically as the estimation becomes quite close to the true value. In this case, it plays the role of filtering against the measurement noise. But in order to see this in action, we need to introduce some measurement noise and to observe the result on a longer time horizon.

The result are shown in Fig. 9.29 where the same scenario is simulated with two different values of the trade-off parameter $\lambda \in \{1, 0.95\}$.

In this case, it can be observed that on the long term, the use of $\lambda < 1$ seems to produce two effects:

(a)

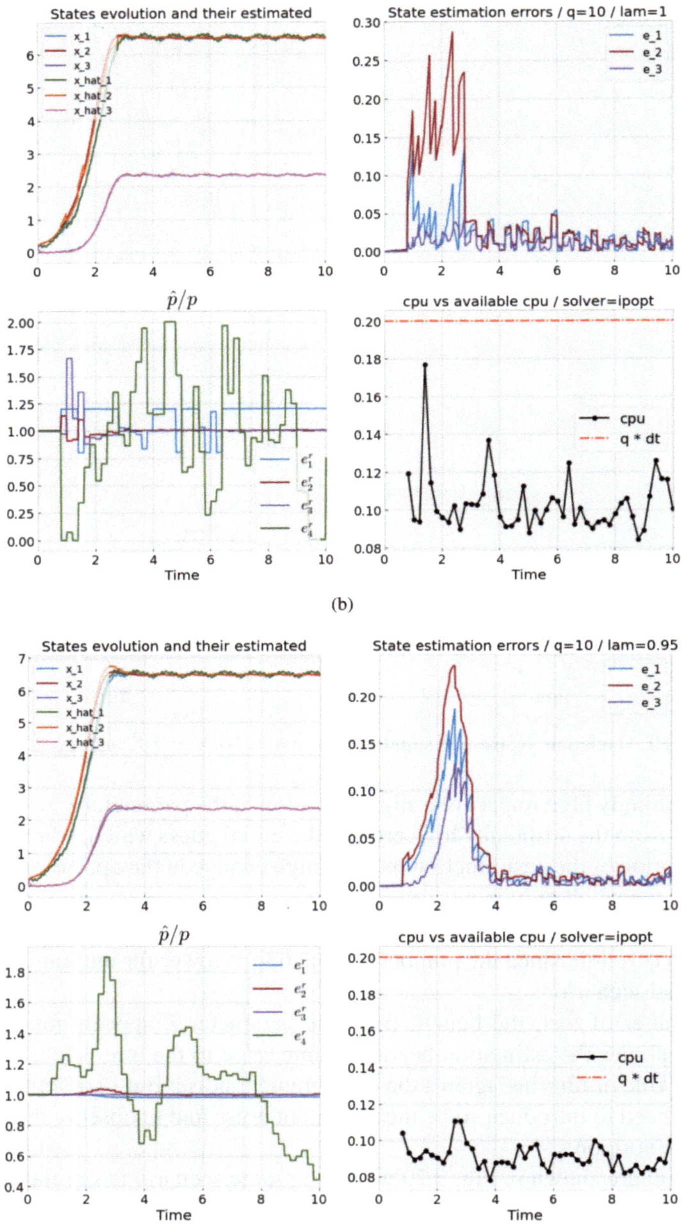

Fig. 9.29 Behavior of the MHE for two different values of the filtering parameter $\lambda = 1$ **a** and $\lambda = 0.95$ **b**

- ✓ The asymptotic error on the state estimation is lower in the presence of measurement noise.
- ✓ The parameter estimation seems to be less nervous when $\lambda < 1$ is used which is expected since when $\lambda = 1$, only the measurement is considered when updating the estimation of the state and parameter and since the noise *moves the location* of the optimal solution, this move is transmitted to the produced estimation without the slightest filtering leading to nervous and more erroneous estimation.
- ✓ The computation time profile seems also to be more *nervous* which is again expected since the filtering makes the optimal solution move more slowly and hence makes the warm start-induced initialization closer to the optimal solution and therefore more quickly reachable hence requiring smaller number of iterations of the `Ipopt` solver.

Notice that the simulations depicted in Fig. 9.29 intentionally start with prefect initial guess and since the measurement based estimation updating only starts after $N_O + q$ sampling periods, the initial interval before that instant corresponds to zero error. It is only when the noisy measurement is used that the errors become to appear in the estimation. As a matter of fact, in this specific case, the best value of λ is 0! But this is obviously an unrealistic situation since in real life, there is no way to know how good is the initial guess.

This being said, the optimal tuning of an observer remains an open question except for the ideal case of linear system corrupted with white noise and provided that the covariance matrices of the state and measurement noise are perfectly known.

It is even possible to adopt a time-varying updating period (time-varying q) depending on the context as it is suggested in [1] for predictive control. This is no surprise since it is all about tuning the gain that one can get from a better solution to the drawbacks that is induced by the presence of noise and uncertainties.

It is also important to keep in mind that the trade-of between the confidence one has in the measurements and in the model coupled to the previous estimation does also depend on the weighting matrices Q_f and P_f that are involved in the definition of the cost function (9.22) that the optimizer attempts to minimize under the associated constraint.

The result of all this *mess* is probably that essay/error tuning is probably the best option from a realistic and pragmatic point of view.

This chapter hopefully gave the basic tools for a first implementation that should be then improved and tuned for each specific use-case!

A final remainder might be necessary. All the above mentioned python codes are by no mean neither optimal nor the most elegant implementation of the extended state parameter MHE. But I tried to make it the quite general in order to be used in many circumstances. Readers who are good coders should feel free to use the code as an inspiration for a better offer in terms of generic MHE!

It is also wise to browse the Internet[19] for the last proposed packages that pops-up on a monthly basis here and there in an active and great community of researchers.

9.7 Conclusion

Building on-line state and parameter estimators lies in the heart of many strategies that aim at addressing the presence of uncertainties in the controlled and the diagnosed dynamical systems.

Mastering at least a specific version of such a state/parameter on-line estimator is mandatory for any algorithm's designer who is in charge of controlling or monitoring uncertain systems with partial measurements.

In this chapter, the foundations and the associated algorithms for the design and the implementation of optimization-based MHE have been explained in detail.

References

1. Alamir, M.: On adaptive measurement inclusion rate in real-time moving-horizon observers. IFAC Proceed. **47**(3), 552–557 (2014)
2. Alamir, M.: Monitoring control updating period in fast gradient based nmpc. In: 2013 European Control Conference (ECC), pp. 3621–3626 (2013). https://doi.org/10.23919/ECC.2013.6669122

[19] Or just ask ChatGPT if it is still there!

Scalable Sparse Identification of Nonlinear Relationships

10

10.1 Chapter Content at a Glance

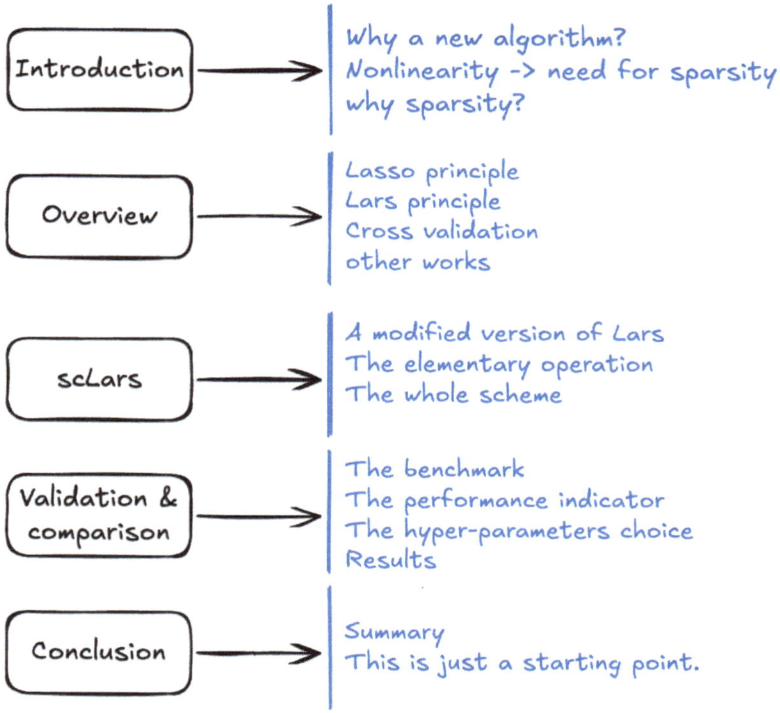

M. Alamir, *Nonlinear Control of Uncertain Systems*,
https://doi.org/10.1007/978-3-031-93287-8_10

10.2 Introduction

Identifying nonlinear relationships lies in the heart of the topic of this book. This is because ignoring the relationships that govern the interdependence of the trajectories of some quantities of interest, for whatever reasons, might be viewed as the ultimate form of the presence of uncertainty.

Indeed, compared to the context of parametric uncertainty where the structure of a parametric model is known but the real values of the parameters are uncertain, here, the mathematical structure of the relationships is not known and needs to be approximated via appropriate mathematical forms that is to be guessed and quantitatively determined from the collected data.

It is however important to underline the following fact:

! This Chapter Is Not About Identification in the Control-Related Sense

Although the algorithms studied in this chapter can be used to derive specific algorithms that address the standard *control*-related identification problem,[1] this is not rigorously the object of the present chapter. In other words, this chapter is about the identification of nonlinear relationships and not about the identification of nonlinear dynamical systems.

This chapter is about an algorithm that enables to identify, **in a parsimonious and scalable** way, the relationship that enables to predict a *label* as a function of a vector of *features*. The chapter studies the algorithm and compares its performance to the state-of-the-art alternatives addressing the same problem.

Links to Other Chapters

In Chap. 11, two examples are given to show how the algorithm, developed in this chapter, can be used to derive a solution to the problem of **partial observation-target reconstruction** in the sense defined in Sect. 2.10.

As a matter of fact, these two examples show how it is possible to address the ambitious goal of partial estimation in the presence of uncertainties.

But let us first answer some high-level questions, namely:

1. Why a new algorithm?
2. Why do we need scalability?
3. Why do we need sparsity?

Some brief answers to the above three questions are provided in the following section.

[1] Which consists in finding the minimum number of measured output and their dynamic relationships so that the system can be simulated.

10.2.1 Why a New Algorithm?

The main keywords in the title of this chapter are:

scalable and nonlinear.

As a matter of fact, it is the fact that we target the objective of identifying nonlinear relationships that makes scalability a crucial issue.

Indeed, the motivation for developing a new algorithm was a simple consequence of the failure of the existing ones!

Indeed, in the beginning, I was in a *user-role* and I was confident that the existing sparse identification algorithms were able to address my nonlinear identification problems involving **real industrial data**. After all, these algorithms are widely used and extremely challenged by thousands of users and developers over the world which strongly suggests that there was no reason to think that something else is necessary or even possible to come out with. It is only when these algorithms started to *crash* for some instances of my problems that I began to consider having a deeper look on the situation.

So let us first examine where does the need for **scalability** come from and why it is intimately linked to **nonlinearity** before we recall the generic relevance of **sparse solutions** in engineering.

These two items are recalled in the following two sub-sections.

10.2.2 Why Nonlinearity Induces the Need for Scalability?

Let us assume that one disposes of time-series that are generated by n_s different sensors and let us denote the values of these sensors at instant k by:

$$s_k := \begin{bmatrix} s_k^{[1]} \\ \vdots \\ s_k^{[n_s]} \end{bmatrix} \in \mathbb{R}^{n_s}. \tag{10.1}$$

Assume also that some scalar quantity[2] ℓ is also measured and that one seeks a relationship between ℓ and s.

An instance of this problem has been already seen in Chap. 2 where the expression of an observation target[3] ξ in terms of the previous measurement $y_k, \ldots, y_{k-N_O}$ was

[2] Referred to as the label.

[3] This can be a component of the state vector for instance.

searched for:

$$\xi_k \approx \texttt{ML}\left(\begin{bmatrix} y_k \\ \vdots \\ y_{k-N_O} \end{bmatrix}\right) := O\left(\mathbf{y}_k^{(-)}\right). \tag{10.2}$$

Another instance of this problem appears when the dynamic of a nonlinear system is to be identified in terms of the measured variables leading to the following instantiation of the label ℓ:

$$\ell_k = \Delta s_k^{[j]} := s_{k+1}^{[j]} - s_k^{[j]},$$

where $j \in \{1, \dots, n_s\}$ is a specific sensor index.

The corresponding relationship one is hence looking for takes the following form:

$$\Delta s_k^{[j]} \approx F(\underbrace{s_k, \dots, s_{k-d}}_{z_k}). \tag{10.3}$$

Consequently, there is a general feature that holds when dynamic systems are involved that we can state as follows:

The Need for Delayed Instances of the Sensors' Values

In identification problems involving dynamical systems, the relationship that correctly matches the value of some *label*[4] ℓ at instant k, namely y_k might involve not only s_k but the previous instances of s over some observation window of length[5] n_d leading to a vector of features z_k that is defined by:

$$z_k := \begin{bmatrix} s_k \\ \vdots \\ z_{k-n_d} \end{bmatrix} \in \mathbb{R}^{n_z} \quad \text{where} \quad n_z := n_s(n_d + 1). \tag{10.4}$$

Once the feature vector z is defined, solving the identification problem requires a finite dimensional representation of the unknown map $F(z)$ that provides a good solution to the least-squares problem:

$$y \approx F(z). \tag{10.5}$$

A common option is to consider the structure according to which F is a linear combination of a set of n_{feat} basis functions denoted by:

$$\left\{\phi_i\right\}_{i=1}^{n_{\text{feat}}}.$$

[4] In a Machine Learning vocabulary.

[5] In the case of the estimation problem recalled above, $N_O = n_d$.

By so doing, solving the identification problem amounts at solving a linear least-squares problem in which the decision variable is the vector of coefficients:

$$\left\{c_i\right\}_{i=1}^{n_{\text{feat}}},$$

that provides the *best* approximation of the targeted label y over the cloud of instants k, namely:

$$y(t_k) \approx \sum_{i=1}^{n_{\text{feat}}} c_i \phi_i(z(t_k)) = \Phi(z(t_k)) \cdot c, \tag{10.6}$$

where t_k, $k = 1, \dots, n_t$ are the measurements acquisition instants while $\Phi(z)$ is defined by:

$$\Phi(z) := \begin{bmatrix} \phi_1(z) \\ \vdots \\ \phi_{n_{\text{feat}}}(z) \end{bmatrix}^T \in \mathbb{R}^{1\times n_{\text{feat}}}. \tag{10.7}$$

Different underlying nonlinear basis functions ϕ_i can be used such as radial [1], polynomial [2], wavelet-like [3], multi-variate Fourier series [4] to cite but few examples.

! This Is Not the Only Option!

Notice that such a structure that is linear in the parameters are not the only option. In particular, Neural Networks (NNs) do not fall into this class and this is the reason why they are not *fitted* by solving linear least-squares problems. Instead, gradient back-propagation descent is used.

While this is not the only option, it is the one explored in this chapter.

The reason of this choice is precisely that the above structure allow sparse identification algorithms providing sparse solutions which is of a tremendous importance in **industrial** data-driven context.[6]

As an example, let us consider a targeted polynomial relationship that one might use to fit the label using n_d delayed versions of the sensors. Assume that a multi-variate polynomial of degree d is used in the identification.

This results in a number of *features* (or monomials in this case), namely n_{feat} that is given by:

$$n_{\text{feat}} := \binom{n_s(n_d + 1) + d}{d}. \tag{10.8}$$

In order to give an idea of the resulting numbers, let us consider the two following realistic examples:

[6] This importance is not sufficiently acknowledged while being quite obvious to grasp which might be one of the side-effects of DNN dominant culture in the current scientific community.

Two examples of the number of unknowns for a multi-variate polynomial of degree d involving n_s sensors and their n_d delayed instances:

$$(n_s, n_d, d) = (10, 2, 5) \quad \Longrightarrow \quad n_{\text{feat}} = 324632.$$

$$(n_s, n_d, d) = (6, 3, 6) \quad \Longrightarrow \quad n_{\text{feat}} = 593775.$$

Notice that the above considered triplets (n_s, n_d, d) are quite common in the industrial context where many sensors are available (the used values are even much lower than in the reality) and where the need for some delayed versions of the time-series is mandatory to incorporate the hidden internal dynamics one tries to represent.

As for the degree d of the polynomial, it is important to acknowledge that quite high values might be necessary to *blindly*[7] represent non polynomial relationships by polynomial development. The same *explosion* in the number n_{feat} of features can be easily observed when using other nonlinear basis functions such as the ones mentioned earlier.

One of the contexts where the need for a large number of features materializes is the context of **data-driven unsupervised anomaly detection** in industrial time-series [5–7]. Indeed, the data-scientists addressing this problem commonly dispose of a high number of long time-series coming from the recordings of on-board sensors. Moreover, these recordings quite frequently come without any a priori provided knowledge regarding the underlying relationships.

On the top of that, the provided time-series might come from sensors that are not necessarily linked to the potential sources of anomaly leading to a priori **huge amount of useless information**. The reason is that the provider of the time-series simply ignores why they are used for and whether they are useful or not in solving the underlying problem anyway.

The task of the data-scientist is then to come out with **scalable** algorithms that provide a *blind* characterization of the so-called normality. This characterization can be exploited in order to design relevant and preferably **explainable**[8] alarms indicating that the supervised equipment goes outside the normality domain.

10.2.3 Why Sparsity Is Needed?

The relevance of the sparsity in this case comes from the fact that the training data generally represents only a subset of all possible *contexts-of- use*. Consequently, new

[7] Without knowing in advance which original sensor appears in a non polynomial relationship.
[8] Explainability has much to gain from sparsity!

contexts that are not present in the training data might trigger an inadmissibly high rate of *False alarms*.

This is where *parsimonious* identification of dynamical models governing the normal behavior can be of a tremendous help since it provides a characterization of the normality that is likely to be be **context-independent**.

As an example, such a characterization is for instance given by the concept of transfer function or a state-space model for dynamical systems that are governed by affine relationships. Indeed, these models are independent of the excitation signals used to create the learning data[9] and the form of reference trajectories that are used in the building of the training data.

Now let us go back to our scalability problem while having in mind the fact established earlier according to which, for a set of quite realistic instances of the nonlinear identification problem, the number of decision variables might reach hundreds of thousands or even millions.

It seems quite obvious that in the presence of such an extremely high number of features n_{feat}, the following fact needs to be kept in mind:

> **! Sparse Identification Is the Only Reasonable Option!**
> A **standard least-squares fitting is not an option** when hundred of thousands of decision variables is to be handled as it automatically leads to the so-called **over-fitting** phenomenon [8] that has been already illustrated in Sect. 3.4.3 where some reminders regarding Machine Learning concepts are provided.

Let us briefly recall that over-fitting results in the optimal parameter vector c being finely tailored to *fit the slightest detail in the training data* inducing a bad extrapolation power when unseen samples that lie outside the domain covered by the samples used in the training step are presented to the predictor.

The over-fitting phenomenon is illustrated in Figs. 10.1 and 10.2. Both figures show two fitted polynomial models of order 1 and 3 using four training points.[10] Figure 10.1 shows the results when no measurement noise is present while Fig. 10.2 shows the results when a small measurement noise is present. Obviously the model with the higher number of parameters show higher sensitivity to measurement noise because it tries to fit the smallest details in the training data, including those that are not meaningful. In particular, here the three-degree polynomial fitted the small displacements of the four training points (since there are sufficient degrees of freedom to do so) which induced a quite high errors at those regions of the features space (here scalar horizontal axis) that are unseen during the fitting.

---

[9] As far as the excitation is sufficiently rich.

[10] For the sake of illustration.

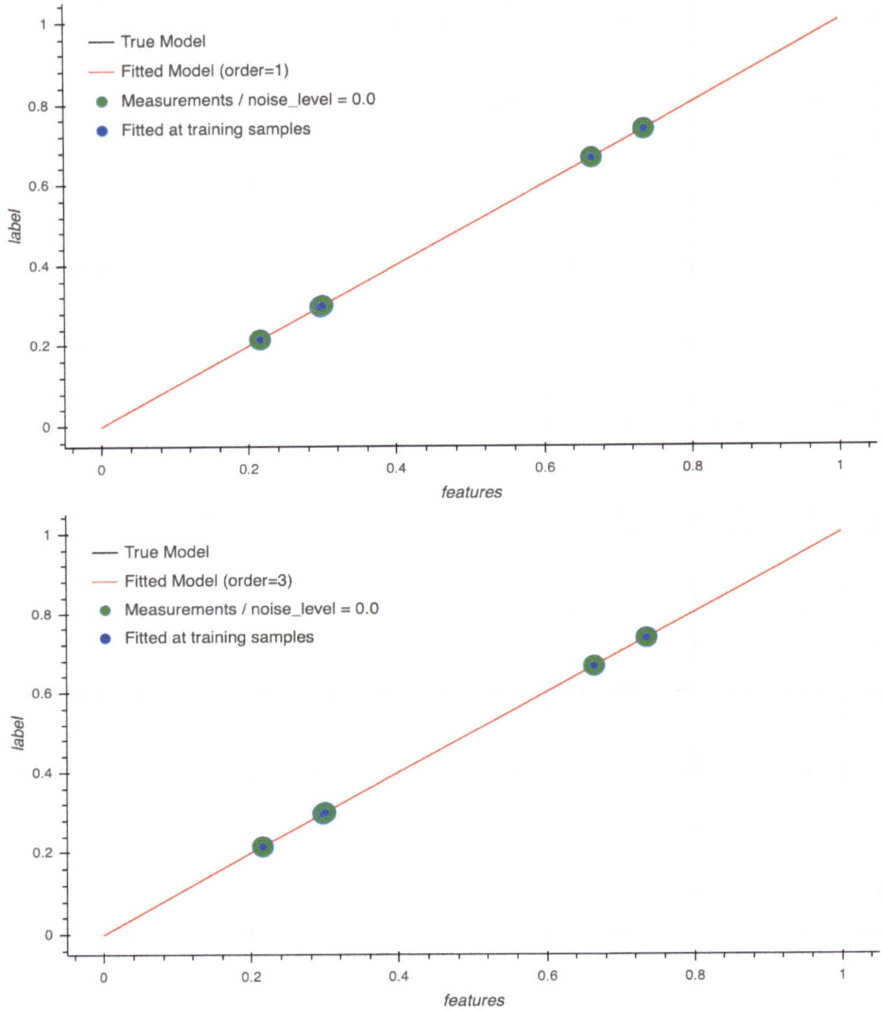

Fig. 10.1 Illustrating the over fitting issue: Fitting results in the absence of measurement noise using polynomials of order 1 and 3

On the contrary, the one degree polynomial barely changed due to the measurement noise since it possesses no enough degrees of freedom to accommodate for the noise-induced tiny changes.

Notice also that memory-induced problems make the situation even more critical with large number of features as the number of *samples* (lines) in the dataset needs to be reduced in order to avoid memory issue leading to even less lines making the number of samples almost comparable or even lesser than the number of columns (features) which enforces the risk of over-fitting.

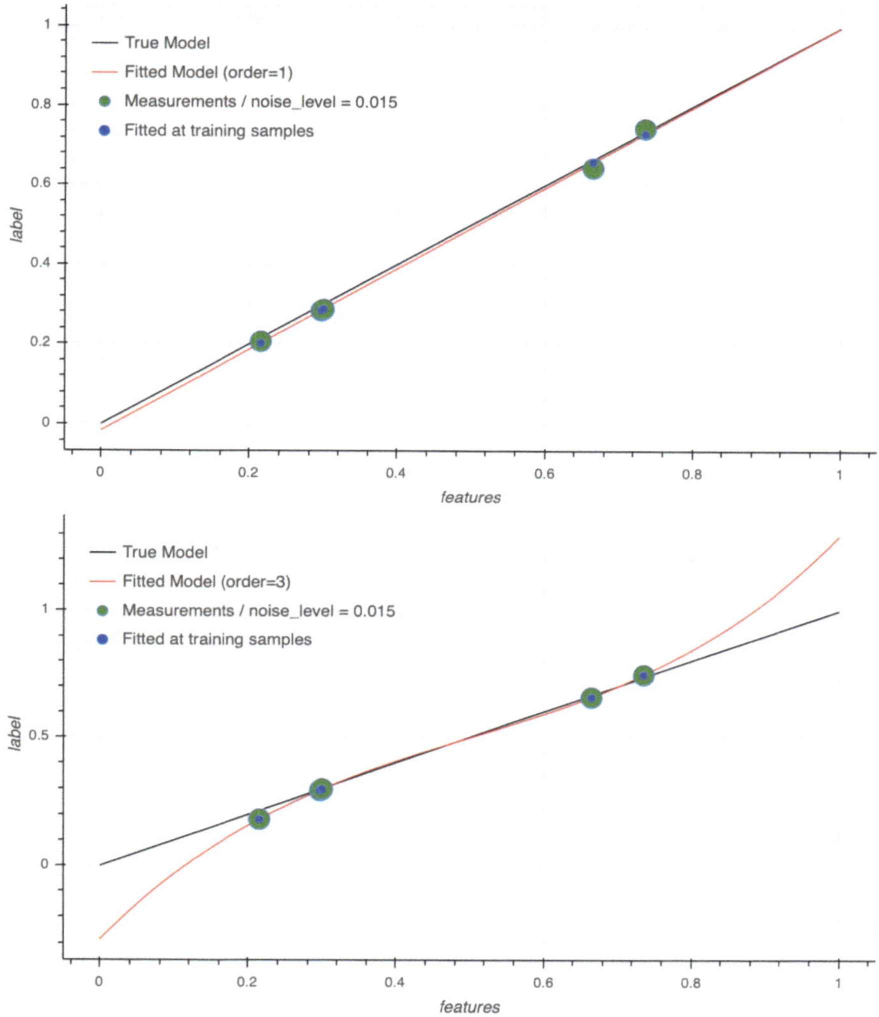

Fig. 10.2 Illustrating the over fitting issue: fitting results in the presence of a small measurement noise using polynomials of order 1 and 3

The previous discussion can be summarized as follows:

> **To Summarize**
> The appropriate framework to solve (10.6) with very large n_{feat} is the one involving the concept of **sparsity** where solutions containing a largest number of zero components in the decision variable c are privileged.

In the next section, the two most common approaches that are used to induce sparse solutions are discussed.

10.3 Overview of Main Sparse Least-Squares Solving Algorithms

Two main approaches have been considered to address this problem which are briefly described in this section.

10.3.1 LASSO: Least Absolute Shrinkage and Selection Operator

The first approach that can be used to induce zero components in the optimal solution is denoted as[11] *Least Absolute Shrinkage and Selection Operator* **LASSO** [9]. This approach consists in adjoining to the original cost function, an additional term representing ℓ_1 penalty on the vector of decision variables c.

This leads to the following modified cost function:

$$\alpha\|c\|_1 + \sum_{k=1}^{n_t} \|\Phi(z(t_k))c - y(t_k)\|_2^2 , \tag{10.9}$$

as this *enhances* zeroing the components of c as α increases. The reason for this impact is schematically illustrated in Fig. 10.3 where two modified costs are shown with their associated resulting optimal solutions for some value of the penalty coefficient α. The latter *tunes* the size of the iso-values of the penalty terms.

Notice how the form of the iso-value curves of the ℓ_1 penalty induces higher probability of the optimal solution being positioned at one of the summits of the polygon (which means that some components are rigorously zero).

On the contrary the spherical shape of the iso-value curves corresponding to the ℓ_2 norm generically leads to dense optimal solutions where all the components are not rigorously equal to 0.

10.3.2 LARS: The Least-Angle Regression Approach

The second widely used approach to induce sparse solutions is referred to as the *Least-Angle Regression* **LARS** approach [10]. The reference to the concept of *least angle* stems from the fact that the features are progressively selected based on the angle they make with the current residual. More precisely, the algorithm follows the steps below:

1. The residual is initially set to be equal to the `label` vector.
2. Then the feature that shows the **least angle** with the residual is selected.
3. The step in the direction of this *best* feature is set to be the one for which the corresponding new updated residual would become *more aligned* with another new feature.

[11] Do not try to infer any understanding from this *weird* name!

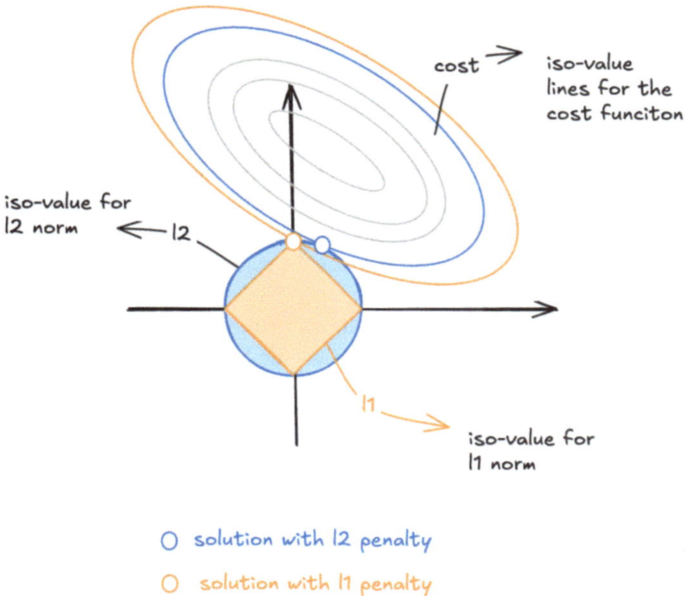

Fig. 10.3 Illustration of the lasso principle: The ℓ_1-norm used in the penalty induces higher probability that the optimal solution lies at one of the summits of the iso-value polygone corresponding to rigorously zero values on some components of the decision variable

4. The process of features selection continues until the number of allowed features is reached.

Notice that each time a feature is added, the contribution of each of the already selected features is updated.

When a feature is added, the coefficients of all the already selected features are appropriately updated. Notice that as mentioned in step 4, above, the number of so added features is a tuning parameter of the algorithm.

10.3.3 About the Cross Validation

Notice that each of the two methods sketched above involves a **tuning parameter**. Indeed, the `Lasso` is parameterized by the penalty coefficient α used in (10.9) in the ℓ_1 added term. As for the `Lars` method, the parameter might be the number of successively introduced *least-angled* features during the previously recalled process. Another stopping condition that can replace this number is the level of fitting error on the training data.

Consequently, for both algorithms, each value of the parameter induces a specific balance between the precision of the fit on training data and the number of non zero parameters in the delivered solution.

As it is mentioned in Chap. 2 where recalls on machine learning are briefly given, one way to come out with a parameter-free design is to use the *cross-validation* procedure [11].

The *cross-validation* procedure consists in splitting the training data into n_{cv} subsets, to learn the model, for a given value of the parameter on $n_{cv} - 1$ subsets and examine the quality of the fit on the remaining subset. Repeating this process n_{cv} times.[12] enables to compute a **robustness score** for that specific value of the parameter by averaging the fitting errors obtained by the n_{cv} so-fitted models. The one finally adopted is the one that minimizes the regression error (averaged over the n_{cv} possible splits).

Performing this process for a set of values of the parameters enables to choose the `optimal` value which is the one that shows the best extrapolation capabilities on unseen data.

10.3.4 The `scikit-learn` Cross-Validated Version of the Algorithms

As a matter of fact, both above presented algorithms are implemented in their cross validation versions in the widely used `scikit-learn` Python library [12] through the `LassoCV` and the `LarsCV` classes of the `linear_models` module. Moreover, the same module provides a mixed version involving ideas from both algorithms that is implemented through the `LassoLarsCV` class.

10.3.5 Some Other Works Addressing Scalability

Without pretending to be exhaustive, let us mention some representative works that attempt to address the scalability issue of sparse least-squares problems' solvers.

Several works addressed the *scalability* of parsimonious solvers in terms of **the number of samples** (lines) while keeping moderate the number of features (columns) [13, 14].

Some other works addressed specifically the number of features as in [15]. Nevertheless, although large numbers of features were addressed, they never exceed 30, 000 which is far beyond the targeted numbers discussed earlier in the beginning of this chapter.

In [16], a Franck–Wolfe implementation of the LASSO algorithm is proposed that is impressively scalable in the number of features.[13] Sadly enough, the results of [16] *only* show that the lasso path can be generated *more efficiently* than in previously available implementations but as it is shown in the remainder of this chapter, it happens that in terms of the ability to retrieve the true features involved in the ground-truth nonlinear relationships, the lasso-based solutions lead to quite bad results.

---

[12] In each of which a different subset is left for the validation.

[13] Problems with 4 millions of features are addressed.

This obviously renders the issues of scalability and computation time of marginal interest.[14]

The remainder of this chapter proposes a simple modified **scalable** version of the Least-Angle Regression algorithm, named `scLars`, which addresses the above mentioned scalability problem.

Moreover, the chapter provides a benchmark dataset for sparse identification of multi-variate polynomial relationships that is used to compare the performances of the proposed algorithm to those of the commonly used algorithms mentioned above which are available through the `scikit-learn` library.

As it is mentioned earlier, the algorithm is then used in **Chap.** 11 in order to address the problem of partial observation targets reconstruction in the presence of uncertainties on the parameters of the model.

Let us start with the presentation of the algorithm.

10.4 scLars: A Modified Scalable Version of Lars

The algorithm is built on a main elementary operation that is first described.

10.4.1 The Elementary Operation

The elementary operation consists in iteratively selecting a set of features that are associated to a *slice* of the data set. This selection is based on the same principle of least-angle measure that is used in the original Lars algorithm with some differences that are fundamental in terms of their impact on scalability.

The Elementary Operation
The elementary operation that is extensively used in the `scLars` algorithm consists in **selecting a set of features** associated to a given slice of the training dataset. The slice consists in w consecutive samples that are randomly selected.

Anticipating the following presentation, let us *reveal* that several slices will be used later and the associated results are aggregated to deliver the solution. But for the time being, let us focus on the description of the elementary operation which is schematically shown in Fig. 10.4.

[14] We need scalable solutions with good results.

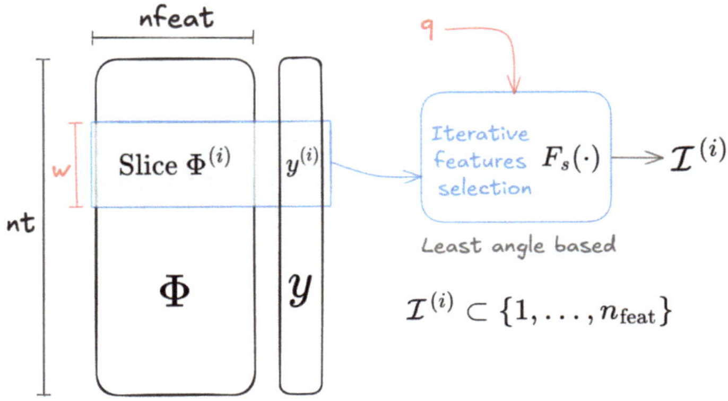

Fig. 10.4 Schematics of the elementary operation consisting in iteratively selecting q features via least-angle principle applied to a specific slice, i, containing w samples of the training dataset

More precisely, Fig. 10.4 shows a sliced window containing w samples of the training data. This slice is randomly sampled at iteration i (of the elementary operation) leading to the features sub-matrix $\Phi^{(i)} \in \mathbb{R}^{w \times n_{\textbf{feat}}}$ and the associated label vector $y^{(i)} \in \mathbb{R}^w$.

The resulting pair:

$$\left(\Phi^{(i)}, y^{(i)}\right),$$

is used to select q features indices (columns indices of the matrix of features), denoted by $\mathcal{I}^{(i)}$, namely:

$$\mathcal{I}^{(i)} := \text{The set of indices of selected columns} \subset \{1, \dots, n_{\text{feat}}\},$$

which makes q a hyper-parameters of the method as it is shown in Fig. 10.4.

This features selection elementary operation is symbolically denoted hereafter as follows (see Fig. 10.4):

$$\mathcal{I}^{(i)} = F_s(\Phi^{(i)}, y^{(i)}, q), \tag{10.10}$$

in which F_s denotes the iterative features selection operation representing the task achieved in the elementary operation for the specific slice leading to the definition of the pair $(\Phi^{(i)}, y^{(i)})$ that is used to form the first two calling arguments to F_s.

More concretely, the operation performed inside F_s is rigorously described in the following algorithm:

The Features Selection Map $F_s(\Phi, y, q)$ of the Elementary Operation

Input: $\Phi \in \mathbb{R}^{m\times n_{\text{feat}}}$, $y \in \mathbb{R}^m$, $q \in \mathbb{N}$
Output: Indices of selected features $I \subset \{1, \dots, n_{\textbf{feat}}\}$.

1: $r \leftarrow y, I = \emptyset$ ▷ Initialize the residual and I.
2: $\bar{\Phi} \leftarrow \Phi/\texttt{norm}(\Phi, \texttt{axis} = 0)$ ▷ Normalize columns
3: **while** $\texttt{card}(I) \le q$ **do** ▷ Iterate over the set of remaining columns indices
4: $\quad \theta \leftarrow r^T\bar{\Phi}$ ▷ Compute r-projections on columns.
5: $\quad i^\star \leftarrow \texttt{argmax}(|\theta|)$ ▷ Index of the most aligned column.
6: $\quad I \leftarrow I \cup \{i^\star\}$ ▷ Add the best index to I.
7: $\quad r \leftarrow y - \texttt{lstsq}(\theta_{i^\star}, r)$ ▷ update the residual.
8: **end while**

While this algorithm looks like the original `lars` in that it selects q features that are the successive most aligned columns with the successive values of the continuously updated residual, it shows some differences that are underlined below:

(1) First of all, it is important to remember that the objective of the computation involved in the elementary operation is not the derivation of the ultimate solution to the problem. It is rather about selecting a set of q features associated to the current slice of the dataset at iteration i. For this iteration, the algorithm is called with the corresponding arguments $\Phi^{(i)}$ and $y^{(i)}$. The amount of data contained in this slice (denoted by w in the presentation above) is generally far smaller than that of the training dataset.[15]

(2) When selecting the successive features (et step 5: of the algorithm), contrary to the LARS algorithm, the contribution of the previously selected features are not updated upon finding a new selected feature. On the other hand, when a new feature (column) is found to be the most aligned one to the current residual, the step in the direction of this most aligned column is computed in order minimize the fitting of the current residual, through the scalar least-squares optimization problem involved in step 7 of the algorithm.

On the contrary, in the LARS algorithm, the step is the sense of the most aligned feature is taken to be the one that induces a residual for which another column becomes more aligned with the corresponding residual.

(3) This last aspect, in the original LARS algorithm, induces rather heavy computation when the number of features gets very high since all the angles of all the remaining features need to be monitored for the candidate values of the step size.

Now that we have defined the elementary features selection map F_s let us see how it is integrated in a whole solution's scheme.

---

[15] Typically the training dataset might have millions of samples while a slice has typically some thousands of samples.

10.4.2 `scLars`: The Whole Scheme

After randomly sampling N different slices[16] and collecting the resulting set of selected indices of features:

$$\mathcal{I}^{(i)} \quad \text{for } i = 1, \dots, N. \tag{10.11}$$

A dictionary of all the so-collected indices is created which associates to each index the number of times this index appeared among the N-times the elementary operation is performed. This number of occurrences is termed below the index's cardinality. More precisely:

$$\forall j \in \mathcal{I} := \bigcup_{i=1}^{N} \mathcal{I}^{(i)}, \quad \kappa_j := \texttt{card}\big(\{i \mid j \in \mathcal{I}^{(i)}\}\big). \tag{10.12}$$

More clearly,

> **The Cardinality of a Collected Features**
> For each collected feature $j \in \mathcal{I}$, the notation
>
> $$\kappa_j \in \{1, \dots, N\},$$
>
> refers to the number of times the j-th feature has been selected during the N different features selection rounds of execution of the elementary step.[17]

At this stage, it might be interesting to precisely distinguish between the two terms used in the discussion, namely: the `selected` set of features and the `collected` set of features, namely:

- Features are *selected* at each execution of the elementary operations. Since each execution of the elementary operation involves a different slice of the training data, the same feature might be selected several times at several execution of the elementary operation.
- Once the pre-determined number N of elementary operations is executed, the N corresponding sets of features $\mathcal{I}^{(i)}$ are available. Each of which contains the *selected* features of a different elementary operation that has been executed on a specific slice, i, of the training data.
  This enables to understand that a features might have been *selected* in more than a single one of these sets which precisely defines its cardinality.

---

[16] This is another hyper parameters of the algorithm.

[17] Notice that since the number of times the elementary operation is performed is N, κ_j is bounded above by N. Moreover, since the feature j has been collected at least once, $\kappa_j \geq 1$.

Here starts the *collection* step, where each of the selected features is brought in a single dictionary together with its associated cardinality (the number of times it has been selected).

Now based on the results of the N selection rounds, one disposes of a number of collected features (members of $\mathcal{I}$), each with it own cardinality. This enables to define N different sets of eligible features defined through a threshold on the cardinality attribute κ_j, namely:

$$\mathcal{S}_\eta := \Big\{ j \in \mathcal{I} \mid \kappa_j \geq \eta \Big\} \qquad \eta \in \{1, \ldots, N\}. \tag{10.13}$$

In other words,

Meaning of $\mathcal{S}_\eta$

$\mathcal{S}_\eta$ is the set of collected indices of features that have been selected at least η times. Notice that for some $\eta > 1, \mathcal{S}_\eta$ might be empty. Moreover, the following property holds true:

$$\mathcal{S}_N \subset \mathcal{S}_{N-1} \subset \cdots \subset \mathcal{S}_1 \quad , \quad \text{card}(\mathcal{S}_1) \leq qN. \tag{10.14}$$

Notice that each of these sets $\mathcal{S}_\eta$, $\eta = 1, \ldots, N$ can be used to define a feature matrix Φ^r_η by taking the associated features' columns in Φ corresponding to elements contained in $\mathcal{S}_\eta$, namely:

$$\Phi^r_\eta := \Big[\Phi[:, j]\Big]_{j \in \mathcal{S}_\eta} \in \mathbb{R}^{n_t \times r_\eta}, \tag{10.15}$$

where n_t is the total number of samples in the dataset while r_η is the cardinality of the set of indices $\mathcal{S}_\eta$. Notice that by definition, one has:

$$q \leq r_\eta \leq qN. \tag{10.16}$$

Notice that the maximum value qN corresponds to the case where totally distinct q features have been selected at each round of execution of the elementary operation.[18] Similarly, the minimum value q corresponds to the situation where exactly the same q features are selected at each run of the elementary operation.

Now since the matrices Φ^r_η contain a drastically reduced number of columns that are defined on the whole dataset, it is possible to define a standard regularized least-squares problem based on the following standard quadratic cost defined by:

$$J_\eta(c_\eta) := \alpha \|c_\eta\|^2 + \Big\| \big[\Phi^r_\eta\big] c_\eta - y \Big\|_2^2, \tag{10.17}$$

where the penalty α can be determined using the same cross-validation technique recalled earlier in this chapter. This is done hereafter using the

[18] Which is not likely to occur in real data.

`sklearn.linear_model RidgeCV` algorithm meaning that quadratic penalty coefficient α has not to be tuned.

Upon solving the associated least-squares problems for all $\eta \in \{1, \dots, N\}$ for which the set $\mathcal{S}_\eta$ is not empty, one disposes of the set of optimal costs and optimal solutions that are denoted by:

$$\left\{J_\eta^\star\right\}_{\eta=1}^{\bar\eta} \quad ; \quad \left\{c_\eta^\star\right\}_{\eta=1}^{\bar\eta}, \tag{10.18}$$

where $\bar\eta \le N$ is the largest value of $\eta \in \{1, \dots, N\}$ for which $\mathcal{S}_\eta \neq \emptyset$. At this stage, it is important to keep in mind that:

! η-Dependent Sparsity Rate
The solutions $(J_\eta^\star, c_\eta^\star)$ generally involve different numbers of active features. More precisely, the higher η, the lower is the number of features.

The above described process that starts from the availability of the set of indices $\mathcal{I}^{(i)}$ and ends by delivering the quantities invoked in (10.18) is summarized in Fig. 10.5.

The remaining part of the algorithm is dedicated to a rational choice of the design parameter η. This computation targets the minimum number of features that does not significantly deteriorates the quality of the fitting.

More precisely, the best fitting result, regardless of the number of features used is first determined based on:

$$J^\star := \min_{\eta \in \{1, \dots \bar\eta\}} J_\eta^\star.$$

Then the highest[19] η that corresponds to a cost that is not greater than $(1+\varepsilon)J^\star$ is selected to be the optimal value:

$$\eta^\star := \max\left\{\eta \in \{1, \dots, \bar\eta\} \;\mid\; J_\eta^\star \le (1+\varepsilon)J^\star\right\}, \tag{10.19}$$

which delivers the final regression error and the *optimal* vector of coefficients, namely $J_{\eta^\star}^\star$ and $c_{\eta^\star}^\star$.

Notice that in the previous discussion, ε is a precision-related parameter that is chosen a priori.

This ends the presentation of the algorithm. But before we proceed with some comparisons and applications, let us summarize the hyper parameters of the algorithm:

---

[19] Notice that the highest η is the lowest is the number of eligible features to participate in the computation of the solution since η is a threshold on the number of times the features has been selected during the N rounds of computation of the elementary operation on the different slices of the training data.

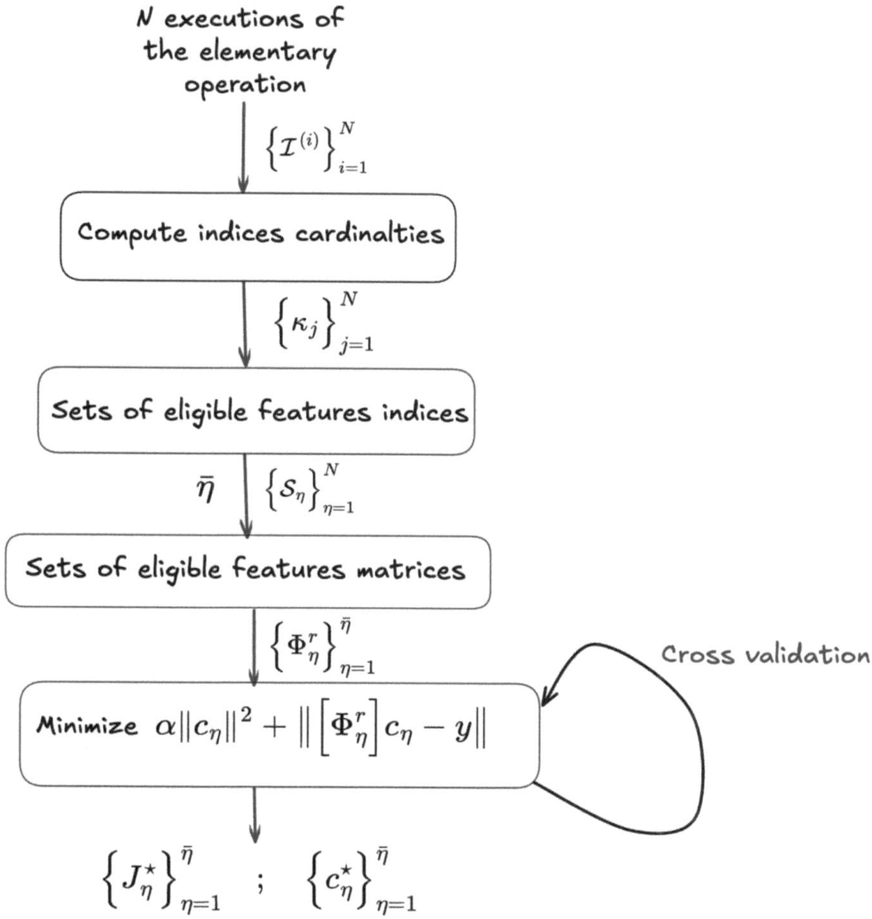

Fig. 10.5 Schematic view of the process starting by the set of indices $\mathcal{I}^{(i)}$, $i = 1, \dots, N$ produced by N successive rounds of execution of the elementary operation and delivering the optimal costs and solutions $J_\eta^\star$ and $c_\eta^\star$ for $\eta \in 1, \dots, \bar{\eta}$

> **The scLars Algorithm's Hyper-parameters**
> The scLars algorithm described above shows the following hyper-parameters:
>
> w the width of the window that is used in slicing the training data,
> N the number of the slices used,
> q the number of selected features for each elementary operation, and
> ϵ the precision threshold used in (10.19)

while tuning these parameters using cross validation is a good practice, notice that

$$\epsilon = 0.05,$$

might always be a reasonable choice. As for w, values such as 200, 500 and 1000 are typical options:

$$w \in \{200, 500, 1000, 5000\}.$$

Obviously, the appropriate values of these last parameters heavily depend on the functions basis $\{\phi_i\}$ being used in (10.6). This is the designer's choice and is heavily problem-dependent. The `scLars` algorithm presented above is used once such a choice is done by the designer. The `scLars` algorithm handles already specified linear least-squares problems that might involve whatever nonlinear functions basis.

For instance, in the next validation and comparison section, the data set used in the comparison is built on a multi-variate polynomial basis. This basis is used to generate the label so that the adequacy of the basis to the label is not questioned. This enables to focus on the comparison between the efficiency of the solvers and not on the adequacy of the designer choice in terms of the relevance of the functions basis used in the formulation of the problem.

10.5 Validation and Comparison

As mentioned above, we shall compare the performance of the above described `scLars` algorithm with the competing algorithms from the `scikit-learn` python data-science library [17] that incorporate a cross validation mechanism. But before we do that, let us first describe the benchmark used in the comparison.

10.5.1 The Dataset Used as Benchmark

The benchmark used in this section in order to examine the performance of the `scLars` algorithm and to compare it to the performances of other alternatives is based on the **generation of a high number of instances of the least-squares problem** expressed by (10.6).

These instances of the problem are generated using functions basis elements $\phi_i(z)$ that are multi-variate monomials of the form:

$$\phi_i(z) = \prod_{j=1}^{n_z} z_j^{p_{ij}} \quad i \in \{1, \dots, n_{\text{feat}}\}, \tag{10.20}$$

in which:

- n_z is the dimension of the z variable which is the common argument of the basis functions.

- n_{feat} is the total number of eligible monomials that might participate in the definition of the label through a relationship of the form (10.6).
- p_{ij} is the integer power of the j-th component of z in the i-th monomial ϕ_i. These integer powers is denoted by:

$$p_i := \begin{bmatrix} p_{i1} \\ \vdots \\ p_{in_z} \end{bmatrix}. \tag{10.21}$$

As an example, let us consider the following instantiation:

$$n_z = 3 \quad ; \quad p_i = (3, 2, 1) \quad \Longrightarrow \quad \phi_i(z) = z_1^3 z_2^2 z_3, \tag{10.22}$$

which is a monomial of degree 6.

Having the definition of the basis function ϕ_i, it is possible to understand that the generation of an instance of the least-squares problem is based on the following process that is presented in a condensed way before a more detailed description of each step is provided later on:

Generation of a Single Least-Squares Problem in the Benchmark
Each least-squares problem results from the choices involved in the following ingredients:

1. Choose the dimension n_z of the argument z.
2. Generate a random matrix $Z \in \mathbb{R}^{n_t \times n_z}$ of normalized values. Each line represent a grid point in the space of values of z.
3. Choose the maximum number of eligible features $n_{\text{feat}}^{\text{max}}$ in the problem.
4. Compute the **maximum** degree $d \leq 9$ of the multi-variate polynomial in n_z unknowns that induces a number of monomials that is lower than $n_{\text{feat}}^{\text{max}}$.
5. The value of the above chosen/determined values of n_z and d leads to the following number of monomials:

$$n_{\text{feat}} = \binom{n_z + d}{d}. \tag{10.23}$$

6. Choose the integer number, denoted by card, among the set of n_{feat} eligible features (function basis) to be used in the definition of the multi-variate polynomial. This number (card) represents the number of non zero components of the coefficient vector c involved in (10.6).

7. Choose the **different** indices of these card basis functions randomly inside the set $\{1,\dots,n_{\text{feat}}\}$. Denote the sampled set by:

$$\mathcal{J} := \left\{\zeta_j\right\}_{j=1}^{\texttt{card}} \quad ; \quad \zeta_j \in \{1,\dots,n_{\text{feat}}\}. \tag{10.24}$$

8. The values of the non zero coefficients are randomly chosen in the set:

$$c_{\zeta_j} \in \{-1,+1\} \quad ; \quad j \in \{1,\dots,\texttt{card}\}.$$

All the other components of $c \in \mathbb{R}^{n_{\text{feat}}}$ are set to 0.

9. Using all the previously set quantities, it is possible to generate the label vector y according to:

$$y := \sum_{i=1}^{n_{\text{feat}}} c_i \prod_{j=1}^{n_z} Z[:,j]^{p_{ij}}. \tag{10.25}$$

10. The pair $(\Phi(Z), y)$ totally defines the least-squares problem.

Notice that in the above presentation of the process of generating a single problem of the benchmark used hereafter, there still are some details and explanations to be given before the benchmark is completely defined. This is done in the following discussion.

- The values of n_z in step 1. are taken in the following set of possible values:

$$n_z \in \left\{5, 10, 20, 50\right\}.$$

- The set of samples Z is generated using the `numpy.random.randn` generator, more precisely the following script is used:

```
nt = 200000
X = 1.2*np.random.randn(nt, nz)
```

meaning that $n_t = 200000$.

- The maximum number of features is taken in the following set of admissible values:

$$n_{\text{feat}}^{\max} \in \left\{2, 5, 10, 20, 30, 50, 100, 150, 200, 250, 400\right\} \times 1000$$

meaning that the $n_{\text{feat}}^{\max}$ ranges from 2000 up to 400000.

- The number `card` of monomials participating in the definition (having non zeros coefficients c_{ξ_j}) spans the following set of admissible values:

$$\texttt{card} \in \left\{3, 5, 8, 10, 15, 20, 30\right\}. \tag{10.26}$$

- The multi-variate polynomial degree invoked in item 4. of the above generation process is taken lower than 10.

Based on the choices, definitions and process, for each value of n_z and each *targeted* number of polynomial features $n_{\text{feat}}^{\max}$, the polynomial degrees in n_z variables that correspond to a number of polynomial features lower than the targeted one are computed and the largest one is selected.

For the resulting (n_z, d), 7 polynomials are created corresponding each to one value of the number `card` of *active* monomials where `card` is taken within the above mentioned 7-values set (10.26).

Finally the values of the non zero coefficients is randomly generated in the set $\{-1, +1\}^{\texttt{card}}$ as mentioned above. The reason for which this set is used to sample the values of the coefficients of the monomials is that it makes easier the derivation of a relevant evaluation of the ability of the algorithm to reconstruct the coefficients being involved.

This generation process produces 308 experiments[20] whose characteristics are detailed in Table 10.1.

In the following sections, this set of problems is used to compare the performance of the `scLars` algorithm presented in this chapter with the main utilities of the `scikit-learn` library that addresses the same problem of solving parsimonious least-squares problems.

10.5.2 Results and Comparison

Since this section is about a comparison between several algorithms, let us first precisely defined the performance indices being investigated.

[20] This is the produce of the number of values of n_z (4) times the number of values of $n_{\text{feat}}^{\max}$ (11) times the number of values of `card` (7) leading to $4 \times 11 \times 7 = 308$.

Table 10.1 Characteristics of the generated least-squares problems serving in the benchmark

n_{feat}	n_z	d	# of instances
1001	{10}	{4}	7
1287	{5}	{8}	7
1326	{50}	{2}	28
1771	{20}	{3}	21
3003	{5, 10}	{10, 5}	77
8008	{10}	{6}	7
10626	{20}	{4}	21
19448	{10}	{7}	14
23426	{50}	{3}	42
43758	{10}	{8}	7
53130	{20}	{5}	21
92378	{10}	{9}	14
184756	{10}	{10}	21
230230	{20}	{6}	14
316251	{50}	{4}	7

10.5.2.1 The Performance Indicators

Let us first underline that when it comes to evaluate the performance of the algorithms, we focus on two indicators:

- The **regression error** indicator that depends on the quality of the predicted label $\hat{y}$ using the fitted relationship in terms of its distance to the original label y used in the call for the solver.
  This distance is defined hereafter as follows:
$$e_y := \frac{\texttt{percentile}(|y-\hat{y}|, 95)}{\texttt{percentile}(|y|, 50)}. \tag{10.27}$$
  Namely, the error is defined to be the ratio between the 95% percentile of the absolute value of the regression error and the median of the absolute value of y. This measure enables to compare the error to the *content* of the signal (relative error) and to avoid taking into account outliers that might be present in the measurement vector itself.
  Notice however that considering only the regression error would totally ignore the parsimony concern. Indeed, it is obvious that by involving all the columns (feature) of the features matrix, one can make the regression error artificially extremely small. Here is where the second performance indicator comes into light.
- The **coefficient reconstruction error** that is related to the difference between the solution $c^\star$ provided by the algorithm and the true vector of coefficients c that is used to produce the label using the columns of the features matrix through the definition (10.25).

This error is defined by:

$$e_c := \frac{1}{\|c\|_0}\|c^\star - c\|_1. \tag{10.28}$$

Recall that the term $\|c\|_0$ refers to the number of non identically zero components in the vector c which is equal to the parameter `card` as far as the experiments used in this specific benchmark is concerned.

10.5.2.2 The Hyper-Parameters Setting

All the results shown hereafter are produced while setting the hyper-parameters of the proposed algorithm to the following values:

$$\varepsilon = 0.05 \ ; \ N \in \{1, 10, 20\} \ ; \ q = \texttt{card} + 5 \ ; \ w = 2000,$$

where it is to be recalled that:

- ε is the precision threshold used in the parsimony enhancing at the final ℓ_2 least-squares solution involving all the collected features.
- N is the number of times the elementary operation is executed.
- q is the number of features selected at each elementary operation.
- w is the window used to define the angle and hence to select the different features.

As for the `scikit-learn` algorithms, namely:

`lassocv`, `larscv`, `lassolarscv`,

the computation is done using a randomly selected slices of total length of 20000 which corresponds to the same amount of data as the proposed algorithm in which $N = 10$ is used (since $w = 2000$).

! Split Train-Test

It is important to recall that each of the 308 experiments used in the benchmark consists of 200000 lines while the fitting is done using 20000 lines (for the `scikit-learn` algorithm and using 2000, 20000 or 40000 for the `scLars` algorithm with $N = 1, 10$ and 20 respectively.

This naturally include a train / test split where the `test_size` is equal to 90% for the `scikit-learn` algorithm and to respectively 99%, 90% and 80% for the different settings of the `scLars` algorithm.

This is the reason why no reference to train/split procdure in the previous presenation of validation process.

Figure 10.6a, b show respectively the percentiles of the two above defined performance indicators for the different competing solvers.

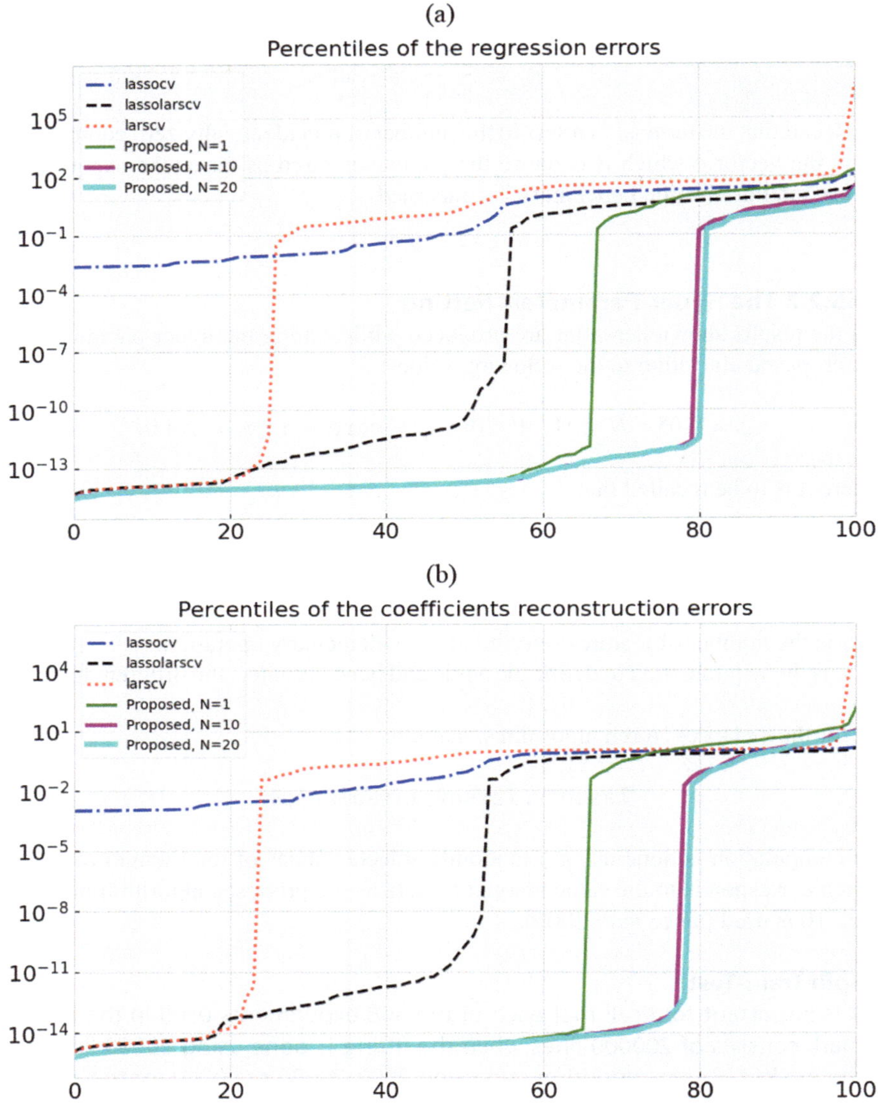

Fig. 10.6 Statistics of **a** the regression errors and **b** the coefficients reconstruction errors. The histograms are computed on the only experiments where the competing solvers were able to produce results (see Fig. 10.7 showing the presence of results as a function of the number of features involved in the least-squares problems)

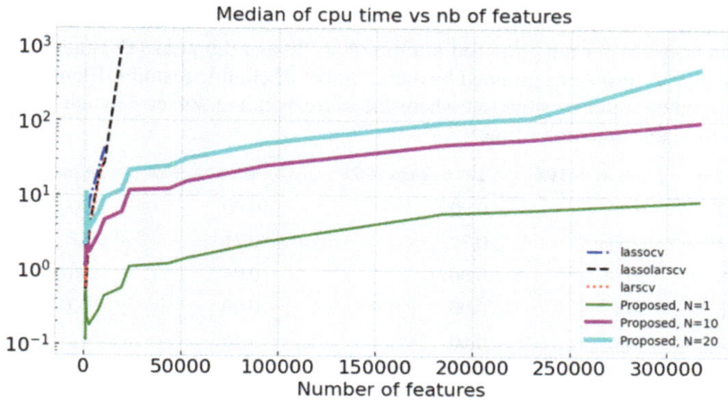

Fig. 10.7 Median of cpu-computation times for the different solvers using a `MacBook Pro, M3, 18Go`

This figure clearly shows that, based on the above defined metrics of performance, **only the mixed `lassolarscv` algorithm stands the comparison**. Nevertheless it remains true that it is clearly outperformed by the the `scLars` algorithm discussed in the present chapter.

As far as the studied benchmark of problems is concerned, he `lassocv` and the `larscv` show quite mediocre results

The results also show the benefit from the multiple selection ($N = 10, 20$) compared to the single selection option ($N = 1$). They also suggest that the choice $N = 20$ offer only a minor improvement in terms of performance suggesting $N = 10$ cycles of the elementary operation enables to capture the relationships for almost all the instances of the problem contained in the dataset.

In terms of scalability, Fig. 10.7 shows the median of the computation times and the superiority of the proposed `scLars` algorithm in tackling problems with very high number of features. This figure shows in particular that the `scikit-learn` algorithms were not able to produce results for problems with a number of features that goes over around 30000 features columns.[21]

Table 10.2 shows a comparison of the coefficients reconstruction error obtained under the proposed `scLars`/$N = 20$ algorithm and the `scikit-learn` `lassolarscv` algorithm as the latter seemed to be the only one that *deserves* to be compared in more details (see the above discussion).

This table shows that the madian values of the `scLars` solver over the 308 experiments are systematically zero which is not the case for the `lassolarscv` solver. Moreover, this remains the case for a bunch of experiments with quite large number of features, namely all the 42 experiments with a number of features in the set $\{53130, 230230, 316251\}$.

[21] It is likely that something could have been done to make this possible but as far as the computer I was using (`MacBook Pro, M3, 18Go`) is concerned, the computation times got insanely increasing as it is shown in the Figure.

Table 10.2 Sparsity Performance Comparison: Median and 95% percentile of the coefficients reconstruction errors e_c for the proposed `scLars`/$N = 20$ and the `scikit-learn lassolarscv` over the benchmark's problems grouped by their number of eligible features. Elements of the table with the '–' symbols refer to situation where the corresponding solver was not able to produce results in a reasonable amount of time

n_{feat}	# exp	`scLars-50%`	`lassolars-50%`	`scLars-95%`	`lassolars-95%`
1001	7	0.00	0.00	0.00	0.00
1287	7	0.00	0.39	0.71	0.51
1326	28	0.00	0.00	0.06	0.06
1771	21	0.00	0.00	0.00	0.00
3003	77	0.00	0.60	5.76	0.82
8008	7	0.00	0.00	0.02	0.17
10626	21	0.00	0.00	0.00	0.00
19448	14	0.00	0.24	0.07	0.24
23426	42	0.00	–	0.00	–
43758	7	0.00	–	0.34	–
53130	21	0.00	–	0.00	–
92378	14	0.00	–	0.65	–
184756	21	0.00	–	1.94	–
230230	14	0.00	–	0.00	–
316251	7	0.00	–	0.00	–

Table 10.3 Computation times: Comparison of statistics of computation times (in sec), over the 308 problem of the benchmark, between the proposed `scLars` solver with $N = 20$ and the `scikit-learn lassolarscv`. The times are obtained on `MacBookPro, M3, 18Go`

n_{feat}	# exp	`scLars-50%`	`lassolars-50%`	`scLars-95%`	`lassolars-95%`
1001	7	2.31	0.36	6.20	0.41
1287	7	2.28	1.44	5.12	1.46
1326	28	10.64	0.55	30.18	0.57
1771	21	5.08	0.87	14.07	0.89
3003	77	3.63	3.76	8.04	3.97
8008	7	6.11	19.50	12.72	21.69
10626	21	9.25	27.27	22.71	28.61
19448	14	11.68	1102.73	25.87	1102.73
23426	42	21.71	–	52.77	–
43758	7	24.72	–	47.34	–
53130	21	31.09	–	62.24	–
92378	14	49.36	–	98.16	–
184756	21	93.68	–	200.73	–
230230	14	111.51	–	223.18	–
316251	7	488.62	–	1981.79	–

Finally, Table 10.3 shows the comparison of the median and 95% percentile needed by the `scLars` solver studied in the present chapter and the `scikit-learn`, `lassolarscv` solver.

Links to Other Chapters
The `sclars` developed in the present chapter is used in Chap. 11 to address the problem of designing a partial state estimator of nonlinear systems with uncertain parameters.

10.6 Conclusion

In this chapter, a scalable algorithm has been examined that addresses the problem of finding parsimonious solutions to linear least-squares problems involving a very large number of features (several hundreds of thousands).

The reason for which this algorithm is included in a book dedicated to handling uncertain nonlinear problems is that the parsimonious identification of unknown nonlinear relationships might be done through linear least-squares problems through the use of nonlinear basis functions.

Moreover, the relevance of parsimonious identification of relationships in addressing several engineering problems involving incomplete and/or uncertain data has been discussed and hopefully sufficiently highlighted.

! Keep in Mind
Finally it is important to underline the fact that the place given to this algorithm in the present book does not reflect its high relevance and potential impact it might have in addressing many of the problems that is addressed in different ways and algorithms throughout the book.
The reason for that is that the development of this algorithm was done simultaneously with the process of writing of the book. Many new results and combinations are still to come. Keep tuned!

References

1. Laurain, V., Tth, R., Piga, D., Zheng, W.X.: An instrumental least squares support vector machine for nonlinear system identification. Automatica **54**, 340–347 (2015) https://doi.org/10.1016/j.automatica.2015.02.017, https://www.sciencedirect.com/science/article/pii/S0005109815000709
2. Kekatos, V., Giannakis, G.B.: Sparse volterra and polynomial regression models: Recoverability and estimation. IEEE Trans. Sig. Process. **59**(12), 5907–5920 (2011). https://doi.org/10.1109/TSP.2011.2165952
3. Sureshbabu, N., Farrell, J.: Wavelet-based system identification for nonlinear control. IEEE Trans. Autom. Control **44**(2), 412–417 (1999). https://doi.org/10.1109/9.746278
4. Boyd, J.P.: Chebyshev and Fourier spectral methods. Courier Corporation (2001)

5. Mejri, N., Lopez-Fuentes, L., Roy, K., Chernakov, P., Ghorbel, E., Aouada, D.: Unsupervised anomaly detection in time-series: An extensive evaluation and analysis of state-of-the-art methods. Expert Systems with Applications, p. 124922 (2024)
6. Hsieh, R.J., Chou, J., Ho, C.H.: Unsupervised online anomaly detection on multivariate sensing time series data for smart manufacturing. In: 2019 IEEE 12th conference on service-oriented computing and applications (SOCA), pp. 90–97. IEEE (2019)
7. Kim, B., Alawami, M.A., Kim, E., Oh, S., Park, J., Kim, H.: A comparative study of time series anomaly detection models for industrial control systems. Sensors **23**(3), 1310 (2023)
8. Pothuganti, S.: Review on over-fitting and under-fitting problems in machine learning and solutions. Int. J. Adv. Res. Electr. Electr. Inst. Eng. **7**, 3692–3695 (2018). https://doi.org/10.15662/IJAREEIE.2018.0709015
9. Gauraha, N.: Introduction to the lasso: A convex optimization approach for high-dimensional problems. Resonance **23**(4), 439–464 (2018)
10. Khan, J.A., Van Aelst, S., Zamar, R.H.: Robust linear model selection based on least angle regression. J. Am. Stat. Assoc. **102**(480), 1289–1299 (2007)
11. Sweet, L.-b, Müller, C., Anand, M., Zscheischler, J.: Cross-validation strategy impacts the performance and interpretation of machine learning models. Artif. Intell. Earth Syst. **2**(4), e230026 (2023)
12. Pedregosa, F., Varoquaux, G., Gramfort, A., Michel, V., Thirion, B., Grisel, O., Blondel, M., Prettenhofer, P., Weiss, R., Dubourg, V., Vanderplas, J., Passos, A., Cournapeau, D., Brucher, M., Perrot, M., Duchesnay, E.: Scikit-learn: Machine learning in Python. J. Mach. Learn. Res. **12**, 2825–2830 (2011)
13. Stražar, M., Curk, T.: Approximate multiple kernel learning with least-angle regression. Neurocomputing **340**, 245–258 (2019)
14. Fraley, C., Hesterberg, T.: Least angle regression and lasso for large datasets. Stat. Anal. Data Mining: ASA Data Sci. J. **1**(4), 251–259 (2009)
15. Yamada, M., Saha, A., Ouyang, H., Yin, D., Chang, Y.: n^3-lars: Minimum redundancy maximum relevance feature selection for large and high-dimensional data. arXiv preprint arXiv:1411.2331 (2014)
16. Frandi, E., Nanculef, R., Lodi, S., Sartori, C., Suykens, J.A.: Fast and scalable lasso via stochastic frank-wolfe methods with a convergence guarantee. Mach. Learn. **104**, 195–221 (2016)
17. Pedregosa, F., Varoquaux, G., Gramfort, A., Michel, V., Thirion, B., Grisel, O., Blondel, M., Prettenhofer, P., Weiss, R., Dubourg, V., Vanderplas, J., Passos, A., Cournapeau, D., Brucher, M., Perrot, M., Duchesnay, E.: Scikit-learn: Machine learning in Python. J. Mach. Learn. Res. **12**, 2825–2830 (2011)

11 Partial State Estimator Design for Nonlinear Uncertain Systems

11.1 Chapter Content at a Glance

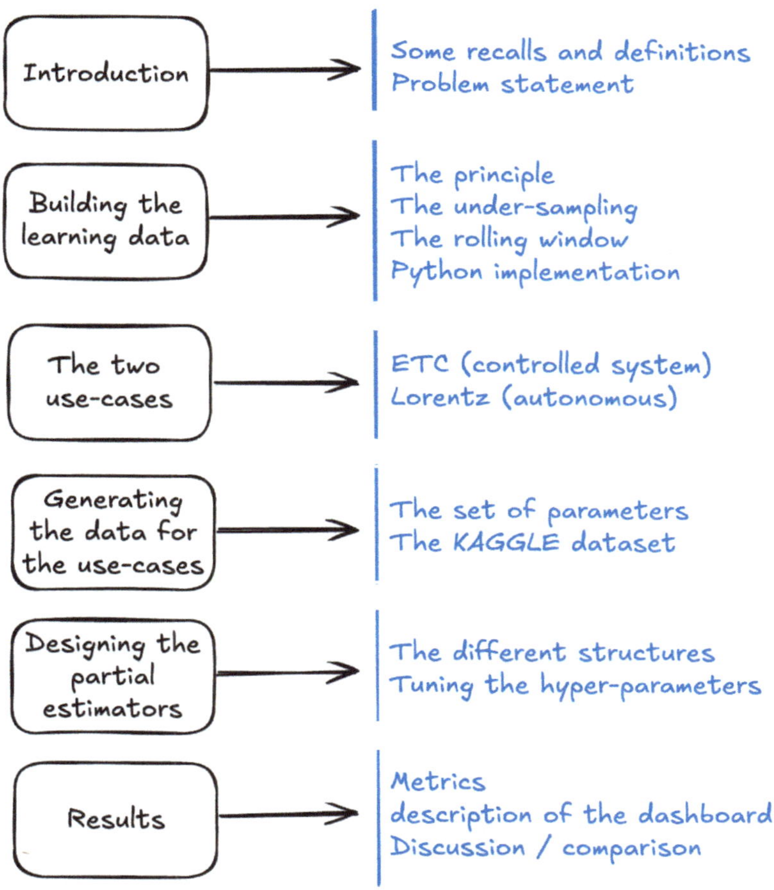

M. Alamir, *Nonlinear Control of Uncertain Systems*,
https://doi.org/10.1007/978-3-031-93287-8_11

11.2 Introduction and Problem Statement

This chapter introduces a framework for the derivation of partial state estimation algorithms. The concept of partial state estimation is introduced in Sect. 2.10 which is dedicated to an overview of the main concepts discussed in the book.[1]

More precisely, the problem of state estimation has been introduced in Chap. 2, first in the standard setting (Sect. 2.10.1), also referred to as the full state estimation setting, where the whole state vector is to be reconstructed and then in its partial form (Sects. 2.10.2 and 2.10.3) where one is interested in the reconstruction of some expression[2] z of the state and control. The reconstruction of the latter does not necessarily require the successful estimation of all the components of the state vector.

For the sake of easiness of reading of the present chapter and in order to clearly state the problem, the concept of partial observation target's reconstruction is recalled here briefly.

Let us consider a parametric dynamical system that is given in the following Ordinary Differential Equation (ODE) form:

$$\dot{x} = f(x, u, p), \tag{11.1a}$$

$$y = h(x, u, p), \tag{11.1b}$$

where

- $x \in \mathbb{R}^{n_x}$ is the state vector;
- $u \in \mathbb{R}^{n_u}$ is some exogenous signal[3]; and
- $y \in \mathbb{R}^{n_y}$ is the vector of measured quantities.

It is assumed in this chapter that the functions f and h **are known** although the values of the parameters gathered in p might be only known with some **uncertainties**. On the contrary it is assumed that the statistics of the parameters dispersion are known.[4]

We are interested in designing an algorithm that uses the measurements of the system in order to reconstruct the evolution of some **non-measured quantity**, denoted by z that depends on the triplet (x, u, p) through a relationship T of the form:

$$z = T(x, u, p), \tag{11.2}$$

where T is supposed to be known.

! z Is Supposed to Be a Scalar

It is assumed here, **without loss of generality**, that the observation target z is a real scalar. In the case where more than one observation target are to

[1] It might be helpful to take a quick reminder of the content of Chap. 2.

[2] Referred to as the *observation target*.

[3] This has not necessarily to be a control input.

[4] So that relevant random sampling might be used that yields a representative set of realizations.

be reconstructed, the process described in this chapter needs simply to be duplicated to each one of them.[5]

Recall that in the notation introduced in Chap. 2, the vector of past measurements at instant k is denoted by[6] [see (2.93)]

$$\mathbf{y}_k^{(-)} := \begin{bmatrix} y_k \\ \vdots \\ y_{k-(N_O-1)} \end{bmatrix} \in \left[\mathbb{R}^{n_y}\right]^{N_O}, \tag{11.3}$$

and that the very possibility to reconstruct the observation target z from the past measurements relies on the implicit assumption according to which there exists some map $\mathcal{O}$ such that the following relationship holds true during the system's evolution:

$$z_k \approx \mathcal{O}\big(\mathbf{y}_k^{(-)}\big). \tag{11.4}$$

Notice however that the existence of $\mathcal{O}$ is only an ideal basic working assumption in the sense that it supports the relevance of searching for a map that approximately matches the relationship even if such a map does not rigorously exist. In other words, we say that the assumption of the existence of $\mathcal{O}$ is *not constructive*. But it is always important to keep in mind what are the leading assumptions (or conjectures) behind any engineering attempt to find a solution to a problem.

Based on the above recalls and definitions, the problem addressed in the present chapter can be stated as follows:

> **Problem Statement**
> Given the following items:
>
> - the dynamics and the measurement maps f and h involved in (11.1a) and (11.1b);
> - an observation target map T invoked in (11.2);
> - relevant subsets $\mathbb{X}$, $\mathbb{U}$ and $\mathbb{P}$ in the state, exogenous signal and parameter spaces; and
> - generators of relevant instances of state, exogenous signal and parameters.
>
> **Design a framework** for the derivation of **partial estimators** of z that takes explicitly into account the uncertainty level affecting the knowledge of the model's parameters and the level of noise affecting the measurements.

[5] Notice that the process of generating the learning data might be common to all the instances of z. Only the fitting process needs to be repeated.

[6] N_O is referred to as the observation horizon.

In the following section, the general process of building the training data for the learning process is described.

11.3 The General Process of Building the Learning Data

In this section, we start by sketching the principles guiding the process of generating the learning data before a concrete Python implementation is discussed.

11.3.1 The Principle

The very existence of the partial observation map invoked in (11.4) might need the observation horizon's length to be taken sufficiently large. This is because large observation horizons provide more information on the system's trajectory that helps determining the current value of the observation target z (see [1] for an illustration of this intuitive fact).

On the other hand, the measurement acquisition period is not necessarily *tuned* to its appropriate value as it is very often chosen depending on the measurement device capability and/or the sampling period that is used for the integration of the system's equations which might be taken quite small for accuracy reasons that are not linked to the reconstruction goal or needs.

These situations might induce unnecessarily small measurement acquisition periods. In such cases, the use of large observation horizons leads to a high number of samples in the definition of $\mathbf{y}_k^{(-)}$ involved in the relationship (11.4) one is trying to identify.

This is the reason for which we introduce the under-sampled version of $\mathbf{y}_k^{(-)}$ that is depicted in Fig. 11.1.

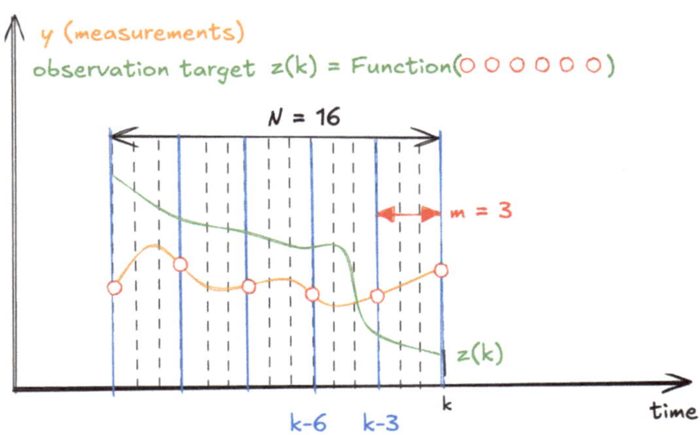

Fig. 11.1 Illustration of the under-sampling used to define the measurement profile used as the argument of the observation map

More precisely, an under-sampling integer parameter m is introduced leading to the past under-sampled measurement profile defined by[7]

$$\mathbf{y}_k^{(-)} := \begin{bmatrix} y_k \\ y_{k-m} \\ \vdots \\ y_{k-\lfloor N/m \rfloor m} \end{bmatrix}, \tag{11.5}$$

where $\lfloor \xi \rfloor$ stands for the integer part of ξ. This definition leads to the input argument of the observation map O being of dimension $\lfloor N/m \rfloor$.

In order to generate the dataset needed to learn the hidden relationship linking z to $\mathbf{y}_k^{(-)}$, it is mandatory to dispose of many simulated scenarios of the dynamical system (11.1a) under a sufficiently representative set of *scenarios*.

Notice that a scenario is defined by the necessary information to simulate the system over some time interval. Therefore, a scenario is defined by the triplet:

$$(x_0, U, p),$$

representing:

- the initial state x_0;
- the sequence U of values of the exogenous signal u; and
- the value of the parameter vector p.

In order to be able to dispose of a representative set of scenarios, it is assumed that we dispose of three associated `generators` that can be called using the following syntax:

$$\texttt{x0}, \texttt{U}, \texttt{p} \leftarrow \texttt{Generate\_sc}(),$$

which generate a scenario by sampling the initial state, the sequence of exogenous inputs and the vector of parameters according to the underlying sets $\mathbb{X}$, $\mathbb{U}$ and $\mathbb{P}$ and the associated randomness laws. This mechanism is problem-dependent and should be carefully thought of so that the training dataset encompasses representative samples of these quantities.

Obviously, each scenario leads to a simulation that lasts over as many sampling instants as there are elements in the sequence U, namely `card`(U). From such a scenario and using a rolling window of length N, it is possible to extract a number of instances of the pair:

$$(\mathbf{y}_k^{(-)}, z_k),$$

[7] Notice that the same notation is kept for the sake of simplicity as the past measurement profile will be used for a given pair (N, m).

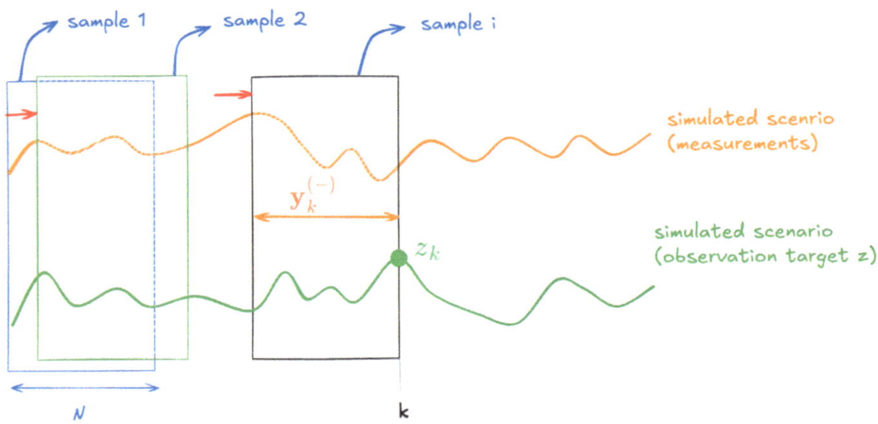

Fig. 11.2 Illustration of the process of rolling window used to collect learning samples from a simulated scenario

that is equal to:

$$\texttt{card}(U) - N.$$

Now repeating the process for N_{sc} scenarios leads to a training data with a number of lines equal to:

$$\texttt{nLines} := N_{sc} \times (\texttt{card}(U) - N),$$

that can be used in the learning process.

This process is illustrated in Fig. 11.2.

Let us implement this principle using a dedicated Python class.

11.3.2 Python Implementation

In order to come out with a generic class, let us first define the Syst structure that is used in the remainder of this chapter in order to summarize all the information regarding a specific system and a specific partial reconstruction problem.

The structure Syst is declared in the following script:

```
# The definition of the namedtuple system object.

from collections import namedtuple

Syst = namedtuple('Syst',
            [
                'n_states',      # Number of states
```

```
            'n_control',    # Number of exogenous input
            'n_meas',       # Number of measurements
            'fun_ode',      # Function defining the ODE
            'fun_meas',     # Function defining the measurments
            'param_dict',   # Dictionary of the model's parameters
            'p_nom',        # Nominal values of the parameter
            'dt',           # Sampling period
            'nsub_steps',   # Under-sampling parameter
            'U_gen'         # Generating of exogenous sequences.
        ])
```

Notice that the `namedtuple` is an **immutable** type that enables to refer to each of the attributes of an instance using the "." notation. For instance, this is an example of use of the above definition:

```
sys = Syst(...) # provide the input arguments
print(sys.n_states)
```

The attributes of the `Syst` type are the following:

`n_states`	the number of states,
`n_control`	the number of exogenous signals,
`n_meas`	the number of measurements,
`fun_ode`	the ODE function (f),
`fun_meas`	the measurement function (h),
`param_dict`	the dictionary of parameters (see p. 443 for an example),
`p_nom`	the nominal value of the parameter vector,
`dt`	the sampling integration period,
`nsub_steps`	the under-sampling parameter (m), and
`U_gen`	the generator of exogenous signal profiles.

Examples of the creation of instances of `Syst` are given in Sect. 11.4 where specific instances of the partial estimation problem are studied for two supporting examples.

For the time being let us focus on the generic part of the code.

As mentioned above, the concept of scenario is extensively used in the generation of the learning data. So let us define the associated `Python` class:

```
class Scenario:
    '''
    -----------------------------------------------------------
    An instance of this class defines a simulation scenario.
    Namely, an initial state, a control input sequence (if any)
    and a vector of parameters.
    -----------------------------------------------------------
    '''
    def __init__(self, x0, U, p):

        self.x0 = x0
        self.U = U
        self.p = p
```

The main class that enables to generate the learning data is called `Obspars`.[8]

In order to create an instance of this class (via the `init` method), a single argument is needed that is an instance of the `namedtuple Syst` mentioned above. The beginning script for the definition of the class and its instantiation method is given below:

```
def __init__(self, syst):

        # Reading the fields of the dynamical System
        self.nx = syst.n_states
        self.nu = syst.n_control
        self.ny = syst.n_meas
        self.f = syst.fun_ode
        self.h = syst.fun_meas
        self.param_dict = syst.param_dict
        self.p_nom = syst.p_nom
        self.np = len(syst.p_nom)
        self.dt = syst.dt
        self.U_gen = syst.U_gen
        self.nsub_steps=syst.nsub_steps
        self.tau = syst.dt * syst.nsub_steps

        ... to be continued!
```

[8] For Parsimonious Observer as the relevance of the parsimonious relationship's identification is the main message of this chapter.

Notice that in this first part, only a copy of the attributes of the Syst instance is performed. Notice also that the attribute tau plays the role of the large sampling period that takes into account the under-sampling associated to the field named nsub_steps. More precisely, tau is intended to represents the measurements acquisition period.

In the following part of the init method, some CasaDi-based utilities are created which are necessary for an efficient simulation of the required high number of scenarios that might be necessary for the building of the learning dataset.

This starts by creating the CasaDi version of the ODE, the integration-related one-step ahead map and then the estimation-related large one-step ahead operation as shown in the script below:

```
# The __init__ function (continued)

        #---- Create the casadi ode
        x = MX.sym('x', self.nx)
        u = MX.sym('u', self.nu)
        p = MX.sym('p', self.np)
        xdot = vertcat(*self.f(x, u, p))
        self.f_casadi = Function('F1', [x,u,p], [xdot])

        #---- Create the small one step casadi
        k1 = self.f_casadi(x,u,p)
        k2 = self.f_casadi\index{Casadi}(x+0.5*k1*self.dt,u,p)
        k3 = self.f_casadi(x+0.5*k2*self.dt,u,p)
        k4 = self.f_casadi\index{Casadi}(x+1.0*k3*self.dt,u,p)
        x_plus = x + self.dt / 6 * (k1 + 2 * (k2 + k3) + k4)
        self.F1_casadi\index{Casadi} = Function('F1', [x,u,p], [x_plus])

        #--- Create the large one-step casadi
        xact = 1.0 * x
        def F(x,u,p):
            xnext = self.F_casadi(x,u,p)
            xnext = xnext.full()[:,0]
            return xnext
        self.F = F
```

Notice that the sequence of exogenous inputs U mentioned above is defined in terms of the large sampling period. This is the reason why the simulation of the scenario involves the use of the last created CasaDi map F as it is shown in the following script:

```
# The __init__ function (continued)

        #-- Create the multi-step simulator
        def sim(x0, U, p):
            X=[x0]
            for u in U:
                X.append(self.F(X[-1],u,p))
            t = np.array([self.tau * i for i in range(len(X))])
            X = np.array(X)
            Uout = np.array(U + [U[-1]]).reshape(-1,self.nu)
            return t, X, Uout
        self.sim = sim
```

The last created map enables to take the fields of a scenario as input and returns the components of the system's trajectory.

The previously defined `sim` method is used in the following `perform_one_scenario` map shown below in order to create the state trajectory for a given scenario, then performs a **rolling window** spanning of the resulting time-series (see Fig. 11.2) in order to create the lines of the learning data that are steered from the specific scenario.

This is done using the following script which is a part of the `Obspars` class instantiation method `init`:

```
# The __init__ function (continued)


    def perform_one_scenario(self, sc, N, m):
            '''
            Simulate a scenario defined by (x0,U,p)
            and extract the features/label results
            using a rolling window of window and nJump
            for under-sampling.

            The output is a tuple of features, label matrices.
            '''

            # Extract the scenario and simulate it

            x0, U, p = sc.x0, sc.U, sc.p
            t, X, U = self.sim(x0, U, p)
```

```
        # Create the output trajectory
        y = []
        for i in range(len(X)):
            x, u = X[i], U[i]
            y.append(self.h(x,u,p))
        y = np.array(y)

        # Roll the window along the scenario's trajectories

        features = np.array([y[i:i+N:m, :].flatten()
                             for i in range(len(y)-N)])

        label_x =  np.array([X[i:i+N:m, :][-1]
                             for i in range(len(y)-N)])

        return features, label_x
```

Notice that in the previous script, the function returns several labels' columns corresponding each to a component of the state vector. This enables to avoid uselessly repeating the computation for each of the components of the state.

! **About the Returned Labels**

Notice that the previous script considers as labels, all the components of the state. This is appropriate when the observation targets only depends on the state.

It is possible (and easy) to render the class more general to make it appropriate for observation targets that depend on the state, the parameter and the exogenous signal, by adding as arguments to the currently used arguments, a map `z=T(x,u,d)` which generates all the specific observation targets. This is possible because the values of `x`, `u`, and `p` for the specific scenarios are available when the returned `label_x` is computed. The reader is strongly encouraged to write the associated more general code for his/her own need.

The script above enables to handle a single scenario that is defined by the values of x_0, U and p. As mentioned above, the learning data is built using a large set of scenarios in order to be relevant in terms of representativity. Therefore we need to precisely determine how the cloud of values of the above three items has to be created. This is explained in the sequel.[9]

[9] Obviously, the presented instantiation of the generation of scenarios is not intended to cover all the possibilities of relevant generation for specific use-cases. It should be taken as an inspiring instance.

Let us start by the method that generates a list of values of the vector of parameters p involved in (11.1a). This method takes the following arguments:

`designated`	The list of parameters for which dispersion is to be created.
`n`	The number of created samples.
`mode`	The dispersion mode (`normal`, `uniform`).
`sig`	The dispersion relative amplitude.

This leads to the script shown below:

```
# A method of the class Obsepars

def generate_P(self, designated, n, mode, sig):
        '''
        Generate a list of n instances of the parameter vector.
        if mode = "normal" then Gaussian relative dispersion
        with standard deviation sig is used, otherwise, a uniform
        dispersion is applied.
        Notice that dispersion affects only the list of designated
        components of the parameter vector.

        The output is a list of n values of the vector of parameters.
        '''
        P = []
        for _ in range(n):
            p = 1.0 * self.p_nom
            for s in designated:
                p_s = 1.0 * p[self.param_dict[s]]
                if mode == 'normal':
                    ecart = np.random.randn() * sig * p_s
                    p[self.param_dict[s]] += ecart
                else:
                    ecart = (-1+2*np.random.rand()) * sig * p_s
                    p[self.param_dict[s]] += ecart
            P.append(1.0 * np.array(p))

        return P
```

The method returns a list of `n` vectors representing different realizations of the vector of parameters p.

The way the `designated` parameters are *dispersed* in the loop generating the cloud of values enables to understand the way the `param_dict` should be defined in the `Syst` structure. Indeed, the keys of the dictionary are the names of the parameters, while the corresponding values are the index of the specific parameter inside the parameter vector p.

As an example, assume that a model involves the following parameters representing the mass `M`, the friction coefficient `mu` and some parameter named `beta`, namely:

$$p := \begin{bmatrix} M \\ \mu \\ \beta \end{bmatrix}, \tag{11.6}$$

then the corresponding `param_dict` dictionary should be defined as follows:

```
param_dict = {'M':0, 'mu':1, 'beta':2}
```

so that, when one needs to examine the dispersion that affects only the parameters M and β it is sufficient to provide the following definition for the `designated` input argument:

```
designated = ['M', 'beta']
```

as this enables to understand the part of the above script that is responsible for the creation of parameters dispersion:

```
        p = 1.0 * self.p_nom
        for s in designated:
            p_s = 1.0 * p[self.param_dict[s]]
```

Indeed, when `s` spans the `designated` list, it takes successively the values `M` and `beta`. When `s = 'M'` the assignment line becomes

```
p_s = 1.0 * p[0],
```

while when `s = 'beta'` the following line is executed:

```
p_s = 1.0 * p[2].
```

Notice that the size of dispersion parameter `sig` plays different roles depending on the choice of the `mode` input. Indeed:

- If `mode=normal` then a normal distribution with relative standard deviation of `sig` is applied.
- On the other hand, if `mode=uniform` is used,[10] `sig` stands for the relative half-width of the admissible domains of values in which a uniform sampling is operated.

Regarding the cloud of instances of the initial state x_0, we consider in this chapter the case where the set $\mathbb{X}$ is a hypercube defined by a vector of lower and upper bounds denoted by `xmin` and `xmax` respectively. Uniform sampling inside this hypercube is done.

Finally, the sequence of exogenous inputs are provided by the function `u_gen` that is an attribute of the `Syst` instances and which becomes an attribute of the `Obspars` instances as a corresponding attribute is created in the beginning of the `__init__` instantiation method (see p. 438).

The previous choices impacting the cloud generation lead to the following method of the `Obspars` class which takes, in addition to the previously mentioned input arguments `designated`, `n`, `mode`, and `sig` the bounds that define $\mathbb{X}$ as well as the duration of the simulation.[11]

```
# A method of the class Obspars

def generate_scenarios(self, designated, n,
            mode, sig, xmin, xmax, tf, nt_max=200000):


        t = np.array([i * self.tau
                for i in range(nt_max) if i*self.tau<=tf])

        P = self.generate_P(designated, n, mode, sig)

        Dx = np.array(xmax)-np.array(xmin)
        set_x0 = [np.array([xmin[i] + np.random.rand()*Dx[i]
```

[10] As a matter of fact anything other than `normal` is interpreted as `uniform`.

[11] Notice the use of `nt_max` that avoid memory overflow that might be induced by a too large choice of the simulation time `tf`.

```
                        for i in range(self.nx)])
                        for i in range(n)]

        set_U = [self.U_gen(t) for i in range(n)]

        set_of_sc = [
                        Scenario(x0=set_x0[i],
                        U=set_U[i], p=P[i])
                        for i in range(n)
                    ]

        return set_of_sc
```

The above-defined method returns a set of `n` scenarios that can be fed to the previously defined `perform_one_scenario` in order to deliver the associated samples that are then concatenated to form the learning data.

This is done using the following method of the `Obspars` class:

```
def perform_multiple_scenario(self, set_of_sc, N, m):
        '''
        --------------------------------------------------------
        Generate the features matrix and label for the
        the list (set_of_sc) of scenarios using the window and
        nJump parameters.

        The output is a triplet containing:
        - The features matrix
        - The label's matrix
        - the features names' column.
        --------------------------------------------------------
        '''
        X = []
        yx = []
        n = len(set_of_sc)

        # Iterate on the scenarios to append the associated
        # features and label's matrices.
        for i in range(n):
            sc = set_of_sc[i]
```

```
        feat, label_x = self.perform_one_scenario(sc, N, m)
        X.append(feat)
        yx.append(label_x)

    # Create the features columns naming
    L = [[[f'y_{j+1}_{k}'] for j in range(self.ny)]
            for k in range(0,N,m)]

    cols = []
    for l in L:
        for ell in l:
            cols.append(ell[0])

    X = np.vstack(X)
    yx = np.vstack(yx)

    return X, yx, cols
```

Notice how in addition to the creation of the concatenated matrices of features and label, namely `X` and `yx`, the method also provides the vector `cols` of names which describes the contents of the features columns.

These names are determined by the component index in the measurement vector y (on the basis of which the feature vector is defined) and on the corresponding instants that take the form: $0, m, 2m, \ldots$ where 0 correspond to the instant at the beginning of the rolling window, m is the one $m\tau$ instants after and so on.

For instance, a possible names might be:

$$y\_2\_0, \quad y\_1\_3,$$

which respectively translate to

$$y_2(t_{k-\lfloor N/m \rfloor m}), \quad y_1(t_{k-(\lfloor N/m \rfloor -3)m}).$$

This naming process is important since it enables to reconstruct the mathematical expression of the function that gives the estimated observation target in terms of the original measurements.

About Naming the Features

Obviously, this symbolic reconstruction is not mandatory for the reconstruction itself, however, it can be interesting in the case where a high number of sensors are present in order to identify which ones of these sensors and which past instants of the associated measurements are necessary to reconstruct a specific observation target z.

This ends the presentation of the generic class `Obspars` that facilitates the generation of the learning data. Recall that the instantiation of this class needs the definition of the problem-dependent object `syst` which is an instance of the `namedtuple` object `Syst` that is to be defined in each specific instances of the partial estimation problem.

In the next section, the two use-cases that are studied in this chapter in order to illustrate the way a partial observation target reconstruction can be designed are presented together with their associated instance of the `Syst` type.

11.4 Presentation of the Two Use-Cases

In this section, the two examples used to illustrate the design of partial estimators design are described and the associated fields of the `Syst` structure used to represent them are also provided for the sake of illustration. The reason for which two examples are shown is that they enable to illustrate different settings of the partial estimator design parameters.

11.4.1 The Electronic Throttle Controlled (ETC) System

This is the same dynamical controlled system that is studied in Sect. 4.3 in order to illustrate standard control related methods and in Chap. 9 in order to illustrate the full state Moving-Horizon Estimator (MHE). Nevertheless, it is briefly recalled here for the sake of easiness of reading. A schematic view of the ETC system is shown in Fig. 11.3.

This system describes the control of the angle $x_1 = \theta$ that moderates the admission of the airflow entering the combustion engine. Two parts are clearly distinguishable in Fig. 11.3, namely the mechanical part and the electric part. The role of the electric part is to act on the mechanical part through an electro-motor force $x_3 := e_a$ that is controlled through the current $i_a = u$ which directly controls the electro-motor force $x_3 := e_a$. The latter induces a torque that acts on the rate of change of the angular velocity $x_2 = \dot{x}_1 = \dot{\theta}$ so that the latter follows some reference value.

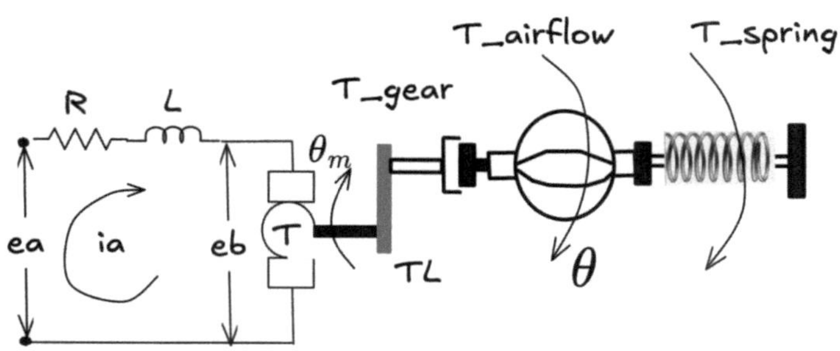

Fig. 11.3 The ETC automotive unit

The mathematical equations describing the ODEs governing the dynamics of the above-described components of the state are given by

$$\dot{x}_1 = x_2, \tag{11.7a}$$

$$\dot{x}_2 = \frac{1}{N_m^2 J_m + J_g}\Big[\phi(x,p) + N_m K_t x_3\Big], \tag{11.7b}$$

$$\dot{x}_3 = \frac{1}{L_a}\Big[-N_m K_b x_2 - R_a x_3 + i_a\Big], \tag{11.7c}$$

in which the map $\phi(x,p)$ is given by

$$\phi(x,p) := -K_{sp}(x_1 - \pi/2) - (N_m^2 b_m + b_t)x_2 - \pi R_p^2 R_{af}\Delta(x_1)\cos^2(x_1) \tag{11.8}$$

where

$$\Delta(x_1) := 2P_{\text{atm}} \times (1 - x_1/\pi),$$

is the over-pressure depending on the opening angle $x_1 = \theta$ of the device.

Using the above equations, it is possible to start the problem-dependent module:

`etc._user_defined.py`

by defining the parameters dictionary invoked in the previous section using the following script:

```
# The etc_user_defined.py

#-------------------------------------
# The parameters' dictionary
#-------------------------------------
param_dict = {
    'Rp': 0,
    'Raf': 1,
    'Ksp': 2,
    'P_atm': 3,
    'J': 4,
    'theta0': 5,
    'Kf': 6,
    'N': 7,
    'Kt': 8,
    'La': 9,
    'Kb': 10,
    'Ra': 11
}
```

where the parameters are numbered for later use as explained above.

Once the parameters are ordered following this dictionary, one can define the ODE's function using the following script:

```
# The etc_user_defined.py (continued)

def etc(x,u,p):
    '''
    -----------------------------------------
    Definition of the ODE for the ETC system.
    -----------------------------------------
    '''
    term0 = p[0]*np.cos(x[0])
    term = 2*np.pi*p[1]*term0*term0
    term *= p[3]*(1-x[0]/np.pi)
    xdot1 = x[1]
    xdot2 = 1.0/p[4]*(-p[2]*(x[0]-p[5])-p[6]*x[1]+p[7]*p[8]*x[2]-term)
    xdot3 = 1/p[9]*(-p[7]*p[10]*x[1]-p[11]*x[2]+u[0])
    xdot = (xdot1,xdot2,xdot3)

    return xdot
```

which simply implements the ODEs defined by (11.7).

Since the vector of nominal values of the vector of parameters p is one of the attributes of the `Syst` structure to be provided to the `Obspars` class, the following script is added to `etc._user_defined.py`:

```
# The etc_user_defined.py (continued)

#-------------------------------------
# Nominal vector of parameters
#-------------------------------------

bm = 0.03
bt = 3.4e-3
Jg = 0.005
Jm = 0.0004
Kb = 0.1051
Ksp = 0.4316
Kt = 0.1045
```

```
La = 0.003
N = 4
Ra = 1.9
Raf = 0.002
Rp = 0.0015
Jg = 0.2
theta0 = np.pi/2
P_atm = 1.01325e5
#----
J = N ** 2 * Jm + Jg
Kf = (N ** 2) * bm + bt

#----- the nominal value of the vector of parameters.

p_nom = np.array([
    Rp, Raf, Ksp, P_atm, J, theta0, Kf, N, Kt, La, Kb, Ra
])
```

Notice that the same order as the one used in the definition of the `param_dict` dictionary is used.

In this study, we shall assume that the measurement vector y for the ETC system is defined as follows:

$$y = h(x, u, p) = \begin{bmatrix} x_1 \\ i_a = u \end{bmatrix} \in \mathbb{R}^2. \tag{11.9}$$

Therefore, the measurement map can be defined by the following script:

```
# The etc_user_defined.py (continued)

#-------------------------------------
# The measurement function
#-------------------------------------
def h(x,u,p):
    return np.array([x[0], u[0]])
```

As for the partial estimation problem, the following two observation targets are studied:

$$z = x_2 \quad \text{and} \quad z = x_3. \tag{11.10}$$

Now we need to provide the generator of the exogenous sequences U that is required in the Syst structure, namely the function:

$$U\_gen(t),$$

that is a part of the field of the Syst structure.

In order to define the generator of the control profiles, it is suggested here to consider that the partial estimator design step has to be done prior to any control design. This is by no means the only possible choice. Indeed, we can imagine using the system's model in order to design a feedback following the same methodology that is proposed in Chap. 4 and to simulate the resulting closed-loop trajectories and use them to build the learning data. Here it is suggested to use a different approach for the sake of illustration.

More precisely, we shall consider a damped oscillating **Open-Loop** (OP) excitation sequences that are defined as follows:

$$u(t) := u_0 \times \sin(\omega t)e^{-\lambda t}, \tag{11.11}$$

where the triplet (u_0, ω, λ) is randomly generated inside the following admissible sets of values:

$$u_0 \in [-50, +50]; \quad \omega \in [1.0, 10.0]; \quad \lambda \in [0.1, 1.0].$$

Three typical possible resulting scenarios are shown in Fig. 11.4.

The function that generates a sequence of exogenous OP current i_a based on the above definition completes the user-defined module etc._user_defined.py:

```
# The etc_user_defined.py (terminated)

def U_gen(t):
    '''
    ----------------------------------------
    Definition of the sequence of actions
    for the ETC system.
    ----------------------------------------
    '''
    lam_max = 1e-0
    lam_min = 1e-1
    omega_min = 1
    omega_max = 10.0
    u_max = 50
    u_min = -50
```

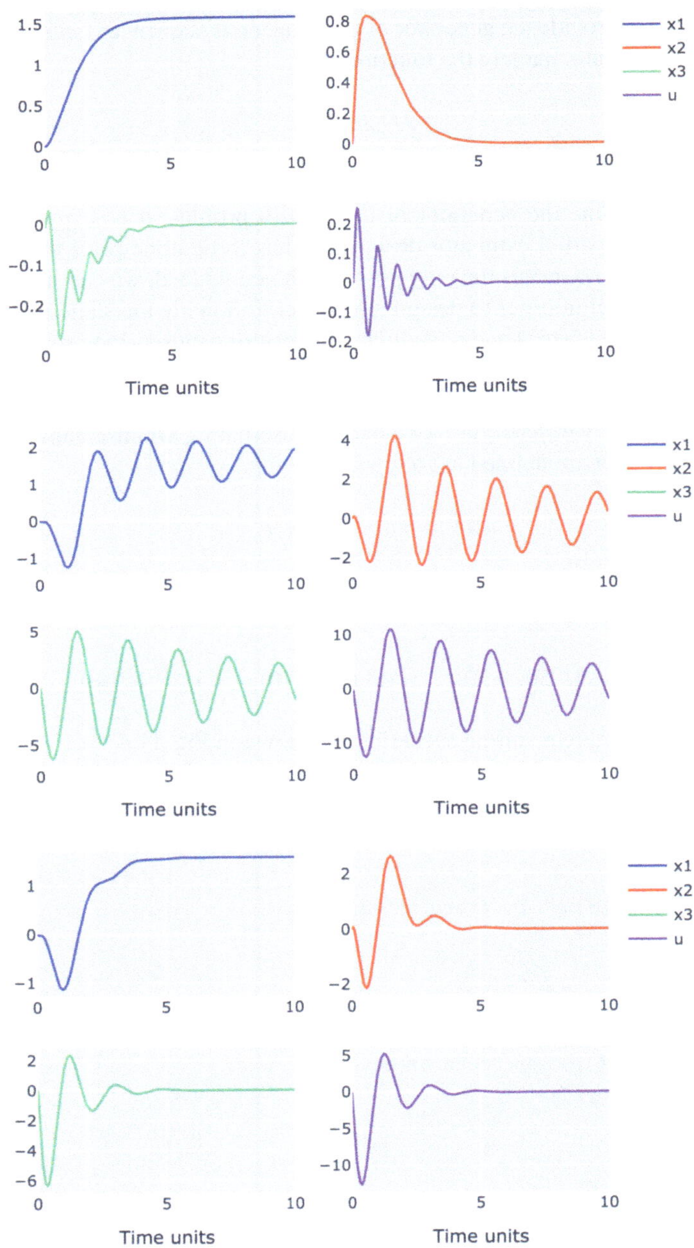

Fig. 11.4 Tree typical possible scenarios of excitation given by the control sequence resulted from the proposed generator of U-OL profile

```
    lam = lam_min + np.random.rand() * (lam_max - lam_min)
    omega = omega_min + np.random.rand() * (omega_max - omega_min)
    u = u_min + np.random.rand() * (u_max - u_min)
    U = list(u * np.sin(omega * t) * np.exp(-lam * t))

    return U
```

By now we have all that we need to create the ETC instance of the `Syst` structure. This is done in the following script:

```
# Import the needed classes and functions
#-------------------------------------------------------
from obspars import Syst, Obspars
from etc_user_defined import h, etc, param_dict, p_nom, U_gen

# Declare the Syst instance representing the ETC system.
#-------------------------------------------------------
sys = Syst(n_states=3,
           n_control=1,
           n_meas=2,
           fun_meas=h,
           fun_ode=etc,
           param_dict=param_dict,
           p_nom=p_nom,
           dt=1e-3,
           nsub_steps=10,
           U_gen=U_gen)
```

notice that the elementary sampling period that is needed to be quite small for the purpose of integration stability is taken equal to dt=10^{-3} while the measurement sampling period is taken equal to 0.01 by setting the attribute `nsub_steps` to 10.

11.4.2 The Lorentz Oscillator

Here is another three-dimensional system that is autonomous (no exogenous signals are involved) and that depends on three parameters that are introduced using the following `param_dict` (dictionary) that lies in the beginning of the user-defined file named:

`user_defined_lorentz.py` :

```
# The user_defined_lorentz.py file

param_dict_lorentz = {
    'sigma': 0,
    'rho': 1,
    'beta':2
}
```

This system is governed by the following ODEs that are written directly in `Python` form since we are now used to the syntax of the `Syst` object:

```
# The user_defined_lorentz.py file (continued)

def lorentz(x,u,p):

    xdot1 = p[0] * (x[1]-x[0]) + x[2] * u[0]
    xdot2 = x[0]*(p[1] - x[2])-x[1]
    xdot3 = x[0]*x[1]-p[2]*x[2]
    xdot = (xdot1, xdot2, xdot3)

    return xdot
```

notice that although the system does not involve exogenous signals in its original and widely considered version, a control related term is introduced in the first line of the ODEs for possible use of excited version.[12] This enables to avoid some technicalities in the code that would be required in order to handle the difference between excited and autonomous systems. This also enables to keep the focus on the main observer-related key aspects.

As for the measurement map, it is assumed here that only the first state is measured. This enables to end the `user_defined_lorentz.py` file with the following script that introduces the nominal value of the parameter vector, the output map and the trivial zero-valued fictitious exogenous signal used in the definition of the `U_gen` generator.

[12] In the forthcoming results, the excitation signal is set to 0 for the sake of comparison with potential existing attempts to build state estimators for this system.

```
# The user_defined_lorentz.py file (completed)

p_nom_lorentz = np.array([10, 28, 8/3])

def h_lorentz(x,u,p):
    return np.array([x[0]])

def U_gen(t):
    return list(0 * np.sin(t))
```

This provides all that we need to create the instance of the `Syst` structure that represents the Lorentz oscillator system. This is done in the following script:

```
# import the needed classes and functions.
#------------------------------------------
from obspars import Syst, Obspars
from lorentz_user_defined import lorentz, param_dict_lorentz
from lorentz_user_defined import p_nom_lorentz, U_gen, h_lorentz

# Create the corresponding instance of Syst
#------------------------------------------
sys = Syst(n_states=3,
           n_control=1,
           n_meas=1,
           fun_meas=h_lorentz,
           fun_ode=lorentz,
           param_dict=param_dict_lorentz,
           p_nom=p_nom_lorentz,
           dt=1e-2,
           nsub_steps=1,
           U_gen=U_gen)
```

This ends this section that is dedicated to the presentation of the two illustrative examples that are used in the remainder of this chapter to examine different aspects of the partial estimator design via Machine Learning (ML).

For each of the use-cases, we dispose now of a corresponding instance of the `Syst` type that we can use to instantiate the `Obspars` class in order to generate the different sets of learning data, each for a given level of the model's parameters dispersion.

This is precisely done in the following section.

11.5 Generating the Sets of Learning Data for the Two Use-Cases

In this section, the parameters that are used in the generation of the learning data for the two use-cases are first given before the resulting publicly available dataset is briefly described.

11.5.1 The Set of Parameters Used in the Generation

For both use-cases, once the corresponding instance `sys` of the `Syst` type has been created, it is possible to create an instance of the `Obspars` class using the script:

```
o = Obspars(syst=sys)
```

By so doing, it is possible to use the method `generate_scenarios` of the `Obspars` class in order to generate a set of scenarios. Recall that this method needs the following input arguments to be provided:

`designated`, `n`, `mode`, `sig`, `xmin`, `xmax` and `tf`.

Table 11.1 gives the different arguments used for each of the use-cases.

The table shows that three levels of relative standard deviation are used, namely

$$\texttt{sigma\_p} = 0, 0.05 \text{ and } 0.1$$

for each use-case.

Table 11.1 Calling parameters for the generation of `n` scenarios for the two use-cases. The parameters `N` and `m` used later in the simulation are also given

Parameter	ETC	Lorentz
Designated	All parameters	All parameters
`n`	1000	250
`Mode`	`Normal`	`Normal`
`sig`	$\in \{0, 0.05, 0.1\}$	$\in \{0, 0.05, 0.1\}$
`xmin`	$[-0.5, -0.5, -0.5]$	$[-1, -1, -1]$
`xmax`	$[0.5, 0.5, 0.5]$	$[1, 1, 1]$
`tf`	3.0	4.0
`N`	30	50
`m`	2	10

Recall that this impact the random sampling of the values of the model's vector of parameters p involved in the simulated scenarios.

The generation of the set of scenarios is obtained for both cases using the following script in which the input arguments are specific to each scenario as shown in Table 11.1.

```
set_of_sc = o.generate_scenarios(
                designated=designated,
                n=n,
                mode='normal',
                sig=sigma_p,
                xmin=xmin,
                xmax=xmax,
                tf=tf
        )
```

These scenarios can now be simulated using the `perform_multiple_scenario` of the `Obspars` class using the following call:

```
X, yx, cols= o.perform_multiple_scenario(set_of_sc=set_of_sc, N=N, m=m)
```

which requires the two input arguments `N` and `m` and to be defined. The values of these two parameters for both use-cases are given in Table 11.1.

As a matter of fact, at this stage, it seemed interesting to create a public dataset on the data-sharing platform `Kaggle`[13] in order to make the learning data available for a larger public [2].

The resulting dataset is briefly described in the next section before its use in illustrating the design of partial estimators is shown and several choices are compared.

11.5.2 The Resulting `Kaggle`'s Publicly Available Dataset

Based on the generation process described in the previous section, 12 datasets have been created. This is because for each use-case (ETC, Lorentz), there are three levels

[13] https://www.kaggle.com/datasets/mazenalamir/dataset-for-nonlinear-state-estimator-design/data.

of relative parametric dispersion (0, 0.05, and 0.1) and two possible observation targets (x_2 and x_3). This leads to 12 different datasets.

These datasets are gathered in a single dictionary named:

`data.pkl`,

that can be downloaded on the `Kaggle` open data repository [2].

Once the `data.pkl` file is locally available, the following script enables to examine the keys of the dictionary it contains:

```
import pickle
data = pickle.load(open('data.pkl', 'rb')

>> print(data.keys())

[('etc', 0.0, 'x2'),
 ('etc', 0.0, 'x3'),
 ('etc', 0.05, 'x2'),
 ('etc', 0.05, 'x3'),
 ('etc', 0.1, 'x2'),
 ('etc', 0.1, 'x3'),
 ('lorentz', 0.0, 'x2'),
 ('lorentz', 0.0, 'x3'),
 ('lorentz', 0.05, 'x2'),
 ('lorentz', 0.05, 'x3'),
 ('lorentz', 0.1, 'x2'),
 ('lorentz', 0.1, 'x3')]
```

This shows the keys that enables to get the dataset for a specific triplet of system, parametric relative dispersion level and observation target.

More precisely, assuming that `k` belongs to the list of above shown keys of the `data` dictionary, `data[k]` is also a dictionary with the following keys:

`Xtrain`, `Xtest`, `ytrain` and `ytest`.

As an example, the following script enables to examine, for instance, the size of the datasets for both systems and for the observation target $z = x_2$ with no parameter dispersion:

```
print(data[('etc', 0.0, 'x2')]['Xtrain'].shape)
print(data[('etc', 0.0, 'x2')]['Xtest'].shape)
print(data[('lorentz', 0.0, 'x2')]['Xtrain'].shape)
print(data[('lorentz', 0.0, 'x2')]['Xtest'].shape)

>>
(136000, 30) (136000, 30) (44000, 5) (44000, 5)
```

Notice that the number of features in the case of the ETC system is equal to 30 because for this system, there are two measurements and because the values $N = 30$ and $m = 2$ are used (see Table 11.1) leading to 15 columns per measurement.

As for the Lorentz system, the number of features can be explained by the presence of a single measured variable together with the values $N = 50$ and $m = 10$.

The above-shown dimensions are the same for a given system regardless of the level of the parametric dispersion and regardless of the observation target being used in the dictionary key.

Moreover, the splitting of the datasets into *train* and *test* subsets has been done *without shuffle*.[14] More precisely, the split has been done by taking the first half of the set of randomly generated scenarios to serve in the construction of the training dataset while keeping the second half of them for the creation of the test-related datasets.

> **Only 5% of the Training Data Is Used to Fit the Models**
> Because of the need to fit 36 models for each of the set of compared structure of the fitting model (see Sect. 11.6.1), only 5% of the training sets has been used in the fitting process. More precisely, the following slicing set:
>
> `Xtrain[::20]`,
>
> has been used for each appropriate `Xtrain` in order to fit the corresponding model. This also enables the examine the ability of the framework to extract the underlying relationships from small amounts of data.
> To be more specific, the amount of training data corresponds to 6800 samples for the ETC-related case and 2200 samples for the `Lorentz`-related case while the validation is done on datasets of sizes 68000 and 22000 respectively.

It is also important to know that the matrices of features `X` produced by the call for the `perform_multiple_scenario` of the `Obspars` class (see p. 457) have been normalized before they were used to create the features matrices that are included in the `Kaggle` open data platform. This helps better appreciating the amplitudes of the estimation errors for the different partial estimators computed in the remainder of this chapter.

[14] See p. 143 of Chap. 3 for a discussion about this option.

Before we step to the next section that is dedicated to the design of the partial estimators and to the comparison, the issue of measurement noise needs to be briefly discussed.

Recall that first of all, the features matrices contained in the `data.pkl` are constructed based on the simulation of the two systems without adding measurement noise. The reason for this is twofold:

1. First of all, it is quite easy to add the measurement noise to the matrix `X` beforehand with the required level of noise that one would like to use in order to analyze the impact on the results. There is no point in repeating all the simulations for all the required levels of the measurement noise. So the first reason in delivering measurement-free features matrices is linked to the purpose of avoiding computational and memory burden.
2. There is also another reason which is linked to the relevance of comparative studies between different levels of the measurement noise. Indeed, we need to have the same measurement-free features matrices (the same underlying scenarios) in order for the comparison to be relevant. This is because, in order to analyze the impact of measurement noise on the quality of the designed estimators, one needs to compare the results that are computed on the same scenarios with the measurement noise being the only difference between the compared results. Otherwise, the difference might be related to the change in the underlying scenarios and not induced by the level of measurement noise.

To summarize:

Regarding the Measurement Noise

For each choice of the structure of the model used in the partial estimator, three **different** estimators will be designed for each of the 12 datasets.

More precisely, for each one of the 12 datasets, an estimator is designed based on a new feature matrix that is computed by adding normal measurement noise having one of the following three levels[15]:

$$\nu \in \{0.0, 0.025, 0.05\}. \tag{11.12}$$

This means that for each one of the competing `ML` structures, 36 estimators will be designed, each is dedicated to a specific choice of the following items:

- the dynamic system (×2),
- the observation target (×2),
- the standard deviation of the relative parameters dispersion (×3), and
- the level of measurement noise (×3).

[15] Recall that the features matrices are normalized.

11.6 Design of Partial Estimators of Observation Targets

In this section, the following structures are used to derive different families of partial estimators for the 36 above described problems.

11.6.1 The Different Structures Used in Designing the Partial Estimators

- The **`RandomForestRegressor`** of the `ensemble` module of the `scikit-learn` library [3].
- The **`KNNRegressor`** of the `neighbors` module of the `scikit-learn` library. The prediction of this regressor is based on the aggregation of the values taken by the label at the k nearest neighbors of the features vector where the estimation is to be provided.
- The **`XGBRegressor`** of the `eXtreme Gradient Boosting (XGBoost)` package which is quite similar to the Random Forest (RF) in that it builds decision trees although the construction method is different and relies on an advanced mix of gradient boosting and tree pruning (see [4] for more details).
- The **`DNNRegressor`** which is a Deep Neural Network (DNN) structure created using the `torch.nn` module. Notice that the number of possible DNN architecture is infinite and there is no specific effort here to look for an *optimal choice*. This is because of the belief that, anyway, DNN is probably not the appropriate choice for parsimonious design that uses small amounts of data. This structure is added here for the sake of completeness.
- The **`Lassolarscv`** of the `linear_models` module of the `scikit-learn` library. This regressor can possibly be applied to the `polynomial-Features`-based extended features leading to the identification of possibly nonlinear maps. However, the application of this algorithm in this chapter is limited to its simplest linear form in order to appreciate the degree of nonlinearity in the use-cases.
- **`plars:`** The last algorithm that is included in the comparison is the one based on the algorithm proposed in Chap. 10 and applied again to the `polynomial-Features`-based extended features matrices exactly as the previous one. This algorithm is referred to hereafter as the **`plars`** algorithm which is a method in the **`scLars`** module presented in Chap. 10.

Table 11.2 Definition of the different settings used in the cross-validation aiming at choosing the hyper-parameters used in the different structures involved in the design partial estimators

Structure	Hyper-parameters	Comments
scLars.pLars	$\begin{bmatrix} \texttt{w} = 200 \\ \texttt{deg} = \{1, 5\} \\ \varepsilon = 0.01, q = 50, N = 50 \end{bmatrix}$	See Chap. 10
RF	max_leaf$\in \{100, 1000, 5000\}$	
XGBoost	$\begin{bmatrix} \texttt{n\_estimators} \in \{100, 1000\} \\ \texttt{max\_depth} \in \{3, 10, 25\} \\ \texttt{learning\_rate} \in \{0.1, 0.5\} \end{bmatrix}$	
K-Nearest Neighbors (KNN)	n_neighbors$\in \{1, 3, 5, 10\}$	
LassoLarsCV	cv$\in \{3, 7, 10\}$	
DNN	$\begin{bmatrix} \texttt{n1} = 25 \\ \text{5 hidden layer} \\ \text{of 25 nodes each} \\ \text{\#epoch} = 200 \end{bmatrix}$	

11.6.2 Tuning the Hyper-Parameters

When appropriate, the above-mentioned structures have been fitted for different settings (see Table 11.2) of the hyper-parameters **on half of the training data**[16] and the quality of the fit has been examined on the remaining part of the training data. Based on the resulting validation scores, the best setting has been determined and used for the prediction on the test data and the resulting scores are used in the comparison.

Table 11.2 shows the design parameters used for each of the competing structures. When sets of possible values are shown for the parameters, then the product set is used as the list of settings to be examined in order to determine the setting to be adopted in the final comparison.

Notice that for each of the structures mentioned in Table 11.2, one might tune and re-tune the design parameters endlessly. There is no claim to be exhaustive here. Nevertheless, it is likely that the qualitative conclusions of the comparison remain quite valid as far as the problem of designing partial estimators via learning approaches is considered and as far as one deals with smoothly defined nonlinear models.

[16] Recall that the original training data is the first half of the dataset and that only 5% of this half is used leading to what is designated here as the training data. The test data is the remaining second half of the original data.

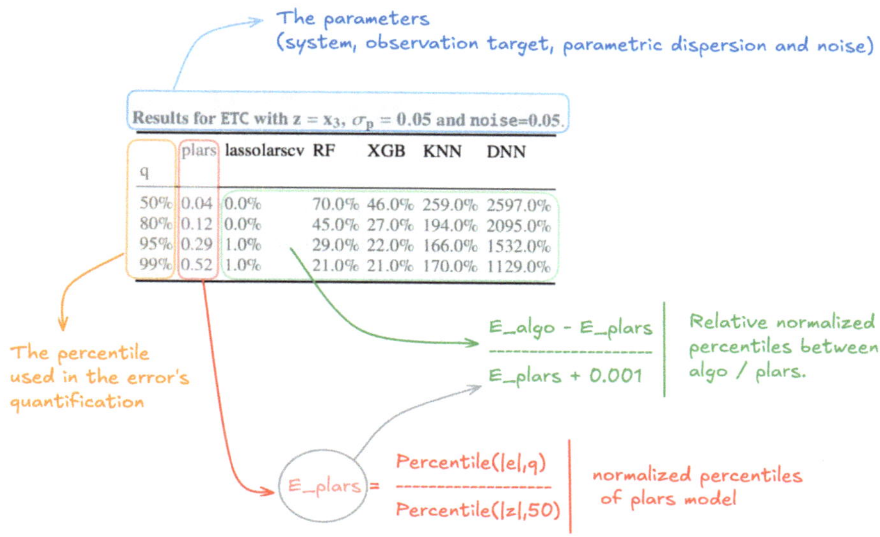

Fig. 11.5 Illustration of the content of the tables showing the results

11.7 Results

The results that are presented in this section gather 40 tables regrouped in 12 Figures. Half of these corresponds to the partial estimation results related the ETC system while the other half corresponds to the results regarding the Lorentz system.

For each system, the partial estimation results are shown when the observation target is x_2 and when it is set equal to x_3. Let first of all describe the content of the tables shown in the remainder of this section in order to compare the precision of the partial estimators for the different use-cases and settings.

11.7.1 Description of the Tables Providing the Comparison Results

This content is described in Fig. 11.5. The following items describe the content of this figure.

- The errors indicators corresponding to the `plars` polynomial model are taken as the reference. The errors indicator associated to the remaining model's structures are normalized by those of the `plars` structure.

- More precisely, the error of an algorithm, say `algo`, is computed as follows:

$$e_{\texttt{algo}}(q) := \frac{\texttt{percentile}\Big(|z(\cdot) - \hat{z}(\cdot)|, q\Big)}{\texttt{percentile}\Big(|z(\cdot)|, 50\Big)}, \tag{11.13}$$

  where z and $\hat{z}$ represent the true observation target's profile and the estimated profile by the `algo` variant. This represents the **normalized percentiles** as all the percentile are divided by the median of the absolute value of the observation target.
- The values $e_{\texttt{plars}}$ for $q = 50, 80, 95$ and 99 are shown in the column named `plars` of the table. This is the only column where the normalized percentiles are shown as defined by (11.13).
- For all the other structures, the corresponding columns of the table show the following relative values:

$$(\text{Column of } \texttt{algo}) = \frac{e_{\texttt{algo}}(q) - e_{\texttt{plars}}(q)}{e_{\texttt{plars}}(q) + 0.001}. \tag{11.14}$$

As there are a lot of figures, let us briefly describe how they are organized.

11.7.2 Organization of the Dashboard of Results

In order to understand the organization of the dashboard of results, let us first recall that each table described in the previous section is determined by the specific instance of the following set of information:

- the system: `ETC` or `Lorentz`,
- the observation target: x_2 or x_3,
- the model's parameters dispersion defined by σ_p, and
- the measurement noise.

Having these items in mind, Table 11.3 summarizes the content of figures that follows.

11.7.3 Discussion

The following items summarize some facts and help focusing on different aspects contained here and there in the tables below:

- In the absence of noise and parameter dispersion, the relationship can be almost perfectly captured (see the top tables on Figs. 11.6, 11.7, 11.12 and 11.13).
- The captured relationships are linear in the case of `ETC` while a multi-variate polynomial of degree 5 is necessary for the `Lorentz` system.

Table 11.3 Overview of the settings concerned by the 12 figures that show the results of the partial estimators

Figures	System	z	σ_p	Noise
11.6	ETC	x_2	0	$\{0, 0.025, 0.05\}$
11.7	ETC	x_3	0	$\{0, 0.025, 0.05\}$
11.8	ETC	x_2	$\{0, 0.05, 0.1\}$	0
11.9	ETC	x_3	$\{0, 0.05, 0.1\}$	0
11.10	ETC	x_2	$\begin{bmatrix} 0.05 \\ 0.1 \\ 0.05 \\ 0.1 \end{bmatrix}$	$\begin{bmatrix} 0.025 \\ 0.025 \\ 0.05 \\ 0.05 \end{bmatrix}$
11.11	ETC	x_3	$\begin{bmatrix} 0.05 \\ 0.1 \\ 0.05 \\ 0.1 \end{bmatrix}$	$\begin{bmatrix} 0.025 \\ 0.025 \\ 0.05 \\ 0.05 \end{bmatrix}$
11.12	`lorentz`	x_2	0	$\{0, 0.025, 0.05\}$
11.13	`lorentz`	x_3	0	$\{0, 0.025, 0.05\}$
11.14	`lorentz`	x_2	$\{0, 0.05, 0.1\}$	0
11.15	`lorentz`	x_3	$\{0, 0.05, 0.1\}$	0
11.16	`lorentz`	x_2	$\begin{bmatrix} 0.05 \\ 0.1 \\ 0.05 \\ 0.1 \end{bmatrix}$	$\begin{bmatrix} 0.025 \\ 0.025 \\ 0.05 \\ 0.05 \end{bmatrix}$
11.17	`lorentz`	x_3	$\begin{bmatrix} 0.05 \\ 0.1 \\ 0.05 \\ 0.1 \end{bmatrix}$	$\begin{bmatrix} 0.025 \\ 0.025 \\ 0.05 \\ 0.05 \end{bmatrix}$

- As far as this benchmark is concerned, **the parsimonious models (`plars` and `lassolarscv`) are the only structures that are capable of a decent generalization to unseen data** in the case of ETC-related use-case.

Results for ETC with z = x_2, σ_p = 0. and noise=0.0.

q	plars	lassolarscv	RF	XGB	KNN	DNN
50%	0.0	-100.0%	17000.0%	10600.0%	17200.0%	93400.0%
80%	0.0	-50.0%	23900.0%	13450.0%	24850.0%	131950.0%
95%	0.0	-80.0%	20140.0%	11160.0%	22200.0%	103140.0%
99%	0.01	-89.0%	22278.0%	11189.0%	23300.0%	83167.0%

Results for ETC with z = x_2, σ_p = 0.0 and noise=0.025.

q	plars	lassolarscv	RF	XGB	KNN	DNN
50%	0.04	-36.0%	383.0%	214.0%	383.0%	2350.0%
80%	0.1	-34.0%	409.0%	197.0%	423.0%	2705.0%
95%	0.24	-33.0%	329.0%	146.0%	363.0%	2197.0%
99%	0.51	-32.0%	298.0%	109.0%	318.0%	1376.0%

Results for ETC with z = x_2, σ_p = 0.0 and noise=0.05.

q	plars	lassolarscv	RF	XGB	KNN	DNN
50%	0.06	-19.0%	203.0%	108.0%	197.0%	1541.0%
80%	0.16	-18.0%	220.0%	99.0%	225.0%	1648.0%
95%	0.38	-15.0%	179.0%	70.0%	199.0%	1381.0%
99%	0.78	-14.0%	169.0%	46.0%	171.0%	943.0%

Fig. 11.6 Partial estimation results (on unseen test data) for the observation target $z = x_2$ of ETC system with no parametric dispersion and different level of noise

Results for ETC with z = x_3, σ_p = 0.0 and noise=0.0.

q	plars	lassolarscv	RF	XGB	KNN	DNN
50%	0.0	nan%	inf%	inf%	inf%	inf%
80%	0.0	-100.0%	11200.0%	5500.0%	34800.0%	279500.0%
95%	0.0	-100.0%	12050.0%	6000.0%	37850.0%	246650.0%
99%	0.0	-100.0%	8100.0%	4900.0%	26340.0%	143600.0%

Results for ETC with z = x_3, σ_p = 0.0 and noise=0.025.

q	plars	lassolarscv	RF	XGB	KNN	DNN
50%	0.01	-15.0%	231.0%	131.0%	915.0%	7662.0%
80%	0.04	-14.0%	198.0%	79.0%	736.0%	6721.0%
95%	0.1	-15.0%	156.0%	61.0%	646.0%	4986.0%
99%	0.18	-15.0%	138.0%	69.0%	629.0%	3776.0%

Results for ETC with z = x_3, σ_p = 0.0 and noise=0.05.

q	plars	lassolarscv	RF	XGB	KNN	DNN
50%	0.02	0.0%	136.0%	82.0%	505.0%	4455.0%
80%	0.07	-1.0%	113.0%	49.0%	401.0%	3838.0%
95%	0.17	-2.0%	88.0%	41.0%	356.0%	2871.0%
99%	0.3	-2.0%	76.0%	43.0%	341.0%	2073.0%

Fig. 11.7 Partial estimation results (on unseen test data) for the observation target $z = x_3$ of ETC system with no parametric dispersion and different level of noise

Fig. 11.8 Partial estimation results (on unseen test data) for the observation target $z = x_2$ of ETC system with no measurement noise and different parametric dispersion levels

Results for ETC with z = x_2, σ_p = 0.0 and noise=0.0.

q	plars	lassolarscv	RF	XGB	KNN	DNN
50%	0.0	-100.0%	17000.0%	10600.0%	17200.0%	93400.0%
80%	0.0	-50.0%	23900.0%	13450.0%	24850.0%	131950.0%
95%	0.0	-80.0%	20140.0%	11160.0%	22200.0%	103140.0%
99%	0.01	-89.0%	22278.0%	11189.0%	23300.0%	83167.0%

Results for ETC with z = x_2, σ_p = 0.05 and noise=0.0.

q	plars	lassolarscv	RF	XGB	KNN	DNN
50%	0.04	-97.0%	377.0%	233.0%	423.0%	2387.0%
80%	0.11	-98.0%	344.0%	184.0%	381.0%	2159.0%
95%	0.24	-98.0%	288.0%	148.0%	337.0%	1762.0%
99%	0.42	-98.0%	291.0%	151.0%	336.0%	1433.0%

Results for ETC with z = x_2, σ_p = 0.1 and noise=0.0.

q	plars	lassolarscv	RF	XGB	KNN	DNN
50%	0.12	-11.0%	118.0%	61.0%	125.0%	775.0%
80%	0.31	-10.0%	105.0%	47.0%	116.0%	737.0%
95%	0.65	-10.0%	98.0%	47.0%	112.0%	653.0%
99%	1.06	-8.0%	102.0%	50.0%	117.0%	579.0%

Fig. 11.9 Partial estimation results (on unseen test data) for the observation target $z = x_3$ of ETC system with no measurement noise and different parametric dispersion levels

Results for ETC with z = x_3, σ_p = 0.0 and noise=0.0.

q	plars	lassolarscv	RF	XGB	KNN	DNN
50%	0.0	nan%	inf%	inf%	inf%	inf%
80%	0.0	-100.0%	11200.0%	5500.0%	34800.0%	279500.0%
95%	0.0	-100.0%	12050.0%	6000.0%	37850.0%	246650.0%
99%	0.0	-100.0%	8100.0%	4900.0%	26340.0%	143600.0%

Results for ETC with z = x_3, σ_p = 0.05 and noise=0.0.

q	plars	lassolarscv	RF	XGB	KNN	DNN
50%	0.03	0.0%	97.0%	59.0%	352.0%	3345.0%
80%	0.1	-2.0%	59.0%	27.0%	253.0%	2789.0%
95%	0.24	-7.0%	33.0%	20.0%	208.0%	2000.0%
99%	0.45	-8.0%	22.0%	20.0%	214.0%	1489.0%

Results for ETC with z = x_3, σ_p = 0.1 and noise=0.0.

q	plars	lassolarscv	RF	XGB	KNN	DNN
50%	0.07	0.0%	33.0%	33.0%	131.0%	1401.0%
80%	0.21	-2.0%	19.0%	17.0%	91.0%	1080.0%
95%	0.53	-1.0%	9.0%	10.0%	62.0%	772.0%
99%	0.97	-2.0%	6.0%	11.0%	54.0%	532.0%

Fig. 11.10 Partial estimation results (on unseen test data) for the observation target $z = x_2$ of ETC system with different levels of parametric dispersion and measurement noise

Results for ETC with z = x_2, σ_p = 0.05 and noise=0.025.

q	plars	lassolarscv	RF	XGB	KNN	DNN
50%	0.06	-19.0%	219.0%	125.0%	246.0%	1539.0%
80%	0.16	-17.0%	202.0%	101.0%	226.0%	1525.0%
95%	0.35	-17.0%	174.0%	80.0%	206.0%	1299.0%
99%	0.59	-17.0%	182.0%	82.0%	215.0%	1081.0%

Results for ETC with z = x_2, σ_p = 0.1 and noise=0.025.

q	plars	lassolarscv	RF	XGB	KNN	DNN
50%	0.1	-19.0%	145.0%	74.0%	164.0%	829.0%
80%	0.28	-19.0%	127.0%	56.0%	144.0%	859.0%
95%	0.58	-18.0%	122.0%	56.0%	139.0%	781.0%
99%	0.93	-17.0%	131.0%	67.0%	143.0%	655.0%

Results for ETC with z = x_2, σ_p = 0.05 and noise=0.05.

q	plars	lassolarscv	RF	XGB	KNN	DNN
50%	0.08	-12.0%	155.0%	88.0%	173.0%	1140.0%
80%	0.2	-11.0%	141.0%	66.0%	160.0%	1144.0%
95%	0.44	-11.0%	118.0%	50.0%	143.0%	987.0%
99%	0.78	-9.0%	114.0%	47.0%	141.0%	775.0%

Results for ETC with z = x_2, σ_p = 0.1 and noise=0.05.

q	plars	lassolarscv	RF	XGB	KNN	DNN
50%	0.12	-11.0%	118.0%	61.0%	125.0%	775.0%
80%	0.31	-10.0%	105.0%	47.0%	116.0%	737.0%
95%	0.65	-10.0%	98.0%	47.0%	112.0%	653.0%
99%	1.06	-8.0%	102.0%	50.0%	117.0%	579.0%

Fig. 11.11 Partial estimation results (on unseen test data) for the observation target $z = x_3$ of ETC system with different levels of parametric dispersion and measurement noise

Results for ETC with z = x_3, σ_p = 0.05 and noise=0.025.

q	plars	lassolarscv	RF	XGB	KNN	DNN
50%	0.03	-3.0%	78.0%	50.0%	313.0%	3028.0%
80%	0.1	-1.0%	53.0%	28.0%	232.0%	2547.0%
95%	0.26	-2.0%	31.0%	20.0%	191.0%	1867.0%
99%	0.47	0.0%	23.0%	23.0%	199.0%	1377.0%

Results for ETC with z = x_3, σ_p = 0.1 and noise=0.025.

q	plars	lassolarscv	RF	XGB	KNN	DNN
50%	0.06	0.0%	35.0%	33.0%	144.0%	1498.0%
80%	0.2	-1.0%	22.0%	17.0%	100.0%	1165.0%
95%	0.51	-2.0%	10.0%	10.0%	68.0%	813.0%
99%	0.93	-2.0%	9.0%	12.0%	60.0%	574.0%

Results for ETC with z = x_3, σ_p = 0.05 and noise=0.05.

q	plars	lassolarscv	RF	XGB	KNN	DNN
50%	0.04	0.0%	70.0%	46.0%	259.0%	2597.0%
80%	0.12	0.0%	45.0%	27.0%	194.0%	2095.0%
95%	0.29	1.0%	29.0%	22.0%	166.0%	1532.0%
99%	0.52	1.0%	21.0%	21.0%	170.0%	1129.0%

Results for ETC with z = x_3, σ_p = 0.1 and noise=0.05.

q	plars	lassolarscv	RF	XGB	KNN	DNN
50%	0.07	0.0%	33.0%	33.0%	131.0%	1401.0%
80%	0.21	-2.0%	19.0%	17.0%	91.0%	1080.0%
95%	0.53	-1.0%	9.0%	10.0%	62.0%	772.0%
99%	0.97	-2.0%	6.0%	11.0%	54.0%	532.0%

Fig. 11.12 Partial estimation results (on unseen test data) for the observation target $z = x_2$ of `Lorentz` system with no parametric dispersion and different level of noise

Results for `lorentz` with $z = x_2$, $\sigma_p = 0.$ and `noise=0.0`.

q	plars	lassolarscv	RF	XGB	KNN	DNN
50%	0.0	700.0%	500.0%	300.0%	575.0%	25125.0%
80%	0.01	600.0%	450.0%	300.0%	520.0%	11660.0%
95%	0.03	1152.0%	303.0%	355.0%	362.0%	4500.0%
99%	0.09	876.0%	309.0%	546.0%	180.0%	2757.0%

Results for `lorentz` with $z = x_2$, $\sigma_p = 0.0$ and `noise=0.025`.

q	plars	lassolarscv	RF	XGB	KNN	DNN
50%	0.03	135.0%	38.0%	27.0%	50.0%	3765.0%
80%	0.05	140.0%	38.0%	25.0%	47.0%	2109.0%
95%	0.09	272.0%	52.0%	52.0%	55.0%	1326.0%
99%	0.18	416.0%	180.0%	174.0%	53.0%	1393.0%

Results for `lorentz` with $z = x_2$, $\sigma_p = 0.0$ and `noise=0.05`.

q	plars	lassolarscv	RF	XGB	KNN	DNN
50%	0.05	111.0%	13.0%	11.0%	15.0%	2091.0%
80%	0.09	117.0%	13.0%	12.0%	16.0%	1235.0%
95%	0.15	137.0%	15.0%	16.0%	15.0%	801.0%
99%	0.27	270.0%	65.0%	44.0%	21.0%	755.0%

Fig. 11.13 Partial estimation results for the observation target $z = x_3$ of `Lorentz` system with no parametric dispersion and different level of noise

Results for `lorentz` with $z = x_3$, $\sigma_p = 0.0$ and `noise=0.0`.

q	plars	lassolarscv	RF	XGB	KNN	DNN
50%	0.01	321.0%	-7.0%	0.0%	29.0%	321.0%
80%	0.03	342.0%	12.0%	19.0%	46.0%	365.0%
95%	0.05	460.0%	35.0%	58.0%	56.0%	540.0%
99%	0.07	994.0%	101.0%	93.0%	96.0%	976.0%

Results for `lorentz` with $z = x_3$, $\sigma_p = 0.0$ and `noise=0.025`.

q	plars	lassolarscv	RF	XGB	KNN	DNN
50%	0.02	228.0%	6.0%	6.0%	17.0%	222.0%
80%	0.04	229.0%	9.0%	11.0%	26.0%	214.0%
95%	0.06	356.0%	19.0%	34.0%	34.0%	307.0%
99%	0.09	733.0%	55.0%	62.0%	49.0%	666.0%

Results for `lorentz` with $z = x_3$, $\sigma_p = 0.0$ and `noise=0.05`.

q	plars	lassolarscv	RF	XGB	KNN	DNN
50%	0.02	146.0%	4.0%	4.0%	8.0%	163.0%
80%	0.05	145.0%	6.0%	9.0%	11.0%	172.0%
95%	0.08	254.0%	9.0%	13.0%	17.0%	330.0%
99%	0.12	560.0%	27.0%	32.0%	27.0%	570.0%

Fig. 11.14 Partial estimation results (on unseen test data) for the observation target $z = x_2$ of `Lorentz` system with no measurement noise and different parametric dispersion levels

Results for `lorentz` with $\mathbf{z = x_2}$, $\sigma_{\mathbf{p}} = \mathbf{0.0}$ and `noise`=0.0.

q	plars	lassolarscv	RF	XGB	KNN	DNN
50%	0.0	700.0%	500.0%	300.0%	575.0%	25125.0%
80%	0.01	600.0%	450.0%	300.0%	520.0%	11660.0%
95%	0.03	1152.0%	303.0%	355.0%	362.0%	4500.0%
99%	0.09	876.0%	309.0%	546.0%	180.0%	2757.0%

Results for `lorentz` with $\mathbf{z = x_2}$, $\sigma_{\mathbf{p}} = \mathbf{0.05}$ and `noise`=0.0.

q	plars	lassolarscv	RF	XGB	KNN	DNN
50%	0.01	450.0%	400.0%	233.0%	450.0%	16750.0%
80%	0.02	331.0%	350.0%	219.0%	363.0%	7356.0%
95%	0.04	814.0%	234.0%	261.0%	270.0%	2970.0%
99%	0.1	817.0%	296.0%	308.0%	200.0%	2895.0%

Results for `lorentz` with $\mathbf{z = x_2}$, $\sigma_{\mathbf{p}} = \mathbf{0.1}$ and `noise`=0.0.

q	plars	lassolarscv	RF	XGB	KNN	DNN
50%	0.05	111.0%	13.0%	13.0%	20.0%	2017.0%
80%	0.09	114.0%	11.0%	13.0%	19.0%	1300.0%
95%	0.16	141.0%	13.0%	20.0%	18.0%	836.0%
99%	0.28	237.0%	33.0%	62.0%	41.0%	793.0%

Fig. 11.15 Partial estimation results for the observation target $z = x_3$ of `Lorentz` system with no measurement noise and different parametric dispersion levels

Results for `lorentz` with $\mathbf{z = x_3}$, $\sigma_{\mathbf{p}} = \mathbf{0.0}$ and `noise`=0.0.

q	plars	lassolarscv	RF	XGB	KNN	DNN
50%	0.01	321.0%	-7.0%	0.0%	29.0%	321.0%
80%	0.03	342.0%	12.0%	19.0%	46.0%	365.0%
95%	0.05	460.0%	35.0%	58.0%	56.0%	540.0%
99%	0.07	994.0%	101.0%	93.0%	96.0%	976.0%

Results for `lorentz` with $\mathbf{z = x_3}$, $\sigma_{\mathbf{p}} = \mathbf{0.05}$ and `noise`=0.0.

q	plars	lassolarscv	RF	XGB	KNN	DNN
50%	0.02	210.0%	-5.0%	-10.0%	10.0%	230.0%
80%	0.04	272.0%	28.0%	28.0%	42.0%	292.0%
95%	0.07	361.0%	52.0%	58.0%	58.0%	348.0%
99%	0.1	652.0%	65.0%	75.0%	57.0%	599.0%

Results for `lorentz` with $\mathbf{z = x_3}$, $\sigma_{\mathbf{p}} = \mathbf{0.1}$ and `noise`=0.0.

q	plars	lassolarscv	RF	XGB	KNN	DNN
50%	0.03	122.0%	11.0%	7.0%	19.0%	144.0%
80%	0.05	153.0%	13.0%	13.0%	23.0%	168.0%
95%	0.1	227.0%	18.0%	19.0%	24.0%	256.0%
99%	0.16	409.0%	34.0%	29.0%	27.0%	412.0%

Fig. 11.16 Partial estimation results (on unseen test data) for the observation target $z = x_2$ of `Lorentz` system with different levels of parametric dispersion and measurement noise

Results for `lorentz` with $z = x_2$, $\sigma_p = 0.05$ and `noise`=0.025.

q	plars	lassolarscv	RF	XGB	KNN	DNN
50%	0.03	124.0%	59.0%	28.0%	52.0%	3383.0%
80%	0.06	134.0%	60.0%	34.0%	53.0%	1950.0%
95%	0.1	268.0%	63.0%	71.0%	63.0%	1190.0%
99%	0.18	410.0%	121.0%	145.0%	75.0%	1390.0%

Results for `lorentz` with $z = x_2$, $\sigma_p = 0.1$ and `noise`=0.025.

q	plars	lassolarscv	RF	XGB	KNN	DNN
50%	0.03	138.0%	46.0%	31.0%	62.0%	3581.0%
80%	0.05	151.0%	49.0%	38.0%	64.0%	2340.0%
95%	0.1	288.0%	59.0%	61.0%	73.0%	1458.0%
99%	0.18	412.0%	120.0%	130.0%	104.0%	1370.0%

Results for `lorentz` with $z = x_2$, $\sigma_p = 0.05$ and `noise`=0.05.

q	plars	lassolarscv	RF	XGB	KNN	DNN
50%	0.05	112.0%	14.0%	14.0%	20.0%	1988.0%
80%	0.1	120.0%	16.0%	14.0%	21.0%	1155.0%
95%	0.16	144.0%	25.0%	20.0%	22.0%	738.0%
99%	0.27	249.0%	76.0%	64.0%	24.0%	975.0%

Results for `lorentz` with $z = x_2$, $\sigma_p = 0.1$ and `noise`=0.05.

q	plars	lassolarscv	RF	XGB	KNN	DNN
50%	0.05	111.0%	13.0%	13.0%	20.0%	2017.0%
80%	0.09	114.0%	11.0%	13.0%	19.0%	1300.0%
95%	0.16	141.0%	13.0%	20.0%	18.0%	836.0%
99%	0.28	237.0%	33.0%	62.0%	41.0%	793.0%

Fig. 11.17 Partial estimation results (on unseen test data) for the observation target $z = x_3$ of `Lorentz` system with different levels of parametric dispersion and measurement noise

Results for `lorentz` with $z = x_3$, $\sigma_p = 0.05$ and `noise`=0.025.

q	plars	lassolarscv	RF	XGB	KNN	DNN
50%	0.03	130.0%	-15.0%	-15.0%	-4.0%	148.0%
80%	0.05	163.0%	-2.0%	0.0%	8.0%	186.0%
95%	0.08	262.0%	20.0%	29.0%	27.0%	277.0%
99%	0.13	505.0%	38.0%	51.0%	30.0%	525.0%

Results for `lorentz` with $z = x_3$, $\sigma_p = 0.1$ and `noise`=0.025.

q	plars	lassolarscv	RF	XGB	KNN	DNN
50%	0.02	161.0%	0.0%	4.0%	13.0%	178.0%
80%	0.05	191.0%	7.0%	11.0%	17.0%	220.0%
95%	0.09	262.0%	23.0%	26.0%	27.0%	385.0%
99%	0.14	455.0%	31.0%	38.0%	29.0%	476.0%

Results for `lorentz` with $z = x_3$, $\sigma_p = 0.05$ and `noise`=0.05.

q	plars	lassolarscv	RF	XGB	KNN	DNN
50%	0.03	107.0%	-3.0%	0.0%	3.0%	130.0%
80%	0.06	135.0%	4.0%	5.0%	9.0%	158.0%
95%	0.09	234.0%	14.0%	20.0%	23.0%	230.0%
99%	0.14	474.0%	32.0%	39.0%	32.0%	452.0%

Results for `lorentz` with $z = x_3$, $\sigma_p = 0.1$ and `noise`=0.05.

q	plars	lassolarscv	RF	XGB	KNN	DNN
50%	0.03	122.0%	11.0%	7.0%	19.0%	144.0%
80%	0.05	153.0%	13.0%	13.0%	23.0%	168.0%
95%	0.1	227.0%	18.0%	19.0%	24.0%	256.0%
99%	0.16	409.0%	34.0%	29.0%	27.0%	412.0%

In the `Lorentz` use-case only `plars` is able to address the problem conveniently. Among the remaining structures, the `DNN` show the worst results due to the small amount of data used to train the over-parameterized model being used.

- In the case of `ETC`, the `scikit-learn lassolarscv` provides generally better results than the `plars` when it comes to capture the linear relationship between the observation targets and the used features vector. Nevertheless, the difference is generally quite small.[17]
- Even in the presence of rather important measurement noise (standard deviation of 0.05 on standardized data) and 10% of relative parameter dispersion ($\sigma_p = 0.1$), 80% of the partial estimation errors indicators, provided by the `plars`-based partial estimators, remain quite small, namely:

[17] Notice that the percentage of difference is rather meaningless when the `plars` column show almost 0 values.

- <30% of the median of $|x_2|$ in the ETC use-case,
- <20% of the median of for $|x_3|$ in the ETC use-case,
- <10% of the median of for $|x_2|$ in the `Lorentz` use-case, and
- <5% of the median of for $|x_3|$ in the ETC use-case.

11.8 Conclusion

In this chapter, an example of use of parsimonious ML fitting in the design of partial estimators is provided. These estimators address the problem of reconstructing the dynamics of specific observation targets without the need for a full state estimation.

Moreover, the design methodology explicitly incorporates the case where a parametric dispersion affects the model's parameters and where the measurements are corrupted by a non-negligible measurement noise.

Obviously, when the set of observation targets covers all the non-measured states, it might be interesting to use the so-designed partial estimators in a global estimation scheme in which the nominal dynamics of the system is used together with the partial estimator in a standard prediction/correction expression in order to filter the output of the partial estimator that is mainly measurement focused.

Another possibility is to use a rolling averaging window on the output of the partial estimator in order to come out with a filtered estimation.

References

1. Alamir, M.: Partial extended observability certification and optimal design of moving-horizon estimators. IEEE Trans. Autom. Control **67**(7), 3663–3669 (2021)
2. Alamir, M.: Dataset for nonlinear state estimator design (2024). https://doi.org/10.34740/KAGGLE/DSV/10016280, https://www.kaggle.com/dsv/10016280
3. Pedregosa, F., Varoquaux, G., Gramfort, A., Michel, V., Thirion, B., Grisel, O., Blondel, M., Prettenhofer, P., Weiss, R., Dubourg, V., Vanderplas, J., Passos, A., Cournapeau, D., Brucher, M., Perrot, M., Duchesnay, E.: Scikit-learn: machine learning in Python. J. Mach. Learn. Res. **12**, 2825–2830 (2011)
4. Azmi, S.S., Baliga, S.: An overview of boosting decision tree algorithms utilizing AdaBoost and XGBoost boosting strategies. Int. Res. J. Eng. Technol **7**(5), 6867–6870 (2020)

Further Advanced Topics 12

12.1 Introduction

In this chapter, some additional formulations that can address the presence of uncertainties in nonlinear systems are invoked with much less details than in the previous chapters.

This is done for the sake of completeness and in order to keep the present book within reasonable size. Nevertheless, references that enable the readers to go deeply in the topics are systematically provided.

As it is previously mentioned in several contexts here and there in this book, there is no indisputably *better* solution to a specific problem. It is even not necessarily excluded that the designer's preferences impact the decision step regarding which solution to adopt. After all, we all better master what we appreciate and it is sometimes better to perfectly implement a non-necessarily best solution rather than loosely try to implement a non-perfectly mastered alternative.

This is why the set of frameworks and methodologies sketched in this book should be viewed as possible alternatives that might behave more or less appropriately depending on the problem statement, constraints, sizes as well as on the effectively encountered realizations of the uncertainties and the specifically used computational device among many other parameters that condition the outcome of a real-life implementation.

Let us start by mentioning a clustering approach that enables a tractable solution to the problem of **Stochastic Model Predictive Control (SMPC)**.

M. Alamir, *Nonlinear Control of Uncertain Systems*,
https://doi.org/10.1007/978-3-031-93287-8_12

12.2 Uncertainty-Clustering-Based Solutions

This class of solutions intends to address the computational burden associated to the stochastic optimal control problems formulation stated in Sect. 2.6.2. Notice that solving the above problems enables to implement SMPC feedback.

Recall that in this context, we assume that the nonlinear system can be described by a discrete-time state equation of the form:

$$x^+ = F(x, u, w), \tag{12.1}$$

in which $w \in \mathbb{R}^{n_w}$ stands for an uncertain, non-measured and non-predictable term.

Moreover, it is assumed that there is a cost function of the form:

$$J(\mathbf{u}^{(i)} \mid x, \mathbf{w}), \tag{12.2}$$

that expresses the cost of a control trajectory of some length N (the prediction horizon) should the initial state be x and the uncertainties profile be $\mathbf{w}$ over the prediction horizon.

In this section, we shall quickly *visit* two alternatives to address the problem.

12.2.1 The Multi-stage Scenario Tree Decomposition

In this approach [1,2] that has an increasingly large number of *followers*, the basic assumption stated below is considered:

Discrete Uncertainties Set
The possible values of the uncertainty vector w lie in a discrete set of moderate cardinality q that is denoted by

$$\mathbb{W} := \left\{w^{(1)}, \ldots, w^{(q)}\right\},$$

where $w^{(i)} \in \mathbb{R}^{n_w}$ for $i \in \{1, \ldots, q\}$.

This assumption obviously implies that the set of possible uncertainties realizations over a prediction horizon of length N is of cardinality q^N.

As this potentially leads to a combinatoric explosion, the following additional assumption is used:

Frozen Uncertainty Profiles
The uncertainty profile $\mathbf{w}$ becomes constant after N_r steps, referred to hereafter as the **robust horizon**.

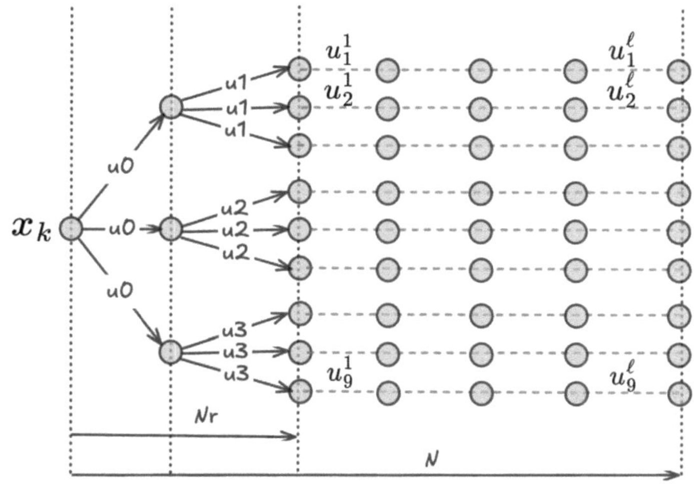

Fig. 12.1 The scenario's tree propagation scheme in the case the cardinality of the set of uncertainty values is equal to $q = 3$. Notice that $N = 6$, $N_r = 2$ and hence $\ell = N - N_r = 4$

This second assumption reduces the number of potential scenarios to

$$\text{(Number of scenarios)} \qquad n_s := q^{N_r}, \tag{12.3}$$

instead of q^N.

Figure 12.1 shows an example of the resulting tree of trajectories in the specific case where the following values are used:

$$N = 6 \quad ; \quad N_r = 2 \qquad \text{and} \quad q = 3,$$

where a control profile is associated to each uncertainties profiles with some constraints as discussed in the remainder of this section. Notice how at each node, three possible new paths ($q = 3$) are generated corresponding to the three possible values of the uncertainty vector at that specific instant in the future. For the time being we do not consider the control vector involved as this is explained later on.

Now in order to proceed with the computation of the stochastically defined cost function J one needs to know the probability ω_i that is associated to each possible value of the uncertainty profile, namely $\mathbf{w}^{(i)}$. This is the object of the third assumption:

Known Probability of Each Scenario
Each one of the possible scenarios, namely

$$\mathbf{w}^{(i)} \qquad \text{where} \qquad i \in \{1, \ldots, n_s = q^{N_r}\},$$

of the uncertainty profile comes with a known probability $\omega^{(i)} \in [0, 1]$.

Notice that this assumption is quite common and has been already encountered in the previous chapter as all the previously proposed uncertainties handling solutions were based on the knowledge of the statistics of the realizations of the uncertain items. Without such a knowledge[1] no explicit and relevant handling of the uncertainties can be imagined.

Based on the above elements, it is possible to compute the expected cost for a given initial state x and candidate control profile $\mathbf{u}$ using the following formulae:

$$\mathbb{E}\big[J \mid \mathbf{u}, x\big] := \sum_{i=1}^{n_s} \omega_i J(\mathbf{u} \mid x, \mathbf{w}^{(i)}).$$

Now we can state the principle of the scenario tree approach to handle the presence of uncertainties in dynamical systems:

The Principle

The principle of the scenario tree approach is to associate, for a given current state x, a specific control profile $\mathbf{u}^{(i)}$ to each possible scenario $\mathbf{w}^{(i)}$.

Now in order to have as many control profiles as there are possible scenario's profiles, the proposed framework introduces constraints on the control profile that can be stated as follows:

The Non-anticipation Constraint

All the control inputs starting from the same state (root) have to be equal (see Fig. 12.1).

The consequence of this constraint is shown in Fig. 12.1. Indeed, the use of the notation u_0, u_1, u_2, and u_3 recalls that since all the trajectories start at the same initial state x_k, all the control profiles have to start with some common value denoted by u_0.

The three different branches issued from the node x_k corresponds to the state trajectories corresponding to the three different possible realizations ($q = 3$) of the uncertainty vector.

This leads to three different states from which different controls can be applied, namely: u_1, u_2, and u_3. Each of these control leads to $q = 3$ different new paths depending on which instance of the uncertainty materializes and so on.

Note also that similarly to the assumption on the frozen uncertainties after $N_r = 2$ time steps, the control is also *frozen* after N_r steps and keeps constant values.

To summarize, the above mentioned assumptions and the associated control parameterization aim at deriving **tractable optimization problems** that depend on decision variables of dimension:

$$n_u \times q^{N_r+1}$$

[1] Or a relevant assumption on the underlying statistics.

and which are **totally deterministic** since the uncertainties are simply handled by enumerating all the presumably restricted number of possible values of the uncertain vector of parameters.

Now one might ask:

About the Finite Number of Scenarios?

How is it possible to consider that there are only q possible values of the uncertainty vectors and how such representative values can be determined?

It is needless to say that no one can assume that the effectively visited values of the uncertainty vector w is truly limited to q values, except for a very limited number of real-life situations if any.

The idea is to consider that there are three representative values that *correctly* span the possibilities of impacts of the uncertainties on the cost function and the constraints satisfaction in a sufficiently representative manner.

As for the concrete computational way to get these q representative values, data clustering approaches can be used either in a non-supervised way (just using similarity metrics) or in a supervised way as it is the case in the framework of the next section.

Notice however that the framework proposed in the next section is not an implementation step of the scenario tree approach described in this section but a totally independent framework.

12.2.2 Stochastic MPC via Supervised Clustering

The framework proposed in [3] is another attempt to derive a tractable approximate version of a stochastic Nonlinear Model Predictive Control (NMPC) setting. As in the previous scenario tree framework, the solution relies on performing an intermediate clustering step with two major differences:

1. First, the clustering step is done repeatedly and not once for all as in the previous section's framework.
2. Moreover, the clustering is explicitly described and is based on a supervised clustering which is defined by solving a family of deterministic Open-Loop (OL) OCP.

These differences and the overall associated framework are briefly described in the present section.

The notation regarding the general model involving uncertainties as well as the cost function is similar to the setting of the previous section. The only difference is that the setting of [3] assumes parametric uncertainties meaning that the vector of uncertainties w is assumed to be unknown but constant along the prediction horizon.

Moreover, it is assumed that the statistical law $\mathcal{W}$ governing the dispersion of w is known so that relevant random samples can be drawn for the sake of tuning the control law.

The principle of the methodology proposed by [3] lies in the following ideas:

1. A learning data $\mathcal{D}_b$ is created and managed over time that contains a set of optimal trajectories that solve each a deterministic OL OCP associated to a specific value of the parameter $w^{(i)}$.
2. This set of learning data is managed over time on a First-In-First-Out (FIFO) basis. Namely, the oldest set of trajectories is forgotten in order to introduce the newly computed one so that the maximum memory associated to the dataset is kept under control.
3. The new set of optimal trajectories is created by solving at each sampling period k a fixed number of newly stated OCP that are associated to the pair

$$(x_k, w^{[j]}),$$

for each $j \in \{1, \dots, \ell\}$. The realizations $w^{[j]}$ are randomly sampled values of the uncertainty vector (following the presumably known statistical law $\mathcal{W}$). Therefore, at each sampling interval k, ℓ new set of *optimal* trajectories of the form:

$$\left(w^{[j]}, \mathbf{u}_\star^{[j]}, J_\star^{[j]}, g_\star^{[j]}\right)_k \quad j \in \{1, \dots, \ell\}, \tag{12.4}$$

is obtained, where

- $\mathbf{u}_\star^{[j]}$ is the optimal control profile associated to the initial state x_k and the vector of parameters $w^{[j]}$.
- $J_\star^{[j]}$ is the associated optimal open-loop cost.
- $g_\star^{[j]}$ is the optimal solution's constraint's satisfaction indicator (in case soft constraints are involved in the formulation).

4. As the time index progresses, these sets of quadruplets are concatenated in the so-called learning dataset $\mathcal{D}_b$ as shown in Fig. 12.2.
5. It is assumed that the maximum size of the learning data is an integer multiple of the number of OCPs that are solved at each sampling period, namely:

$$n_b = q \times \ell.$$

Notice that the reference to *Threads* in Fig. 12.2 refers to the possibility to solve the set of sampled OCP using parallel solvers that might be executed using several instances of some replicated application on the cloud.
6. Recall that in the beginning of this chapter, it has been mentioned that the main idea is to create clusters over which the statistics of the cost function can be approximated in a tractable manner.
This is precisely where the so-collected learning data is used. As a matter of fact:

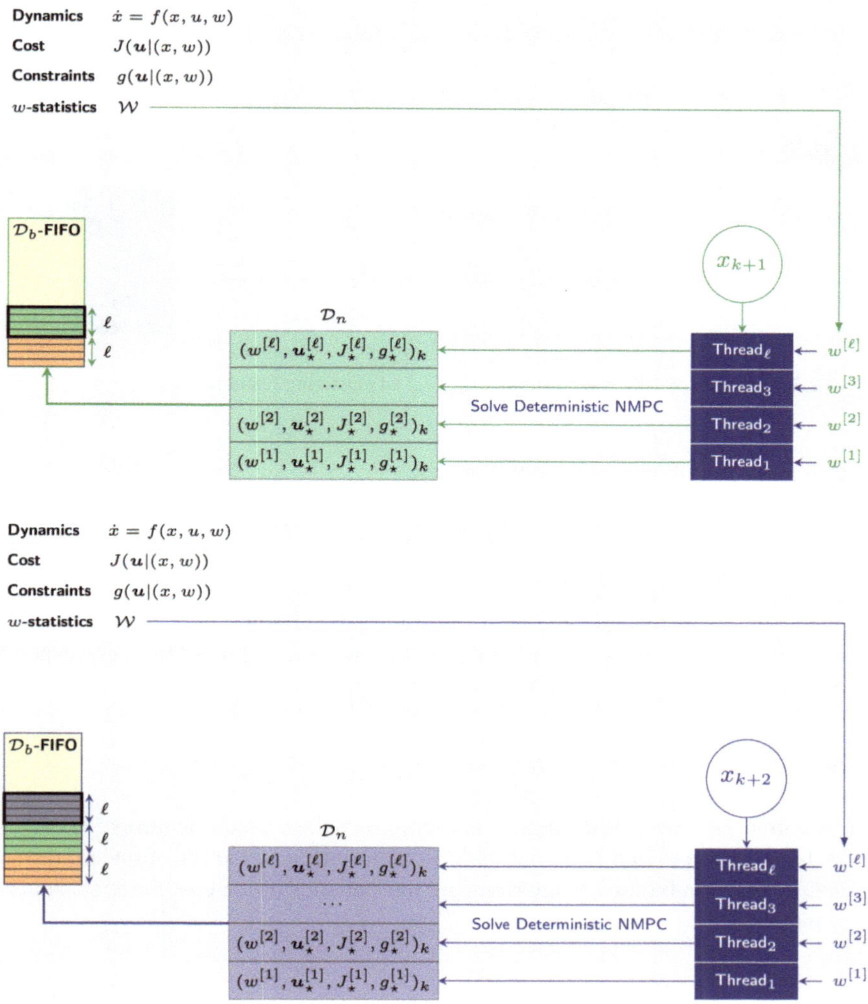

Fig. 12.2 Stochastic NMPC via clustering: creating the learning data. Two successive batches of open OCP solving to feed the clustering features

> **Optimal Solutions as Clustering Features**
> The columns of the learning data $\mathcal{D}_b$ shown in Fig. 12.2 are used as features in order to cluster the different solutions of the randomly generated OCP problems.

More precisely, assuming that the OCP are solved without constraints violation, all the $g_\star^{[j]}$ would be systematically 0 leaving as non-trivial features for clustering the

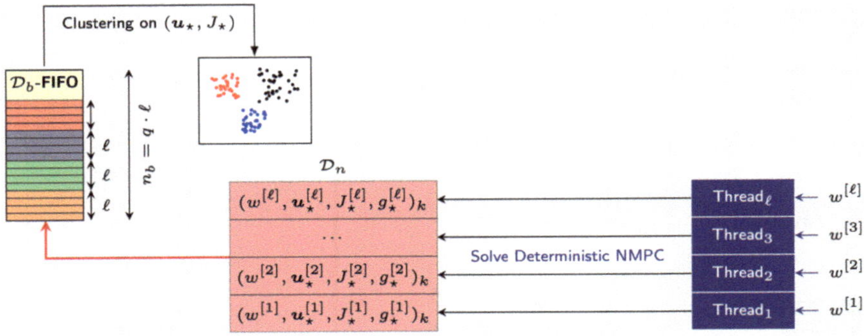

Fig. 12.3 Stochastic NMPC via clustering: $(u_\star, J_\star)$ ad clustering features

pair of control profiles and the optimal cost:

$$(\mathbf{u}_\star, J_\star).$$

This is shown on Fig. 12.3.

Now obviously, the outcome of the clustering materializes in creating a partition of the lines numbers in associated clusters and since each of these lines corresponds to a specific value of the uncertainties, it results that:

> Supervised Clustering of Uncertainties
> The **unsupervised** clustering of the optimal control related features $(\mathbf{u}_\star, J_\star)$ induces a **supervised** and control-related clustering of the set of uncertainty values that can be used to approximate the statistics of the cost functions and constraints.

This is shown in Fig. 12.4.

It is important to underline the subtleties of this clustering process. Indeed, recall that the quantities we want to cluster originally are the possible values of the uncertainty vectors w.

Clustering algorithms are based on a similarity criteria in order to separate the set of values to be clustered. Consequently, if one runs a non-supervised clustering algorithm such as `Kmeans`, `DBSCAN`, to cite but few examples that are freely available in the `scikit-learn` `Python` library [4], the resulting clusters would be based on the similarity in the space of possible values $\mathbb{W}$ without any consideration to the impact of these similarities on the control problems involving these uncertainty realizations as parameters of the problem to be solved.

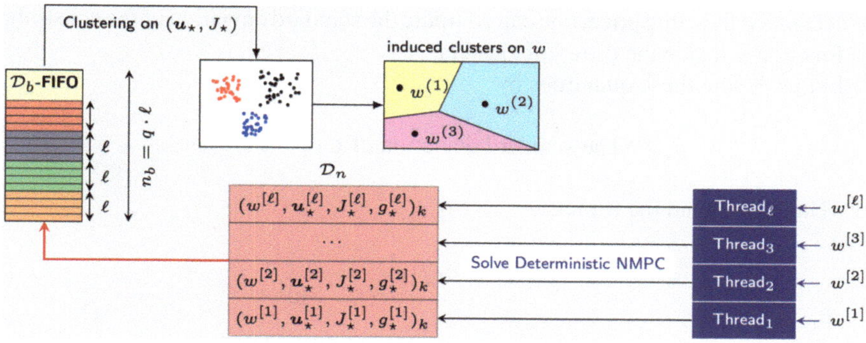

Fig. 12.4 Stochastic NMPC via clustering: induced clusters on the uncertainties. The non-supervised clustering of the control-related features induces a supervised clustering on the set of uncertainty values

On the contrary, clustering the features built on the optimal control trajectories and the cost function enables to **put in the same class all these lines that corresponds to uncertainty values $w^{(i)}$ that can be accommodate for using the same kind of control profiles and/or leading to comparable levels of the cost function.**

The above process might consider as *close* some values of the uncertainty vector that might be quite far when viewed in the *isolated* space of uncertainties without connection to the control problem at hand.

7. Now assume that the number of clusters[2] is equal to n_{cl} and let us denote the centers of the resulting clusters by (see Fig. 12.4):

$$\left\{w^{(i)}\right\}_{i\in\{1,\dots,n_{\text{cl}}\}}.$$

8. Assume also that the relative population sizes (number of lines belonging to clusters) of the different clusters indexed by

$$i = 1\dots, n_{\text{cl}},$$

are denoted by ω_i such that:

$$\sum_i \omega_i = 1.$$

Then obviously, the values ω_i can be used as estimations of the probabilities that a randomly sampled uncertainty vector belongs to the associated clusters.

[2] Either chosen or automatically determined by the clustering algorithm being use in the clustering process.

9. At *almost* the same price, one can compute the standard deviation of the constraint function g inside the different clusters.
   Let us denote these quantities by

$$\sigma_g^{(i)} \text{ : The standard deviation of } g \text{ inside cluster } i. \tag{12.5}$$

10. Having at hand the triplets:

$$\left\{(w^{(i)}, \omega_i, \sigma_g^{(i)})\right\}_{i\in\{1,\ldots,n_{\text{cl}}\}},$$

   of uncertainty vector values (centers of the clusters) and their associated probabilities and standard deviations, one can reformulate an approximated version of the constrained stochastic OCP that underlines the implementation of the stochastic NMPC as follows:

The Approximated Stochastic OCP
For each state value x, find the control profile $\mathbf{u}$ and the slack variable $\mu \geq 0$ so as to minimize the deterministic cost function:

$$\min_{\mathbf{u},\mu} \sum_{i=1}^{n_{\text{cl}}} \omega_i J(\mathbf{u}\,|x, w^{(i)}) + \rho\mu, \tag{12.6}$$

under the soft constraint

$$\forall i \in \{1,\ldots,n_{\text{cl}}\} \quad g(\mathbf{u}\,|x, w^{(i)}) + \frac{1-\varepsilon}{\varepsilon}\sigma_g^{(i)} \leq \mu \tag{12.7}$$

where the slack variable μ is used to induce always solvable problems while ε is used together with the standard deviations $\sigma_g^{(i)}$ to manage the risk of constraints violation that might be induced by the non-perfect representativity of the possible uncertainty values via the centers of the clusters.

More precisely, if $\varepsilon = 1$ no consideration is given to the inter-cluster dispersion while for small ε the risk is taken into account in an increasing way as ε decreases.

It is worth underlying that the above presentation includes some simplification steps compared to the original complete version of [3]. For the details of the complete implementation, the reader can refer to the above cited work where the framework is applied to the problem of drug dosage in the combined therapy of cancer.

12.3 Learning-Based Sensitivity Analysis and Feedback Design

When facing the challenging task of designing feedback control for uncertain non-linear systems with a high number of uncertain parameters with partial measurement of the state, the problem of determining which are the components of the state and parameter vectors that highly impact the outcome of the closed-loop is an important question.

This paradigm can be referred to using the term *global sensitivity analysis* by opposition to sensitivity analysis that is based on local gradient estimation which holds only locally.

Such an estimation of the sensitivity of the results to different impact items might be beneficial for different reasons that are briefly discussed hereafter:

1. Determining the key components of the vector of parameters that highly impact the outcome might be useful in reducing the dimension of the extended state in an extended observer design framework.[3]
   More precisely, if such a global sensitivity analysis shows that a high number of model's parameters seems to be of almost no impact (as far as their dispersion domain around their presumable nominal values is concerned), then these parameters can be frozen to their nominal values leading to a decision variable (in the extended observer formulation) that is of much lower dimension.[4]
2. It is shown hereafter that a by-product of the sensitivity analysis is the derivation of a low-dimensional vector of features that enables to fit a robust model for the success/failure of the parameterized feedback design.
   This classifier enables to predict the outcome of a control strategy over a wide region of the extended state/parameters vector. Coupling this information with the presumably known statistics of the parameters dispersion, one can get a dashboard of success/failure which provides an overview of the possible outcomes for all initial states within a region. This can be a valuable decision-making tool.
3. Moreover, since a subset of the parameters involved in the previous study involves **control design parameters**, it goes without saying that having the previously mentioned dashboard provides, for the same price, a way to design a **state-dependent** adaptive control which increases the rate of success and the quality of the closed-loop.

It is worth recalling that the presentation of the present section provides the main guidelines of the method. For the full details and a precise instantiation of the steps on the quite involved case of the 6-states/24-parameters model of **triple therapy**

[3] See Chap. 9.

[4] For example, it is shown in [5] that using the methodology of this section, the results show that among the 24 parameters that are involved in the model, only 4 parameters really impact the outcome of the closed-loop. As for the state components, only 2 over the 6 states should be available for the output feedback design.

of cancer involving chemotherapy, vaccine, and immunotherapy, interested readers might consult [5].

So let us first state the problem and the associated working assumptions that are needed for the framework to be applied. This is the aim of the next section.

12.3.1 Working Assumptions

It is assumed that we dispose of a dynamic model of the form

$$x^+ = f(x, u, p_{\text{model}}), \tag{12.8}$$

where $p_{\text{model}} \in \mathbb{R}^{n_p}$ is the vector of parameters involved which are highly uncertain. $u \in \mathbb{R}^{n_u}$ is the vector of control inputs (manipulated variables).

The second assumption is that one disposes of a parameterized feedback law of the form:

$$u = K(x, p_{\text{ctr}}), \tag{12.9}$$

that is designed based on a deep understanding of the system logic and the role assigned to the components of u in the expression of the dynamical system. The vector p_{ctr} involved in (12.9) is a vector of control design parameters.

Regarding the Control Design Vector p_{ctr}

Notice that the vector of control design parameters can include any possible tunable parameters such as gains, thresholds, set-points, constraints-related weights to cite but few examples.The state of mind should be the following:

There should be an appropriate combination of values of these parameters to yield good results,finding these values is to be done in a next later step!

Notice that an example of such a feedback design has already been shown in Chap. 4 regarding the control of the Electronic Throttle Controlled system (see Sect. 4.3). Two other examples can be examined in [6,7] regarding two different model of multi-drugs cancer therapy.

For the time being, we keep the presentation at a general and conceptual level by assuming that a problem-dependent control design step involving a vector of parameters p_{ctr} has been achieved leading the state feedback law expressed by (12.9).

Another example of such design parameters can be the saturation level on the control in case the dimension of the actuator is not yet decided and where the currently studied framework is used in the purpose of determining the actuator sizing.

By applying this feedback law to the dynamical system (12.8), the following autonomous (closed-loop) dynamical system is obtained:

$$x^+ = f(x, K(x, p_{\text{ctr}}), p_{\text{model}}). \tag{12.10}$$

Notice that for a given pair $(p_{\text{model}}, p_{\text{ctr}})$ of model's parameters and control design vector, and a given initial state x_0, one can compute the resulting closed-loop trajectories of the state and control over some simulation window.

The duration of the closed-loop trajectory can be determined a priori (fixed final time) or defined through a terminal state-dependent condition.

The Experiment

Regardless of the type of terminal condition used in the above mentioned closed-loop simulation, the latter is referred to as an *experiment*.

In both cases, it is assumed that the resulting trajectories depending on the initial state x_0 and the vectors p_{model} and p_{ctr} can be characterized as successful or not through a computable map of the form:

$$C(x_0, p_{\text{model}}, p_{\text{ctr}}) \in \{\texttt{true}, \texttt{false}\}, \tag{12.11}$$

where for instance, `true` refers to a *successful* while `false` stands for an *unsuccessful* scenario.

Obviously, the boolean function can be defined such that

$$C(x_0, p_{\text{model}}, p_{\text{ctr}}),$$

is `true` if a set of constraints defined on the trajectories are satisfied.

Some of these constraints can be defined for all instants on the trajectories while others might concern the satisfaction of some inequality that is defined in terms of the terminal state.

Another example might be associated to the integral of the used control (the total amount of drugs injected) or the sum-of-square of some variable (constrained on some economic budget), etc.

Obviously the number of possible instantiations is infinite and no specific structure is required.

Finally, it is also assumed that there is some vector of measurable quantities:

$$Y(x_0, p_{\text{model}}, p_{\text{ctr}}) \in \mathbb{R}^{n_Y}, \tag{12.12}$$

whose components are quantities for which one would like to examine or to analyze the statistical behavior under the parameterized control law.

Examples of such quantities might be

- The quantity of used drug in a drug delivery's optimization problem.
- The produced quantity in a reaction control problem.
- The consumed energy during the experiment.
- The time duration before some condition is satisfied (as in anesthesia control problem).
- etc.

Before going further in the presentation of the framework, let us summarize the working assumptions discussed so far:

The Working Assumptions
It is assumed that we dispose of the following items:

1. A dynamic parameterized model (12.8).
2. A parameterized state feedback control law (12.9).
3. A boolean function expressing the success/failure of a simulated scenario (12.11).
4. A vector of measurable quantities (12.12) defined over closed-loop trajectories for which some properties need to be optimized.

Based on the above items, it is shown in the next section how a learning data is constructed based on a given assumption regarding the statistics of dispersion of the model's parameter vector p_{model}. This dataset is later used in order to derive the different elements of the sensitivity analysis, the dashboard creation, and the feedback adaptive design.

12.3.2 Building the Sensitivity-Oriented Learning Data

The construction of the learning data is based on the execution of the following steps:

1. Choose an integer $n_R \in \mathbb{N}$ representing the number of samples in the learning data.
2. Use the presumably valid statistical law of the dispersion of the model's parameters in order to draw n_R samples of the vector of parameters p_{model}. The resulting set of values is denoted hereafter by

$$\mathbb{P}_{\text{model}} := \left\{p_{\text{model}}^{(1)}, \ldots, p_{\text{model}}^{(n_R)}\right\}. \tag{12.13}$$

3. Generate a set $\mathbb{X}$ of n_R initial states following a potentially problem-dependent argument regarding the importance of such or such domain of State Space (SS). If no such an argument is obvious, take a uniform sampling of initial states inside the hypercube of interest delimited by relevant lower and upper bounds of the state components.
   This leads to the set of states of interest denoted by $\mathbb{X}$:

$$\mathbb{X} := \left\{x_0^{(1)}, \ldots, x_0^{n_R}\right\}. \tag{12.14}$$

4. For each triplet of the form:

$$\left(x_0^{(i)}, p_{\text{model}}^{(i)}, p_{\text{ctr}}^{(i)}\right) \quad i \in \{1, \ldots, n_R\}, \tag{12.15}$$

simulate the corresponding experiment and compute the associated indicators of success/failures:

$$C^{(i)} \in \{\texttt{true}, \texttt{false}\},$$

and the quantities

$$Y^{(i)} \in \mathbb{R}^{n_Y}.$$

5. This provides a vector of boolean (two-classes) label[5]:

$$C := \begin{bmatrix} C^{(1)} \\ \vdots \\ C^{(n_R)} \end{bmatrix} \in \{\texttt{true}, \texttt{false}\}^{n_R}, \tag{12.16}$$

as well as n_Y different real label vectors defined by[6]:

$$(\forall j \in \{1, \dots, n_Y\}) : \qquad Y_j := \begin{bmatrix} Y_j^{(1)} \\ \vdots \\ Y_j^{(n_R)} \end{bmatrix} \in \mathbb{R}^{n_R}. \tag{12.17}$$

6. While the two previous labels stand for the *label* parts of two different Machine Learning (ML) fitting problems, the associated features matrix (containing n_R lines) is defined by the following matrix X:

$$X := \begin{bmatrix} x_0^{(1)} & p_{\text{model}}^{(1)} & p_{\text{ctr}}^{(1)} \\ x_0^{(1)} & p_{\text{model}}^{(2)} & p_{\text{ctr}}^{(2)} \\ \dots & \dots & \dots \\ x_0^{(n_R)} & p_{\text{model}}^{(n_R)} & p_{\text{ctr}}^{(n_R)} \end{bmatrix} \in \mathbb{R}^{n_R \times (n_x + n_p + n_{ctr})}, \tag{12.18}$$

where n_{ctr} is the number of design parameters included in p_{ctr}.

The triplet:

$$(X, C, Y),$$

constitutes the learning data that is exploited in the following sections.

---

[5] This vector of labels is used below to fit a binary classifier for *success/failure* prediction.

[6] These labels are used to fit a multi-variable regressor in order to predict the outcome of the experiment quantitatively regarding the different quantities involved.

12.3.3 Global Sensitivity Analysis

The sensitivity analysis we are interested in here aims to answer the following question:

Sensitivity Analysis?

Among all the components of the features vector including:

- the components of the initial state,
- the model's parameters, and
- the degrees of freedom involved in the control law,

which ones strongly impact the success/failure of an experiment?

Answering this question is crucial knowing that the number of features might be quite large even for rather *small* problems.

In particular, the possibility to perform this selection in a rationale way frees the designer from the burden of censoring the possible options and ideas that might come to his/her mind during the early control design step.

In other words, as soon as the designer thinks that some design tunable parameter is potentially interesting to consider as one of the degrees of freedom used in the control law, he/she can just add it leaving to the forthcoming systematic selection task to take place following the steps described hereafter:

- Use the (X, C) learning data to **fit a binary classifier** using a Random Forest (RF) model. For instance, the `Python` model named:

  `sklearn.ensemble.RandomForestClassifier`

  of the `scikit-learn` package [4] can be used.
- The advantage of using the above suggested model is that it comes with an incorporated algorithm that ranks the involved features in a decreasing order according to their importance in the underlying fitting algorithm.
  More precisely, as a RF model is based on successive *splitting* operations each of which is based on a single feature, it seems reasonable to conclude that if a feature is never involved as a splitting variable, one can infer that its value has no or little impact on the *success/failure* of the experiment.
- For instance, assuming that the fitted classifier in the first step is denoted by `cl`, the following attribute:

  $L \leftarrow$ `cl.features_importance_`

  provides the decreasingly ordered list of features.

- Now it is possible to iterate on the ordered list L removing progressively the least relevant features and checking the quality of a newly fitted success/failure prediction model based on the only remaining features in order to come out with the least number of features that correspond to an acceptable meaningless degradation of the classification quality.[7]

Depending on which class of parameters are removed during this selection step, the consequences are different. More precisely:

- If the removed components belong to the set of control design parameters, this removal facilitates the potential use of on-line state-dependent adaptation of the controller parameters. Indeed, since this step involves on-line search for the better p_{ctr} to be used, reducing the cardinality or the dimension of this set is of a great interest.
- On the other hand, if the removed components belong to the set of components of the model's parameters vector p_{model}, this enables to reduce the number of decision variables in a possibly used extended state/parameter estimator as these removed components can be frozen to their nominal values.
- Finally, if the removed components are components of the initial state vector x_0, this enables to reduce the dimension of the *success/failure* dashboard. More precisely, it is easier to grasp success/failure conclusion looking at two-dimensional dashboard rather than at a six-dimensional one.

Here again, it is highly recommended that the interested reader takes a deeper look on the use-case analyzed in [5] in order to get a concrete realization of the above conceptual ideas.

12.3.4 Learning-Based Adaptive Feedback Design

In this section, we shall assume that the previous sensitivity analysis step provides a reduced dimensional set of control design parameters, denoted hereafter by

$$\theta := \{p_{\text{ctr}_j}\}_{j\in\mathbb{J}\subset\{1,\dots,n_{ctr}\}},$$

where $\mathbb{J} \subset \{1, \dots, n_{ctr}\}$ is the subset of indices that are kept after the previous sensitivity-based dimensionality reduction step.

Similarly, let us denote by x_0^r and x_0^c the retained components and the complementary (removed) components of the state. More precisely, x_0^r gathers the relevant and important features while x_0^c gathers the remaining (*not important features*).

The problem we are interested in here can be stated as follows:

[7] All the procedure is obviously conditioned by the fact that the original initial classifier is very good. Otherwise, the foundations that justify the material of this section are not valid.

> **Problem Statement**
> Given the reduced dimensional initial state x_0^r, design an uncertainty aware algorithm that delivers the best controller's parameters $\theta^\star$ to be used.

Notice here again the benefit of reducing the dimension of the relevant initial state x_0^r. Indeed, assuming that the above problem is solved, then one can build a *look-up table* in the reduced dimensional set of values of x_0^r by solving the above problem on a grid of values of the reduced initial state x_0^r. The outcome of this procedure is a tabulated adaptation of the feedback law without any on-line optimization.

This procedure would obviously be intractable for higher dimension of x_0^r.

Closing the above parenthesis, let us sketch the process of solving the above stated problem using the probabilistic certification framework introduced in Sect. 2.11.3.

We have already defined the design vector θ. Regarding the uncertainty vector, the following definition is used:

$$w := \begin{bmatrix} x_0^c \\ p\_\text{model} \end{bmatrix}. \tag{12.19}$$

Namely, we shall sample random instances of the remaining state x_0^c and the vector of model's parameters.[8]

The success/failure map that is mandatory in the probabilistic certification procedure in order is already defined by the expression:

$$g(\theta, w \mid x_0^r) := \texttt{cl}\Big((x_0, p_\text{model}, p_\text{ctr})|_{(\theta, x_0^r, w)}\Big), \tag{12.20}$$

where $\texttt{cl}$ is the classifier that is fitted in Sect. 12.3.3.

Notice that the notation

$$(x_0, p_\text{model}, p_\text{ctr})|_{(\theta, x_0^r, w)},$$

in the r.h.s of (12.20) refers to the features vector which is build by assigning the values of θ and x_0^r to their corresponding *important* features while using the information in w to assign the values of the remaining *less important* features.

As for the cost function to be used in the certification, it can be defined using the fitted regressor, denoted by $\texttt{reg}$, based on the label Y defined above which obviously

[8] Notice that the reduction in the dimension of the vector of parameters of the model is not needed here as the complexity of the probabilistic certification does not depend on the dimension of the uncertainty vector w.

takes the standard form:

$$J(\theta \mid x_0^r) = \frac{1}{N}\sum_{i=1}^{N} \texttt{reg.predict}(\theta, w^{(i)}).$$

Consequently, we have all the required ingredients for the randomized optimization that leads to the computation of the optimal design:

$$\theta^\star(x_0^r).$$

Notice that the number N of samples to be drawn depends on the targeted precision and confidence level following the setting already discussed in Sect. 2.11.3.

References

1. Lucia, S., Finkler, T., Basak, D., Engell, S.: A new robust NMPC scheme and its application to a semi-batch reactor example. IFAC Proc. **45**(15), 69–74 (2012)
2. Mart, R., Lucia, S., Sarabia, D., Paulen, R., Engell, S., de Prada, C.: Improving scenario decomposition algorithms for robust nonlinear model predictive control. Comput. Chem. Eng. **79**, 30–45 (2015)
3. Alamir, M.: On the use of supervised clustering in stochastic NMPC design. IEEE Trans. Autom. Control **65**(12), 5392–5398 (2020)
4. Pedregosa, F., Varoquaux, G., Gramfort, A., Michel, V., Thirion, B., Grisel, O., Blondel, M., Prettenhofer, P., Weiss, R., Dubourg, V., Vanderplas, J., Passos, A., Cournapeau, D., Brucher, M., Perrot, M., Duchesnay, E.: Scikit-learn: machine learning in Python. J. Mach. Learn. Res. **12**, 2825–2830 (2011)
5. Alamir, M.: Learning-based sensitivity analysis and feedback design for drug delivery of mixed therapy of cancer in the presence of high model uncertainties. J. Theor. Biol. **568**, 111508 (2023)
6. Alamir, M.: Learning against uncertainty in control engineering. Ann. Rev. Control **53**, 19–29 (2022)
7. Alamir, M.: On probabilistic certification of combined cancer therapies using strongly uncertain models. J. Theor. Biol. **384**, 59–69 (2015)

Part III
Applications

In this last part, several examples are given in order to illustrate the use of the tools and concepts introduced in the first two parts to solve control problems where uncertainties play a key role.

The objective of this last part is not to give a detailed presentation of the solutions. Instead, focus is made on the following aspects:

- the problem statement and the associated challenges,
- the role and the difficulties induced by the uncertainties and/or the disturbances,
- the key elements of the adopted solutions and the links to the methodologies and solutions discussed in the previous chapters, and
- the kind of results that might be achieved.

For each of the examined use-cases, references are given where interested readers can find more detailed exposition of the implementation, results as well as potential possible alternative solutions.
From a more general perspective, the objective of this part is to provide a quick overview of the wide domain of applications and the wide variety of problem settings that the tools discussed so far might address in a rather efficient way.
Beside this general overview, the main message of this part dedicated to the applications is to give instances of the following general statement:

> **There Is No Universal Solution!**
> There is no such a thing as a universal solution to a class of problems. Each problem comes with its own specific settings, difficulties and constraints. Consequently, the optimal solution for a specific problem is rarely obtained by taking an off-the-shelf algorithm that helped somebody else solving another problem.

Having this in mind, this part should be considered as a panorama of challenging use cases that go beyond the simple toy examples that are sometimes used in scientific papers on nonlinear systems and to show how the tools and algorithm presented in the first two parts can be deployed in holistic and problem-specific tailored solutions.

The best one can expect from such a presentation is to provide the reader with an accelerated experience regarding the diversity of possibilities so that when it comes to addressing his/her own specific problem, appropriate choices can be made based on ideas and combinations of tools that are encountered here and there in the following (and the preceding chapters).

This being said, it is worth mentioning that the use-cases presented in this part relate to a part of the author's personal experience which spreads over three decades. Consequently, some of the solutions presented hereafter are not fully up-to-date given the tools and concepts that have been developed since the solutions have been adopted and *sometimes* deployed.

In other words, for some of the related use-cases, different and probably *better* solutions would have been found that might address the same problem in a more appropriate way.

Nevertheless, only a part of those solutions that are still of interest and which are still relevant beyond the specific instance discussed here have been inserted in this part. It is only the full optimality of the adopted solutions that is not claimed.

Finally, it is worth underlying that some of the work mentioned in some of the following chapters are mainly due to some of my former Ph.D. students I had the pleasure to supervise. In such cases, references to their relevant publications on the topics are systematically inserted for a deeper examination by those readers who are willing to go beyond my own short rewriting of the stories!

Control of Automotive Automated Manual Transmission

13

13.1 Problem Statement

The Automated Manual Transmission (AMT) system is a technical device that enables to transmit the driver's will to change the gearbox ratio to the real transmission system and to perform smoothly the needed action so that the changes can be realized *smoothly* and without disagreable oscillations.

As such, the AMT operates similarly to the old manual transmission except that it does not require direct clutch actuation or gear shifting by the driver. The latter only has to express the desired gear ratio to the controlled system which performs the needed work in order to achieve the task while satisfying some comfort and transparency conditions.

Figure 13.1 shows a schematic view of the AMT system.

On this view, the following variables can be noticed that play a crucial role in the statement of the problem:

- The driver's torque demand which is expressed via the pedal position denoted by

$$\text{(The Pedal Position)} \qquad X_{\text{pedal}},$$

  that is translated into a torque demand via a precomputed and given dedicated *interpretation* table.
- The resulting engine torque which is denoted by

$$\text{(The engine torque)} \qquad T_e.$$

  This is the result of the action of the engine control unit.
- The rotational speed of the engine, denoted by ω_e.
- The rotational speed of the main shaft, denoted by ω_c.

M. Alamir, *Nonlinear Control of Uncertain Systems*,
https://doi.org/10.1007/978-3-031-93287-8_13

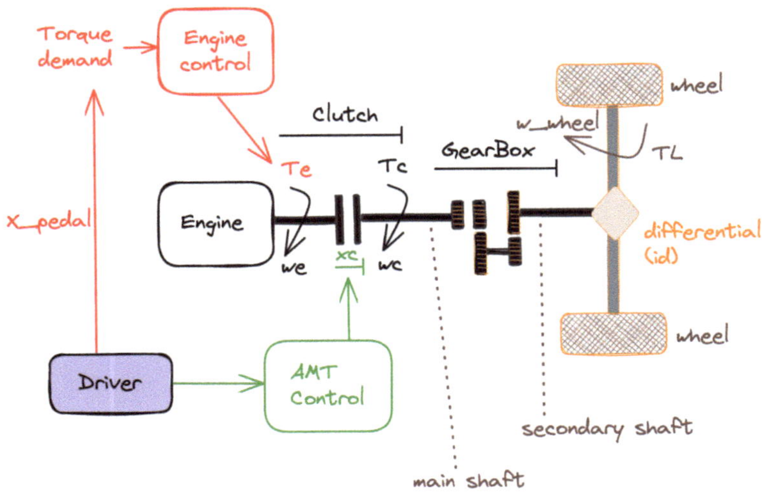

Fig. 13.1 Schematic view of the automated manual transmission system

- The transmission torque, denoted by

$$\text{(The transmission torque)} \qquad T_c,$$

  between the crankshaft (with speed ω_e) and the main shaft (with speed ω_c) is realized via the clutch unit whose state is determined by the distance x_c.
- The transmission between the main shaft and the secondary shaft (the one connected to the wheel shaft via the differential i_d) is realized via the gear-box which has its own actuators.

The main objective we are interested in here is to control the clutch distance x_c so that the changes of the gearbox ratio required by the driver can be obtained *smoothly*.

More precisely, each time a change is operated, the two rotational speeds ω_e and ω_c are instantaneously decoupled after being equal and the system is required to bring them back to a common value, with the new required gear ratio and this without too many oscillations:

$$\text{(Smooth transition):} \qquad \omega_e - \omega_c \to 0.$$

Moreover this has to be done while meeting the so-called *transparency* condition meaning that the engine speed ω_e should be compatible with the desired torque, denoted by[1]:

$$T_e^d(X_{\text{pedal}}, \omega_e),$$

[1] This is to be interpreted as follows: There exists a table named T_e^d and available to the control algorithm, that associates to each pair of pedal position $X_{\mathbf{pedal}}$ and an engine speed ω_e a corresponding value of the driver's will that is expressed in term of engine torque.

as interpreted from the pedal's position $X_{\mathbf{pedal}}$ as set by the driver at that same instant and for the current value of the engine's rotational speed ω_e.

Mathematically, for the sake of the control design, the desired torque delivered by the table T_e^d can be transformed into a set-point on the engine rotational speed that is defined by

$$\omega_e^{\text{ref}} := \max\Big\{\omega_e^0, \mathcal{T}^{-1}(T_e^d(X_{\text{pedal}}, \omega_e))\Big\}, \tag{13.1}$$

in which $\mathcal{T}^{-1}$ is a kind of *inverse map* that delivers the value of the engine speed that is compatible with the desired torque (as translated from the pedal position X_{pedal} using the relationship T_e^d).

From the above description, it comes out that two regulated variables are involved, namely the engine speed ω_e and the main shaft speed ω_c. These two variables are coupled through the value of the distance x_c between the two surfaces of the clutch mechanism.

The objective of the control is two-fold:

1. Keeping the transparency of the transmission which is viewed as a tracking problem in which the regulated variable is ω_e while the reference is given by ω_d^{ref} defined by (13.1).
2. Performing smooth transmission when gearbox ratio changes are required by the driver. This is a regulation problem in which the regulated variable is the difference $\omega_e - \omega_c$ that needs to be *smoothly* steered 0 with almost no oscillations.

This multi-variable control problem has to be handled using a sampling period that is **less than 1 ms** for obvious reactivity reasons.

This is one of the challenges of the control problem together with the one associated to the high level of uncertainties that is discussed in the next section.

Before we describe the sources of uncertainties involved in the problem stated above, let us first summarize the latter as follows:

The Statement of the AMT Control Problem
Design a feedback that delivers the appropriate value of the clutch distance x_c **at an updating rate of** 1 **msec** using the measurements ω_e and ω_c such that the following two tasks are achieved:

- The engine torque follows the reference value defined by (13.1) driven by the pedal position X_{pedal}.
- The slipping speed:

$$\omega_{sl} := \omega_e - \omega_c,$$

  is *smoothly* steered towards 0 with almost no oscillations.

In the next section, the origin of the uncertainties that impact the behavior of the dynamical system involved in the above defined problem is first examined before the technical solution that has been adopted to cope with them is briefly sketched.

13.2 The Origin and the Handling of the Uncertainties

In order to localize and understand the role of the uncertainties in the control problem described above, it is worth to have a look on the equations that govern the dynamical system one is attempting to control.

This is done in the following section.

13.2.1 The Equations Governing the System

The AMT system is governed by the following set of ODEs:

$$J_e\dot{\omega}_e = T_e - \texttt{sign}(\omega_{sl}) \cdot T_c(x_c), \tag{13.2a}$$

$$[J_c + J_{eq}(i_g, i_d)]\dot{\omega}_c = \texttt{sign}(\omega_{sl}) \cdot T_c(x_c) - T_{cw}, \tag{13.2b}$$

$$J_w\dot{\omega}_w = T_{cw} - T_L, \tag{13.2c}$$

$$\dot{\theta}_{cw} = \frac{\omega_c}{i_g i_d} - \omega_w, \tag{13.2d}$$

$$T_{cw} := \frac{1}{i_g i_d}\Big[k_{tw}\theta_{cw} + \beta_{tw}\Big(\frac{\omega_c}{i_g i_d} - \omega_w\Big)\Big], \tag{13.2e}$$

in which:

- The terms involving J's designate inertia.
- The terms involving T's are torques.
- The subscripts e, c, m, t, and w refer to engine, clutch, main shaft, transmission, and wheel respectively.
- k_{tw} and β_{tw} are elasticity and friction coefficients respectively.
- J_{eq} is the equivalent inertia of the transmission which depends on i_g and i_d associated to the gear-box.
- $\theta_{cw} := \theta_c - \theta_w$ where θ_c and θ_w denote the angular position of the main shaft and the wheel respectively.
- T_L is the load torque which is generally viewed as an unknown disturbance that includes aerodynamic resistance, tire-road frictions, etc.

Using the above equations, we can now discuss the impact of the uncertainties on the control problem we are about to address.

13.2.2 Discussion Regarding the Uncertainties

The system of Eqs. (13.2a)–(13.2e) involves many sources of uncertainties which are

- The load torque that depends on many (road/aerodynamic)-related inaccessible sources of uncertainties.
- The friction and elasticity coefficients k_{tw} and β_{tw} that are time-varying and/or badly known.

Therefore, one can imagine an approach in which bounds on the domain of variation of these quantities are first defined. Then an advanced, uncertainty aware-approach that explicitly handles the presence of these uncertainties within the prescribed bounds can be deployed.

Any of the approaches discussed in the previous chapters can be imagined, including the different approaches to implement Stochastic Model Predictive Control (SMPC) (scenario tree, on-line clustering, etc.).

This would obviously lead to intractable formulations given the available computation time (1 ms).

! Danger!
Using advanced learning-based methodologies here would have been a risky (a hopeless in fact) task! This is particularly true in for the problem described above given the prescribed control updating rate 1000 Hz!

Here is the use of a dynamic observer, as it is suggested in Chap. 4, can be of great help since it enables to totally suppress the uncertainties from the picture. The way this is done in the specific problem under interest is shown in the next section.

13.2.3 Removing the Uncertainties via State Estimation

It is always worth to examine the situation before taking a decision regarding the best control strategy to adopt. In order to orient the examination, it is always crucial to recall what is the control objective and what are the key variables that are involved in the problem statement.

Recall that in the use-case we are interested in here, we need to control the two regulated variables, ω_e and $\omega_{sl} := \omega_e - \omega_c$.

Having in mind this fact, we can focus on the dynamical equations (13.2a)–(13.2b) in which, the **only source of uncertainties is the term** T_{cw}!

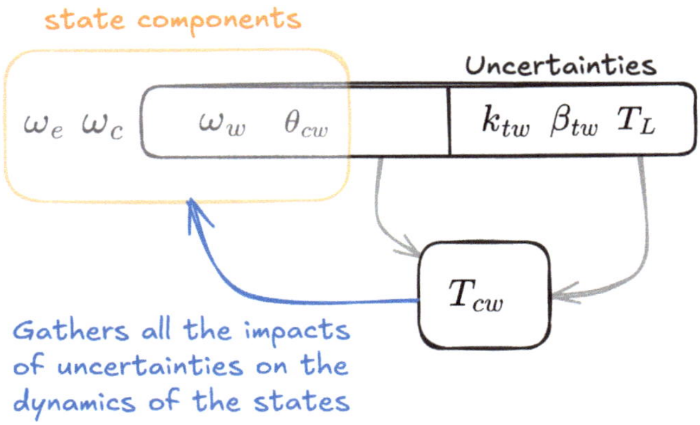

So let us focus on these two equations that we re-write for the sake of clarity:

$$J_e\dot{\omega}_e = T_e - \mathtt{sign}(\omega_{sl}) \cdot T_c(x_c), \tag{13.3a}$$

$$[J_c + J_{eq}(i_g, i_d)]\dot{\omega}_c = \mathtt{sign}(\omega_{sl}) \cdot T_c(x_c) - T_{cw}. \tag{13.3b}$$

By so-doing, we get an uncertainty-free problem provided that we can build an extended observer for the resulting *reduced* dynamics following the philosophy presented in Chap. 4. This is discussed shortly in the sequel.

As a matter of fact, we can even go further once the observer hint found its way to our minds!

Indeed, it is worth noticing that the terms T_e and $T_c(x_c)$ in the above equations are the true engine and clutch torques that are delivered by the actuators and the underlying systems.

As a matter of fact, these values are not rigorously our manipulated (control) variables as they are the outputs of other controlled subsystems. They simply result from the underlying controlled subsystems and algorithms and their respective set-points.

In addition to the above remark, it is worth noticing that measuring torques is not a common practice and the associated sensors (should they exist) might be prohibitive.

All these reasons make the following move even *more attractive*!

In fact while one cannot manipulate neither measures these torques, one can certainly manipulate and measure the associated set-points we shall denote hereafter by $u_1 = T_e^{\text{sp}}$ and $u_2 = T_c^{\text{sp}}$ respectively.

Therefore, our truly manipulated variable is defined by the torques set-points:

$$u := \begin{bmatrix} T_e^{\text{sp}} \\ T_c^{\text{sp}} \end{bmatrix}. \tag{13.4}$$

Notice that there are unavoidable differences between the set-points and the actual values of the torques which are simply the tracking errors of the respective closed-loop.

Denoting by δ_e and δ_c the corresponding tracking errors, it comes that

$$\delta T_e = T_e^{\text{sp}} - T_e \qquad \textbf{and} \qquad \delta T_c = T_c^{\text{sp}} - T_c^{\text{sp}}.$$

Injecting this notation in the simplified equations (13.3a)–(13.3b) leads to the following equations:

$$J_e\dot{\omega}_e = u_1 - \texttt{sign}(\omega_{sl}) \cdot u_2 - \overbrace{(\delta T_e + \texttt{sign}(\omega_{sl})\delta T_c)}^{\delta_e}, \tag{13.5a}$$

$$[J_c + J_{eq}(i_g, i_d)]\dot{\omega}_c = \texttt{sign}(\omega_{sl}) \cdot u_2 - \underbrace{(T_{cw} + \text{sign}(\omega_{sl})\delta T_c)}_{\delta_c}, \tag{13.5b}$$

where two blocks of unmeasured and unpredictable uncertainties have been highlighted and denoted by δ_e and δ_c leading to the following simplified model involving the two unknown and unpredictable quantities δ_e and δ_c:

$$J_e\dot{\omega}_e = u_1 - \texttt{sign}(\omega_{sl}) \cdot u_2 - \delta_e, \tag{13.6a}$$

$$[J_c + J_{eq}(i_g, i_d)]\dot{\omega}_c = \texttt{sign}(\omega_{sl}) \cdot u_2 - \delta_c. \tag{13.6b}$$

But these two uncertain terms can be estimated through a dedicated observer using the following extended system to build the observer:

$$\dot{\omega}_e = \frac{1}{J_e}\Big[u_1 - \texttt{sign}(\omega_{sl}) \cdot u_2 - \delta_e\Big], \tag{13.7a}$$

$$\dot{\omega}_c = \frac{1}{[J_c + J_{eq}(i_g, i_d)]}\Big[\texttt{sign}(\omega_{sl}) \cdot u_2 - \delta_c\Big], \tag{13.7b}$$

$$\dot{\delta}_e = 0, \tag{13.7c}$$

$$\dot{\delta}_c = 0. \tag{13.7d}$$

In this particular case where the computation time is extremely restricted, a moving-horizon observer such as the one discussed in Chap. 9 would be inappropriate because of the computation time and the need to embed quite involved solvers on board.

Fortunately, as it can be observed, the dynamics is almost linear for each pair (i_g, i_d) and each given sign of ω_{sl}. Both quantities are constant during a relevant maneuver so that a linear simple Luenberger-like estimator can be designed. Obviously even simpler observer can be derived on a line by line basis.

As mentioned in the introduction of this part dedicated to the application use-case, the details of the implementation are skipped in order to focus on the part dedicated to uncertainty handling tricks which is the objective of this book.

As far as this aspect of the design is concerned, the *Take-Home* message from this application example can be stated as follows:

> **Remove Uncertainties via State Observers**
> When achieving a regulation/tracking control objective, it might be wise to focus on the dynamics of the regulated variables and to reconstruct the uncertainty-induced group of terms, appearing in the lines describing their derivatives, via dedicated state observers.
>
> This enables to transform the uncertain problem into a deterministic problem in which some of the terms are delivered by the extended state/parameters estimation algorithm.

13.2.4 A Word Regarding the Control Design

Although a detailed presentation of the control design can be found in [2], it is worth giving some highlights on the control design and how it is connected to the tools and concepts described in the previous chapters.

First of all, the adopted control law is based on the principle of the Model Predictive Control (MPC) design introduced in Chap. 2. However, for the sake of real-time implementability and the on-board installation (no heavy numerical libraries allowed), some specific tricks have been deployed in order to cope with the above stringent requirements.

More precisely, the following items summarize the specific MPC implementation:

- Note first of all that as soon as the estimated values $\hat{\delta}_e$ and $\hat{\delta}_c$ are delivered by the state estimator at a given instant k, a short horizon **deterministic** optimal control problem can be formulated based on the first two Eqs. (13.7a)–(13.7b) in which the estimated terms are injected. This open-loop optimal control problem is then used in the definition of the MPC feedback to be implemented on line.
- Assuming that during the maneuver, i_g, i_d, and the sign of ω_{sl} are constant (which happens to be true in closed-loop!), these two equations are linear and hence analytical expressions of the trajectories of the variable ω_e and ω_c can be derived.
- Using these expressions, an analytical solution of the optimal trajectories that end with the satisfaction of the terminal constraints:

$$\omega_{sl}(N) = 0,$$

  where N is the NMPC's prediction horizon length can be obtained.
- **The prediction horizon length** N **is chosen to be the decision variable**. Indeed, using short prediction horizons might lead to the violation of the saturation constraints on the actuator while long prediction horizons can certainly be achieved while keeping the actuator levels inside their admissible range.
- This enables to perform **binary search** to find the shortest prediction horizon at each decision instant that delivers the optimal solution.

Obviously this is a *non-standard* MPC implementation! But it perfectly addresses the stringent requirements of this challenging use-case.

In the next section, some real-life results are shown which have been experimentally obtained on the real system in order to illustrate the efficiency of the proposed control design.

Regarding the MPC design and implementation, a second *Take-Home* message from this use-case might be the following one:

MPC Design and Implementation Should Be Tailored
Under stringent updating frequency and/or on-board implementability constraints, it might be necessary to use tailored MPC formulations that exploit the structural properties of the control-dedicated model. It is even wise to tailor the control-dedicated model so that it leads to efficient computation. Finally, sometimes, it is necessary to accept simplifying assumptions that have every reason to be satisfied in real-life although they might be theoretically not always guaranteed although they can be possibly certified using probabilistic certification.

Notice that this is not the first time the importance of choosing the NMPC formulation is highlighted. Indeed, this feature has been already discussed in Chap. 2.

13.3 Experimental Results

Figure 13.2 shows a first gearing up maneuver where the gearbox ratio is changed according to:

$$1 \to 2 \to 3.$$

The figure shows two y-scales. In the left one, the rotation speed ω_e and ω_c are shown in `rpm` (rotation per minute) while in the right scale, the linear speed of the vehicle is shown in `km/h`.

Notice that the maneuver lasts less than two seconds and that oscillations are barely visible after the *meeting* instant[2] between the two speeds ω_e and ω_c.

On the other hand, Fig. 13.3 shows the opposite maneuvers corresponding to the changing ratio defined by:

$$3 \to 2 \to 1,$$

where the same two observations can be made regarding the maneuver's duration and the quasi-absence of oscillations. Finally, Fig. 13.4 shows the cpu execution time associated to the different tasks where the one associated to the control algorithm

[2] This is the instant where the two speeds are equal and the clutch is closed.

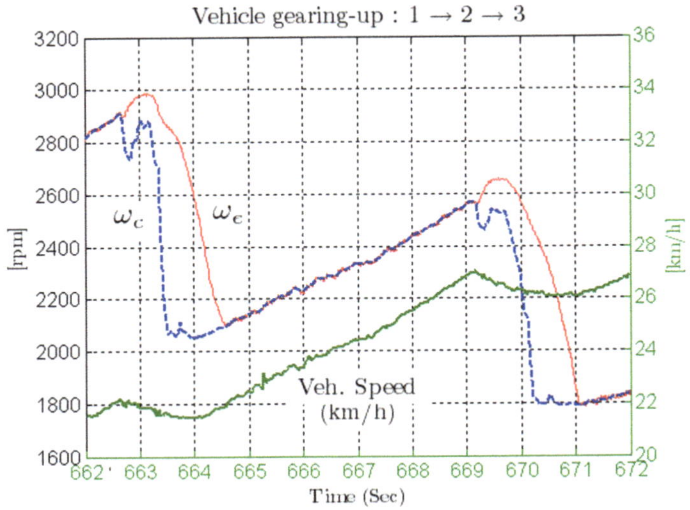

Fig. 13.2 Gearing Up maneuver 1 → 2 → 3. Note the satisfaction of the assumption regarding the unique sign of the sleep speed during a maneuver. Notice also the absence of noticeable oscillations at the speed meeting instant

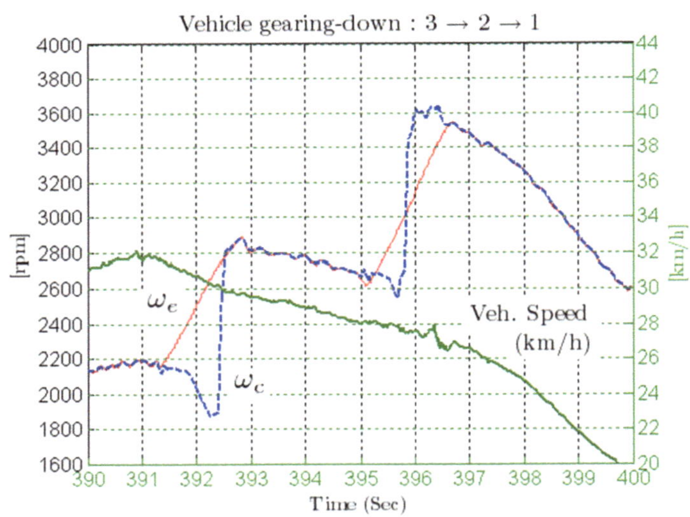

Fig. 13.3 Gearing Up maneuver 3 → 2 → 1

(including the MPC and the estimation tasks) is referred to as task 2. It is worth mentioning that the computation times correspond to an implementation that dates back to 2006!

The figure clearly shows that the time needed by the control algorithm for the computation at each sampling instant is systematically lower than 500 μ-s which is obviously real-time compatible with the required updating rate of 1 kHz.

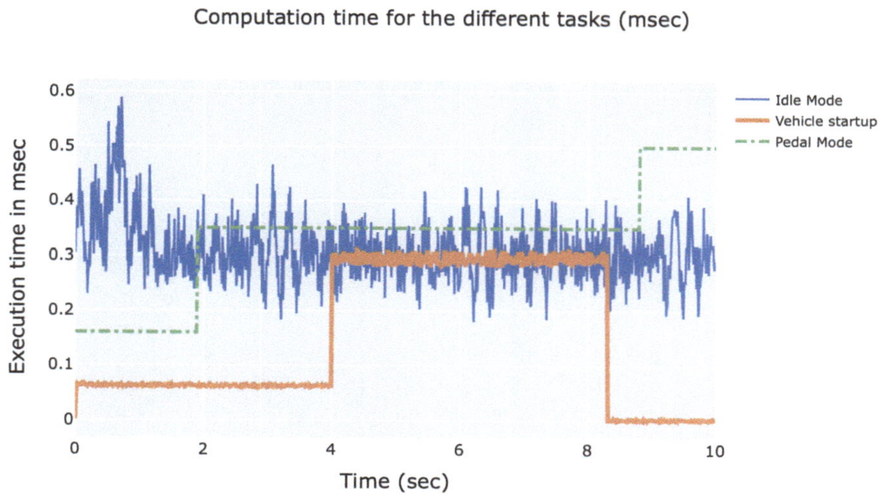

Fig. 13.4 cpu-time for the different tasks inside the control central unit. Task 2 is the one that is associated to the control algorithm. It shows that all the computation instances are completed within <500 μ-s

Acknowledgements The material of this chapter is based on a joint work with my former Ph.D. student Rachid Amari and my colleague and Friend Paolo Tona from the IFP Renewable Energies (IFPEN) who co-supervised the Ph.D. work [1]. While the concepts involved in the finally adopted solution were elaborated collectively, all the tedious implementation and validation works have been exclusively done by Rachid Amari.

References

1. Amari, R.: Commande pr dictive unifi e de groupes motopropulseurs complexes. Ph.D. thesis (2010). http://www.theses.fr/2010INPG0140. Th se de doctorat dirig e par Alamir, Mazen et Tona, Paolino Automatique et productique Grenoble INPG 2010
2. Amari, R., Alamir, M., Tona, P.: Unified MPC strategy for idle-speed control, vehicle start-up and gearing applied to an automated manual transmission. IFAC Proc. Vol. **41**(2), 7079–7085 (2008)

Power Management in an EV-Charging Station

14

14.1 Introduction and Problem Statement

Electric Vehicles (EVs) integration in the energy network is one of the major topics to be addressed by modern societies. With the increasing adoption of EVs, it is expected that the Electric Vehicles Charging Stations (EVCSs) will spread widely.

Large EVCSs might represent a quite important consumer node in a micro-grid besides building districts and factories. The power demand of these consumers (and sometimes producers at the same time) is sometimes managed by a unique **Distributed Systems Operator** (DSO), in charge of managing the power in micro-grids in a reliable and economically profitable way.

Figure 14.1 shows the DSO in its environment including the EVCS among other entities such as renewable sources of energy, storage devices, residential buildings as well as factories and of course the electricity grid.

All these actors share interest in a rational energy management that optimizes the use of renewable energy sources by optimizing the use of resources along the day in a way that reduces the amount of energy to buy from the electric grid. A key issue in energy management is the availability of a reliable prediction of the needs of all the actors as well as a reliable prediction of the production to come from the renewable sources.

There are several models and possibilities regarding the energy market in the presence of several consumers/producers as shown in Fig. 14.1. But in some of the instantiations of such a market and as far as the EVCS is concerned, it might be required that each EVCS provides **a day-ahead** commitment expressed in terms of a **maximum power profile along the day** as shown in Fig. 14.2.

More precisely, at day $j-1$, a detailed prediction of the maximum power time-series $P_{\text{lim}}(\cdot)$ is computed as a function of the time at day j together with a commitment that the truly used power $P(t)$ by the EVCS at any instant t, of day j, is lower

M. Alamir, *Nonlinear Control of Uncertain Systems*,
https://doi.org/10.1007/978-3-031-93287-8_14

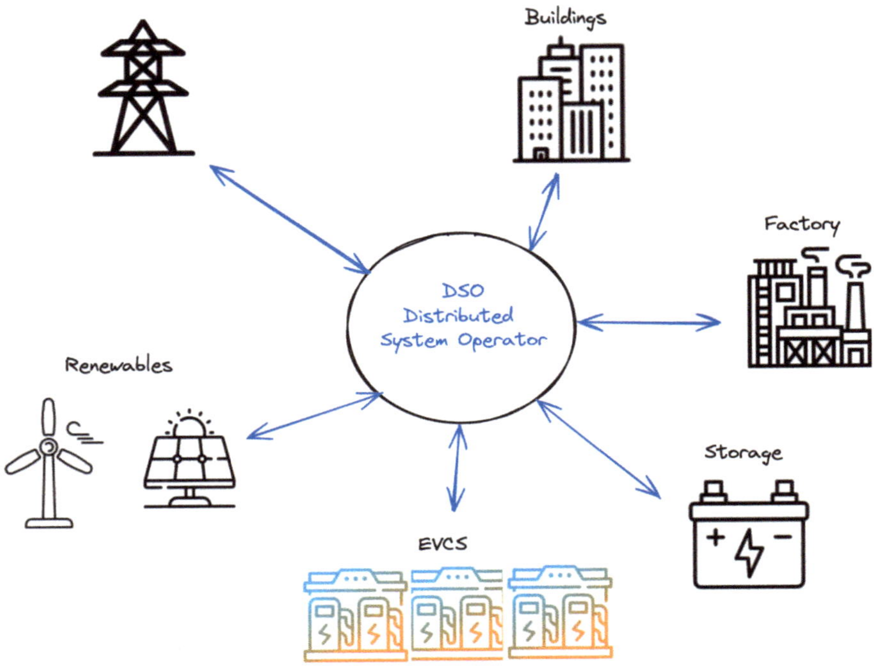

Fig. 14.1 Schematic view of a distributed systems operator in a micro-grid

than $P_{\text{lim}}(t)$:

$$(\forall t) \qquad P(t) \leq P_{\text{lim}}(t). \tag{14.1}$$

Notice that in the market's model we are discussing, the maximum power profile P_{lim} cannot be revised during day j. Only the newly sent profile for day $j+1$ can be re-defined at day j based on the previous day's outcomes.

Now the question is

What Determines P_{lim}?

What are the criterion for the determination of the maximum power profile by the EVCS manager?

Obviously, taking high values of P_{lim} enables to guarantee the satisfaction of the contractual constraints (14.1). But this might be expensive since the higher P_{lim}, the more expensive it costs to the EVCS which has to buy this power from the grid the day before.

Consequently, P_{lim} should not be too high, but just sufficient for the EVCS performance to be appreciated by its clients, namely, the EVs owners expecting their connected EV to be charged in a reasonable time and to a reasonable rate.

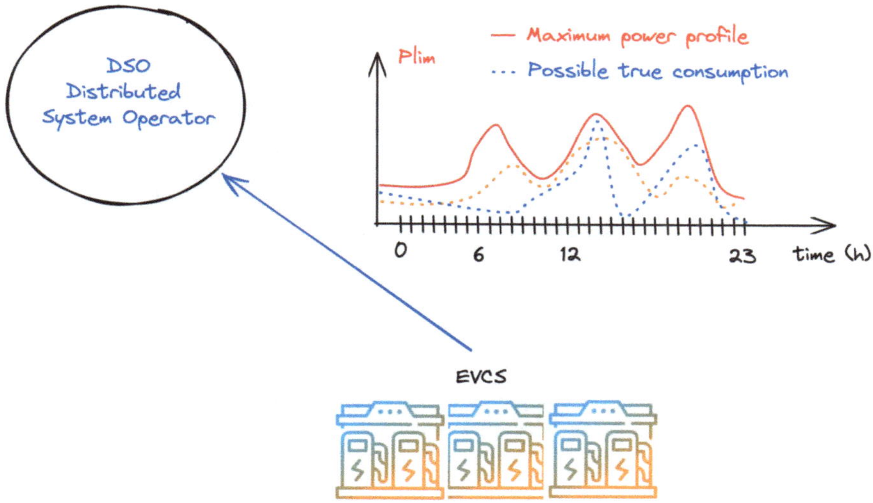

Fig. 14.2 EVCS sends the maximum power profile for the next day to the DSO

Consequently, in order to derive rationale optimization, the client's satisfaction criterion should be clearly defined through a *mathematical* formulae that can be used in tractable algorithms.

In the next section, a formulation of the EVs user's satisfaction regarding the service provided by the EVCS is proposed to be later used in the development of the power management algorithm and the associated certification.

14.1.1 The Client's Satisfaction Criterion

In [2], a satisfaction criterion has been defined that involves the following quantities:

- The ratio between the quantity of energy E_{charged} received by the EV and the amount of energy E_{req} that would be necessary to fully charging the EV:

$$\frac{E_{\text{charged}}}{E_{\text{req}}}.$$

- This ratio needs to be above some threshold c_{satis}, say 90%, should the EV remain connected more than a specific amount of time T_{th}, say 3 h.

This leads to the following condition:

$$\frac{E_{\text{charged}}}{E_{\text{req}}} \geq c_{\texttt{satis}} \times \min\left\{1, \frac{t_{\text{park}}}{T_{\text{th}}}\right\}, \tag{14.2}$$

that can be illustrated as follows:

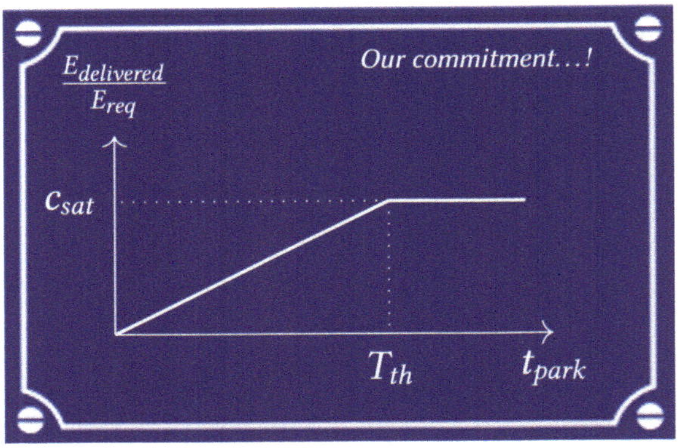

Notice that if for whatever reason, the client removes his EV from the charging station after a duration that is less than T_{th}, the satisfaction threshold defined in the r.h.s of (14.2) decreases proportionally since it is not the EVCS which is responsible for this decision!

> **A Single EV Satisfaction**
> Keep in mind that the condition (14.2) represents the satisfaction of a single EV instance. In other words, in a probabilistic certification framework, it labels the associated *experiment* for this specific vehicle as **successful**!

Obviously, the reader might already felt that the appropriate design setting here is the probabilistic certification discussed in Sect. 2.11. Regarding this framework, we just defined the criterion for the success/failure of a single *experiment*. It remains to clarify the following issues:

Remaining Issues (For Probabilistic Certification)
In order to build a complete probabilistic certification framework, we need to clarify the following items:

1. What is the random process involved and how to define the generator that enables to draw consistent random instances for the certification?
2. What is the design variable θ that is to be tuned in order to come out with an optimal design?
3. What is the cost function to be minimized by tuning the decision variable θ?

Providing answers to the items mentioned above is the road-map for the resolution of the problem of the energy management in EVCSs considered in this chapter.

Based on the above discussion, the problem can be stated as follows:

> **Problem Statement**
> Define the items mentioned above in order to come out with a probabilistic certification framework that enables the EVCS manager to provide the appropriate maximum power profile P_{lim} that minimizes the cost while satisfying the EVCS's client with a probability greater than $\eta = 95\%$!

It is important to distinguish between the 90% that was used to define the individual EV satisfaction (the threshold on the ratio between the provided energy and the required energy within the duration T_{th} on one hand and the $\eta = 95\%$ which is the precision to be used in the probabilistic certification on the other hand.

Please refer to Sect. 2.11.3 for a reminder of the meaning of the precision parameter η in the probabilistic certification framework.

In the following section, a particular solution to the problem stated above is provided following the work by [2].

14.2 Probabilistic Certification-Based Design for EVCS

In this section, the items mentioned above are successively addressed. Let us start by analyzing the random process that justifies the use of the probabilistic certification paradigm.

14.2.1 Analysis and Modeling of the Random Process

It is needless to say that the uncertainties involved in this use-case are linked to following uncertain items:

- the arrival times of EVs,
- the parking duration of the different EVs,
- the State of Charge (SoC) of their batteries upon arrivals, and
- the number of EV to be expected during the day j.

Obviously, one can use assumptions and common-sense probabilities to define the sampling process of the above mentioned quantities but nothing is better than real-life data to capture the randomness of the underlying processes.

Fortunately, thanks to `Schneider Electric` Company, we had access to the `e-car` experimental EVCS situated at the `Euref` campus in `Berlin`, Germany.

These statistics are shown in Fig. 14.3 which displays the histograms of the four variables

$$t_{\text{arr}}, t_{\text{park}}, E_{\text{req}}, n_{\text{arr}},$$

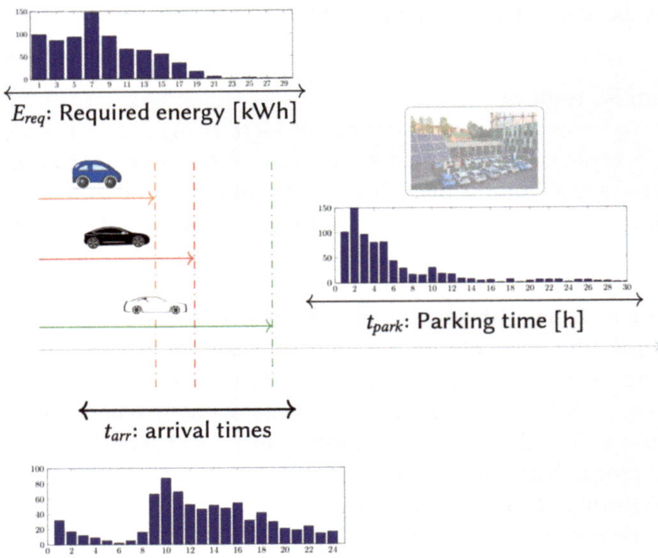

Fig. 14.3 The statistics of arrival and parking times as well as the required energy E_{req} for the Euref-campus, Berlin, Germany

that are required to build the scenario corresponding to a single car's experiment with the EVCS.

Notice that the variable n_{arr} stands for the number of vehicles that attempt a connection to the EVCS. Samples of this variable are obtained over a high number of days.

Now by using the original raw data (not the corresponding histogram) of these three scalar random variables, it is possible to use freely available tools that enable to fit **non-parametric probability distribution** such as the `Python scipy.stats` module's `Gaussian_kde` that can be imported and used as shown below:

```
# import the module
from scipy.stats import gaussian_kde
# fit the non parametric probability distribution
kde = gaussian_kde(data)
# use the identified object to draw new samples
num_samples = 1000
new_samples = kde.resample(num_samples)
```

where the vector `data` stands for any of the available arrays of values t_{arr}, t_{park}, E_{req} or n_{arr} we need to fit a probability distribution for.

Figure 14.4 shows comparisons between histograms of real measured data and the histograms of data generated using the identified non-parametric probability distributions for the three stochastic variables t_{arr}, t_{park}, and E_{req}.

The figure clearly shows that the conclusions that can be inferred from the future probabilistic certification step would be relevant in terms of the representativity of the sampling process.

Once the probability distributions of all the four variables are obtained, one can generate a day scenarios by

1. first sampling the number of EV in the day;
2. for each EV instance, the values of t_{arr}, t_{park}, E_{req} can be randomly sampled using the distribution laws previously identified.

This completely defines the random generator of the EV behavior along a day and answers the first of the three questions that remained to be addressed.

In the following section, the definition of the design variable is explained.

14.2.2 Definition of the Vector of Design Parameters θ

Recall that the objective of the design is to determine the maximum power profile that is to be sent at day $j-1$ by the EVCS to the DSO which defines the maximum power profile that the EVCS is allowed to use from the grid at day j.

Consequently, it is expected that the design parameter θ encompasses the information that enables this profile to be determined. Figure 14.5 shows the parameterization used hereafter which involves a vector of four parameters $\theta \in \mathbb{R}^4$ which defines a two-Gaussian mixture that are defined by the positions and the values of two maximums.

Notice that the central values of the components of the design parameter vector θ are determined by averaging the power profiles that are obtained by simulating over a high number of random scenarios (sampled using the above mentioned identified statistics) a **simple rules-based charging strategy which consists in charging the EVs as soon as they arrive with the nominal power of 3.3 kW**.

Moreover in addition to the first four parameters appearing in Fig. 14.5, a saturation level θ_5 has been added to enrich the parameterization.

! θ_5-Saturation Is Used

Notice that in addition to the four parameters shown in Fig. 14.5, a fifth parameter $\theta_5 > 0$ is used that enables a forced saturation on the original two-modes Gaussian mixture.

The resulting nominal setting is then used to define some search domain around the nominal values for use in the probabilistic certification algorithm.

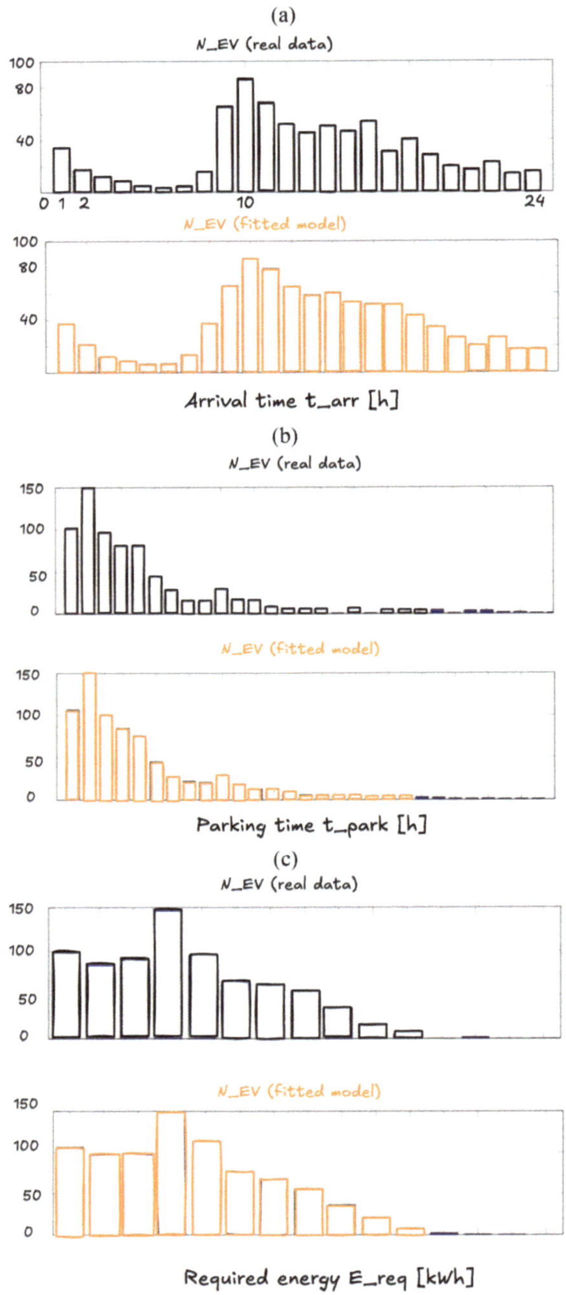

Fig. 14.4 EVCS: comparison between histograms of real and identified probability distribution. **a** t_{arr} **b** t_{park} **c** E_{req}

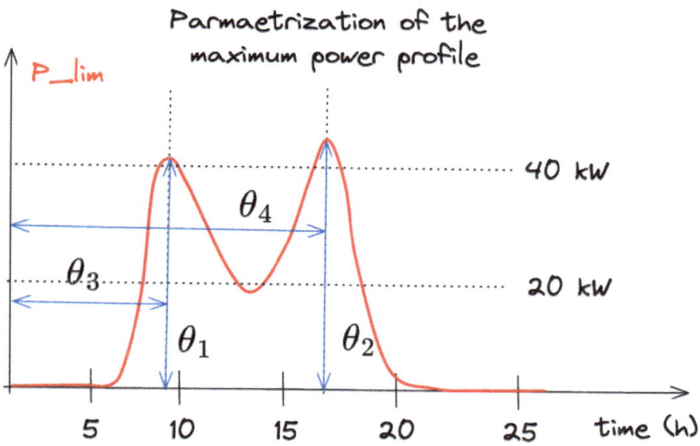

Fig. 14.5 The unconstrained parameterization of the maximum power profile to be delivered at day $j-1$ by the EVCS to the DSO for day j. A fifth parameters θ_5 is then added to saturate the original profile

The maximum power profile corresponding to a given value of the vector of parameters θ is hereafter referred to as

$$P_{\text{lim}}(\cdot, \theta),$$

where the first argument refers to the time along the day.

The same simple rules-based algorithm is applied when a candidate profile $P_{\text{lim}}(\cdot, \theta)$ is used to predict its consequence on the satisfaction of the cloud of EVs sampled using the above identified distributions.

For the sake of completeness, this simple rule-based strategy is described in the following statement:

> **K_θ-Strategy: Charging Strategy Given $\mathbf{P_{lim}}(\cdot, \boldsymbol{\theta})$**
> At each minute t_k, distribute uniformly the available power, namely $P_{\text{lim}}(t_k, \theta)$ on the connected EVs while managing the maximum accepted power per EV.
>
> In case of remaining power, distribute the remaining power to the EVs starting by those showing the lowest SoC values.

This fully defines the vector of design parameters θ and the way it impacts the power management for each experiment in the set of experiments to be used in the probabilistic certification process.

14.2.3 Definition of the Cost Function to Be Minimized

The cost function to be minimized in the probabilistic certification algorithm is represented by the pre-allocated energy declared by the EVCS manager to the DSO, namely:

$$J(\theta) := \int_{t=0}^{24} P_{\lim}(t, \theta)dt. \tag{14.3}$$

Notice that it would have been possible to use an economic cost under non-uniform power cost $c(t)$ that would be defined on a day-time basis. The corresponding cost function would be given by:

$$J(\theta) := \int_{t=0}^{24} c(t) \times P_{\lim}(t, \theta)dt, \tag{14.4}$$

without noticeable changes in the underlying methodology.

This ends the definition of all the algorithmic components that are needed to perform the optimal design using the probabilistic certification.

The concrete implementation is summarized in the next section.

14.2.4 The Probabilistic Certification Algorithm

Recall that the framework of probabilistic certification has been discussed in Sect. 2.11. It might be useful to refer to this discussion for a smooth reading of the current section.

First of all, a discrete-set Θ of candidate values of the vector of parameters θ is defined. This set is built based on different possible ratios to the central values that have been computed by averaging the power profiles corresponding to the simple rules-based strategy:

$$\Theta := \{0.9, 1.0, 1.1\}^3 \times \{0.9, 1.0, 1.1, 1.2\} \times \{0.9, 0.95, 1, 1.05\}, \tag{14.5}$$

which corresponds to a number of candidate values that equals

$$n_\Theta = 3^3 \times 4^2 = 432.$$

More precisely, if the setting $(0.9, 1.1, 1.05) \in \Theta$ is used then the value used for θ_1 is 90% of the nominal value, θ_2 is 10% higher than the nominal value while θ_3 is 5% higher than the nominal value.

Regarding the precision η, the confidence parameter δ and the number of unsuccessful samples m, the following values are used:

$$\eta = 0.05; \quad \delta = 0.05; \quad m = 20,$$

which together with $n_\Theta = 432$ when injected in the formulae[1] giving the number of samples to be drawn to get the certification leads to $N = 920$.

Since each of these scenarios needs to be simulated for each of the candidate values of the vector of parameters, the number of simulated scenarios is given by

$$N \times n_\theta \approx 4 \times 10^5,$$

which is totally tractable given the extremely infinitesimal computation time of a single *simulation*.

Recall that by *a scenario*, it is meant a day of operation in the EVCS and a one day experiment is considered to be a success if 95% of the EVs are satisfied in terms of the criterion defined above.

Now for a design parameter value θ to be admissible, less than $m = 20$ unsuccessful days should be obtained.

Among the admissible values, the one corresponding to the lowest cost function is selected. In the use-case under consideration, the optimal vector of design parameter is given by

$$\theta^\star = \begin{bmatrix} 0.9 \\ 1.0 \\ 1.0 \\ 1.2 \\ 0.95 \end{bmatrix}. \tag{14.6}$$

In the next section, this relevance of this optimal solution is evaluated from different points of views.

14.3 Validation of the Designed Solution

In this section, different aspects of the results of applying the above mentioned strategy are examined.

14.3.1 The Relevance Compared to the Rules-Based Strategy

First of all, let us examine the advantage of using the strategy described in the previous section in terms of the micro-grid market's commitment mentioned above.

More precisely, let us examine if the application of the strategy enables the EVCS to cope with its commitment regarding the maximum level of power consumed during day j compared to the application of the *blind* simple rules-base strategy consisting in charging the EVs with the nominal *direct charging* power as soon as they connect to the EVCS facility.

[1] See Sect. 2.11.

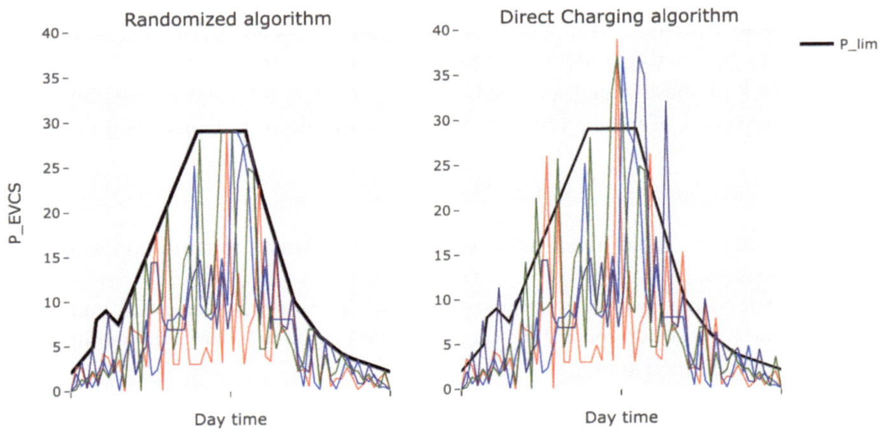

Fig. 14.6 Typical comparison between the probabilistic certification-based strategy and the direct charging strategy for three different days

The results are shown in Fig. 14.6. More precisely, three random scenarios are used under both strategies to show that the proposed strategy enables a perfect respect of the power profile sent to the **DSO** while under the direct charging strategy the actually used power profiles clear violate the specified bound.

On the other hand, although a more systematic check of the client's satisfaction rate and the tight fulfillment of the maximum allowable power is undertaken in the next section, it is worth mentioning that for the specific random scenarios involved in Fig. 14.6, bounding maximal power sent by the EVCS to the DSO is fully respected.

14.3.2 Checking the Statistics of Quality of Service (QoS)

In this section, the conformity of the QoS's statistics to the one expected by the probabilistic certification is validated through extended simulations.

More precisely, 5000 days scenarios have been randomly generated and the proposed strategy has been simulated on each of them.

The number of unsuccessful scenarios was found to be equal to 1.1% of total number of scenarios. This is lower than the theoretical one used in the probabilistic certification design ($\eta = 5\%$).

Figure 14.7 shows the cloud of satisfaction of the connected EVs in the appropriate coordinates that enable to show the boundary of the satisfaction condition expressed by (14.2).

More precisely, each circle represents the outcome of charging a single EV. The EV is satisfied if the circle lies below the dashed blue line.

Notice that only from this plot, it can be noticed that the majority of the EV leaves the EVCS fully charged while an extremely tiny number of EV's outcomes violate the satisfaction constraints.

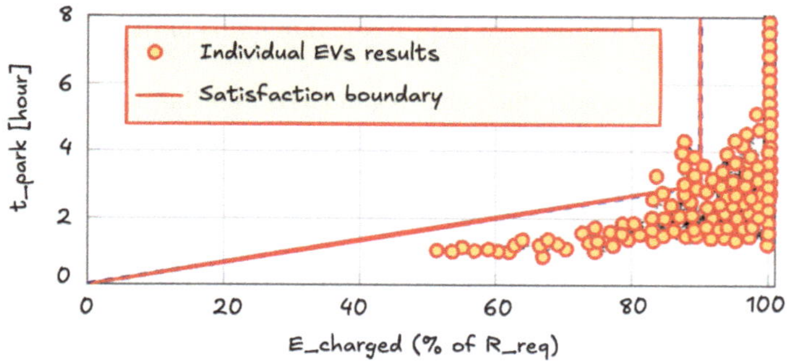

Fig. 14.7 Cloud of satisfaction of the EVs under the proposed probabilistically certified charging strategy. Each circle represents the outcome of charging a single EV. The EV is satisfied if the circle lies below the dashed bounding line

Notice that only the experiments of 20 days among the 5000 drawn days experiments are shown on Fig. 14.7.

14.3.3 The Over-Estimation Rate

Recall that the counter-part of high EV's satisfaction rate is the over estimation of the power to be bought from the grid at day $j-1$n to be used in day j. That is the reason why the rate of over-estimation of the power is a key information to be examined together with the satisfaction statistics in order to get the full picture of the quality of the strategy. To say it differently, a full satisfaction can always be obtained at very high price corresponding to a high security margin in terms of the over-estimation of the needed power.

A detailed study regarding the power's over-estimation issue is analyzed in [2]. Interested readers can examine this detailed study. Let us simply mention that this study shows that when the number of charging spots is equal to $n_{CP} = 100$, the power over-estimation is around 30% which can be considered to be rather *tight* for the targeted application.

14.3.4 The Computation Time

Using an Intel Core i7-3540M machine at 3-GHz, the total computation time needed for the computation of the probabilistically certified solution is around 40 min. This clearly means that more ambitious strategies can be computed given that this computation is done once for all.

As a matter of fact, a possibility that might be used in order to improve the quality of the solution might be to take a more rich parameterization of the maximum power profile which has been taken here to be a *saturated bimodal Gaussian* profile with few (five) parameters. A more complicated parameterization can be imagined that might enable to grasp few percentages of improvement in terms of the over-estimation of the power profile.

Another option is to keep the same parameterization but to use more eligible values in the discrete set Θ. This induces higher cardinality n_Θ and hence requires a larger number of samples for the probabilistic certification framework that can be managed at the price of a much longer computation time.

14.4 Conclusion

In this chapter it is shown that using the probabilistic certification framework, it is possible to come out with solutions that address quite non-conventional and complicated requirements.

Notice that while the parameterization of the cost function, the satisfaction constraint, and the choice of the decision variables are clearly problem-specific, the methodology can be replicated to many use-cases as it is shown in different chapters in this book.

It remained to be kept in mind that for the whole methodology to be applicable, it is important that the computation time required for the simulation of each single scenario be kept computationally cheap in order to scale in terms of the number of examined scenarios. This generally requires a set of clever and problem-dependent choices regarding the parameterized control strategy to be used.

This is another illustration of the seemingly unavoidable *no-free-lunch* theorem.

Acknowledgements The material of this chapter is based on a joint work with my former Ph.D. student Peter Pflaum and my former Ph.D. Yassine Lamoudi from the Schneider Electric Industrie Company, who co-supervised Peter's Ph.D. work [1]. While the concepts involved in the finally adopted solution were elaborated collectively, all the tedious implementation and validation works have been exclusively done by Peter Pflaum.

References

1. Amari, R.: Commande prédictive unifiée de groupes motopropulseurs complexes. Ph.D. thesis (2010). URL http://www.theses.fr/2010INPG0140. Thèse de doctorat dirigée par Alamir, Mazen et Tona, Paolino Automatique et productique Grenoble INPG 2010
2. Pflaum, P., Alamir, M., Lamoudi, M.Y.: Probabilistic energy management strategy for EV charging stations using randomized algorithms. IEEE Trans. Control Syst. Technol. **26**(3), 1099–1106 (2017)

Feedback with Probabilistic Certification for Propofol Injection During Anesthesia

15

15.1 Introduction

Using feedback strategies during surgery on humans is a crucial and delicate topics. It is *delicate* because of the generally high level of uncertainties in the underlying models and because it is almost impossible to faithfully reconstruct of values of the parameters involved in these models.

This impossibility is a consequence of the comprehensively low excitation of the involved signals. This lack of excitation is incompatible with a good reconstruction of the generally high number of parameters involved in the underlying dynamical representations of the physiological phenomena.

As a matter of fact, unlike human designed technological systems such as robots, electrical motors, engines and so, which can be tightly modeled using well established first principles of mechanics, electricity, thermodynamics and so, modeling the dynamics of biological processes is much more complicated as the underlying mechanisms are only very partially understood and even less, precisely quantified.

Even when sufficiently faithful functional structures are derived (which is rather rare), the parameters that such models involve are highly uncertain and strongly differ among individuals and even change with time and during surgery for the same patient.

The consequence of these facts is that, any candidate model-based feedback strategy that is intended to rationalize the drug's delivery needs the robustness of its performances to be assess, against these unavoidable and particularly high levels of uncertainties in the parameters.

Designing feedback control strategies for anesthesia makes no exception.

Nevertheless, the challenge is worth since successful automated process might be of great interest since it enables to:

M. Alamir, *Nonlinear Control of Uncertain Systems*,
https://doi.org/10.1007/978-3-031-93287-8_15

1. Avoid the dramatic consequences of a possible *transient* lack of attention of the anesthetist.
2. Reduce the sensitivity to patient-related estimation of the responsiveness to the drug which is mainly made dependent on some set of high level factors such as: age, sex and weight to cite few of them while the variety of individuals is too rich to be summarized by such a short list of features.

In this chapter, it is shown how the methods and ideas developed throughout the preceding chapters can be used to derive a simple and yet certifiable feedback control design for the deep anesthesia via Propofol injection.

More precisely, as there is no direct measurement of the depth of anesthesia, the Bi-spectral Index (**BIS**) measurement which is provided via commercially available EEG-based index is used as a *surrogate* measurement for the depth of anesthesia.

It is however important to underline that [1]:

! BIS Is Not Alone!

The **BIS** indicator is one among a set of multiple indicators that the anesthetist keeps monitoring. Similarly, the depth of anesthesia is not the only variable to be regulated.

The consequence of the above warning is that the topics of this chapter should be viewed as an example of a methodology that can be adapted to more complicated settings whose complete handling might go beyond the scope of the present general-scope book.

A successful anesthesia operation is characterized by the following considerations:

1. The BIS indicator should definitively enters in the interval $[40, 60]$ after few minutes (50 being the targeted value) without leaving it, except during specific transients that are fired by the surgeon's intervention (The latter are not considered in the present use-case).
2. During the operation, the BIS never goes below 40 including in the starting transient operation that steers the BIS to the above mentioned interval.

Notice that the first requirement concerns the steady state phase while the second one mainly concerns the potential transient excursions.

It goes without saying that, whatever is the proposed solution, a validation step should be undertaken using some specific model for which a cloud of parameters has to be sampled according to some statistics of dispersion. However, the detailed feedback design should be independent of the detailed information regarding the model used in the validation as this renders more credible the extrapolation of the closed-loop performance over a wider class of models.

As for the tools and concepts involved in the solution, the following items are used:

- the simple output feedback design presented in Sect. 4.4,
- the state estimation paradigm discussed in Chap. 9, and
- the probabilistic certification framework discussed in Sect. 2.11.

Note however that the observer used in the present chapter is a simple Luenberger linear observer that does not involve the concept of Nonlinear Moving-Horizon Estimator.

The remainder of this chapter is organized as follows:

- In Sect. 15.2, the general structure defining the family of models to which the feedback design is addressed is first defined and the way it is impacted by the presence of uncertainties is discussed.
- Section 15.3 is dedicated to a clear statement of the control problem to be addressed.
- The proposed solution is discussed in Sect. 15.4.
- The setting of the probabilistic certification is discussed in Sect. 15.5.
- The corresponding results are shown in Sect. 15.6.

15.2 The Structure of the Uncertain Dynamic Model

In order to introduce the general structure of a model that governs the dynamics, let us first introduce some key variables that any model of the anesthesia should involve.

To do so, recall that the class of models we are going to use belong to the set of so-called *compartmental models* which is based on the idea of having different communicating compartments between which the key quantities circulates. Each of these compartments is supposed to be homogeneous and represents a specific population of the *same kind*.

In the anesthesia class of models we focus on in this chapter, two compartments are to be distinguished, namely:

- The *effect site* compartment, denoted by the index e so that the concentration of the drug inside this compartment is denoted by:

$$\text{(Effect site drug's concentration)} \qquad C_e \qquad \text{mg/L}.$$

- The *central* compartment (the blood), denoted by the index p so that the concentration of the drug in this compartment is denoted by:

$$\text{(Blood drug's concentration)} \qquad C_p \qquad \text{mg/L}.$$

This is a common compartment-based modeling approach in which, **only the drug present in the *effect site* if effective**.

This is expressed by the fact that the BIS level directly depends on the drug concentration in the *effect site*:

$$\text{BIS} := \Psi_B(C_e \mid p_B), \tag{15.1}$$

in which p_B is a vector of uncertain modeling parameters.

The map Ψ_B is supposed to be invertible for a given value of p_B, namely, it will be possible to write the concentration of the drug in terms of the BIS level as follows:

$$C_e = \Psi_B^{-1}(\text{BIS} \mid p_B). \tag{15.2}$$

The transfer of the drug between the blood and *effect site* is governed by the following first order equation:

$$\dot{C}_e(t) = k_{e0}\Big[C_p(t) - C_e(t)\Big], \tag{15.3}$$

which simply models the diffusion-induced dynamics between the injection of the drug and its effect on the BIS indicator.

As for the dynamics of C_p, it is supposed to be governed by the following general nonlinear equation:

$$\dot{C}_p = \Phi\Big(\xi, C_e, C_p\Big) + v(t), \tag{15.4}$$

where

- $\xi \in \mathbb{R}^{n_\xi}$ is a time-varying **non measured** quantity with unknown dynamics defined by the unknown map Φ. This term gathers many non modeled dynamics and quantities including the *leaks* of drug from the blood to some of the body's tissues.
- v stands for the normalized drug injection intensity (mg/min) which is supposed to belong to the admissible set $\mathbb{V}$:

$$\mathbb{V} := [0, \bar{v}], \tag{15.5}$$

in which $\bar{v}$ stands for the maximum drug's injection rate.

The last component of the general structure of the model is given by the dynamics that governs the internal quantity ξ involved in (15.4). This unknown dynamic is supposed to take the following general form:

$$\dot{\xi} = F(\xi, C_e, C_p, v, p_\phi), \tag{15.6}$$

where $p_\phi \in \mathbb{P} \subset \mathbb{R}^{n_\phi}$ is a vector of parameters. Notice that the last form does not correspond to any specific assumption except the regularity induced by ξ being governed by a differential equation.

The need to explicitly invoke the expression (15.6) is formal so that we can proceed in the presentation of the solution and the simulator that we shall use in the validation step.

The above discussion can be summarized as follows:

Structure of BIS↔Propofol the Dynamics

It is assumed that the dynamics of the BIS, serving as an indicator of the level of anesthesia, under the action of the Propofol drug's injecting (at rate v) is governed by the following system of equations:

$$\dot{C}_e = k_{e0}\Big[C_p - C_e\Big], \tag{15.7a}$$

$$\dot{C}_p = \Phi\Big(\xi, C_e, C_p\Big) + v(t), \tag{15.7b}$$

$$\dot{\xi} = F(\xi, C_e, C_p, v, p_\phi), \tag{15.7c}$$

$$\text{BIS} = \Psi_B(C_e \mid p_B), \tag{15.7d}$$

where

- k_{e0}, p_B, p_ϕ are highly uncertain parameters,
- Φ, F are unknown functions, and
- Ψ_B is a known function involving the unknown vector of parameters p_B.

It is important to underline that (15.7) provides only the structure of the model, almost all the ingredients of this structure are unknown including the dimension n_ξ and the definition of the internal *sub-state* ξ and obviously, its governing dynamics.

On the other hand, it is quite clear that this means that the above structure can represent a wide range of possible nonlinear complicated dynamics that might be used to derive a faithful model that might fit a large number of specific patients reaction to the drug administration.

Nevertheless, it is important to underline the previously discussed implicit assumption:

! **Underlying Representativity Assumption**

The forthcoming developments are relevant to the problem only if any possible patient reaction to drug administration can be captured by the above model's structure for a specific choice of the unknown maps and parameters.

In what follows we consider that an instantiation of the model is given if instantiations of the unknown maps Φ and F are defined together with sampling statistics for the unknown remaining parameters, namely:

Model's Instantiation
An instance of the class of models described by (15.7) is obtained once the following quantities are provided:

1. instances of the unknown maps Φ and F,
2. a probability distribution $\mathcal{P}_\phi$ that enables to randomly draw samples of the unknown vector of parameters p_ϕ, and
3. two probability distributions $\mathcal{P}_B$ and $\mathcal{P}_{e0}$ that enables to get random realizations of p_B and k_{e0} involved in Eqs. (15.7d) and (15.7a) respectively.

Having at hand all the above mentioned definitions, notation and assumptions, we have all that we need to state the problem addressed in the present chapter.

This is done in the next section.

15.3 Problem Statement

The objective of the control design is to provide a dynamic output feedback that uses the only available information represented by the BIS indicator (EEG-based estimation of the anesthesia depth) in order to compute the rate with which the drug (Propofol) has to be injected leading to statistically certified outcomes.

When reading the following problem statement, it is very important to appreciate the complexity of the stated problem compared to standard regulation/tracking control problems that the control community was working on during decades.

To be more specific, the difference does not lie in the complexity of the equations or not even their size, it is all about the expression of what is expected from the control design to achieve:

> **Problem Statement**
> Assuming that:
>
> - The BIS is measured.
> - An instance of the model is considered together with associated statistics of dispersion of the parameters.
> - A duration T for the surgery is given.
>
> The objective of the design is twofold:

1. Design a dynamic output control law of the form:

$$\dot{z} = g(z, \text{BIS}) \tag{15.8a}$$
$$v = K(z, \text{BIS}) \tag{15.8b}$$

   that steers the BIS inside the interval

$$I = [50 - \delta_B, 50 + \delta_B],$$

   after a finite (short) time

$$\tau_r >, 0$$

   and maintain it inside I over the remaining interval $[\tau_r, T]$ while **never crossing the lower bound** $50 - \delta_B = 40$.
2. Provide an acceptably small upper bound $\bar{\tau}_r$ on the settling time τ_r mentioned above such that the following condition holds regarding the probability of τ_r being lower than $\bar{\tau}_r$:

$$\text{Pr}\Big[\tau_r \leq \bar{\tau}_r\Big] \geq 1 - \eta. \tag{15.9}$$

In the following section, the guidelines of the proposed solution are sketched and commented.

For full details regarding some technicalities, interested reader might refer to [2].

15.4 The Proposed Solution

Following the problem statement, there are two steps in the road map leading to the control design, namely, the definition of the structure of the feedback (very problem-dependent) on one hand and the certification step on the other hand.

More precisely, first we shall examine the structure of a feedback law that involves some parameters that remain to be tuned. This tuning is done through probabilistic certification process that is then discussed.

Notice that this feedback design that is problem-dependent recall the spirit of the design that we examined in Chap. 4 dedicated to the promotion of standard control methodologies that better fit a class of problems in which the regulation under uncertainties is the main goal. As far as the use-case of the present chapter is concerned, such a design represents only a step within a holistic design that incorporate different other modules, including the probabilistic certification.

So let us first examine the kind of *parameterized* feedback law that might address the control problem as stated above. This is the aim of the next section.

15.4.1 Feedback Design

In this section, the following notation are used to refer to the measured and regulated BIS and the sub-state of interest:

$$y = \text{BIS}; \quad x := \begin{bmatrix} C_e \\ C_p \end{bmatrix}.$$

Using this notation, Eqs. (15.7a)–(15.7b) can be rewritten as follows:

$$\dot{x}_1 = k_{e0}(-x_1 + x_2) \qquad k_{e0} \text{ uncertain,} \tag{15.10a}$$
$$\dot{x}_2 = \Phi + v \qquad \Phi \text{ unknown,} \tag{15.10b}$$
$$y = \Psi(x_1 \,|\, p_B). \tag{15.10c}$$

Recall that Φ is a time-varying term that is driven by a dynamically varying term ξ through the relationship (15.7c) that involves a vector of unknown parameter p_ϕ.

On the other hand, the evolution of the measured quantity y = BIS involves the unknown parameters that are gathered in the following notation:

$$p_c := \begin{bmatrix} p_B \\ k_{e0} \end{bmatrix}.$$

In what follows, the notation p is used to denote all the uncertain parameters involved in the model, namely:

$$p := \begin{bmatrix} p_c \\ p_\phi \end{bmatrix}. \tag{15.11}$$

Since the instantiation of the model includes the information regarding the statistics of dispersion of all the above mentioned parameters through $\mathcal{P}_B$, $\mathcal{P}_{e0}$ and $\mathcal{P}_\phi$, it is possible to induce statistical law for the sampling of the whole vector of parameters p. This sampling law is denoted hereafter by $\mathcal{P}$.

Notice that these statistics come with *nominal* values of the vector of parameters p_c, namely:

$$p_c^{\text{nom}} := \begin{bmatrix} p_B^{\text{nom}} \\ k_{e0}^{\text{nom}} \end{bmatrix}. \tag{15.12}$$

Using the nominal value p_c^{nom}, it is possible to use (15.2) in order to produce a *virtual sensor* for the quantity x_1, namely

$$\hat{x}_1 = \hat{C}_e = \Psi_B^{-1}(y \,|\, p_B^{\text{nom}}).$$

This virtual measurement can be used together with the state equation (15.8a) to build a state estimator that enables to reconstruct an estimation $\hat{x}_2$ of x_2.

More precisely, this state estimator takes the following standard form:

$$\dot{\hat{x}} = \begin{bmatrix} -k_{e0}^{\text{nom}} & k_{e0}^{\text{nom}} \\ 0 & 0 \end{bmatrix} \hat{x} + \begin{bmatrix} 0 \\ v \end{bmatrix} + \begin{bmatrix} L_1 \\ L_2 \end{bmatrix} \left[\Psi_B^{-1}(y|\ p_B^{\text{nom}}) - \hat{x}_1 \right], \tag{15.13}$$

where the matrix

$$L = \left[L_1\ L_2 \right]^{\top},$$

is a standard Luenberger observer gain [3].

Notice that in this state estimator, the assumption $\dot{\Phi} = 0$ is used, which can still lead to acceptable observer performance provided that sufficiently high gain L is used as already explained in Chap. 9. This is a standard practice in extended-state-based observer.

The previous discussion can be summarized as follows:

The State Components x_1 and x_2 can Be Approximately Reconstructed
Using the nominal values of k_{e0} and p_B it is possible to build an **approximate** reconstruction of the values of the state (x_1, x_2) of the dynamical system (15.8a)–(15.8b).

Building on this *nominal* extended observer-based reconstruction, we can focus on designing a **state** feedback for the controller system (15.8a)–(15.8b).

But before going further, the reader might legitimately ask:

! But These Are Only Nominal Values!
It is legitimate to think that the so-reconstructed values of the states x_1 and x_2 are not correct given that the observer is built on the nominal values that are by definition different from the true values of any specific patient!

So how can one be sure that anything that might be built on these incorrect estimations can be relevant?

The question is legitimate and the answer is simply: *Nobody can be a priori sure of anything! It is the probabilistic certification step that will give us the answer!*

This is precisely why this certification framework is a true *game changer* in the realm of control design as it totally *frees the imagination* during the step of building heuristics that combine the know-how of a control designer and leaves the question of fine tuning and the proof (probabilistic proof) to the last phase.

Upon terminating the certification step, in case the certification is not satisfactory, the designer can go back to such or such step in the design and take another option and then try if its consequences on the closed-loop provide better certified results.

Following this road map, let us then assume that the state $x = (x_1, x_2)^T$ is available and let us design a feedback that delivers the appropriate injection rate v.

Recall that our aim is to *regulate* $x := (x_1, x_2)$ around the *desired* value given by:

$$x^d := \begin{bmatrix} 1 \\ 1 \end{bmatrix} \Psi_B^{-1}(50 \mid p_B^{\text{nom}}), \tag{15.14}$$

as this value is the steady state of x that is compatible with

$$y = \text{BIS} = 50.$$

It is now that we need to handle the high degrees of uncertainties that one faces regarding the unknown quantities k_{e0}, p_B and Φ that are involved in the dynamical system one needs to control.

To address this issue, we shall use a known control-trick that will transform the above defined control problem into two control problems on systems of the form that has been studied in Sect. 4.4 regarding the control of uncertain scalar systems with relative degree 0.

The design of the control law uses a technique that is inspired by the well known *back-stepping* approach. In order to better understand the idea, let us rewrite (15.8a)–(15.8b) in the following form:

$$\dot{x}_1 = k_{e0}[-x_1 - e_2 + x_2^{\text{ref}}], \tag{15.15a}$$

$$\dot{x}_2 = \Phi + v, \tag{15.15b}$$

where $e_2 = x_2^{\text{ref}} - x_2$ in which x_2^{ref} is some reference value for x_2 to be designed in the sequel.

In the *back-stepping* approach, one views x_2^{ref} as a control signal that forces x_1 to track x_1^d and views v as a control signal that forces x_2 to track x_2^{ref}.

By so doing, the control problem can be split into two scalar control problems in each of which the dynamic scalar controlled system takes the following form:

$$\dot{x}_i = \alpha_i[u_i - h_i], \qquad i \in \{1, 2\}, \tag{15.16}$$

where α_i and h_i are given by:

$$(\alpha_1, \alpha_2) = (k_{e0}, 1); \quad (h_1, h_2) = (x_1 + e_2, -\Phi), \tag{15.17}$$

while the control inputs are given by:

$$u_1 = x_2^{\text{ref}}; \quad u_2 = v. \tag{15.18}$$

Notice that the form of Eq. (15.16) is precisely the one [see (4.14)] that has been studied previously in which h_i is totally unknown while α_i is unknown but guaranteed to have a strictly positive lower bound $\underline{\alpha}_i$.

Therefore we can use the same dynamic output feedback design studied in Chap. 4 in order to compute the expression of the *control*-like terms, namely, x_2^{ref} and v respectively.

Now according to methodology described in Sect. 4.4, this leads to the following control structure (Only the structure is given here to focus on the principle of the design):

$$\left|\begin{array}{ll} x_2^{\text{ref}} &= K_1(x_1, z_1) \\ \dot{z}_1 &= g_1(z_1, x_1) \end{array}\right. \qquad \left|\begin{array}{ll} v &= K_2(x_2, z_2) \\ \dot{z}_2 &= g_2(z_2, x_2) \end{array}\right. , \tag{15.19}$$

where

- At the left, the structure of the feedback law regulating x_1 around x_1^d (that is omitted in the notation for the sake of simplicity as it is simply a constant value that never changes) with the control given by x_2^{ref} and the internal state z_1.
- At the right is represented the feedback law for the regulation of x_2 around its desired value x_2^d [omitted in the notation] with the control v and the internal state z_2.

It is important to underline here that these feedback laws do not involve the use of any nominal value of the remaining parameters. Rather, they have been conceived to overcome the need for the knowledge of the parameters while leading to nice regulation performances.

It is also important to underline that despite the fact that the reconstructed value of x_2 be erroneous, the reconstruction error can be compensated for thanks to the dynamic output feedback defined above.

But again, only the probabilistic certification results can determine whether the so-designed feedback structure is able to achieve acceptable certified results.

The above discussion explained the structure and the genesis of the feedback control law. The exact instantiation of the related expressions is not mandatory. Nevertheless, it is given hereafter for the sake of completeness.

As explained in Sect. 4.4, the different instances of the feedback laws-related functions K_i, g_i involved in (15.19), need saturation levels on the *control signals* to be adopted.

As far as the virtual control x_2^{ref} is concerned, the bounds are given by:

$$x_2^{\text{ref}} \in [\underline{\gamma} x_2^d \;,\; \bar{\gamma} x_2^d], \tag{15.20}$$

where $\underline{\gamma} \in [0, 1]$ and $\bar{\gamma} \in [1, 2]$ are later used as components in the design parameter θ used in the probabilistic certification.

As for the drug injection rate control v, the obvious saturation interval $[0, \bar{v}]$ is used.

This leads to the feedback law and the associated vector of control-related design parameters θ_c described below in which the saturation map S_1 and S_2 are defined using the interval of saturation designed above:

The θ_c-Parameterized Feedback Law
The dynamic output feedback law is given by:

$$\dot{z}_1 = \lambda_{f1}\Big[S_1\big(\lambda_1(x_1^d - x_1) + z_1\big) - z_1\Big], \tag{15.21a}$$

$$\dot{z}_2 = \lambda_{f2}\Big[S_2\big(\lambda_2(x_2^{\text{ref}}(x_1, z_1) - x_2) + z_2\big) - z_2\Big], \tag{15.21b}$$

$$v = S_2\big(\lambda_2(x_2^{\text{ref}}(x_1, z_1) - x_2) + z_2\big), \tag{15.21c}$$

$$x_2^{\text{ref}}(x_1, z_1) := S_1\big(\lambda_1(x_1^d - x_1) + z_1\big), \tag{15.21d}$$

and the vector of controller design parameters to be used in the probabilistic certification is given by:

$$\theta_c := \big[\lambda_1\ \lambda_2\ \lambda_{f1}\ \lambda_{f2}\ \underline{\gamma}\ \bar{\gamma}\big]. \tag{15.22}$$

Notice that θ_c is only a part of the total design parameter θ that is used in the certification procedure. Other remaining designed parameters will be added as described later.

The last modification that is needed is the one that enables to take into account the uncertainties on the parameter p_B.

Indeed, while the dynamic output feedback laws enables to compensate for the bad knowledge of the parameters, they cannot compensate for an error on the set-point!

Such an error comes from the fact that we use a set-point on the state x_1 as a *proxy* to reach the desired value on the only output we are interested in which is precisely the BIS.

But the proxy desired value x_1^d defined by (15.14) involves the nominal value of the vector of parameters p_B and this introduces a bias on the set-point for the original $y =$ PIB.

This is the reason why a kind of *integrator action* should be introduced in order to avoid systematic errors on the set-point around which the regulation is performed via the above laws.

This can be achieved by modifying the definition of the desired state x^d from the original form (15.14) to the new form:

$$x_1^d := \Psi_B^{-1}(50 + e, |p_B^{\text{nom}}), \tag{15.23}$$

where the dynamics of the correction term e is defined by:

$$\dot{e} = \begin{cases} \lambda_I(50 - y) & \text{if } y \in \mathcal{I}, \\ 0 & \text{otherwise,} \end{cases} \tag{15.24}$$

where $\lambda_I > 0$ is the integrator constant.

Notice that in the definition of the dynamics of e, the correction is only applied inside the terminal region $\mathcal{I}$. This conditional correction enables the integrator state to be updated only in the vicinity of the desired BIS and renders the outcome less sensitive to the value of λ_I.

Nevertheless, this parameter can now be added to the definition of the design parameter θ which becomes:

$$\theta := \begin{bmatrix} \lambda_I \\ \theta_c \end{bmatrix} \in \mathbb{R}^7. \tag{15.25}$$

This makes us ready to address the setting of the probabilistic certification step. As a matter of fact, we need to clarify the following issues:

1. The instance of the model to build the certification on. This includes the choice of the unknown maps as well as the statistical rules under which the uncertain parameters are to be randomly sampled.
2. The success/failure condition as well as the optimization criterion if any.

This is done in the following section.

15.5 The Probabilistic Certification Step

This section tackles the two steps described at the end of the previous chapter, namely, the instantiation of the general model and the definition of the conditions under which a scenario is declared to be successful.

15.5.1 Model's Instance and Associated Statistics

It is important to recall that, instantiating the model amounts at defining the unknown maps Φ and F respectively given by (15.7b) and (15.7c), but also at providing the definition of the admissible sets of the parameters p_c and p_ϕ and their probability distributions to be used in the sampling process of the certification procedure.

Notice that while a specific commonly used model is instantiated hereafter for the derivation of the certification results, any other complicated, high-dimensional model involving complex nonlinear relationships can be used in the framework without any change in the feedback design since the latter exploits no structural properties of the model. Only the quantitative results of the assessment would be different.

As an instance of the general model, let us consider the widely used compartmental model describing the dynamics of the drug in the human body [see [4,5] and the references therein].

In this model, the following two-states dynamics is used to represent the unknown functions Φ and F:

$$\dot{\xi}_1 = k_{12}x_2 - k_{21}\xi_1, \tag{15.26a}$$

$$\dot{\xi}_2 = k_{13}x_2 - k_{31}\xi_2, \tag{15.26b}$$

$$\Phi = -(k_{10} + k_{12} + k_{13})x_2 + k_{21}\xi_1 + k_{31}\xi_2, \tag{15.26c}$$

where Eqs. (15.26a)–(15.26b) define the function F invoked in (15.7c) while the coefficients k_{ij} are given by nominal expressions that depend on the gender (G), age (A), height (H) and weight (W) of the patient:

$$k_{ij}^{\text{nom}} := F_{ij}(G, A, H, W). \tag{15.27}$$

Interested readers can consult the above cited references for the explicit expressions of the functions F_{ij}.

Recall that in the previous discussion, the vector gathering these parameters is the one denoted by p_ϕ, namely:

$$p_\phi := \begin{bmatrix} k_{10} \; k_{12} \; k_{21} \; k_{13} \; k_{31} \end{bmatrix}^\top \in \mathbb{R}^5. \tag{15.28}$$

As for the probability distribution ($\mathcal{P}_\phi$) associated to this vector of parameters, it is defined by the following procedure to generate random samples:

1. First the gender parameter G is randomly chosen by equally probable Male/Female outcome.
2. Then, the continuous parameters

   $$A \in [20, 60], \quad H \in [140, 200], \quad W \in [55, 90],$$

   are uniformly sampled inside the above defined intervals of age, weight and height. Then the corresponding parameter vector p_ϕ^{nom} is computed by (15.27).
3. The results of the two previous steps are then randomly *disturbed* by multiplying each parameter p_{ϕ_i}, $i = 1, \dots, 5$ independently by a factor β_i that is uniformly randomly chosen in the interval $[\underline{\beta}, \bar{\beta}]$ where $\underline{\beta} \in (0, 1)$ and $\bar{\beta} > 1$:

   $$p_{\phi_i} = \beta_i \times p_{\phi_i}^{\text{nom}}(G, A, H, W) \quad \beta_i \in [\underline{\beta}, \bar{\beta}]. \tag{15.29}$$

The values

$$\underline{\beta} = 0.5 \quad \text{and} \quad \bar{\beta} = 1.5,$$

are used in the following investigation. Note that this means that the amplitudes of the uncertainties can be described as follows:

The Uncertainties Level
Each parameter can randomly take any value between 50 and 150% of its nominal value. These nominal values are themselves randomly obtained through random samples of $G, A, H,$ and W.

The static map between the BIS and x_1 is classically defined by [4]:

$$\text{BIS} = \Psi_B(x_1|\ p_B) := 100\left[1 - \frac{x_1^\gamma}{x_1^\gamma + E_{c50}^\gamma}\right], \tag{15.30}$$

which is a model that is defined by the parameter vector:

$$p_B := (\gamma, E_{c50}),$$

for which, the following nominal value is used:

$$p_B^{\text{nom}} := (2.39, 5.6). \tag{15.31}$$

Regarding the random sampling (distribution $\mathcal{P}_B$), a uniform sampling in the interval defined by $\pm 15\%$ is performed around the nominal values of the components.

Similarly the nominal value $k_{e0}^{\text{nom}} = 0.474$ is used for k_{e0} involved in (15.7a) while the *real* value used in the simulation is uniformly randomly chosen in the $\pm 30\%$ interval around the nominal value. This defines the sampling probability $\mathcal{P}_{e0}$.

This fully defines the instantiation of the models to be simulated as well as the statistics of random sampling that are used to represent the dispersion of the model's parameters.

Now let us precisely define the success/failure condition of a *scenario* to be used in the probabilistic certification.

This is the aim of the next section.

15.5.2 The Probabilistic Certification Setting

In this section, the definition of the constraints that determine the success of an operation on a randomly sampled patient (sampled according to the process described in the previous section) is given.

Based on the problem statement given above (see p. 529), we are looking for an approximate solution of the following optimization problem:

$$\min_{\bar{\tau}_r} J(\bar{\tau}_r) := \bar{\tau}_r \quad | \quad \Pr_{\mathcal{P}}\begin{bmatrix} \tau_r(p) \le \bar{\tau}_r \\ \text{AND} \\ \min\limits_{t\in[0,T]} y(t|p) \ge 40 \end{bmatrix} \ge 1 - \eta. \tag{15.32}$$

As it is explained in Sect. 4.4, this formulation is intractable because of the difficulties in evaluating the multi-variate probability function involved.

The essence of probabilistic certification is to replace the above problem by the following more tractable one:

$$\min_{\bar{\tau}_r} J(\bar{\tau}_r) := \bar{\tau}_r \quad | \quad \sum_{i=1}^{N} I(\bar{\tau}_r, p^{(i)}) \leq m, \tag{15.33}$$

in which:

- $p^{(i)}$, $i = 1, \ldots, N$ are N random samples of p which are obtained according to the distribution $\mathcal{P}$.
- $I(\bar{\tau}_r, p^{(i)})$ is the constraint violation indicator:

$$I(\bar{\tau}_r, p^{(i)}) := \begin{cases} 1 \text{ if } \tau_r(p) > \bar{\tau}_r \text{ or } \min_t y(t|p) < 40, \\ 0 \text{ otherwise.} \end{cases} \tag{15.34}$$

- m is some integer such that

$$\frac{m}{N} \leq \eta. \tag{15.35}$$

Roughly speaking, the condition (15.33) states that only m samples inside the population of N samples violate the constraint ($\tau_r(p) \leq \bar{\tau}_r$ and $\min_t y(t|p) \geq 40$).

Now, intuitively, one would expect that if a value $\bar{\tau}_r$ is found such that for a *sufficiently high* number of samples N, only a small number m satisfying (15.35) violates the constraint then one can say that the probability that the constraint is violated is *close* to η.

The problem lies obviously in the terms (*sufficiently high* N) and (*close to* η). That is where the rigorous formalism of probabilistic certification[1] [6] comes into action.

More precisely, this formulation computes the number of scenarios to simulate in order to be able to make a statement that is correct with a probability $1 - \delta$ where δ is called the confidence parameter.

More precisely, the theory gives for a priori chosen m, a priori desired η and a priori defined confidence parameter δ, the number of samples N that one has to use. Note that the confidence parameter δ is the probability with which (15.33) can hold while (15.32) is violated.

In other words, $1 - \delta$ is the probability that one satisfies the intractable inequality (15.32) while checking only the tractable inequality (15.33).

In the remainder of this section, we consider a given setting of the design parameter θ defined by (15.25) and (15.22) leaving as the unique remaining design parameter the value of the settling time $\bar{\tau}_r$ that remains to be tuned.

---

[1] See Sect. 2.11.

Table 15.1 Evolution of the sample size N (number of samples needed to achieve the certification) as a function of the precision η and the cardinality n_r of the design parameter set $\mathbb{T}$ (confidence parameter $\delta = 10^{-3}$ and $m = 1$ are used)

n_r	$\eta = 0.1$	$\eta = 0.05$	$\eta = 0.01$	$\eta = 0.001$
1	132	264	1317	13164
5	154	308	1536	15354
10	163	326	1628	16280
100	193	386	1930	19299
1000	223	445	2225	22249
10000	252	503	2515	25148

Under these assumptions, the number N of scenarios to be sampled depends on the number of candidate values of this remaining design parameter.

In what follows, we consider the following set of candidate values:

$$\bar{\tau}_r \in \mathbb{T} := \left\{\bar{\tau}_1, \ldots, \bar{\tau}_{n_r}\right\}. \tag{15.36}$$

In this case, formulae (15.37) leads to the following number of scenarios:

$$N \geq \frac{1}{\eta}\left(m + \ln\left(\frac{n_r}{\delta}\right) + \left(2m \ln\left(\frac{n_r}{\delta}\right)\right)^{1/2}\right). \tag{15.37}$$

The corresponding values of N for $m = 1$ and the confidence parameter $\delta = 10^{-3}$ are given in Table 15.1.

15.6 Results

In this last section, some results are given regarding the performance of the proposed feedback structure on a large population of the patients that are sampled using the above mentioned statistics.

The probabilistic certification scheme is applied to the following setting of the controller design:

$$\begin{aligned}
&\lambda_1 = 40;\;\; \lambda_2 = 10;\;\; \lambda_{f_1} = 0.01\lambda_1;\;\; \lambda_{f_2} = 0.1\lambda_2,\\
&\underline{\gamma} = 0.7;\;\; \bar{\gamma} = 1.3;\;\; \bar{\nu} = 100.
\end{aligned}$$

The integrator constant is taken equal to $\lambda_I = 0.1$ in (15.24). The sampling period for the control is taken equal to $\tau = 6$ s.

Before the certified performance is stated, let us have a look on the behavior of the closed-loop over three randomly sampled patient's parameters.

The objective here is to show that the same control feedback law that only uses the measured quantity $y =$ BIS is able to realize a tight regulation without a steady error of the regulated output despite the high level of uncertainties being involved.

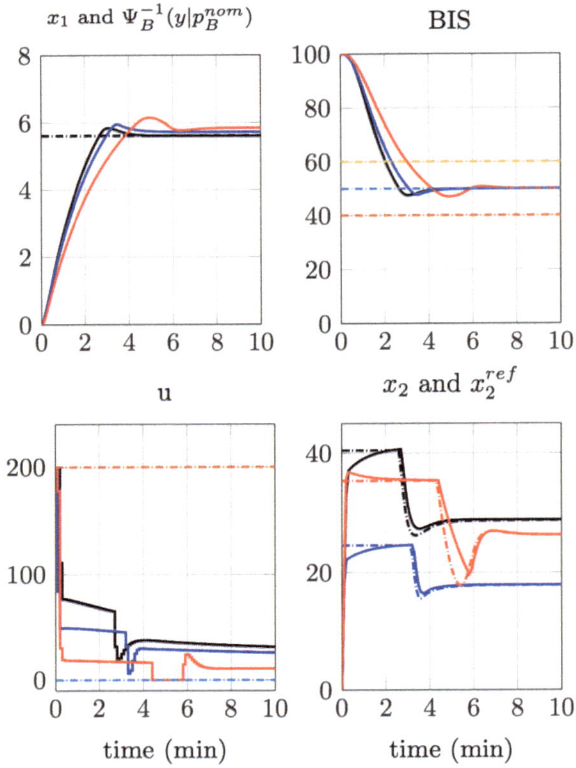

Fig. 15.1 Closed-loop simulations of three randomly sampled patients with the uncertainty levels given by (15.38)

More precisely, three patients are considered with the following relative discrepancies w.r.t. the nominal values of the parameters (the expressions have to be interpreted component-wise):

$$\frac{p_\phi - p_\phi^{\text{nom}}}{|p_\phi^{\text{nom}}|} \in \left\{ \begin{bmatrix} 43\% \\ 5\% \\ 29\% \\ 12\% \\ 11\% \end{bmatrix}, \begin{bmatrix} -45\% \\ -1\% \\ -18\% \\ -11\% \\ -10\% \end{bmatrix}, \begin{bmatrix} 3.5\% \\ -13.3\% \\ -22\% \\ -2\% \\ 12\% \end{bmatrix} \right\}. \tag{15.38}$$

The BIS profile and the drug's injection rate for those patients under the same design dynamic output feedback are shown in Fig. 15.1.

One can easily notice how different are the Propofol injection profiles for the three randomly sampled patients despite of the model-free design of the feedback law.

The role of the integrator variable e introduced in (15.24) can also be observed on the fact that the stationary values of x_1 are corrected and hence become patient-dependent so that the truly regulated BIS is steered to the desired value despite the unknown parameters of the specific patient.

Let us now examine the probabilistic certification step. To do so a number $n_r = 100$ of candidate values of the response time have been taken, which are uniformly

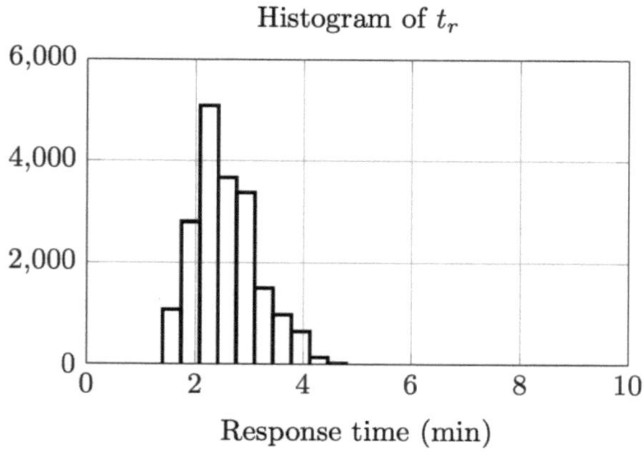

Fig. 15.2 Histogram of the response times for the 1930 sampled patients used in the computation of the probabilistically certified response time

distributed on the interval $[1, 20]$ min. The precision $\eta = 10^{-2}$ and the confidence parameter $\delta = 10^{-3}$ are used.

From Table 15.1, it comes that the above choice leads to a number of samples $N = 1930$ that is necessary to give the certification result.

Note that since a single simulation needs <1 ms, the certification of a given value of the design parameter vector p needs <2 s. Figure 15.2 shows the histogram of the response time corresponding to the randomly sampled population of patients. From this figure, it comes clearly that a certifiable response time can be taken equal to $\bar{\tau}_r = 6$ min. Let us state the result more clearly:

Probabilistic Certification Statement
As far as the instantiated model described in Sect. 15.5.1 is relevant, under the proposed feedback, **the probability** that more than 1% of a patient's BIS indicator remains outside the admissible domain after 6 min or gets their BIS lower than 40 **is lower than 0.1%**.

It is important to recall that a more detailed and computationally intensive probabilistic optimization would have been possible by keeping free not only the settling time τ_r as it is done above but all the components of the vector θ of design parameters.

Such holistic optimization would have probably enabled shorter settling time to be certified at the price of much longer computation time.

Nevertheless, there are two reasons for which the choice has been made to stick to a single *hand-tuned* parameterization of the control feedback, namely:

1. It is always a good practice to check the individual components of a designed feedback law progressively as the design process goes. For instance, the different steps of the design of the two controllers of the scalar systems, the design of the Luenberger state estimator as well as the tuning of the integral action.

When following this progressive checking and tuning process, the designer naturally comes with a set of parameters that is likely to be appropriate.
This setting represents a natural and intuitive candidate for the certification procedure and this is how the used setting has been derived. This does not work all the time! Indeed, in some complicated architecture, while each sub-controller achieves its own task, the interaction between the individually controlled subsystems forces the designer to re-tune each controller so that the global task is appropriately achieved.
2. When the number of design parameter grows, the number of samples to be used in the certification process becomes quite high. On the other hand, one needs to define the set of admissible values for each of the components of the design parameter. This limitation leads anyway to a degree of sub-optimality if one needs to avoid unreasonably high number of samples.

Obviously, if the resulting settling time $\bar{\tau}_r$ has been found to be too large, design tuning should have been repeated with more degrees of freedom (by making free other components of the design parameter than the only $\bar{\tau}_r$). Since the so-obtained performance is compatible with the time practically and commonly encountered by practitioners, the design can be considered to be largely acceptable.

15.7 Conclusion

In this chapter it has been shown that the probabilistic certification framework comes extremely handy when addressing the seemingly challenging problem of drug dosage in anesthesia via Propofol injection.

It is important to underline also that the solution of the problem involves many control-related *tricks* that were crucial in achieving the success of the resulting solution, namely:

1. the use of state estimation paradigm,
2. the use of generic dynamic output feedback design studied in Chap. 4, and
3. the use of the integral action to compensate for errors on the set-points.

All these tools are in the heart of *standard* control concepts and solutions that one should keep in mind.

This strengthens an idea that has been already mentioned elsewhere in this book: A crucial step in the probabilistic certification always relies on a problem-dependent *angle of attack* that condenses the *know-how* of the control specialist who is behind the scene!

Acknowledgements The material of this chapter is based on a joint work with my coleagues Mirko Fiacchini (Gipsa-lab), Sophie Tarbouriech and Isabelle Queinnec (LAAS) and Michel Mazerolles (Toulouse University Hospital).

References

1. Charbati, S., Bracco, D., Mathieu, P.A., Hemmerling, T.M.: Comparison of four different display designs of a novel Anaesthetic monitoring system, the 'integrated monitor of Anaesthesia. Br. J. Anaesthesia **103**(5), 670–677 (2009)
2. Alamir, M., Fiacchini, M., Queinnec, I., Tarbouriech, S., Mazerolles, M.: Feedback law with probabilistic certification for propofol-based control of BIS during anesthesia. Int. J. Robust Nonlinear Control **28**(18), 6254–6266 (2018)
3. Antsaklis, P., Michel, A.N.: A Linear Systems Primer. Birkhauser, Boston (2007)
4. Ionescu, C.M., Keyser, R.D., Torrico, B.C., Smet, T.D., Struys, M.M., Normey-Rico, J.E.: Robust predictive control strategy applied for propofol dosing using BIS as a controlled variable during anesthesia. IEEE Trans. Biomed. Eng. **55**(9), 2170–2191 (2008)
5. Zabi, S., Queinnec, I., Tarbouriech, S., Garcia, G., Mazerolles, M.: New approach for the control of anesthesia based on dynamic decoupling. In: Proceedings of the 9th IFAC Symposium on Biological and Medical Systems, Berlin, Germany, pp. 511–516 (2015)
6. Alamo, T., Tempo, R., Camacho, E.F.: Randomized strategies for probabilistic solutions of uncertain feasibility and optimization problems. IEEE Trans. Autom. Control **54**(11), 2545–2559 (2009)

16 Learning Optimal Energy Management in Hybrid Vehicles

16.1 Problem Statement

We consider a hybrid car equipped with a battery with a State-of-Charge (SoC $= x$) governed by the following ODE:

$$\dot{x} = a_1\Big[a_2 - \sqrt{a_2^2 + a_3 w_1 + a_4 w_2 u}\Big], \tag{16.1}$$

where

- x is the state of charge of the battery that should be kept inside the interval $[0.2, 0.8]$, namely:
$$x \in [\underline{x}, \bar{x}] = [0.2, 0.8]. \tag{16.2}$$
- v is the speed of the vehicle.
- $w_1(v, \dot{v})$ and $w_2(v, \dot{v})$ are measured but unpredictable terms that are functions of the speed and acceleration of the car as required by the driver.
- u is the control charging signal that is constrained to remain in an admissible charging interval:
$$u \in [u^{\min}, u^{\max}] = [0, 150]. \tag{16.3}$$
- a_i for $i = 1, 2, 3, 4$ are constant **known** (identified) parameters.

Regarding the Uncertainties w_1 and w_2
Note that the way the terms $w_i(v, \dot{v})$, involved in (16.1), depend on their arguments $(v, \dot{v})$ is quite well known but it is precisely because the evolution of the latter argument is not predictable that makes the terms w_i behave as unpredictable disturbances.

M. Alamir, *Nonlinear Control of Uncertain Systems*,
https://doi.org/10.1007/978-3-031-93287-8_16

Equation (16.1) states that depending on the pair $(v, \dot{v})$, the total power needed to realize the kinetics required by the driver can be provided while charging or discharging the battery.

This trade-off depends on the value of the control signal u.

More precisely, the optimal choice of the control profile leading from the initial position $p_0 = 0$ to the final position p_f is the one that minimizes the following cost function:

$$J(\mathbf{u}|\mathbf{v}, \dot{\mathbf{v}}) := \int_{t=0}^{t_f} \Big[m_0(t) + m_1(t)u(t) + m_2(t)u^2(t)\Big]dt, \tag{16.4}$$

where, again, by overloading the notation, the terms m_is take the following form:

$$m_i(t) = m_i(v(t), \dot{v}(t)) \quad \text{for} \quad i = 0, 1, 2, \tag{16.5}$$

with **known functions** of the speed v and the acceleration $\dot{v}$ at instant t.

Here again, in spite of the knowledge of the maps m_i, the fact that the evolution of their arguments $(v, \dot{v})$ during an *experiment* leads to a problem where uncertainties impact the very definition of the cost function.

Notice however that the optimization of the above cost (16.4) needs to be done while meeting the saturation constraints (16.2) and (16.3) as well as the final constraints on the position and the state of charge of the vehicle's battery, namely:

$$p(t_f) = p_f; \quad x(t_f) = x_f = 0.5. \tag{16.6}$$

Obviously, the computation of the *exact* optimal control profile denoted by

$$\mathbf{u}^\star(\mathbf{v}, \dot{\mathbf{v}}),$$

requires the knowledge of the speed profile $\mathbf{v}$ which is not available since it depends on the driver's actions that might differ, even for the same path between the initial and the final position p_0 and p_f, because of the state of the traffic and its own variability of behavior.

This is concretely shown in Fig. 16.1 where nine different real-life speed profiles are recorded leading from the same initial position to the same final position on the same track that is addressed overall the chapter.

Notice that the duration of the different profiles are different since all the speed profiles should lead to the increment of position defined by $p_f - p_0$ meaning that slower profiles (in average) need more time to reach the same final position when starting from the same initial position.

This can be clearly seen in Fig. 16.2 where the evolution of the position during the scenarios are plotted showing that they all end at the same terminal value p_f.

As a matter of fact, we dispose of eleven scenarios of driver's behavior when executing the driving cycle linking the positions p_0 and p_f.

Since the relations giving the profiles $\mathbf{w}_1$ and $\mathbf{w}_2$ as well as $\mathbf{m}_0$, $\mathbf{m}_1$, and $\mathbf{m}_2$ in terms of the kinetic profiles $\mathbf{v}$ and $\dot{\mathbf{v}}$ are known, it is assumed here that these

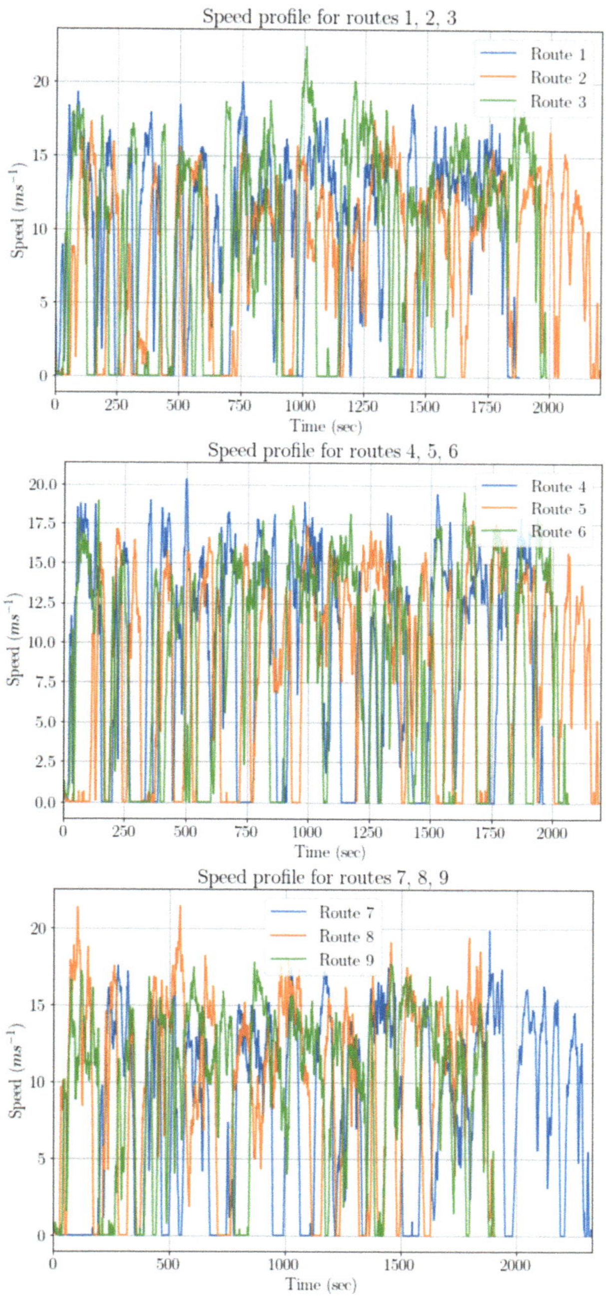

Fig. 16.1 Nine different speed profiles **v** leading from the same initial position to the same final position

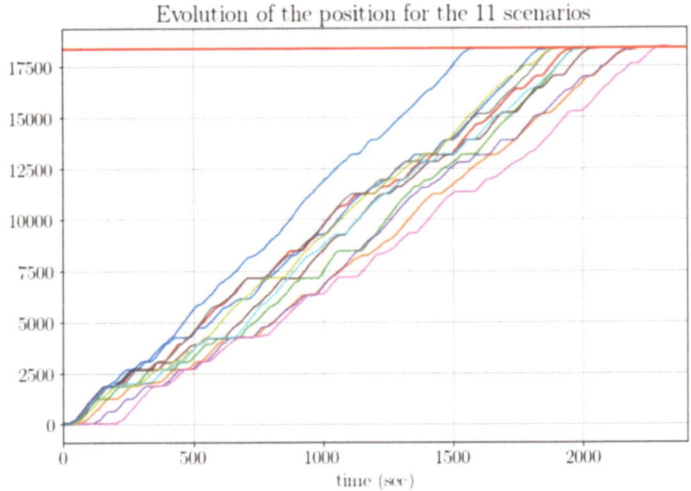

Fig. 16.2 Evolution of the position for the different driving cycles showing that the terminal value is shared by all of the 11 driving scenarios

	t	position	speed	acc	m0	m1	m2	w1	w2
0	0.0	0.0000	0.231500	0.231500	0.616612	0.000583	8.491631e-09	58.307717	0.870083
1	0.2	0.0463	0.272656	0.184171	0.606640	0.000684	1.169367e-08	62.839941	1.021035
2	0.4	0.0926	0.313811	0.136842	0.596879	0.000784	1.537755e-08	65.611543	1.170870
3	0.6	0.1389	0.354967	0.089513	0.587327	0.000884	1.953215e-08	66.622705	1.319593

Fig. 16.3 Head of a typical dataframe for a driving scenario. Note that the sampling period for measurement acquisition is $\tau = 0.2\,\text{s}$

relationships are used in order to derive the following data representing the eleven scenarios:

$$\mathcal{D} := \left\{(\mathbf{w}_1^{(s)}, \mathbf{w}_2^{(s)}, \mathbf{m}_0^{(s)}, \mathbf{m}_1^{(s)}, \mathbf{m}_2^{(s)})\right\}_{s=1}^{n_s}; \quad n_s = 11. \tag{16.7}$$

Each of the above-mentioned scenarios is represented within a `csv` file containing a `dataframe` that takes the form shown in Fig. 16.3 (only the head of the `dataframe` is represented).

Note that the basic sampling period of the measurement acquisition is equal to 0.2 s. Based on the above discussion, the problem we are interested in here can be stated as follows:

> **Problem Statement**
> Use the available data $\mathcal{D}$, given by (16.7), in order to design a feedback control law of the form:
>
> $$u_k = K(y_{k-1}, \ldots, y_{k-n}) =: K(\mathbf{y}_k^{(-)}), \tag{16.8}$$

where

$$y = (x, p, v, a),$$

is the vector gathering the battery SoC, position, speed, and acceleration, such that the resulting cost is close to optimal and the resulting trajectories satisfy the saturation constraints (16.2) and (16.3) as well as the terminal constraint

$$x(t_f) = 0.5,$$

on the final state of charge.

There are several comments that are worth giving regarding the above statement:

- Note that the notation:

$$u_k = u(k\tau),$$

is used where $\tau = 0.2$ is the basic sampling time.
- The control u_k to be applied during the sampling period:

$$[k\tau, (k+1)\tau[,$$

is computed based on the previously gathered information $\mathbf{y}_k^{(-)}$. This stems from the fact that the information regarding the speed and the acceleration cannot be known in advance.
- The *use of the available data* $\mathcal{D}$ should follow the correct rules regarding Machine Learning (ML). This means that if the feedback control law (16.8) is to be learned from data-related computation, it is mandatory that the working data be split without shuffling (see Sect. 3.4.3) into learning and validation datasets in order to check the generalization power of the so-derived feedback solution.
- It is also important to mention that while the available 11 scenarios correspond to different speed and acceleration profiles required by the driver, each of these scenarios does not totally define an optimal control profile since the latter not only depends on the speed/acceleration profiles but it also depends on the **the initial state of charge** x_0. In other words:

! Many Different Scenarios for Learning
Using each one of the eleven drivers' profiles, by starting from different initial states, it is possible to create new experiments.

We will see this feature in a more precise way in the forthcoming sections, but it is necessary to keep in mind that having these 11 kinetic scenarios and the many possible initial values of the state of charge $x_0 \in [0.2, 0.8]$ we dispose of a quite rich set of possibilities to create the learning data and to validate the learned feedback law in a quite confident way.

In the next section, the road map for the solution of the problem stated in this section is discussed before its steps are implemented successively.

16.2 Overview of the Proposed Solution

A short way to summarize the content of the present chapter is to state that

Application of the Design Method Explained in Chap. 7
The solution to the above stated problem follows the methodology sketched in Chap. 7 which implements a tractable approximation of stochastic model predictive control.

Nevertheless, for the easiness of reading, the principle of the approach is sketched again in the present section.

Recall that in the heart of the feedback design detailed in Chap. 7 lies a ML-like fitting problem. But any ML problem requires the precise definition of the features vector as well as the targeted label. The definition of these two ingredients for the problem stated above is described in the next section.

16.2.1 Definition of the Features and Label

The main idea already implicitly invoked in the problem statement of the previous section is to learn a map K that takes as **input argument** the previous measurement $\mathbf{y}_k^{(-)}$ of state of charge, position, speed and acceleration[1] over some past interval involving n previous measurements ($n = 5$ in the figure below) and provides **as output**, the charging control u_k to be applied on the next sampling period.

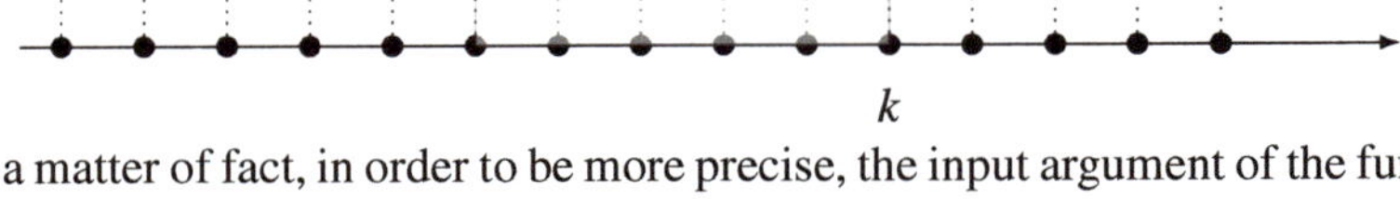

As a matter of fact, in order to be more precise, the input argument of the function K to be learned will precisely be **the averaged values** of

$$y = (x, p, v, a),$$

[1] This is the `features` vector in ML language.

over the previous window of n basic sampling periods.

More precisely, using the language of ML, the vector of features used at instant k is given by

$$\texttt{feature vector}_k = \begin{bmatrix} \frac{1}{n}\sum_{i=1}^{n} x_{k-i} \\ \frac{1}{n}\sum_{i=1}^{n} p_{k-i} \\ \frac{1}{n}\sum_{i=1}^{n} v_{k-i} \\ \frac{1}{n}\sum_{i=1}^{n} a_{k-i} \end{bmatrix} \in \mathbb{R}^4. \tag{16.9}$$

On the other hand, the label to be targeted is defined as follows:

The Label

The targeted label at instant k is defined to be the value of the optimal solution **should the whole profiles of speed and acceleration over the scenario under consideration be completely known,** namely:

$$\texttt{label}_k = u_k^\star|_{\text{perfect knowledge}}. \tag{16.10}$$

In order to build the learning data for the above ML problem, the steps described in the following section are applied.

16.2.2 Building the Learning Data

1. A cloud of pairs of scenarios numbers s and an initial conditions x_0 are sampled[2]:

$$\left\{s, x_0^{(j)}\right\}_{(j,s)} \in \{1, \dots, 10\} \times \{1, \dots, 11\},$$

   leading to 110 different scenarios.
2. For each of the previous scenarios, the deterministic optimal control problem (assuming the exact knowledge of the kinetics over the scenario) is solved to get the optimal control profile

$$\mathbf{u}^\star(s, x_0^{(j)}).$$

3. A sliding window of length[3] n is used to collect the feature vectors and the labels over this scenario as defined through (16.9) and (16.10).
4. The resulting data is divided (without shuffling) into a training data and a test data. The way the data is divided leads to a separation by scenarios meaning that the solution will be learned (fitted) on a subset of scenarios (50%) leaving the remaining scenarios for the validation.

---

[2] $x_0^{(j)}$ are sampled inside $[0.2, 0.8]$.

[3] Several values will be tried for the window size n.

! Data Split on Drivers Basis
Splitting is done on a scenario basis and not on time-series inside a scenario for the sake of avoiding data leakage.

Another way to describe the splitting philosophy is to state that the split is done on drivers basis. The same drivers cannot appear in both the training and the test datasets.

5. A model is used to fit the data using the training data before it is used on the validation data to check its extrapolation power on unseen data.

In the following section, the way one can solve the deterministic optimal control problem for a given initial state x_0 and a given scenarios of kinetics is explained.

16.3 Solving the Deterministic OCP

There are two main difficulties in solving the deterministic optimal control problems when one attempts to create the learning dataset following the steps mentioned at the end of the previous section:

- The first difficulty stems from the size of the problem in terms of the number of decision variables that the optimal solution, namely $\mathbf{u}^\star(s, x_0^{(j)})$, contains.
  Indeed, Fig. 16.4 shows the number of basic sampling periods (of duration $\tau = 0.2$) for each of the eleven available scenarios. This would be precisely the dimension N of the decision variable should one consider that the basic sampling period is also used in the time-structure of the control profile (no control parameterization is used).
  Solution: It will be shown hereafter that a **larger control sampling period** (under-

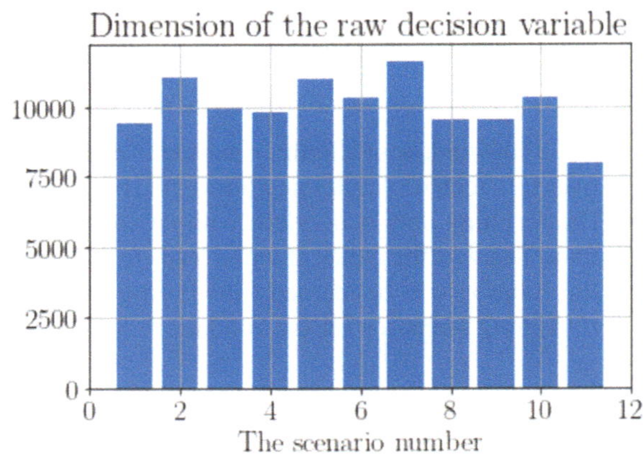

Fig. 16.4 The dimensions N of the decision variable for the different scenarios

sampling) which is set to 5, 10, or even 15 times the basic sampling period leads to almost quasi-optimal solutions (see below).

- The second difficulty in solving the exact optimal control problems comes from the nonlinearity in the state equation (16.1) which, coupled with the previous point, induces quite involved optimal control problems.
  Solution: An averaging-based linearization is proposed and proved to be almost without harm on the quality of the result. This is obviously a specific solution that is tailored and adapted to the specific use-case.
  The reader would probably has already notice that this is a kind of *pattern* in the development of efficient real-life solutions, namely, the need at some stage of the development to design a tailored solution that addresses the specific difficulty of the problem under consideration.

It is shown hereafter that by combining the two above-mentioned tricks, solving each of the deterministic optimal control problems bowls down to solving a moderate dimensional **Quadratic Programming (QP)** problem that can be efficiently done using the powerful `cvxpy` `Python` package [1].

More details regarding this package are given once the underlying problem has been reformulated.

This is the object of the next section.

16.3.1 Reformulating the Deterministic Problem as a QP Problem

Let us choose some integer[4] $n \in \mathbb{N}$ that is used as a parameter that defines the under-sampling rate that helps reducing the number of decision variables.

More precisely, we shall formulate an optimal control problem in which the **control profile is piece-wise constant with a control sampling period of** $n\tau$.

By so-doing the number of decision variables becomes equal to n_ℓ where

$$n_\ell = \left\lfloor \frac{N}{n} \right\rfloor \tag{16.11}$$

where the right-hand side of the equation represents the integer part of the ratio N/n.

Based on this assumption and denoting by

$$\bar{\mathbf{u}} \in \mathbb{R}^{n_\ell},$$

the new lower dimensional vector of decision variables, the cost function (16.4) can be approximated using the following formulae[5]:

---

[4] Typically, $n = 5, 10, 15$ or so!

[5] Recall that the profiles $\mathbf{m}_i$ are supposed to be perfectly known.

$$J_n(\bar{\mathbf{u}}) := \sum_{\ell=0}^{n_\ell-1} [\bar{m}_{0,\ell}] + [\bar{m}_{1,\ell}]\bar{u}_\ell + [\bar{m}_{2,\ell}]\bar{u}_\ell^2, \tag{16.12}$$

$$\text{where} \quad \bar{m}_{i,\ell} := \int_{\ell n\tau}^{(\ell+1)n\tau} m_i(t)dt. \tag{16.13}$$

Note how the terms m_i have been replaced by their averaged values over the new larger sampling periods of duration $n\tau$.

By so-doing the new approximate cost function can be written in the following compact standard quadratic form:

The Cost Function Reformulated

In terms of the new decision variable $\bar{\mathbf{u}}$, the cost function becomes

$$J_n(\bar{\mathbf{u}}) = c + q\bar{\mathbf{u}} + \frac{1}{2}\bar{\mathbf{u}}^T P\bar{\mathbf{u}}, \tag{16.14}$$

$$\text{where} \quad \left| \begin{array}{l} c = \sum_{\ell=0}^{n_\ell-1} \bar{m}_{0,\ell}, \\ q = [\bar{m}_{1,0}, \ldots, \bar{m}_{1,n_\ell}], \\ P = 2\text{diag}(\{\bar{m}_{2,\ell}\}_{\ell=0}^{n_\ell-1}). \end{array} \right. \tag{16.15}$$

Note that the dimension n_ℓ of the decision variable $\bar{\mathbf{u}}$ depends on the integer n used to define the under-sampling rate through (16.11).

It is important to note that the ultimately designed control feedback will use the original sampling period as τ remains the basic feedback sampling period.

The larger sampling period used currently in the formulation of the open-loop optimal control problem is only an intermediate step in the generation of the learning data via simplified deterministic optimal control problem's formulation.

Now a quadratic form of the cost does not simplify the optimal control problem if the underlying dynamics remains nonlinear since the resulting optimization problem would not become a QP problem.

It happens that the model of the system we are interested in admits a specific *treatment* that ultimately induces a linear representation without a significant loss in the relevance of the so-obtained solutions.

This is described in the next section.

16.3.2 Reformulating the Dynamics

Recall that the original dynamics is given by (16.1):

$$\dot{x} = a_1\Big[a_2 - \sqrt{a_2^2 + a_3 w_1 + a_4 w_2 u}\Big]. \tag{16.16}$$

Assuming that the reduced-dimensional control profile $\bar{\mathbf{u}}$ is defined as explained above, it can be argued that the discrete dynamics of x between two successive down-sampling instants $\ell n\tau$ and $(\ell+1)n\tau$ is bounded by the two extreme dynamics, namely, the one obtained for $\bar{u} = u^{\min}$ and the one obtained for $\bar{u} = u^{\max}$.

In order to express these two dynamics, let us use the following short notation:

$$b_j(u) := -a_1\sqrt{a_2^2 + a_3 w_1(j\tau) + a_4 w_2(j\tau)u}. \tag{16.17}$$

Based on this notation, it comes out that if u is maintained at the minimum value $u = u^{\min} = 0$ during n sampling periods, one gets the following dynamics on the state of charge[6]:

$$x_{\ell+1} = x_\ell + n\tau a_1 a_2 + \tau\sum_{j=0}^{n} b_{\ell+j}(0). \tag{16.18}$$

On the other hand, when the maximum $u = u^{\max}$ is used, one gets

$$x_{\ell+1} = x_\ell + n\tau a_1 a_2 + \tau\sum_{j=0}^{n} b_{\ell+j}(u^{\max}). \tag{16.19}$$

Using the above two equations and exploiting the monotonicity of the relation it comes out that for any intermediate value $u \in [0, u^{\max}]$, the dynamics based on the down-sampling defined by n can be approximated by the following formulation[7]:

The Approximated Dynamics

The approximated dynamics that will serve in solving the constrained optimal control problem is given by

$$x_{\ell+1} = \alpha_\ell + \beta_\ell \bar{u} \tag{16.20a}$$

$$\text{where}\quad \left|\begin{array}{l} \alpha_\ell = n\tau a_1 a_2 + \tau\sum_{j=0}^{n} b_{\ell+j}(0) \\ \beta_\ell = \frac{\tau}{u^{\max}}\sum_{j=1}^{n}[b_{\ell+j}(u^{\max}) - b_{\ell+j}(0)]. \end{array}\right. \tag{16.20b}$$

This can be simply checked by injecting successively $\bar{u} = 0$ and $\bar{u} = u^{\max}$ in (16.20) and observing that the results are respectively identical to (16.18) and (16.19) respectively.

The good news is that the new approximated dynamics (16.20a) of the system is now affine in the decision variable $\bar{u}$. This together with the quadratic form of the

[6] Here, the writing of the discrete-time dynamics refer to the longer sampling period $n\tau$ which obviously corresponds to an abuse of notation that we locally accept in order to keep a fluent presentation.

[7] Note that α_ℓ and β_ℓ can be computed once given the kinetic profiles which are supposed here to be known for the creation of the learning dataset.

cost function and the affine nature of the constraints leads to a QP problem that will be easy to solve in almost no time.

This is more clearly and explicitly developed in the remainder of this section.

The dynamics exhibited in the previous relation can be used to predict the successive values of the state of charge x for any possible reduced dimensional control profile $\bar{\mathbf{u}}$, indeed one obviously has

$$\underbrace{\begin{bmatrix} x(0) \\ x(\ell n\tau) \\ \vdots \\ x((n_\ell - 1)n\tau) \end{bmatrix}}_{\bar{\mathbf{x}}} = \underbrace{\begin{bmatrix} \alpha_0 \\ \alpha_0 + \alpha_1 \\ \vdots \\ \alpha_0 + \cdots + \alpha_{n_\ell - 1} \end{bmatrix}}_{\Phi} + \underbrace{\begin{bmatrix} \text{diag}(\beta) \end{bmatrix}}_{\Psi} \bar{\mathbf{u}}, \tag{16.21}$$

which can be shortly written as follows:

$$\bar{\mathbf{x}} = \Phi + \Psi \bar{\mathbf{u}}. \tag{16.22}$$

This can then be used to reformulate the constraints on the state as explained in the next section.

It is worth keeping in mind that by the very definition (16.17) of $b_j(u)$, the matrices Φ and Ψ are dependent on the scenario through the profiles $\mathbf{w}_1$ and $\mathbf{w}_2$.

Notice that the equality constraint (16.22) is a condensed form of the approximate linear dynamics (16.20a) enforced over the whole prediction horizon that last over the duration of the associated scenario.

To this equality constraint representing the dynamics, we have to add the saturation constraints on the decision variable and the equality constraint on the state of charge of the battery in order to cope with all the requirements contained in the statement of the problem.

This is done in the next section.

16.3.3 Completing the Formulation of the Constraints

Based on the last equation, it is possible to rewrite the state constraints as follows[8]:

$$\mathbf{1} \otimes \underline{x} - \Phi \le \Psi \bar{\mathbf{u}} \le \mathbf{1} \otimes \bar{x} - \Phi, \tag{16.23}$$

$$\begin{bmatrix} \Psi[n_\ell, :] \end{bmatrix} \bar{\mathbf{u}} = x_f - \Phi_{n_\ell}, \tag{16.24}$$

where $\Psi[n_\ell, :]$ and Φ_{n_ℓ} denotes the last lines of the matrix Ψ and the last components of the vector Φ respectively. These are affine constraints in the decision variable $\bar{\mathbf{u}}$.

[8] Recall that the bounds and the targeted final value of the state of charges are given by $\underline{x} = 0.2$, $\bar{x} = 0.8$, $x_f = 0.5$.

These constraints can be written in the following standard form:

$$G\bar{\mathbf{u}} \leq h; \quad A\bar{\mathbf{u}} = b, \tag{16.25}$$

with obvious definitions of the associated matrices.

Notice however that for feasibility reasons, the equality constraint is replaced by a high penalty term of the form

$$\rho_f \|A\bar{\mathbf{u}} - b\|^2,$$

that is added to the quadratic form defining the cost function (16.14).

More precisely, this is done by adding the term $\rho_f A_{\text{ineq}}^T A_{\text{ineq}}$ to the matrix P in (16.14) and the vector $\rho_f A_{eq}^T b_{eq}$ to the affine term gain p in (16.14).

This ends the sequence of transformations that leads to a QP problem which, when solved for each scenario (consisting of an initial state and a driver's profiles of v and $\dot{v}$) provides the associated *approximately optimal profile of the control.*

But before we proceed with the use of these reformulated problems and their solutions, let us take a break and discuss in details what has been done in order to get to the reformulated problem and what is the rationale behind the confidence one might have in using it.

This is the object of the next section.

16.3.4 Discussion Regarding the Problem's Reformulation

To summarize, two modifications have been introduced to come up with a tractable solution to the deterministic optimal control problem.

- **The Reduced Parameterization of the Control Profile.**
  The first consists in using a down-sampling of the control profile, namely, by defining a subset of admissible piece-wise profiles that are defined using a larger sampling period (namely, $n\tau$ instead of τ).
  This introduces a degree of sub-optimality as it imposes a reduction in the number of degrees of freedom.
  This first simplification in the optimal problem formulation lies in the fact that the open-loop control remains constant over n sampling periods rather than a single one. As far as the studied instance of the problem is concerned, this simplification has negligible effect on the quality of the computed solution as long as rather moderate values of $n \approx 20$ are used.
  The consequence of this modification is an increase in the cost function when compared to the truly optimal value under the ideal control profile's parameterization in which the control might change each basic sampling period of length τ rather than $n\tau$.

We shall evaluate the amount of pessimism introduced by comparing the resulting approximated optimal costs for different values of $n \in \{5, 10, 15\}$ over some representative scenarios, namely:

$$J_5^\star, \quad J_{10}^\star \text{ and } J_{15}^\star.$$

Notice that very small differences between these values would be a strong indication that the simplification introduced by the down-sampling in the parameterization of the admissible control profile has a very marginal or negligible effect on the optimality of the results.
Anticipating the details of the implementation of the optimal solution computation code, the veracity of the previous statement is illustrated in Fig. 16.6 which shows the relative increase of the cost function:

$$\frac{J_n^\star - J_5^\star}{J_5^\star},$$

as the under-sampling parameter n grows.
It can be noticed that, as far as n keeps moderate values (around $n = 20$), this simplified parameterization of the open-loop control profile has no noticeable effect on the quality of the result.

- **The Averaging-Based Linearization of the State Equation.**
The second simplification introduced above lies in the averaging-based linearization of the state equation. Namely a time-varying linear model is derived such that the increments in the state of charge coincide with the *true model* for the two extreme values of the control, namely $u^{\min} = 0$ and $u^{\max} = 150$.
Now the question is *How to evaluate the impact of the averaging-based linearization on the quality of the results?*
In order to answer this question, it is important to recall that the only impact of the SoC equation on the problem's formulation lies in the computation of the state constraints (16.23)–(16.24). Therefore, the risk of using an erroneous approximation of the dynamics of the state of charge is to see the related constraints violated when applying the so-computed solution on the real dynamics.
Here again, the computations show that this is not the case, more precisely, the constraints are respected with a high fidelity as shown in the typical examples showed in Fig. 16.5. Indeed, from this figure, it appears obvious that the average-based linearization leads to a rather faithful model of the dynamics of the state of charge leading to almost the same trajectory of the state of charge.
Now the idea is not to rely only on the two previous argumentation and computations. The last word will be given to the validation step where a bunch of different scenarios will be used in the training phase and the validation phase. Only the success of the design output feedback on the unseen test scenarios is legitimate for the final validation of the proposed solution.

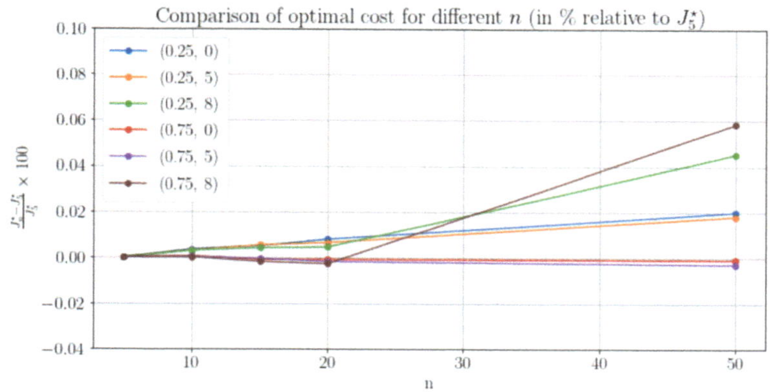

Fig. 16.5 Examination of the impact of the down-sampling of the control profile $\bar{\mathbf{u}}$ on the quality of the optimal solution. The legend shows the pair (x_0, s) of the initial state and the scenario number. Note that for $n = 20$ for instance the relative difference between $J^\star_{20}$ and $J^\star_5$ is less than 0.01%! while for $n = 50$, the difference might reach 0.06% which remains quite small

So let us proceed with the execution of our road-map for the design of the feedback law by coding the solver, generating the learning data and fitting the selected ML model.

The next section shows the `Python` code that is used to formulate and solve the open-loop optimal control problems.

16.3.5 Coding the Solver

In this section, the main part of the code which enables to understand how the QP problem described above is coded and then solved is shown and commented.

The starting point is the set of files containing the scenarios that are present in the folder named `time` (see Fig. 16.7).

An example of the content of these `csv` files has been already shown in Fig. 16.3. Let us reproduce this image here for the sake of easiness of reading of the forthcoming codes.

	t	position	speed	acc	m0	m1	m2	w1	w2
0	0.0	0.0000	0.231500	0.231500	0.616612	0.000583	8.491631e-09	58.307717	0.870083
1	0.2	0.0463	0.272656	0.184171	0.606640	0.000684	1.169367e-08	62.839941	1.021035
2	0.4	0.0926	0.313811	0.136842	0.596879	0.000784	1.537755e-08	65.611543	1.170870
3	0.6	0.1389	0.354967	0.089513	0.587327	0.000884	1.953215e-08	66.622705	1.319593

Below are shown the first lines of the code of the function `solve_avg` that declares and solves the averaging based-optimal control problem once given the following items:

- a scenario file number `ifile` $\in \{0, \ldots, 10\}$,

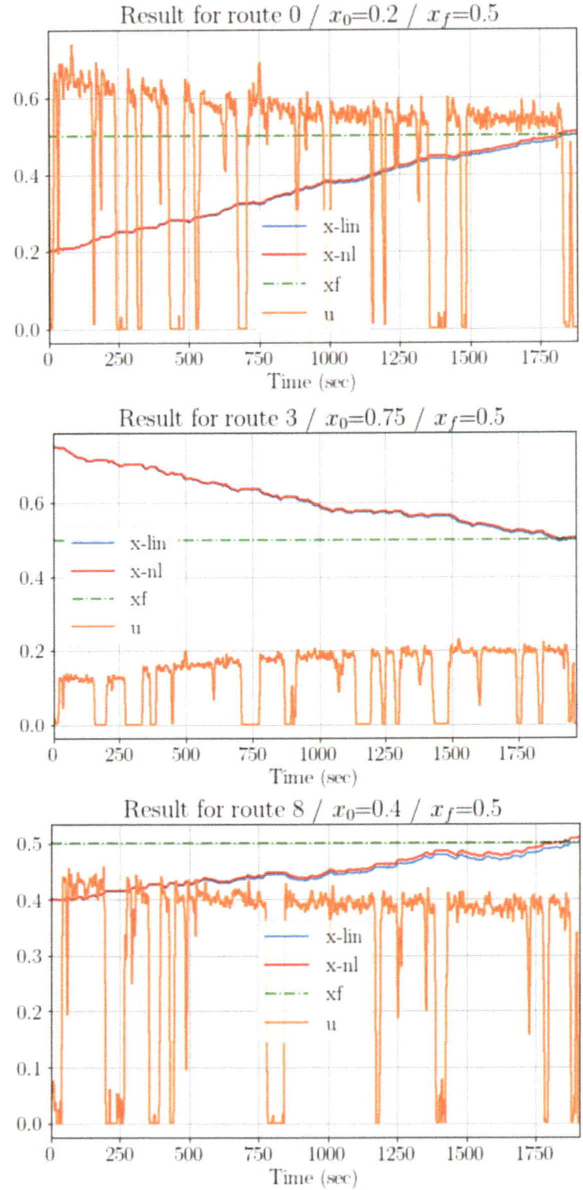

Fig. 16.6 Three typical optimal control trajectories computed with the new formulation of the optimal control problem and applied to the nonlinear and the linear model in order to examine the impact on the evolution of the SoC and hence the satisfaction of the constraints

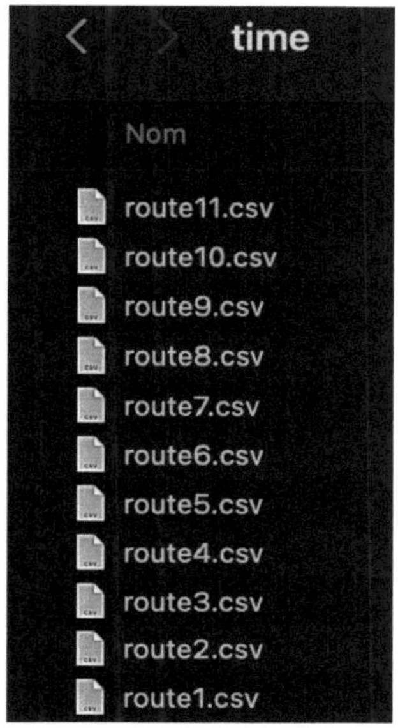

Fig. 16.7 List of files containing the scenarios' data. The folder is named `time`

- an integer `nwindow` (the n-averaging parameter described above and used in the definition of the rate of uncer-sampling,
- an initial state of charge `x0` $\in [0.2, 0.8]$, and
- a targeted final state of charge `xf` $\in [0.2, 0.8]$.

```
def solve_avg(ifile, nwindow, x0, xf):
    #-------------------------------------------------------------
    # find the optimal charging sequence given a route
    # represented by the dataframe df, the decision sampling
    # defined by nwindow (the u is constant over nwindow
    # succesive instants), the initial state of charge x0 and
    # the final desired state of charge xf.
    # ----------------
    # ifile: the file number defining the route to work on
    # nwindow : The averaging window to define the optimal
    #           control problem
    #         (in terms of number of sampling periods).
    #
    # x0 : The initial state of charge
    # xf : the final targeted state of charge
    #-------------------------------------------------------------
```

First, the dataframe is read and the indices of lines that correspond to the under-sampling defined by the parameter `nwindow` are computed through the use of the `numpy.arange` list of indices creator.

```
file = route_files[ifile]
df = pd.read_csv('time/'+file)
# Computes the under sampling indices
ind = np.arange(0,len(df),nwindow)
```

These indices are then used in order to compute the quantities:

$$\bar{m}_{i,\ell},\ \alpha_\ell,\ \beta_\ell,\ \Phi,\ \Psi,$$

described in the previous section regarding the re-formulation of the optimal control problem. This is done in the following script:

```
m0_sum, m1_sum, m2_sum = [], [], []}

dxmin, dxmax = [], []

for k0 in ind[:-1]:
    #-------
    # The sequences containing the INTEGRATED values over the
    # windows of size nwindow

    m0_sum +=  [df.m0[k0:k0+nwindow].sum()*tau]
    m1_sum +=  [df.m1[k0:k0+nwindow].sum()*tau]
    m2_sum +=  [df.m2[k0:k0+nwindow].sum()*tau]

    #-------
    # The profiles of w1 and w2 on the moving window
    w1k = df.w1[k0:k0+nwindow]
    w2k = df.w2[k0:k0+nwindow]

    # The integral of xdot over the window for the extreme values
    # of the inputs umin and umax. These are used to compute the
    # piecewise linear model on each segment through the coefficients
    # alpha and beta.
```

```
    dxmin += [np.sum(a1*(a2-np.sqrt(a2**2+a3*w1k+a4*w2k*umin)))*tau]
    dxmax += [np.sum(a1*(a2-np.sqrt(a2**2+a3*w1k+a4*w2k*umax)))*tau]

nsegments = len(m0_sum)

# compute the Phi and Psi matrices such that
# the trajectory of x = Phi + Psi * U
# ------
alpha = np.array(dxmin)
beta = (np.array(dxmax)-np.array(dxmin))/umax

Phi = x0+np.cumsum(alpha)
Psi = np.zeros((nsegments, nsegments))
for k in range(nsegments):
    Psi[k:,k] = beta[k]
```

Then the definition of the matrices involved in the definition of the cost, namely, the matrix of the purely quadratic term P and the vector involved in the affine term p as defined by (16.15) is implemented (Recall that these quantities will be later modified in order to include the formulation of the soft constraint on the final equality constraints $x(t_f) = x_f = 0.5$):

```
#--------------------------------------------------
# The cost function\index{Cost function} matrices (H, P)
# H is defined with a regularization parameter
# epsilon = 1e-9
#--------------------------------------------------
epsilon = 1e-9
q = np.array(m1_sum)
H = np.diag(m2_sum)
P = H + H.T + epsilon * np.eye(H.shape[0])
```

Notice that a regularization term (involving the `epsilon` penalty) is introduced here in order to derive a QP problem with strictly convex cost function which greatly helps avoiding singular situations without impacting significantly the resulting solution.

The next step is to express the inequality constraints in the form:

$$G\bar{\mathbf{u}} \leq h,$$

which simply expresses the set of constraints described earlier through (16.23):

```
#--------------------------------------------------
# The ineq constraints matrices (G,h)
# Expressing the constaints xmin <= x <= xmax
#--------------------------------------------------
G = np.vstack([Psi, -Psi])
Xmin = xmin * np.ones(nsegments)
Xmax = xmax * np.ones(nsegments)
h = np.vstack([(Xmax-Phi).reshape(-1,1),
               (Phi-Xmin).reshape(-1,1)]).flatten()
```

Finally, the matrix and vector needed to express the terminal equality constraint on the state of charge as well as the lower and upper bound on the decision variable vector are computed[9]:

```
#--------------------------------------------------
# The equality constraints matrices
# in the truly implemented version of the problem
# this equality constraint is replaced by a high
# penalized (via coef_xf) term in the cost function
#--------------------------------------------------
A = Psi[-1,:].reshape(1,-1)
b = np.array([xf - Phi[-1]])
#--------------------------------------------------
# Bounds on the decision variables
#--------------------------------------------------
lb = umin * np.ones(nsegments)
ub = umax * np.ones(nsegments)
```

[9] Recall that in python `Psi[-1,:]` refers to the last line of the matrix `Psi`.

In order to solve the resulting QP problem, we shall use the framework `QPsolver` that is freely available at

https://pypi.org/project/qpsolvers/

which incorporates several possible algorithms developed by different research groups.

The problems addressed by the solver are described by the following transcription of the project website:

Usage

The library provides a one-stop shop "solve_qp" function with a "solver" keyword argument to select the backend solver.
It solves convex quadratic programs in standard form:

$$min_x \quad \frac{1}{2}x^T Px + q^T x \quad \text{subject to } Gx \leq h \quad Ax = b \quad lb \leq x \leq ub$$

vector inequalities apply coordinate by coordinate. The function returns the solution x* found by the solver, or None in case of failure/unfeasible problem. For most solvers, the Matrix P should be positive definite.

which obviously perfectly fits our need.

So let us call one of the available algorithms to solve our problem:

```
#------------------------------------------------
# Solve the quadratic problem
#------------------------------------------------
ubar = solve_qp(P + coef_xf * A.T@A,
                q - coef_xf * A.T@b,
                G, h, None, None,
                lb, ub, solver="scs")
```

Note that in this call, the matrix P and q are modified so that the cost function incorporates a high penalty (`coef_xf=1e7` is used to penalize the violation of the constraint (16.25)).

Notice that the two `None` values used in the call express the fact that there are no **equality constraints** involved in the problem since the only such ones are replaced by a soft version through associated penalties on the cost function.

Recall that the optimal solution $\bar{\mathbf{u}}^\star$ returned by the previous call and named `ubar` in the Python script represents the optimal solution in the reduced dimensional control profile (due to the under-sampling introduced through the integer parameter n (or `nwindow` in the script).

Having this optimal sequence at hand, it is possible to compute the corresponding state trajectory and the associated cost (in the under-sampled framework) using the following script[10]:

```
#-------------------------------------------------
# Resulting trajectory based on the time varying
# linear averaged model.
#-------------------------------------------------
xbar = Phi + Psi @ ubar
tbar = np.array([i * tau for i in ind[:-1]])

# Resutling cost function with the
independent m0-term
J = df.m0.sum()*tau + (q.T @ ubar + 0.5 * ubar.T @ P @ ubar)
```

Moreover, one can use the resulting under-sampled control sequence to reconstruct the associated τ-sampled control sequence and inject it in the exact nonlinear model in order to compute the associated nonlinear-dynamics induced state trajectory and the associated cost since this is the cost and trajectory that really matter.

This can be done using the following script:

```
#-------------------------------------------------
    # simulate the true nonlinear dynamics under
    # the computed control sequence to evaluate the
    # associated true cost.
    #-------------------------------------------------
    ubar_nl = np.array([[ubar[i]] * nwindow
                        for i in range(len(ubar))]).flatten()
```

[10] These computed quantities will be exported as fields in the output object of the function `solve_avg` from which all the previous excerpts are extracted.

```
    xbar_nl = simulate_soc(x0, df.w1, df.w2, ubar_nl)
    tbar_nl = np.array([i*tau for i in range(len(ubar_nl))])

    Jnl = tau * (df.m0.sum() +
                 np.multiply(df.m1.values[:len(ubar_nl)],
                             ubar_nl).sum() +
                 np.multiply(df.m2.values[:len(ubar_nl)],
                             np.power(ubar_nl,2)).sum())
```

Note how the first instruction starting with

```
ubar_nl= ...
```

creates a τ-sampled sequence using the $n\tau$-piece-wise control sequence before this reconstructed control sequence is used in the simulation function `simulate_soc` which simulates the nonlinear state of charge equation over the time of the corresponding scenario.

Finally, the associated cost is computed.

For the sake of completeness, the function `simulate_soc` used in the script above to simulate the nonlinear equation governing the evolution of the state of charge is given below[11]:

```
def simulate_soc(x0, w1, w2, U):
#---------------------------------------------------
# Simulate the evolution of the state of charge (SoC)
# correctly using the squared-root term and not the
# linearized one. This enbles to check that the presumably
# optimal results meets the constraints.
#---------------------------------------------------
X = [x0]
for i in range(len(U)):
    inter = a2**2 + a3*w1[i] + a4*w2[i]*U[i]
    X += [X[-1] + tau * a1*(a2-np.sqrt(inter))]
return np.array(X)
```

[11] Note that a simple first-order explicit Euler method is used because the state of charge evolves very slowly making this scheme largely sufficient even for the down-simplified dynamics.

The last part of the function `solve_avg` is used to construct the dataframe named `dfX` regrouping the results and which will be used in the learning process later on.

More precisely, the dataframe `dfX` shows the following columns:

$$v_1, \dots, v_n, \quad p_1, \dots, p_n, \quad a_1, \dots, a_n, \quad u, x, \text{ifile},$$

representing successively the n successive previous speeds, positions, and accelerations followed by the optimal control value, the state of charge, and the scenario file's number.

It is shown below that the kinetic information (position, speed, and acceleration) are averaged leading to a learning scheme in which the vector of features is composed of the averaged information over a moving window of n-successive values:

$$\texttt{feature vector:} \quad x, \bar{p}, \bar{v}, \bar{a}. \tag{16.26}$$

For the time being, let us have a look on the script that enables to reconstruct the dataframe `dfX`:

```
#---------------------------------------------------
# Extraction of the information regarding position,
# speed and acceleration and state of charge on the windows
# in order to create the dataframe dfX
# for the training of the optimal control model.
#---------------------------------------------------
nL = int(np.floor(len(df)/nwindow)*nwindow)
v = df.speed.values[:nL].reshape(-1,nwindow)
p = df.position.values[:nL].reshape(-1,nwindow)
a = df.acc.values[:nL].reshape(-1,nwindow)
#--------
colp = [f'p{i+1}' for i in range(nwindow)]
colv = [f'v{i+1}' for i in range(nwindow)]
cola = [f'a{i+1}' for i in range(nwindow)]

dfX = pd.DataFrame(v, columns=colv)
dfX[colp] = p
dfX[cola] = a
dfX['u'] = ubar
dfX['x'] = xbar
dfX['ifile'] = [ifile] * len(dfX)
```

```
class Container:
    pass
```

Fig. 16.8 This is a simple class that enables to dynamically add the different information computed inside a function as fields in order to export the computed results in a compact form

Once all the results have been computed inside the function `solve_avg`, they are gathered as fields of an instance of the class `Container` shown in Fig. 16.8 and the resulting instance is exported as an output of the function.

The use of the class `Container` serves simply to export many results in a single compact object.

```
# Regroup the results as field in the object output

output = Container()

# Results on the undersampled time interval
# ------------------------------------------
output.tbar = tbar
output.ubar = ubar
output.xbar = xbar
output.J = J
output.t = tbar_nl
output.x = xbar_nl
output.u = ubar_nl
output.Jnl = Jnl
output.ifile = ifile
output.m0_sum = m0_sum
output.m1_sum = m1_sum
output.m2_sum = m2_sum
output.df = df
output.dfX = dfX

return output
```

Now that we dispose of an efficient solver that delivers the quasi-optimal solution for any pair (x_0, s) consisting of an initial state of charge $x_0 \in [0.2, 0.8]$ and a scenario number, we have all that we need to create the learning data that will be used to fit the feedback K required by the problem statement.

The construction of the learning data is the object of the following section.

16.4 Creating the Learning Data

As mentioned in Sect. 16.2, we shall create a cloud of pairs (x_0, s) of initial states of charge and scenario number.

For each of these pairs, the associated problem is solved and the resulting dataframe is associated in the field named `dfX` of the output of the function `solve_avg` described above.

By so-doing, a list of dataframes is obtained. These dataframes are concatenated using the `pandas` function `concat` to obtain a single quite big dataframe that encompasses the learning data.

This is done in the following rather self-explanatory script[12]:

```
# Choose the under-sampling/averaging window size
nwindow = 20
xf = 0.5
# Create the list of dataframe
dfs = []
for ifile in range(len(route_files)):
    for x0 in [0.2, 0.35, 0.5, 0.65, 0.8]:
        # solve the optimal control problem for the pair (x0, ifile)
        R = solve_avg(ifile=ifile, nwindow=nwindow, x0=x0, xf=xf)
        dfs += [R.dfX]

    # Concatenate the resulting dataframes in a single
    # dataframe and save it
    # -----------------------------------------------
    dfX = pd.concat(dfs)
    pickle.dump(dfX, open(f'dfX_{nwindow}.pkl', 'wb'))
```

[12] Note how the scenarios are iterated on in the outer loop which means that when the data will be split without shuffling in training and test data, the two sets will contain different scenarios. For instance, if a train/test split of 50% is used, the training set will contain the scenarios with indices 0, 1, …, 4 while the test set 6, …, 10. The scenario indexed 5 will be split into two different blocks, each belonging to one of the subsets.

Notice that the above script creates the learning data for the specific choice of $n = 20$.

The result is saved in the `pickle` object named:

`dfX_20.pkl.`

The script is then executed using the different choices:

`nwindow = 5, 10, 15,`

in order to get the associated learning dataframes named:

`dfX_5.pkl, dfX_10.pkl, dfX_15.pkl.`

Each of these dataframes is used hereafter to fit the corresponding feedback `K` that approximately solves the stated problem.

16.5 Computing the Feedback by Fitting Machine Learning (ML) Models

As mentioned earlier, there are so many structures of models that can be used to solve regression problems such as the one we are concerned with here.

Indeed, we would like to find a map taking as arguments the mean values (over a previous measurement window of size n) of the position, speed, and acceleration of the vehicle as well as the value of the state of charge of the battery and delivers as output the associated instance of the optimal sequence at the considered sampling instant $k\tau$.

In the sequel, we shall use a very famous model, namely:

`eXtreme Gradient Boosting (XGBoost),`

Chen and Guestrin [2] which performs similarly to the well-known `RandomForest Regressor` model while being faster to compute in average.

The first thing to do is to import the model using the `import` command. By the way, we import the data splitting facilities used later on. Moreover, we fix the train/test split ratio to 50% and define the sequence of values of `nwindow` for which learning data is available following the scripts of the previous section:

```
from xgboost import XGBRegressor
from sklearn.model_selection import train_test_split
```

```
nwindows = [5, 10, 15, 20]
test_size = 0.5

# -------------------------------------------------------
# Create the dictionary of models with nwindow as key
# -------------------------------------------------------
regs = {nwindow:XGBregressor() for nwindow in nwindows}
```

Note that at the end of the previous script, a dictionary of models is created with the keys being the different values of the integer `nwindow` that defines the available rates of under-sampling used in the reduction of the degrees of freedom.

This is because we shall fit a different model for each value of `nwindow` included in the list of values `nwindows` declared in the head of the previous script.

Remember that the value of the dictionary at some key value, namely `nwindow`, will be accessible through the call

`regs[nwindow]`,

which will be an `XGBregressor` instance representing the feedback map we are looking for when the specific `nwindow` value is used.

For the time being, the regressors (one for each value of `nwindow`) are not yet fitted but only instantiated. Once the loop on the values of `nwindow` is executed, the associated training data is created, the different fields of the dictionary, named `regs` in the previous script, will be correctly fitted for later use.

Now we shall iterate on the possible values of `nwindow` and for each value, the training data is used to fit a `XGBoost` regressor and the resulting fitted map is used to predict the control using the instances of features contained in the unseen test data.

More precisely, after loading the data associated to the value of `nwindow`, the columns associated to

- position;
- speed; and
- acceleration

are first averaged to create a new dataframe with only four features, namely:

$$x, \ \bar{p}, \ \bar{v}, \ \bar{a}.$$

This is done in the following part of the script which extracts the pair `(X,y)` associated to the current value of `nwindow`.

This pair is needed to call the fit method of the XGBregressor module:

```
for nwindow in nwindows:

    # load the saved data for this value of nwindow
    dfX = pickle.load(open(f'dfX_{nwindow}.pkl', 'rb'))

    # Extract the label to be used in the fit step.
    y = dfX['u'].values

    # Compute the average of the position, speed
    # and acceleration.
    # -----------------
    colp = [c for c in dfX.columns if 'p' in c]
    colv = [c for c in dfX.columns if 'v' in c]
    cola = [c for c in dfX.columns if 'a' in c]
    df_avg = pd.DataFrame(dfX['x'].values, columns=['x'])
    df_avg['p'] = dfX[colp].mean(axis=1).values
    df_avg['v'] = dfX[colv].mean(axis=1).values
    df_avg['a'] = dfX[cola].mean(axis=1).values

    # This is the feature matrix to be used in the fit step
    X = df_avg.values
```

Now that we dispose of the learning pair (X,y) of features matrix and label's vector, it is time to slit the data into training subset and a test subset.

The corresponding XGBregressor can therefore be fitted, using the training data, and used to predict the labels corresponding to the instances of features that are present in the unseen data.

This is done by the following call[13]:

```
# Split the training data without shuffle
#-----------------------------------------
Xtrain, Xtest, ytrain, ytest = train_test_split(X, y,
```

[13] Remember that this work is done on the data associated to the current value of nwindow.

```
                                    test_size=test_size, shuffle=False)

# Create an instance of the regressor and fit based on the
# training set.
#---------------------------------------
regs[nwindow].fit(Xtrain, ytrain)
yhat = reg.predict(Xtest)
```

Note the use of the argument option `shuffle=False`[14] in the call of the data splitting utility `train_test_split` which prevents a random shuffle of the lines which would increase the probability of having *twin* lines in the training and the test data leading over-optimistic results.[15]

The consequence of that is that only the first 5 scenarios and a tiny part of the sixth one are used in the fitting of the model.

Note also that the fitted model is the field, of the `regs` dictionary, that is associated to the key `nwindow` which is compatible with the fact that the data we are working on corresponds to the current value of the under-sampling/averaging value of `nwindow`.

The comparison of the predicted and the true optimal results is shown in Fig. 16.9 which is plotted thanks to the following script:

```
fig, ax = plt.subplots(figsize=(15, 4), ncols=2)
    ax[0].plot(ytest, yhat, 'ro', alpha=0.05)
    ax[0].plot(ytest, ytest, 'k-', )
    ax[0].grid(True)
    ax[0].set_xlabel('True test values')
    ax[0].set_ylabel('Predicted test values')
    ax[0].set_title(f'Fitting results for nwindow={nwindow}')
    ax[1].hist(ytest, bins=200, alpha=0.3);
    ax[1].hist(ytest-yhat, bins=200, alpha=0.3, log=True);
    ax[1].grid(True)
    ax[1].set_xlabel('Control u');
    ax[1].set_ylabel('counts')
    ax[1].set_title('Histogram of errors vs histogram of u');
    ax[1].legend(['u', 'Error on u-prediction']);
```

[14] See the related discussion and examples already provided in Sect. 3.4.3.

[15] For a detailed discussion on this over-fitting and shuffling issues, the reader might refer to Sect. 3.4.3.

An interesting observation on Fig. 16.9 is related to the fact that the error on the prediction of the optimal control seems symmetrically distributed on both sides of the 0-error which is a quite promising result.

Indeed, one should remember that this prediction map will be called once each $\tau = 0.2\,\text{s}$ (the basic sampling period) which means that successive errors might compensate each other in average.

But fortunately enough, we do not need to rely on such speculations since the efficiency and the degree of optimality of the resulting fitted feedback will be carefully examined on real and exact simulations over the different scenarios and for different values of the initial state of charge leading to unseen instantiations of the problem on which the generalization power and the robustness of the design feedback law are checked.

This is the aim of the following section.

16.6 Validation of the Resulting Learned Feedback Laws

16.6.1 Closed-Loop Simulations

In order to validate the performance of the resulting learned feedback law, we need to simulate the corresponding closed-loop systems.

This is the aim of the function `simulate_closed_loop` that is explained and commented in the following paragraphs:

`simulate_closed_loop(ifile, x0, nwindow).`

This function takes the following arguments:

- the scenario number `ifile`,
- the initial state of charge `x0`, and
- the integer `nwindow` that defines the under-sampling and the averaging as explained in the previous section.

First of all the saved dictionary of models `regs` is loaded and the model (feedback law) corresponding to the input value of `nwindow` is recovered and named `reg`.[16]

This is done using the following script:

```
# Download the dictionary of fitted model and
# select the one that corresponds to the
```

[16] Remember that this is an `XGBregressor`.

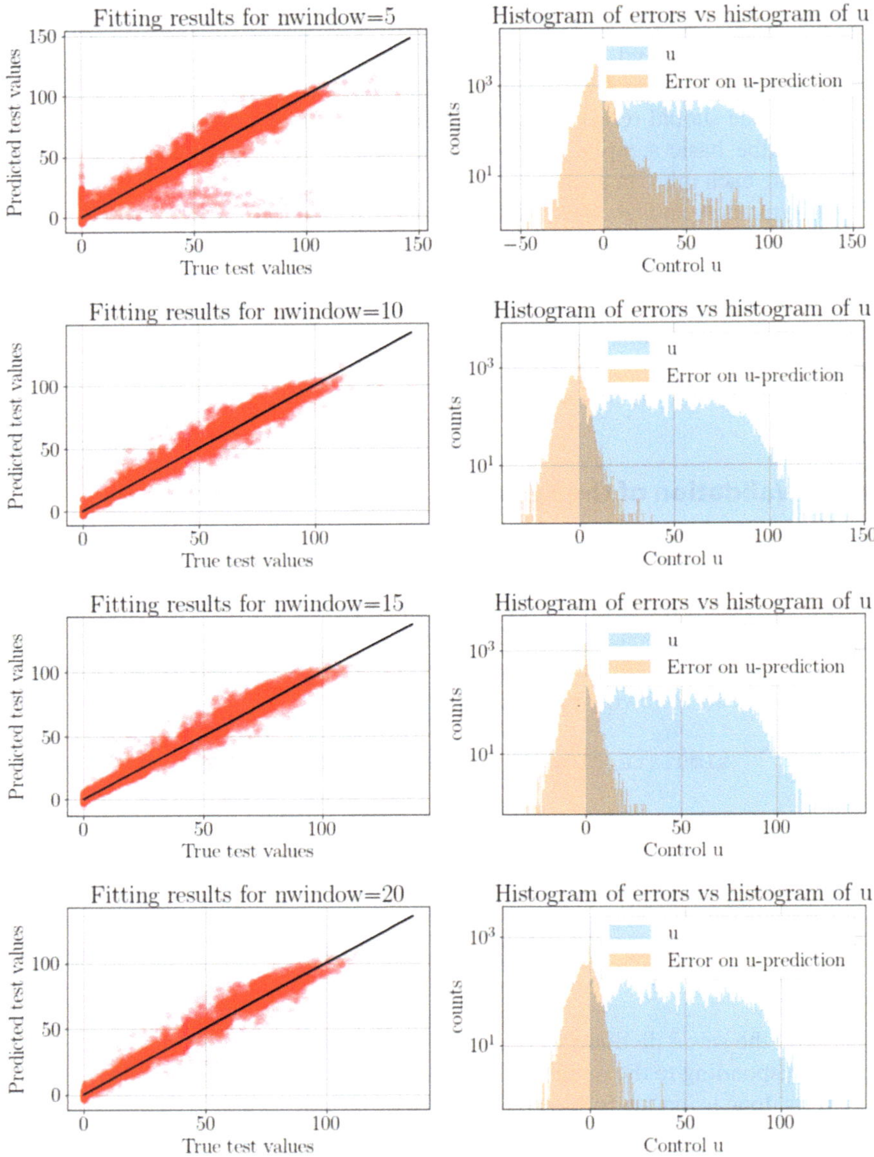

Fig. 16.9 Fitting results for the prediction on the test dataset of the `XGBregressor` identified for the different values of the down-sampling/averaging parameter `nwindow`. Each line corresponds to a given value of `nwindow`. (Left) the comparison between the true optimal values and the predicted ones. (Right) the histogram of the true optimal control and the histogram of the prediction error

```
# argument nwindow.
#---------------------------------------------
regs = pickle.load(open('regs.pkl', 'rb'))
reg = regs[nwindow]
```

Once this is done, the open-loop quasi-optimal solution of the problem defined by the pair `(x0,ifile)` is solved using the function `solve_avg` described earlier. This computation is not needed for the simulation of the closed-loop per se but it is done for the sake of comparison.

The resulting dataframe is named `df`.

On the other hand, the equivalent dataframe in closed-loop is initialized and named `df_cl`.

The column representing the state of charge is created and initialized before its content is updated via the closed-loop computation to come:

```
# Extract the data of the routes
# solving the open-loop problem is just an way extract the data
# so that comparison can be available with the idealistic optimal
# solution.

R = solve_avg(ifile, nwindow, x0, 0.5)
df = R.df
df_cl = df.copy()

# Create the column of closed-loop SoC
#---------------------------------------
df_cl['xcl'] = [x0] * len(df)
```

Now in order to start applying the closed-loop control we need to reconstruct the mean values of the position, speed, and acceleration over a sliding window of width `nwindow`. This is because these averages are the features on which the feedback laws (the `XGBRegressors`) have been fitted and hence, now they are used as the argument of the feedback laws.

The computation of the moving averages is done using the `pandas` utility named `rolling` as shown in the following script:

```
# Create the averaged value so that the prediction model can be
# consistently called. Remove the Nan values. Note that this means
# that the initial state x0 is the value at instant nwindiw * tau

df_cl = df_cl[['xcl', 'position', 'speed', 'acc']].rolling(
                                    window=nwindow).mean()
df_cl = df_cl.iloc[nwindow:]
```

Notice that the last line in the script above is necessary in order to remove the lines containing *non-assigned numbers* (NaN) that always result from the use of the rolling window.

This has little influence on the value of the cost function since we remove no more than 20 lines over the thousands of lines contained in the scenarios.

```
Ucl = [0]
for k in tqdm(range(len(df_cl)-1)):

    # current state of charge & regressor value
    x = df_cl.iloc[k]['xcl']
    X = df_cl.iloc[k].values.reshape(1,-1)

    # compute control using model (to apply at the next period)

    u =  reg.predict(X)[0]
    if u<0: u=0
    w1 = df.w1[k+nwindow]
    w2 = df.w2[k+nwindow]

    # Apply the previously computed control

    term = a2**2+a3*w1+a4*w2*Ucl[-1]
    xplus = x + tau * (a1*(a2-np.sqrt(term)))
    df_cl.iloc[k+1]['xcl'] = xplus

    # save computed control for the next step
    Ucl += [u]
```

In the script above, it is important to notice that the control that is applied to the system at instant k is the one computed in the previous sampling period, namely `U[-1]` while the control computed at instant k through the call to the `predict` method of the regressor `reg`:

```
u =  reg.predict(X)[0],
```

is applied only in the next step, namely over the period $k + 1$.

Notice that the vector of features used in the above call is precisely the k-line of the moving averaged-based dataframe `df_cl`:

```
X = df_cl.iloc[k].values.reshape(1,-1)
```

When the closed-loop simulation is ended, we dispose of the closed-loop control profile `Ucl` that concatenates the values provided by the computed feedback model based on the encountered values of the averaged position, speed, acceleration, and the state of charge during the simulation.

This closed-loop control profile can now be used to compute the associated costs through the following script:

```
m0 = df.iloc[nwindow:]['m0'].values
m1 = df.iloc[nwindow:]['m1'].values
m2 = df.iloc[nwindow:]['m2'].values

ucl = np.array(Ucl)
Jcl = tau * (m0.sum() +
           np.multiply(m1, ucl).sum() +
           np.multiply(m2, np.power(ucl,2)).sum()) * coeff

df_cl['ucl'] = Ucl
```

Note how the `m0`, `m1` and `m2` associated to the chosen scenario are first recovered and how the first `nwindow` lines are removed to be compatible with the time index of the first computed closed-loop control. Then the definition of the cost function is used to compute the resulting cost. Finally a column is added to the closed-loop simulation dataframe that represents the feedback action applied during the closed-loop simulation.

Before ending this study with a more rich and systematic analysis of the quality of the resulting identified feedback control, Fig. 16.10 shows a comparison between quasi-optimal control profiles and the resulting closed-loop profile using the identified (fitted) feedback for different scenarios, different initial SoC and different choices of the down-sampling/averaging window size `nwindow`.

These plots clearly show that the identified (fitted) `XGBregressor` does capture the main dependencies between the *current average situation* in terms of the state of charge and the kinetic variables.

Notice however that these figures showing the control profiles do not tell us that much about the resulting cost nor about the final constraint satisfaction on the state of charge of the battery.

In order to examine these two important features, a more detailed analysis is needed.

This is the aim of the next section.

16.6.2 Validation of the Quality of the Identified Feedback Laws

In order to get a more solid assessment of the quality of the identified feedback laws, 10 closed-loop simulations are performed for each of the eleven available scenarios, each simulation is conducted for a randomly chosen initial state of charge.

Note that an additional column is created that contains the relative difference between the optimal and the closed-loop cost function.

```
nwindow = 20
dfR = []
# iterate over the 11 scenarios
for ifile in range(len(route_files)):

    # for the scenario number ifile
    # generate randomly 10 initial states of charge

    for i in range(10):

        x0 = 0.2 + 0.6 * np.random.rand()
        df_cl, R, Jcl = simulate_closed_loop(ifile, x0, nwindow)
```

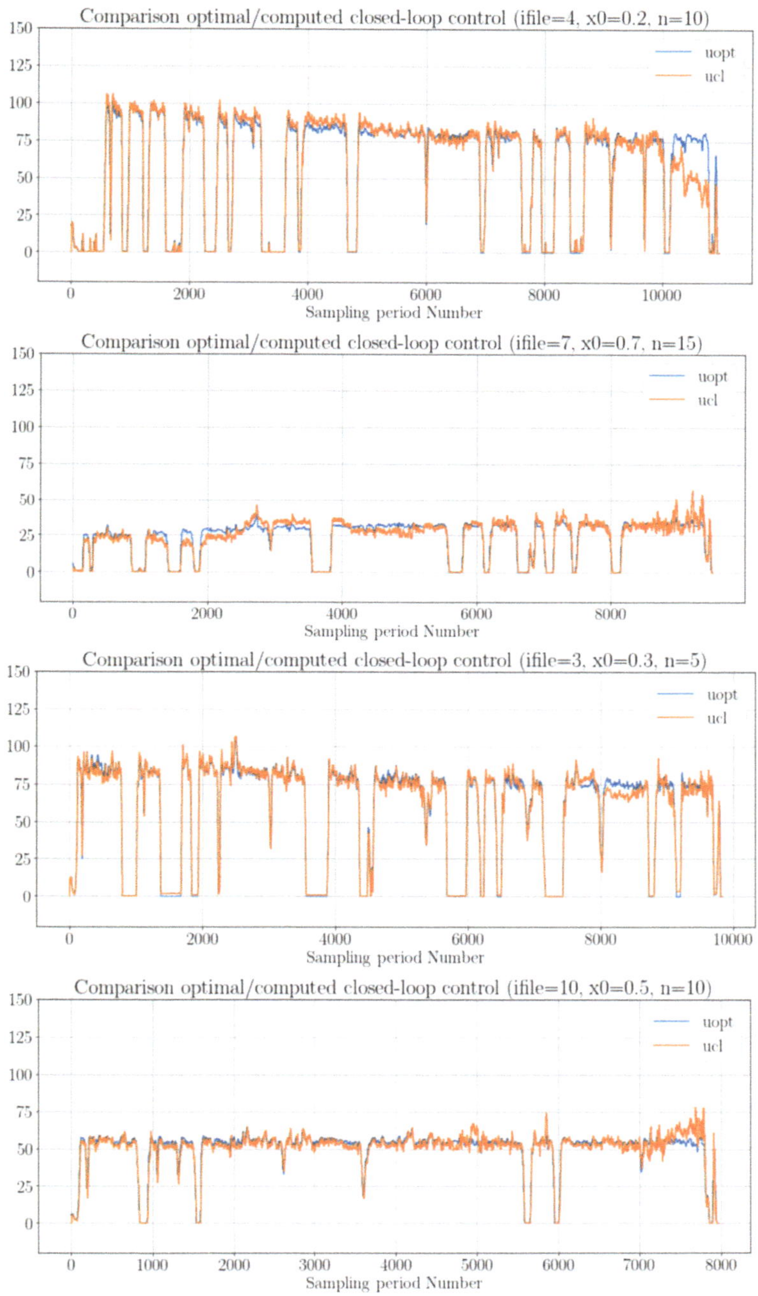

Fig. 16.10 Typical comparison between quasi-optimal control profiles and the resulting closed-loop profiles using the identified (fitted) feedback for different scenarios, different initial state of charge and different choices of the down-sampling/averaging window size

```
        dfR += [pd.DataFrame([ifile, x0, R.Jnl, Jcl,
                              R.xbar[-1], df_cl['xcl'].values[-1]],
                     index=['ifile', 'x0', 'Jopt', 'Jcl',
                            'xf_opt', 'xf_cl']).T]

# Create a single dataframe for this choice of nwindow
dfR = pd.concat(dfR, axis=0)

#compute the relative difference in the cost function
dfR['(Jcl-Jopt)/Jopt'] = (dfR.Jcl-dfR.Jopt)/dfR.Jopt

pickle.dump(dfR, open(f'dfR_{nwindow}.pkl', 'wb'))
```

The whole operation is repeated for the different values of the averaging/down-sampling window size `nwindow`.

This leads to the following four large dataframes:

```
dfR_5, dfR_10, dfR_15, dfR_20.
```

The following figure shows the dataframe `dfR_20`:

	ifile	x0	Jopt	Jcl	xf_opt	xf_cl	(Jcl-Jopt)/Jopt
0	0.0	0.545336	1165.640071	1154.991053	0.499845	0.500010	-0.009136
1	0.0	0.293567	1559.387062	1547.169456	0.499843	0.499672	-0.007835
2	0.0	0.327239	1506.441199	1494.360986	0.499843	0.499838	-0.008019
3	0.0	0.540706	1172.835952	1162.135701	0.499845	0.500007	-0.009123
4	0.0	0.721336	893.315698	886.223601	0.499846	0.500307	-0.007939
...	...	...	...	...	...	...	...
105	10.0	0.297017	1420.015312	1407.829042	0.499842	0.499697	-0.008582
106	10.0	0.753511	708.998460	701.928566	0.499846	0.500288	-0.009972
107	10.0	0.771937	680.649823	674.093510	0.499846	0.500356	-0.009632
108	10.0	0.787685	656.443973	650.459819	0.499846	0.500504	-0.009116
109	10.0	0.787700	656.421003	650.434309	0.499846	0.500502	-0.009120

110 rows × 7 columns

Each line of this dataframe describes the results of a whole closed-loop simulation under the feedback law identified for the value `nwindow` = 20.

So let us check the two main key information, namely:

- the cost function's proximity to the optimal values,
- the satisfaction of the terminal constraint on the final state of charge.

This is done hereafter by using one of the nice methods in the `pandas` package which is the `groupby` method.

This utility enables to perform operations on the lines of the dataframe which are regrouped according to the values of one or several columns.

Here we are going to examine the statistics of the values of the final SoC reached in closed-loop over the 10 possible values of the initial state of charge separately for each scenario number.

This is done using the following script:

```
nwindow=10
dfR = pickle.load(open(f'dfR_{nwindow}.pkl', 'rb'))
dfR.groupby(by='ifile').mean()
```

ifile	E[x0]	E[xf_cl]	E[Jcl-Jopt)/Jopt]	std[x0]	std[xf_cl]	std[Jcl-Jopt)/Jopt]
0.0	0.585965	0.500029	-0.009490	0.174766	0.000229	0.000845
1.0	0.485109	0.502354	-0.005177	0.162113	0.002099	0.003591
2.0	0.505799	0.498510	-0.009189	0.169320	0.003284	0.002499
3.0	0.431917	0.499632	-0.008362	0.140914	0.000326	0.000567
4.0	0.566803	0.501439	-0.005711	0.154465	0.001198	0.002465
5.0	0.545765	0.499829	-0.008045	0.115134	0.000131	0.000619
6.0	0.506520	0.500977	-0.005864	0.210593	0.000607	0.001812
7.0	0.447693	0.499656	-0.008427	0.200293	0.000355	0.000731
8.0	0.503477	0.500491	-0.008333	0.184761	0.000354	0.000740
9.0	0.470792	0.499940	-0.008250	0.164230	0.000140	0.000425
10.0	0.532297	0.499947	-0.010160	0.189628	0.000360	0.000928

Fig. 16.11 Statistics of the closed-loop for `nwindow = 10` for the different scenarios. The columns show the averaged values and the standard deviation over the different initial states of charge used in the simulations. Note that the error on the cost function as well as on the terminal state of charge are quite small

ifile	E[x0]	E[xf_cl]	E[Jcl-Jopt)/Jopt]	std[x0]	std[xf_cl]	std[Jcl-Jopt)/Jopt]
0.0	0.500591	0.499993	-0.008486	0.153418	0.000214	0.000638
1.0	0.544031	0.505510	0.000648	0.178917	0.006592	0.011908
2.0	0.474615	0.497733	-0.009220	0.193126	0.003677	0.002643
3.0	0.463309	0.499219	-0.008363	0.163496	0.001409	0.000792
4.0	0.474636	0.500577	-0.006938	0.142841	0.000810	0.001123
5.0	0.565683	0.499377	-0.007721	0.167512	0.001621	0.000939
6.0	0.520068	0.501239	-0.005823	0.167846	0.000333	0.000873
7.0	0.442710	0.499625	-0.008499	0.154705	0.000283	0.000467
8.0	0.586057	0.501230	-0.007135	0.129339	0.001402	0.003201
9.0	0.493543	0.500065	-0.007813	0.161782	0.000246	0.000520
10.0	0.604774	0.500145	-0.009571	0.209945	0.000307	0.000797

Fig. 16.12 Statistics of the closed-loop for `window = 20` for the different scenarios. The columns show the averaged values and the standard deviation over the different initial states of charge used in the simulations. Note that the error on the cost function as well as on the terminal state of charge are quite small

which leads to the dataframe shown in Fig. 16.11 where the parameter `nwindow = 10` is used.

The same results are shown in Fig. 16.12 for the choice `nwindow = 20`.

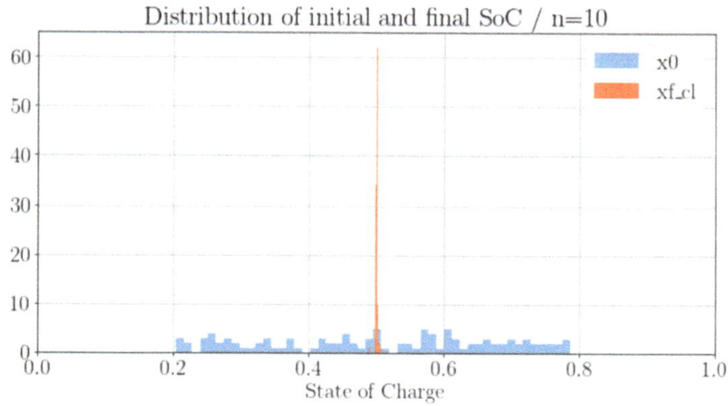

Fig. 16.13 Histogram of initial and closed-loop induced **terminal state of charge** for `nwindow = 10` and for all the 110 scenarios

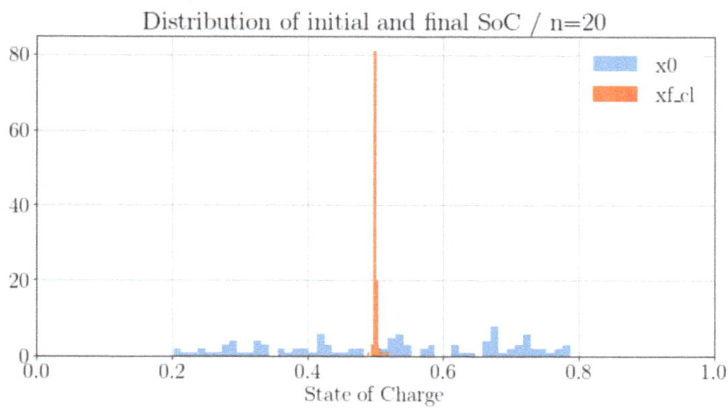

Fig. 16.14 Histogram of initial and closed-loop induced **terminal state of charge** for `nwindow = 20` and for all the 110 scenarios

The results clearly show that the identified feedback map enables to quite precisely enforce the terminal constraint on the state of charge while inducing a cost function that is almost identical to the one obtained via tight approximation of the **full-knowledge-based optimal control problem**.

In order to see the results more clearly, the histogram of the initial state of charge and the terminal closed-loop induced state of charge are shown in Figs. 16.13 and 16.14 for `nwindow = 10` and 20 respectively.

These figures clearly show the extent to which the terminal constraint is respected.

Finally Figs. 16.15 and 16.16 show the histograms of the optimal cost and the difference between the realized closed-loop cost under the identified feedback for `nwindow = 10` and 20 respectively.

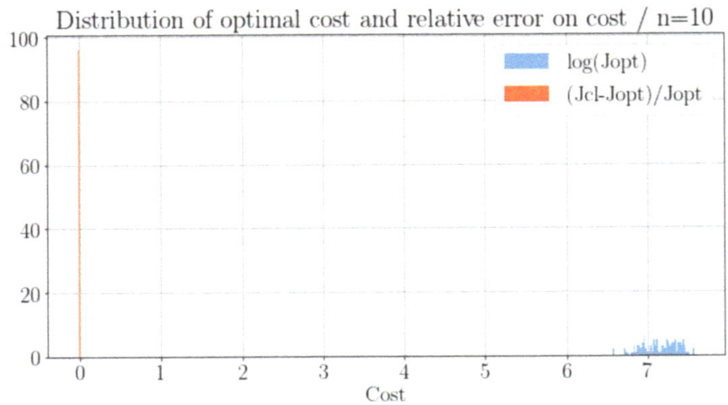

Fig. 16.15 Histogram of optimal cost and the closed-loop realized cost under the identified feedback law corresponding to `nwindow = 10` and for all the 110 scenarios. Note that the optimal cost is shown in a logarithmic scale for sake of readability

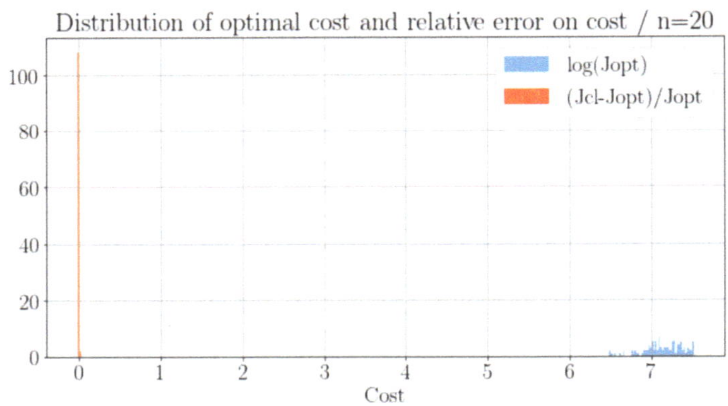

Fig. 16.16 Histogram of optimal cost and the closed-loop realized cost under the identified feedback law corresponding to `nwindow = 20` and for all the 110 scenarios. Note that the optimal cost is shown in a logarithmic scale for sake of readability

16.7 Conclusion

In this chapter, the methodology depicted in Chap. 7 is applied to the real-life problem of power management in hybrid vehicle using real-life collected data.

More precisely, it has been shown that by solving off-line a set of deterministic optimal control problems in which the behavior of the uncertainties are perfectly known, a relevant learning data can be constructed and an associated feedback function can be fit for use in closed-loop so that quasi-optimal results can be achieved while meeting tightly the constraints including the one related to the final SoC of the battery.

However one should keep in mind that the current study considered the same path from the same initial position to the same final position although using different driver's behavior and requirements.

Moreover, this chapter gives another instance of a typical situation, namely: we need to look at the problem's specificity in order to come out with appropriate simplifications and or parameterization of the degrees of freedom.

Acknowledgements The material of this chapter is based on a joint work with my colleague and friend Tielong Shen from the University of Sophia, Tokyo and our colleague Xu Fuguo from Tokyo-City University.

References

1. Diamond, S., Boyd, S.: CVXPY: a Python-embedded modeling language for convex optimization. J. Mach. Learn. Res. **17**(83), 1–5 (2016)
2. Chen, T., Guestrin, C.: XGBoost: a scalable tree boosting system. In: Proceedings of the 22nd ACM SIGKDD Conference on Knowledge Discovery and Data Mining, pp. 785–794 (2016)

17 Explicit Q-Learning-Based Approximation of Stochastic Optimal Control Feedback for Combined Therapy of Cancer

17.1 Introduction

Computing optimal drug delivery based on highly uncertain models is a crucial issue in medical applications that lies in the heart of this book.

Indeed, compartmental models that describe the dynamics governing the tumor cells' proliferation as well as the impact of drug's usage generally involve a high number of uncertain parameters. These parameters strongly depend on the patient as well as on its current state of health.

On the other hand, given the generally low level of excitation and the strong constraints on them together with the high number of parameters, it is obvious that these parameters cannot be adequately *identified* from on line measurements via extended state estimators.

Traditionally, the *applied mathematics*-related literature addresses deterministic optimal control problems based on the nominal values of the parameters involved in the model [1–3].

Although this approach might be relevant when qualitative conclusions regarding the drug delivery regimen (intensive therapy, presence of singular arcs, state-dependent order of drug's delivery and switching conditions to cite but few examples), the so-derived solutions cannot address the problem of the presence of potential high model mismatches.

This is, in particular, true because of the long-range impact of the uncertainties on one hand and the specific character of the control objective that is not a regulation/tracking problem nor is it a pure economic objective since cancer treatment require the *risk management* issue to be seriously addressed (important loss for some patients cannot be compensated by very good results on others).

Regardless of the above mentioned elements, standard works focus on **open-loop** drug delivery profiles without corrective actions that might be based on the on-line acquired measurements of the patients indicators.

M. Alamir, *Nonlinear Control of Uncertain Systems*,
https://doi.org/10.1007/978-3-031-93287-8_17

Such feedback strategies lead to a much lower sensitivity to model mismatches than open-loop application of presumably nominally optimal injection profiles.

Among the many possible feedback design methodologies, Model Predictive Control (MPC) that has been extensively studied in Chaps. 1, 5 and 7 offers the advantages of *smoothly* extending the open-loop optimal control tradition while being a feedback law that corrects the previously computed scheduling based on the latest on-line measurement of patient's state.

However, the performance of such approaches that do not explicitly use the information regarding the model's parameters statistical dispersion can only be observed a posteriori. No a priori certification regarding the closed-loop performances comes out at the design stage.

Another option is to follow the design strategy applied in Chap. 14 in which a parameterized feedback law is first designed based on the understanding of the model's interactions and causalities and then to optimally *tune* the controller's parameters using explicitly the statistical description their dispersion by means of the **randomized optimization framework** studied in Chap. 1. The complete execution of this control design methodology is proposed in [4].

Stochastic Model Predictive Control (SMPC) formalized in Sect. 2.6.2 offers, at least conceptually, an alternative to explicitly address the presence and the statistical laws governing the parameters dispersion. Despite the fact that the computational burden of this approach might be expensive, the available computation time for the specific case of cancer therapy might be sufficient to perform the associated computation (low updating rate and extremely slow dynamics).

This might result in a scalable solution that can be used regardless of the complexity of the underlying mathematical model. On the other hand, this option provides a solution for each initial state but does not provide a global overview of the performance of the strategy on a whole region of the State Space (SS).

Compared to the previous approach, the Stochastic Dynamic Programming (SDP) [5,6], also referred to in the AI literature as the Q-learning approach which has been already sketched in Sect. 2.8 offers the following two appealing features:

1. It leads to an explicit state feedback law that needs no heavy on-line computation and that is valid on a whole region of the SS.
2. It explicitly takes into account any statistical description of uncertainties.

It is no surprise that the above mentioned advantages are obtained at some price, namely, an exponentially increasing complexity of the approach with the state/control dimension since the so called value function function Q (see Sect. 2.8) needs to be represented over the extended space of the state and control. This makes this approach **not scalable** with the model's dimensions.

Moreover, at each step in the minimization of the residual of the SDP equation, one needs to compute statistically defined quantities that involves a high number of randomly sampled instances of the uncertain model's set of parameters in order to get consistent approximations of these quantities.

Notice however that this problem is shared by the SMPC approach as well, except that the SDP-based approach involves only off-line computations that are to be done once for all.

In this chapter, a Q-learning framework that has been already briefly depicted in Sect. 2.8 is used to address the problem of combined therapy of cancer while exploiting some specific properties of the governing model in order to accelerate the underlying computation.

More precisely, the value function Q is approximated by a `Support Vector Machine` (SVM) regression model. As for the *trick* that helps accelerating the fixed-point algorithm, a clustering approach is applied in order to derive fast approximation of the statistical quantities using a reduced number of samples. This induces a tractable approximate solution to the original problem.

Let us first state the control problem to address.

17.2 Problem Statement

17.2.1 The Dynamic Model

Consider the population model used by [7] to describe the dynamics governing the combined immuno-chemotherapy of cancer:

$$\dot{x}_1 = ax_1(1 - bx_1) - c_1x_4x_1 - k_3x_3x_1, \tag{17.1}$$
$$\dot{x}_2 = -\delta x_2 - k_2x_3x_2 + s_2, \tag{17.2}$$
$$\dot{x}_3 = -\gamma_0 x_3 + u_2, \tag{17.3}$$
$$\dot{x}_4 = g\frac{x_1}{h + x_1}x_4 - rx_4 - p_0x_4x_1 - k_1x_4x_3 + s_1u_1, \tag{17.4}$$

where

x_1	the tumor cell population,
x_2	the circulating lymphocytes population,
x_3	the chemotherapy drug concentration,
x_4	the effector immune cell population,
u_1	the rate of introduction of immune cells, and
u_2	the rate of introduction of chemotherapy.

Table 17.1 contains a description of the role of each group of terms in the underlying process.

The nominal values of the parameters involved in the (17.1)–(17.4) are given by:

$$a = 0.25, b = 1.02 \times 10^{-14}, c_1 = 4.41 \times 10^{-10}, g = 1.5 \times 10^{-2},$$
$$h = 20.2, k_2 = k_3 = 0.6, k_1 = 0.8, p_0 = 2 \times 10^{-11}, r = 0.04,$$
$$s_1 = 1.2 \times 10^7, s_2 = 7.5 \times 10^6, \delta = 1.2 \times 10^{-2}, \gamma_0 = 0.9$$

Table 17.1 Signification of the terms involved the dynamic model

Equations	Term	Description
(17.1)	$ax_1(1-bx_1)$	Logistic tumor growth
(17.1)	$-c_1x_4x_1$	Death of tumor due to effector cells
(17.1)	$-k_3x_3x_1$	Death of tumor due to chemotherapy
(17.2)	$-\delta x_2$	Death of circulating lymphocytes
(17.2)	$-k_2x_3x_2$	Death of lymphocytes due to chemo
(17.2)	s_2	Constant source of lymphocytes
(17.3)	$-\gamma_0x_3$	Exponential decay of chemotherapy
(17.4)	$g\dfrac{x_1}{h+x_1}x_4$	Stimulation of tumor on effector cells
(17.4)	$-rx_4$	Death of effector cells
(17.4)	$-p_0x_4x_1$	Inactivation of effector cells by tumor
(17.4)	$-k_1x_4x_3$	Death of effector cells due to chemo

namely, 14 parameters are involved in the dynamic model (17.1)–(17.4).

Note however that the order of magnitude of x_1 is around 10^9 while $h \approx 20$. The consequence of this is that the term h is effective only when the tumor is almost vanishing. However, in the absence of the term h, the effector immune population might keep growing even in the absence of tumor which obviously contradicts its definition being a tumor induced immune population.

Consequently, in the remainder of this chapter, this parameter is supposed to be known and consequently, it is not considered as a part of the set of uncertain parameters. As shown later on, this simplification enables a separable structure of the model that is exploited in the computation [see (17.6)].

Denoting by $p \in \mathbb{R}^{13}$ the vector of all remaining parameters and using simple Euler discretization scheme with some *sufficiently small* sampling period τ, the resulting discrete-time dynamics can be represented in the following condensed form:

$$x^+ = f(x,u,p) \quad (x,u,p) \in \mathbb{R}^4 \times \mathbb{R}^2 \times \mathbb{R}^{13}, \tag{17.5}$$

$$= \Phi(x,u)\Psi(p), \tag{17.6}$$

with straightforward definitions of the functions Φ and Ψ.

The state trajectory of the system that starts at x the control profile $\boldsymbol{u}$ and the instantiation p of the parameter vector is denoted hereafter by $x^{\boldsymbol{u}}(k|x,p)$, namely:

$$x^{\boldsymbol{u}}(k+1|x,p) = f\big(x^{\boldsymbol{u}}(k|x,p), u(k), p\big),$$
$$x^{\boldsymbol{u}}(0|x,p) = x.$$

Now that we have defined the dynamics and the related notation, let us define the cost function and the constraints to be enforced.

17.2.2 The Control Objective and Constraints

Recall that the objective of the combined therapy is to contract the size of the tumor, denoted by x_1, while satisfying the health constraint that consists in keeping the size of the population of lymphocyte cells, denoted by x_2 above some lower bound x_2^{min}.

Moreover, in this chapter and for the sake of simplicity of computation, we assume that this objective has to be satisfied using quantized drug delivery (two extreme values are only allowed for each drug):

$$u \in \mathbb{U} := \{0, u_1^{max}\} \times \{0, u_2^{max}\}, \tag{17.7}$$

while this simplification is done here for the sake of simplicity, it might correspond to an interesting technological simplification in the use of the drug's delivery devices.

In the sequel, an infinite sequence of control values that lie inside the admissible domain $\mathbb{U}$ is denoted by $\mathbb{U}^\infty$.

Based on the control objective, the following stage cost is used which incorporates the soft exact penalty version of the constraint on x_2:

$$L(x, u) := x_1^2 + \rho_c \max\{0, x_2^{min} - x_2\} + \rho_1 u_1 + \rho_2 u_2. \tag{17.8}$$

Therefore, **in the case where the model's parameters are perfectly known**, the following deterministic optimal control problem can be defined for all given initial state x:

$$\mathcal{P}(x|p): \qquad \min_{\boldsymbol{u} \in \mathbb{U}^\infty} J(\boldsymbol{u}|x, p) := \sum_{k=0}^{\infty} \gamma^k L\Big(x^{\boldsymbol{u}}(k|x, p), u(k)\Big), \tag{17.9}$$

where $\gamma \in (0, 1]$ is some discount factor.

Now, because the parameters are supposed to be unknown, a stochastic control formulation should be used.

This is done by replacing the deterministic cost function's version (17.9) by a statistically defined version that is given by:

$$J_\alpha(\boldsymbol{u}|x, \Pi) := \mu(J(\boldsymbol{u}|x, \cdot)) + \alpha\sigma(J(\boldsymbol{u}|x, \cdot)), \tag{17.10}$$

$$=: S_\alpha^{(J)}(\boldsymbol{u}, x), \tag{17.11}$$

in which the terms $\mu(\cdot)$ and $\sigma(\cdot)$ represent respectively the expectation and the variance of their arguments as p is randomly sampled according to some presumably known Probability Density Function (pdf) Π while the term in (17.11) is a short writing of the r.h.s of (17.10).

This leads to the following stochastic optimization problem:

$$\mathcal{P}_\alpha(x|\Pi): \quad \min_{\boldsymbol{u} \in \mathbb{U}^\infty} J_\alpha(\boldsymbol{u}|x, \Pi). \tag{17.12}$$

It is important to underline that in many stochastic control-related works, only the expectation of the cost function is considered by taking $\alpha = 0$ in (17.10). One of the messages of the present chapter is that this is not totally appropriate when cancer therapy is considered.

Indeed, the expectation alone is a quite relevant indicator only when a rather high number of realizations are expected to take place and where losses occurring for some instantiations of the uncertainties can be compensated by profits that is gained when other instantiations take place.

This is the case for instance when optimizing the expectation of energy consumption in Building Management System (BMS) in the presence of bad knowledge of exogenous quantities such as power demands and whether conditions.

In this case, obviously, only the global average is of interest since it corresponds to the bill one has to pay over some period.

In the case of cancer therapy however, a scenario is a patient being treated and a *loss* cannot really be compensated by a potential *win* on another patient.

This means that in the case of cancer, one should explicitly accommodate for the level of risk that **each patient affords** during the therapy. In other words, it is precisely those *bad* scenarios corresponding to a non negligible likeliness that really matter and not uniquely the averaged cost in the sense that this has to do not only with the expectation but also with the associated variance as well.

This justifies the use of $\alpha > 0$ in (17.12).

In the next section, some recalls are proposed on the SDP equations that might be invoked to approximately solve the stochastic optimization problem (17.12).

17.3 Recalls and Preliminary Results

Some of the explanation and results given in this section are also present in Chap. 2. Nevertheless, they are recalled here for the easiness of reading as some minor modifications are necessary to take into account the specific structure of the system.

17.3.1 Stochastic Dynamic Programming (SDP)

SDP is about solving a functional equation in the unknown function $V(\cdot)$ which happens to be the optimal solution of (17.12). More precisely:

> **The Stochastic Bellman Equation**
The stochastic Bellman equation describes the properties of the optimal solution V as a function of the state vector x, namely:

$$V(x) := \min_{u\in\mathbb{U}} Q(x,u) \qquad \text{where,} \tag{17.13}$$

$$Q(x,u) := L(x,u) + \gamma \min_{v\in\mathbb{U}}\Big[\mathcal{S}_\alpha^{(Q)}(x,u,v)\Big], \tag{17.14}$$

$$\text{where } x^+ = f(x,u,p), \tag{17.15}$$

in which $\mathcal{S}^{(Q)}(x,u,v)$ is defined in a similar way as $\mathcal{S}^{(J)}(\boldsymbol{u},x)$ [see (17.11)], namely:

$$\mathcal{S}_\alpha^{(Q)}(x,u,v) := \mu\Big(Q(f(x,u,\cdot),v)\Big) + \alpha\sigma\Big(Q(f(x,u,\cdot),v)\Big), \tag{17.16}$$

where $\mu(\cdot)$ and $\sigma(\cdot)$ are respectively the expectation and the variance of their arguments under p (involved in $f(x,u,p)$) that is randomly sampled according to the presumably known p.d.f Π.

The fact that a function V that meet (17.13) and (17.14) is a solution to (17.12) is simply a direct consequence of the so-called Bellman principle, already invoked in Sect. 2.7.1 together with the the fact that:

$$L = \mu(L) + \sigma(L) = L + 0$$

since $L(x,u)$ is a deterministic term that does not depend on the uncertain parameter vector p.

From the definition above, it appears that **the truly unknown function to be found is** $Q(\cdot,\cdot)$ since the optimal value function $V(\cdot)$ and the optimal control law result from the static optimization problem (17.13) where they appear to be the optimal value and the optimal arguments of the optimization problem.

Now it is quite easy to acknowledge from (17.14) that Q is the solution of a fixed-point iteration that is defined below:

The Fixed-Point Equality

$$Q^{(i+1)} = F\Big(Q^{(i)}\Big) \tag{17.17}$$

where the operator F (commonly called the Bellman operator) is defined by:

$$F(Q)(x,u) := L(x,u) + \gamma \min_{v\in\mathbb{U}}\Big[\mathcal{S}^{(Q)}(f(x,u,\cdot),v)\Big]. \tag{17.18}$$

In the next section, some sufficient conditions are provided for the convergence of the above-defined fixed-point iteration in a general conceptual setting.

Afterwards, a specific implementation is proposed in Sect. 17.4 for the combined therapy of cancer which takes advantages of the particular structure of the mathematical model of the system that is shared by many compartmental models used in the mathematical representation of biological systems.

17.3.2 Fixed-Point Iteration Convergence Analysis

For the sake of easiness of reading, all the rather technical proofs of the results stated in this section can be carefully examined in the paper below:

https://arxiv.org/abs/1905.04937

while the working assumptions and the associated results are comprehensively stated in this section in order to get a high level understanding of the underlying success conditions.

First of all, the following continuity assumption is considered regarding the function used to express the dynamics of the system is used in the proof of convergence.

Obviously, the continuity of the discrete-time dynamics is naturally inherited from the continuity of the original ODEs used in the discretization:

Continuous Dynamics
Assumption 17.1 The map f describing the discrete-time state evolution (17.5) is continuous in its arguments.

The second assumption concerns the boundedness of the set of potential values of the parameters which is obviously satisfied without loss of generality.

Finitely Supported Uncertainties
Assumption 17.2 There exists a compact set $\mathbb{P}$ to which belong all possible realizations of the uncertain vector p.

The next condition that is needed to derive the convergence proof concerns the family of functions that might be used to represent the unknown map Q that takes the pairs (x, u) as arguments.

The following definition states a boundedness condition on the potential excursion of the map when the state, control and parameter vector are bounded.

Bounded Excursion
Definition 17.1 A map G defined on $\mathbb{X} \times \mathbb{U}$ is said to have *B-bounded excursions* w.r.t the dynamics f if the following inequality holds:

$$|G(f(x, u, p), v) - \mu(G(f(x, u, \cdot), v))| \leq B \tag{17.19}$$

for all $(x, u, v, p) \in \mathbb{X} \times \mathbb{U}^2 \times \mathbb{P}$.

This definition states that the potential excursions of the realizations

$$G(f(x, u, p), v),$$

from the mean value

$$\mu\big(G(f(x, u, \cdot), v)\big),$$

are bounded.

Obviously, under Assumption 17.1, such a bound provably exists as soon as the map G is continuous since all the involved bounding sets are compact.

Nevertheless, while the evolution map is supposed to be continuous, we might want to use non continuous approximating functions provided that the above boundedness condition holds. In other words, the continuity of approximating function is a sufficient condition that is not necessary and since we want to keep the possibility of using non continuous Machine Learning (ML) available structures to represent the value function Q, we use here the above stated general boundedness condition.

The following lemma [8] is needed in the statement of the following main result.

> **Fixed-Point Convergence**
> **Lemma 17.1** *Under Assumptions* 17.1 *and* 17.2, *provided that fixed point map* F *defined by* (17.17) *produces only maps* $Q^{(i)}$ *with* B*-bounded excursions, the fixed point iteration converges on*
>
> $$\mathbb{Z} := \mathbb{X} \times \mathbb{U},$$
>
> *for all penalties* α *on the variance-related term in* (17.16) *satisfying the following inequality:*
>
> $$\alpha < \frac{1-\gamma}{2\gamma B}, \tag{17.20}$$
>
> *and for sufficiently small sampling period* τ *that is used to derive the discrete-time dynamics* (17.5). ◇

Notice that the above lemma contains as a particular case, the standard convergence result that holds when no penalty $\alpha = 0$ is put on the variance-related term (see for instance [5,6,9] and the references therein).

The way the boundedness condition invoked in Lemma 17.1 is enforced in the specific solution proposed in this chapter is discussed in the next section.

17.4 The Proposed Computation Framework

Let us first of all focus on the paremeterization of the value function Q.

17.4.1 Parameterization of $Q(x, u)$

In order to derive a concrete computational framework based on the conceptual formulation of the previous section, a finite-dimensional parameterization of the value function Q needs to be chosen. This aspect is considered in the current section.

In this section, the notation:

$$z = (x, u),$$

is used to refer to the extended vector that plays the role of the argument of the value function Q. Moreover, the notation:

$$\mathbb{Z} := \mathbb{X} \times \mathbb{U},$$

is used to refer to the compact set over which the approximation is to be computed.

Consequently, the notation $Q(z)$, $f(z, p)$ become short version of the value function $Q(x, u)$ and the one-step-ahead prediction map $f(x, u, p)$ respectively computed at the pair $z = (x, u)$.

The map representing the finite-dimensional parameterization of the value function Q is defined through the following steps:

- We first consider a fixed grid of values:

$$\mathcal{Z} := \{z^{(i)}\}_{i=1}^{n_g},$$

that is defined on the set $\mathbb{Z}$ of interest.
- The corresponding vector of values of Q at the above set of values is denoted by

$$\boldsymbol{q} := \{q_i\}_{i=1}^{n_g} \quad \text{where} \quad q_i = Q(z^{(i)}).$$

- Let us choose a regression model (Polynomial, Gaussian, Support Vector Machine Regressor (SVMR), etc.) from our favorite ML library and denote it by $\hat{Q}$.
In this chapter, a SVMR with Radial basis is chosen with the default parameters used in the scikit-learn ML (ML) library [10].
- By this choice, it becomes possible to associate to each candidate vector $\boldsymbol{q} \in \mathbb{R}^{n_g}$, the associated fitted function $\hat{Q}(\cdot|\boldsymbol{q})$ defined on $\mathbb{Z}$ which is the one fitted (or *learned* to invoke a ML vocabulary) based on the learning data set defined by $(\mathcal{Z}, \boldsymbol{q})$.

Now consider the following definition regarding the admissibility of the structure of regression model:

Admissible Regression Models

Definition 17.2 A regression model is said to be admissible if and only if for all set $\mathcal{Z}$ of finite grid-points in $\mathbb{Z}$ and any compact set of values $\mathbb{Q} \subset \mathbb{R}^{n_g}$, there exists a finite real number $B > 0$ such that all identified maps $\hat{Q}(\cdot|\boldsymbol{q})$ with $\boldsymbol{q} \in \mathbb{Q}$ admit B-bounded excursions in the sense of Definition 17.1.

Based on the above definitions and choices, we are now able to state the main theoretical result that supports and proves the convergence of the implementation scheme aiming at addressing the above defined control problem.

17.4.2 Main Convergence Result

The following main result can be proved [8] regarding the convergence of the fixed-point iteration scheme in the unknown vector q that parametrizes the approximate value function $\hat{Q}(\cdot)$:

> **Main Result [8]**
> **Proposition 17.1** *Under Assumptions* 17.1 *and* 17.2, *when using an admissible regression model (in the sense of Definition* 17.2*) on the grid*
>
> $$\mathcal{Z} := \{z^{(i)}\}_{i=1}^{n_g},$$
>
> with the initialization
>
> $$q_i = L(z^{(i)}),$$
>
> *the resulting fixed-point iteration* (17.18) *with* $\gamma \in (0, 1)$ *is convergent provided that the penalty* α *on the variance-related term and the sampling period* τ *are sufficiently small.* ◇

With this result, it remains to find an admissible regression model in the sense of Definition 17.1.

Fortunately, many such models are available. Among them, the SVMR is used hereafter with the radial kernel leading to the following structure:

$$\hat{Q}(z|\boldsymbol{q}) := \beta_0(\boldsymbol{q}) + \sum_{i=1}^{n_g} \beta_i(\boldsymbol{q}) G(z^{(i)}, z), \tag{17.21}$$

where it can be proved that the the coefficients $\beta_i(\boldsymbol{q})$ show uniform bound provided that the vector $\boldsymbol{q}$ belongs to some compact set $\mathbb{Q}$.

It is worth recalling that the above discussion and results focused on the convergence of the resulting iteration which is an important issue.

Nevertheless, we should also be careful about the computational burden that might be too heavy.

In order to address this issue, a specific property of the mathematical model governing the combined therapy of cancer is exploited in order to accelerate the computational scheme.

This is the aim of the next section.

17.4.3 Computation of the Expectation and Variance

The implementation of the fixed-point iteration (17.18) requires the computation of the statistically defined quantity:

$$\mathcal{S}^{(Q)}(f(x,u,\cdot),v).$$

For this to be done, the computation of the expectations and variances, that depend on the dispersion law of the vector of parameters p, is required.

This section explains the specific way these quantities are approximated by exploiting the mathematical structure of the model.

Recall that the r.h.s of the equations representing the tumor dynamic in presence of drug injection which is given by (17.6) can be decomposed in two multiplicative terms as follows:

$$x^+ = \Phi(x,u)\Psi(p),$$

in which only the second term does depend on the value of p.

This implies that the computation of the expectation of x^+ depends on the statistics of the term $\Psi(p) \in \mathbb{R}^{14}$ which has to be approximated.

This can be done by exhibiting n_{cl} clusters based on a set $\mathcal{S}_\psi$ of n_s samples of the uncertain vector $\Psi(\cdot)$. These samples can be randomly drawn using the presumably known probability density function (pdf) Π.

Once the clustering computation is done, the centers of the so-computed clusters, denoted hereafter by

$$\Psi^{(j)}, \quad j = 1, \dots, n_{cl},$$

can then be viewed and used as representative of their cluster's population of values.

Moreover, the probability of the clusters can be approximated as the ratio π_j between the sizes (number of samples) of these clusters and the total number of samples n_s. More formally:

$$\pi_j := \frac{1}{n_s}\text{card}\Big\{\psi \in \mathcal{S}_\psi \quad | \quad \texttt{cl}(\psi) = \texttt{cl}(\Psi^{(j)})\Big\}. \tag{17.22}$$

By so-doing, the following expressions that involve the approximated probabilities π_i might be used in the approximation of terms in (17.16):

$$\hat{\mu}\big(x,u,v|\boldsymbol{q}\big) := \sum_{j=1}^{n_{cl}} \pi_j \hat{Q}\Big(\Phi(x,u)\Psi^{(j)}, v|\boldsymbol{q}\Big), \tag{17.23}$$

$$\hat{\sigma}\big(x,u,v,|\boldsymbol{q}\big) := \sum_{j=1}^{n_{cl}} \pi_j \Big[\hat{Q}\Big(\Phi(x,u)\Psi^{(j)}, v|\boldsymbol{q}\Big) - \hat{\mu}\big(x,u,v|\boldsymbol{q}\big)\Big]^2, \tag{17.24}$$

and consequently, the weighted sum (17.16) can be approximated by the following expression:

$$\hat{\mathcal{S}}_{\alpha}^{(Q)}(x,u,v|\boldsymbol{q}) := \hat{\mu}\big(x,u,v|\boldsymbol{q}\big) + \alpha\hat{\sigma}\big(x,u,v|\boldsymbol{q}\big), \tag{17.25}$$

which results in the following updating expressions of the components q_i at the corresponding grid points $z^{(i)}$:

$$q_i^+ := L(z^{(i)}) + \gamma \min_{v\in\mathbb{U}}\Big[\hat{\mathcal{S}}_{\alpha}^{(Q)}(z^{(i)},v)|\boldsymbol{q})\Big]. \tag{17.26}$$

Notice that the low-dimensional optimization problem in v can be solved through a simple enumeration process over the four elements of the admissible set $\mathbb{U}$. This is the reason why the admissible control values have been restricted to the extreme values of the admissible set of drugs injection rates.

The resulting fixed-point iteration on the vector $\boldsymbol{q}$ can therefore be shortly written as follows:

$$\boldsymbol{q}^+ = F(\boldsymbol{q}), \tag{17.27}$$

where the notation F already used in (17.17) is slightly overloaded by using the finite-dimensional parameter $\boldsymbol{q}$.

This is precisely the fixed-point iteration whose convergence has been proved under the assumptions of Proposition 17.1.

In Sect. 17.5 hereafter, the concrete instantiation of the above-described framework on the example of combined therapy of cancer is detailed and the corresponding results are extensively discussed in order to get a better understanding of the impact of the parameters involved in the formulation on the outcome of the simulated closed-loop therapy.

17.5 Application to the Combined Therapy of Cancer

17.5.1 The Parameters Used in the Framework

As far as stochastic controller design is concerned (nominal controller is also investigated for the sake of comparison), the parameters described in the present section have been investigated in the implementation of the above-described framework.

A number of samples that is equal to

$$n_s = 10^4,$$

is randomly drawn and a clustering has been performed on the corresponding set of n_s values of Ψ (see Sect. 17.4.3 hereafter).

The number of *fitted* clusters is set to:

$$n_{cl} = 20.$$

As for the probability density function Π using in the random sampling of the vector of parameters, the following normal relative distribution is used, namely each component value is defined by:

$$p_i = (1+\nu)p_i^{nom},$$

where ν is a random normally distributed value with 0 mean and variance

$$\sigma = 0.4.$$

The grid of arguments $\mathcal{Z}$ is taken to be the Cartesian product of a uniform grid on the state components (having 7 uniformly distributed values, per component, over the normalized intervals [0, 1]) on one hand and the four values that defines the control set previously defined using the extreme values of the admission drugs injection rates $\mathbb{U}$ on the other hand.

This leads to a set of point $\mathcal{Z}$ which contains:

$$7^4 \times 4 = 9604$$

elements.

Notice that the clustering of the Ψ-values has been performed via the

`K-Nearest-Neighbors` (`KNN`),

module of the `scikit-learn` library.

The `scikit-learn` library's `SVMR` regression model is used to represent the approximation of the value function Q with the radial basis kernel and the default values of the remaining parameters.

Regarding the remaining parameters of the problem, notice that the sampling period:

$$\tau = 0.25 \text{ day},$$

is used and the health-related lower bound:

$$x_2^{\min} = 0.05,$$

is used as the normalized value of the minimum allowable size of the population of lymphocyte cells (Notice that prior to any computation, the model has been rewritten in the normalized values of the components of the state using the normalization vector value $(10^9, 10^9, 1, 10^9)$).

17.5.2 The Different Investigated Controllers

In the following discussed results, the following three different controllers are compared:

1. The **nominal controller (Controller #1)** that uses the nominal values of the parameters without any explicit handling of the parameters' dispersion.
   This is done using the same framework in which the probability density function Π is the Dirac distribution centered at the nominal vector of parameters.
2. The **expectation-only-based controller (Controller #2)** for which the distribution of the uncertain parameters is the one described above and where no penalty is applied on the variance term in the cost function's definition ($\alpha = 0$).
3. The **variance-aware controller (Controller #3)** for which a penalty $\alpha = 0.1$ is applied to the variance-related term.

Before we compare the different controllers, let us examine the convergence of the fixed-point iteration when non vanishing values are used for the variance-related penalty α.

This is the aim of the next section.

17.5.3 Convergence of the Fixed-Point Iteration

While the convergence of the fixed-point iteration has been theoretically proved for values of $\gamma \in (0, 1)$ that are strictly lower than 1 and for sufficiently small values of the penalty α and the sampling period τ, the value $\gamma = 1$ has been used since the results showed that the convergence does occur for this ideal value (no discount on the cost function). On the other hand, the penalty $\alpha = 0.1$ is used for the design of Controller #3. An instance of the fixed-point iteration's convergence is shown in Fig. 17.1 where the difference between two successive values that are visited during the iterations, namely:

$$\|\boldsymbol{q}^{(i+1)} - \boldsymbol{q}^{(i)}\|_\infty,$$

is shown.

Notice that, because of the use of $\gamma = 1$, the strictly monotonic behavior that is expected by the theory (not presented here) is not observed. Nevertheless, the convergence eventually occurs and the solution is obtained since the approximated map $\hat{Q}$ is a solution of the fixed-point equation.

Let us now end this chapter by comparing the performances of the three above mentioned controllers. This is the aim of the next section.

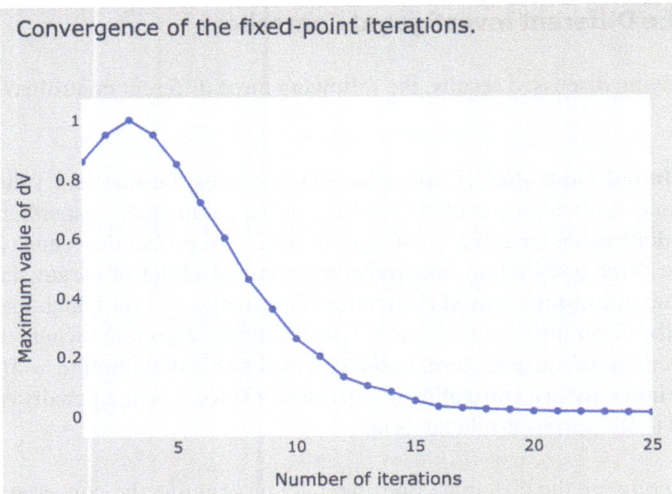

Fig. 17.1 Typical convergence curve for fixed-point iterations: Evolution of the difference $\|\boldsymbol{q}^{(i+1)} - \boldsymbol{q}^{(i)}\|_\infty$ between two successive iterations. $\gamma = 1, \alpha = 0.1$

17.5.4 Advantage of Stochastic Versus Nominal Controllers

Figures 17.2 and 17.3 show the advantage of using the stochastic formulation compared to using the nominally design controller.

This is shown by using:

$$N_{\text{sim}} := 20000,$$

different simulations corresponding to as many randomly drawn samples of pairs of initial states and model's parameters.

The resulting closed-loop costs are computed before statistics are drawn. The 20000 scenarios are obtained by first randomly sampling 100 initial states and for each of these initial states, 200 random samples of the vector of parameters are drawn. Then for each pair $(x^{(i)}, p^{(i)})$, $i = 1, \dots, N_{\text{sim}}$, the following resulting closed-loop cost is computed:

$$J_{\text{cl}}(x^{(i)}, p^{(i)}) := \sum_{k=0}^{N_{\text{sim}}} L(x(k), u(k)), \tag{17.28}$$

where the simulation horizon length $N_{\text{sim}} = 50$ is used leading to a time horizon of (12.5 days) for the simulation while $x(k)$ and $u(k)$ stand for the closed-loop trajectories at instant k starting at initial $x^{(i)}$ and when the model's parameter vector $p^{(i)}$ is used in the model.

Using the associated values of the closed-loop performance, different statistical comparisons are performed in order to compare the three controllers mentioned above, namely, the nominal controller, the stochastic controller that focuses on the

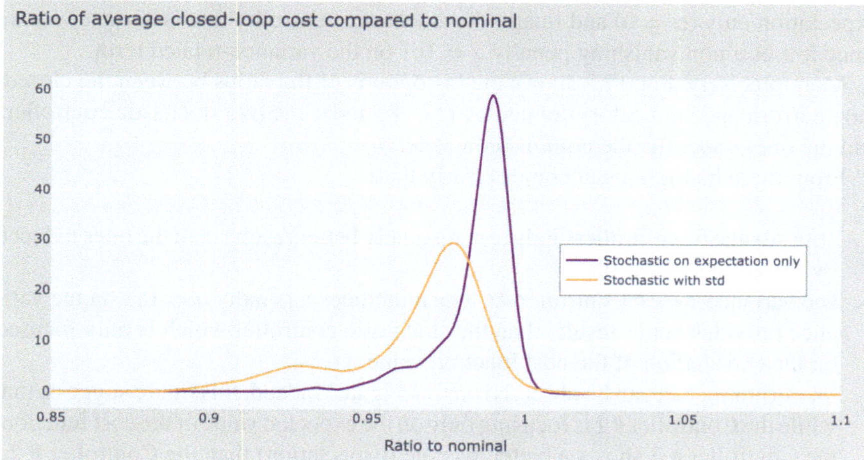

Fig. 17.2 Comparison between closed-loop performance ratios relatively to the performances of the nominal controller for the two stochastic formulations over the 20000 randomly sampled initial states and model's parameter vectors

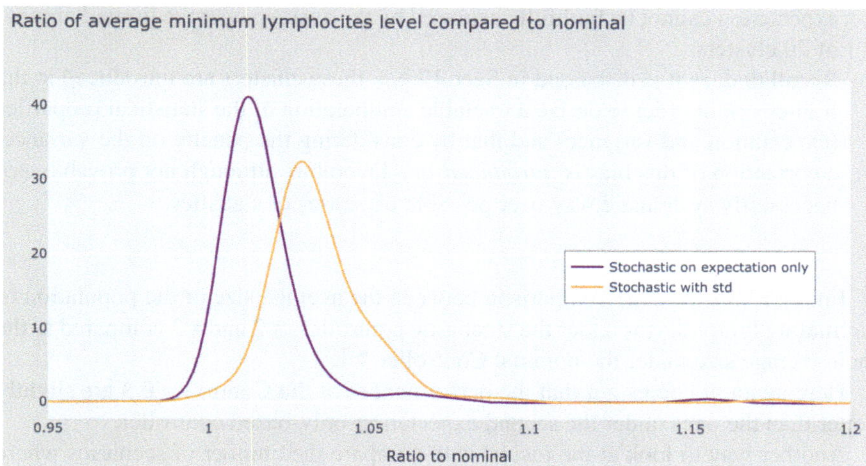

Fig. 17.3 Comparison between the averages of the lymphocytes population level's ratios under the stochastic controllers compared to their average size under the nominal controller for the 20,000 randomly sampled initial states and parameter vectors

expectation only ($\alpha = 0$) and finally the stochastic controller that is variance-aware since it uses a non vanishing penalty $\alpha = 0.1$ on the variance-related term.

More precisely, Fig. 17.2 shows the histograms of the ratios between the closed-loop performance indicators defined by (17.28) under the two stochastic controllers and the ones under the the nominal controller.

From these histograms, it comes clearly that:

1. Both stochastic controllers induce statistically better results than the ones induced by the nominal controller.
2. The variance-aware Controller #3 which includes a penalty $\alpha = 0.1$ on the variance provides better results than the stochastic controller which is only focused on the expectation of the cost function value.
3. An intriguing fact can be observed on this Figure! Indeed, this figure suggests that while the Controller # 2 is focusing only on the expected value of the cost function, the Controller # 3 shows a better average (expectation) than the Controller # 2. In a first thought, one might consider that this observation contradicts the fact that Controller # 2 is supposed to minimize this expectation and hence should show better results in terms of the average closed-loop cost.
   In order to explain this apparent contradiction, it can be conjectured that the expectation cannot be faithfully captured by the average over the limited number of 20 clusters.
   Recall that, as it is discussed in Sect. 17.4.3, these clusters are introduced in the framework in order to derive a tractable computation of the statistical properties (expectation and variance) and that by considering the penalty on the variance, a correction of this bias is *introduced* in a favorable, although not provable, nor necessarily systematic way over possible use-cases or statistics.

Figure 17.3 shows the comparison between the average size of the population of circulating lymphocytes under the stochastic controllers # 2 and # 3 compared to the their average size under the nominal Controller # 1.

Here again, it comes out that the performances of the Controller # 3 are slightly better than the ones under the second expectation-only-based controller.

Another way to look at the results is to compare the number of scenarios where the health constraint

$$x_2 \geq x_2^{\min},$$

is fulfilled.

This is done in Fig. 17.4 that shows the percentage of improvement due to the use of stochastic controllers compared to the nominal controller.

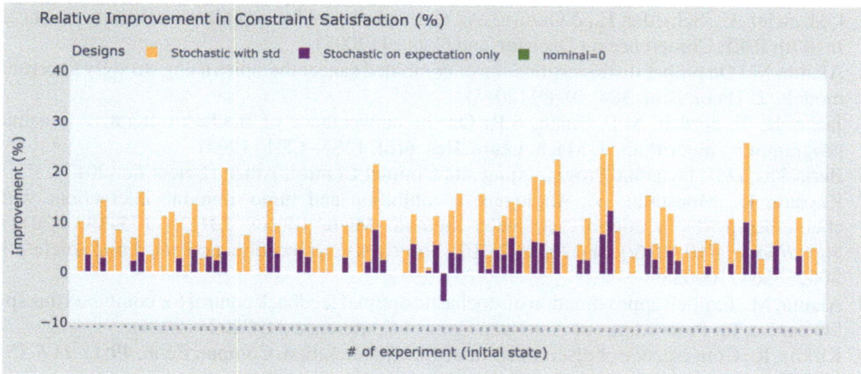

Fig. 17.4 Safety constraints satisfaction improvement via stochastic design. Abscissa stand for the initial state's number (100 are randomly selected) and the values refer to the percentage of improvement regarding the safety over the 200 instances of the vector of parameters that are simulated for the same initial state. For instance, 20% improvement means that there are 20% less instances of the vector of parameters that lead to constraints violation compared to the nominal controller for the corresponding initial state

17.6 Conclusion

In this chapter, it has been shown that for control problems of moderate size, the paradigm of SDP, via its Q-learning version, can be used to derive an explicit feedback that leads to statistically meaningful improvement of the closed-loop outcome when compared to nominal controller that is based on the nominal value of the model's vector of parameters.

Moreover, the proposed method can be easily implemented by means of easy-to-use and freely available ML modules and model's structure.

Moreover, sufficient conditions on the formulation settings are provided for the convergence of the underlying fixed-point iteration.

It is important to underline the fact that similar solutions might have been obtained using the Reinforcement Learning (RL) paradigm and efficient implementation versions might be derived using the `Pytorch` framework in order to benefit from the acceleration provided by the Graphical Processing Unit (GPU) environment through the `Google-Colab` notebook facilities.

References

1. Swan, G.W.: General applications of optimal control in cancer chemotherapy. IMA J. Math. Appl. Med. Biol. **5**, 303–316 (1988)
2. DePillis, L.G., Radunskaya, A.E.: A mathematical tumor model with immune resistance and drug therapy: an optimal control approach. J. Theor. Med. **3**, 79–100 (2001)

3. Ledzewicz, U., Schattler, H., d'Onofrio, A.: Optimal control for combination therapy in cancer. In: 47th IEEE Conference on Decision and Control (2008)
4. Alamir, M.: On probabilistic certification of combined cancer therapies using strongly uncertain models. J. Theor. Biol. **384**, 59–69 (2015)
5. Jaakoola, T., Jordan, M.I., Singh, S.P.: On the convergence of stochastic iterative dynamic programming algorithms. J. Mach. Learn. Res. **6**(6), 1185–1201 (1994)
6. Bertsekas, D.P.: Dynamic Programming and Optimal Control. Athena Scientific (2017)
7. Kassara, K., Moustafid, A.: Angiogenesis inhibition and tumor-immune interactions with chemotherapy by a control set-valued method. Math. Biosci. **231**(2), 135–143 (2011). https://doi.org/10.1016/j.mbs.2011.02.010, http://www.sciencedirect.com/science/article/pii/S0025556411000289
8. Alamir, M.: Explicit approximation of stochastic optimal feedback control for combined therapy FO cancer. In: Proceedings of American Control Conference ACC2024 (2024)
9. Kirkby, R.: Convergence of discretized value function iteration. Comput. Econ. **49**(1), 117–153 (2017)
10. Pedregosa, F., Varoquaux, G., Gramfort, A., Michel, V., Thirion, B., Grisel, O., Blondel, M., Prettenhofer, P., Weiss, R., Dubourg, V., Vanderplas, J., Passos, A., Cournapeau, D., Brucher, M., Perrot, M., Duchesnay, E.: Scikit-learn: machine learning in Python. J. Mach. Learn. Res. **12**, 2825–2830 (2011)

Decentralized Frequency Control in Micro-grids

18

18.1 Problem Statement

The control of microgrids is a very important and vast topic that it deserves an entire dedicated book.

In this chapter, we simply show how the simple design methodology described in Sect. 4.5 and dedicated to the control of systems showing unknown internal destabilizing dynamics can be applied to a crucial microgrids control problem.

More precisely, we are interested in the decentralized frequency control in a micro-grid involving several generators serving a power consuming *load*. A simple example of such a micro-grid that is used hereafter in order to illustrate the framework is shown in Fig. 18.1.

Figure 18.1 shows an example of a *small* micro-grids involving two generators, denoted by G_1 and G_2 providing power to a load.

The guiding principle that governs the majority of works in grid's frequency control is the one stipulating that the dynamic of the frequency is due to the *load unbalance*, namely to the difference between the produced power P_p and the consumed power P_ℓ.

This can be expressed by the following very simple equation:

$$\dot{f} = \alpha\Big[P_p - P_\ell\Big], \tag{18.1}$$

where α is a coefficient that is inversely proportional to the *micro-grid inertia*.

Notice also that, for the use-case we are interested in the produced power P_p being the sum of the produced powers P_1 and P_2 of the two generators connected to the micro-grid:

$$P_p := P_1 + P_2. \tag{18.2}$$

At the beginning of his Ph.D. work, Jean first tried to elaborate a control design that is based on (18.1) together with the framework of Sect. 4.4 which addresses exactly uncertain systems of the form (18.1).

M. Alamir, *Nonlinear Control of Uncertain Systems*,
https://doi.org/10.1007/978-3-031-93287-8_18

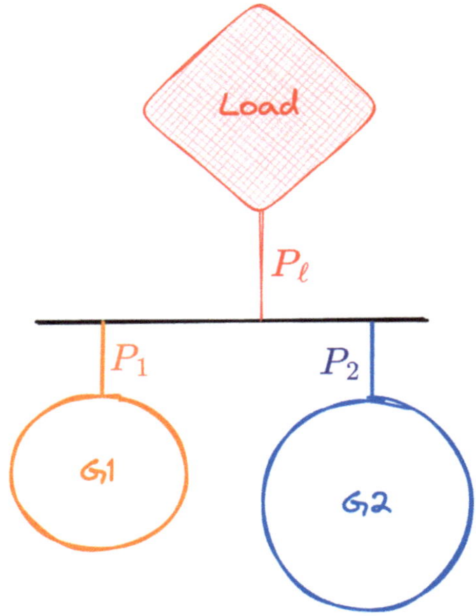

Fig. 18.1 Example of a micro-grids involving two generators serving a load

More precisely, with regards to the above mentioned control framework (see Sect. 4.4), he considered the following correspondences:

- $y \leftrightarrow f$ is the regulated variable,
- $u \leftrightarrow P_p$ is the total produced power viewed as the control input,
- $P_\ell \leftrightarrow h$ is the unknown quantities serving as the totally unknown term. This is practically true as the load is not expected to *send* the power it requires to the frequency controller, and
- $\alpha \leftrightarrow \alpha$ is the non vanishing gain that is bounded below by some known bound.

Based on this mapping, Jean tried to follow the theory[1] depicted in Sect. 4.4 to design a control law of the form:

$$\dot{z} = \lambda_f\Big[S\Big(\lambda(f_0 - f) + z\Big) - z\Big], \tag{18.3a}$$

$$P_p = S\Big(\lambda(f_0 - f) + z\Big), \tag{18.3b}$$

[1] The designed control algorithm would be used to provide the total power P_ℓ needed to stabilize the frequency, then the sharing rule for each generator is based on their respective maximal capacity for instance and is not mentioned here since it is not the part of the control design that is linked to the uncertain quantity P_ℓ. Any sharing rule that produces the same required amount of power would lead to the stabilization of the frequency.

in which $f_0 = 50$ is the desired frequency, z is the internal state of the controller while (λ, λ_f) are the two controller's parameters to be tuned following the framework depicted in of Sect. 4.4.

By so-doing, Jean figured out that a too aggressive control (corresponding to high values of λ and λ_f) induced instability of the resulting closed-loop behavior, when applied to a precise digital twin of the micro-grid, which contradicts the theory expressed by the widely used *swing equation* (18.1).

The only explanation is that the simple equation (18.1) fails to faithfully represent the underlying dynamic **under aggressive control**.

This suggests that an aggressive control might *excite internal dynamics* that are not represented by this simple equation.

As a matter of fact, the existence of such a *hidden* dynamic is no surprise since the produced powers P_1 and P_2 giving rise to the total produced power P_ℓ are the outputs of rather complicated dynamics representing the generators behavior in closed-loop. But we know that the regulated output of a controlled system can be replaced by the corresponding set-point only as far as the dynamics of the latter is sufficiently small for the tracking to be quasi-perfect.

On the contrary, when fast set-points are fed to the controller, the imperfect tracking produces regulated outputs that differ from the set-points. If the latter is designed to control another variable, this discrepancy might lead to undesired oscillation or even to instability.

In the remainder of this chapter, it is shown that using standard identification tools, it is possible to come out with a more faithful model of the underlying dynamics which takes another general form that is also studied in Chap. 4, namely the one precisely analyzed in Sect. 4.5.

This enables the underlying design methodology to be successfully applied to address the problem of frequency control in micro-grids.

18.2 Identification of the Frequency Dynamics

While the results shown in this section are obtained under the assumption of a micro-grid composed of two generators (see Fig. 18.1), the same methodology can be applied to larger micro-grids with potentially higher orders for the resulting dynamic models.

Jean disposed of a detailed power electronic simulator of the micro-grid in which the complete models of the generators are represented together with simple control laws which are aware of the consumption profile P_ℓ inducing imperfect tracking of the total produced power.

This enables to generate the internal dynamics mentioned above which however remain stable. This also provides the produced power profile P_p that is used hereafter in the identification process together with the load power P_ℓ of Fig. 18.2.

Regarding the identification process and the use of the above mentioned high fidelity simulator, the following comment is worth to keep in mind:

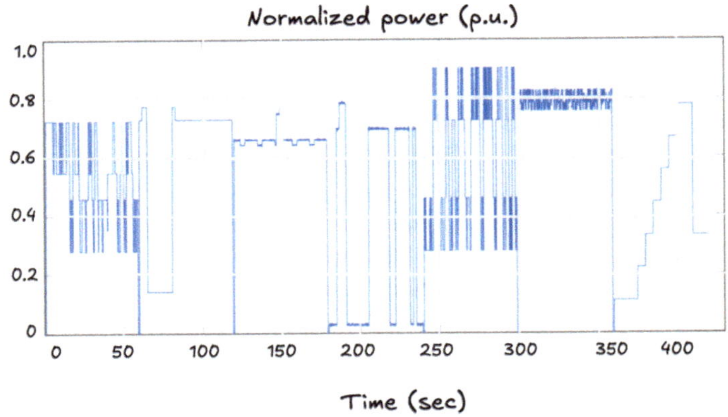

Fig. 18.2 The profile of the load normalized (by 145kVA) power P_ℓ used in the identification of the dynamics of the frequency of the two-generators micro-grids

The Identification Step
It is important to underline that the consumed power is not supposed to be known in operational mode. This is the raison d'être of the whole study. However, for the sake of identifying the dynamics and hence to be able to derive a feedback control law that can operate in the absence of this information, the high fidelity simulator is used here in the identification step.

Now we dispose of the time series

$$P_\ell,\ P_p,\ f,$$

where the frequency profile is delivered by the high fidelity simulator,[2] it is possible to proceed with the identification step.

First of all, the best parameter α was found such that the frequency behavior can be described by:

$$\dot{f} = \alpha(P_p - P_\ell) + e. \tag{18.4}$$

This can be done by considering e to be the residual of an optimization problem in which, the decision variable is α while the cost function is defined by the least-squares problem:

$$\dot{f} \approx \alpha(P_p - P_\ell). \tag{18.5}$$

The next step is to stick to the general form (4.55) for which the detailed expressions of the quantities involved in the parameterization of the feedback law proposed in Sect. 4.5 is used.

[2] It is important to keep in mind that, reconstructing the frequency of a time-series is not that straightforward and can be obtained through an involved signal processing algorithm that we skipped here in order to focus on the main topic.

In order to do so, the error e involved in (18.4) should be the output of a linear system of appropriate order, namely:

$$\dot{\eta} = A\eta + B_1\dot{P}_p + B_2\dot{P}_\ell, \tag{18.6a}$$
$$e = C\eta. \tag{18.6b}$$

Indeed, if a good fit is obtained with the above structure for some order, then the frequency equation would be given by:

$$\dot{f} = \alpha(P_p - P_\ell) + C\eta, \tag{18.7a}$$
$$\dot{\eta} = A\eta + B_1\dot{P}_p + B_2\dot{P}_\ell, \tag{18.7b}$$

which is precisely the general form of the uncertain systems involved in Sect. 4.5.

Notice that the reason for which it is attempted to identify a linear dynamical system having the derivatives of the powers as control input lies in the fact that the reason for which this high order dynamics is searched for is that the swing equation (18.1) seems to fail in the presence of high dynamics on these variables.

The identification of such state-space linear model can be done using a `Python` equivalent module to the known MATLAB®/`n2sid` identification function.

An example of such `Python`-package is `sippy`[3] which can be used as shown in the following script assuming that `u` and `y` are the input and output matrices, namely:

$$u := \left[\dot{P}_p, \dot{P}_\ell\right] \in \mathbb{R}^{n_t\times 2} \quad \text{and} \quad y = f \in \mathbb{R}^{n_t},$$

```
from sippy import subspace_ident

model_order = 3
model = subspace_ident(u, y, SS_det='N4SID', SS_order=model_order)
A, B, C, D = model.A, model.B, model.C, model.D
```

In order to compare the different models, the following metric is used to measure the fit between the model output (frequency) and the measured (*simulated*) frequency:

$$\texttt{fit} := \left[1 - \frac{1}{nt}\sum_{t=1}^{n_t}\frac{|\dot{f}(t) - \hat{\dot{f}}(t)|}{|\dot{f}| + \varepsilon}\right] \times 100, \tag{18.8}$$

where $\hat{\dot{f}}(t)$ is the value predicted by the model.

Table 18.1 shows the fitting residual as defined above for the different orders of the state-space identified model.

[3] `pip install sippy.`

Table 18.1 Micro-grid frequency dynamics identification: fitting residuals for the different model's orders

Identified model's order	Fitting residual (%)
3	73
4	82
5	96
6	91

Notice that the original error e with 0-order model (the one based on (18.1)) corresponds to `fit = 51%`.

Notice that the fitting levels are evaluated on a new scenario that is different from the one used in the identification step.

This explains why the higher 6th order is worse than the 5th order as this can be viewed as an instance of the side effect of over-fitting.

Now that we dispose of a dynamic model of the general form (18.7a)–(18.7b), the computation framework of Sect. 4.5 can be used to determine the best pair of (λ, λ_f) to be used in the feedback law:

$$P_p = \texttt{Sat}\Big[\lambda(f_0 - f) + z\Big] \tag{18.9a}$$

$$\dot{z} = \lambda_f(P_p - z) \tag{18.9b}$$

where $f_0 = 50\,\text{Hz}$ is the desired frequency while `Sat` is the saturation maps that is defined for prescribed bounds $P_{\min}$ and $P_{\max}$ on the total produced power.

It is important to underline that the above dynamic output feedback enables to define a **decentralized control**. This is because the internal dynamics defined by (18.9a)–(18.9b) only depends on items that are available to all the producers in the micro-grid, namely

- the frequency f of the grid for which each producer can *run* the same estimation algorithm,
- the desired set-point frequency f_0 which is known to all the actors in the micro-grid, and
- the internal state z.

This means that, locally at each producer level, it is possible to reconstruct the current instantaneous value of the total power P_p to be produced.

Based on this value and depending on the ratio between each producer's capacity and the total installed capacity, each producer deduces the level of its own power, P_i, that it is supposed to inject into the micro-grid.

Figure 18.3 shows the tracking error on the frequency under the control law (18.9a)–(18.9b) in which the computed $\lambda = 14{,}600$ and $\lambda_f = 350.4$ are used.

Notice that the detailed computation of the intermediate parameters that enables to compute the optimal pair (λ, λ_f) are skipped here and can be consulted in [1].

Figure 18.3 also displays the theoretically predicted bound on the frequency tracking error which is both respected and appear to be not so pessimistic.

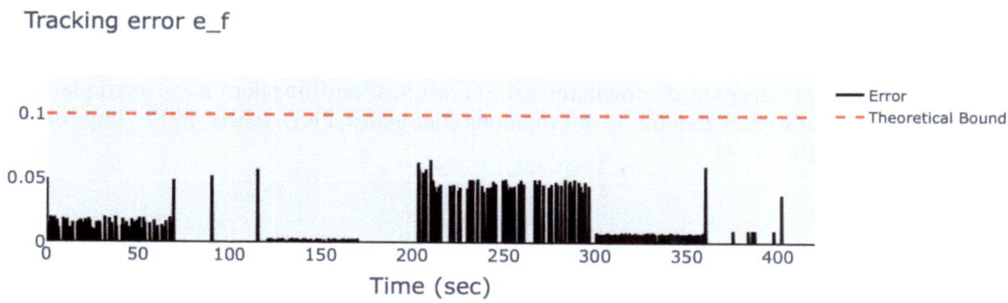

Fig. 18.3 Frequency tracking error in closed-loop under the control law given by (18.9a)–(18.9b) with $\lambda = 14{,}600$ and $\lambda_f = 350.4$ together with the theoretically computed upper bound on the tracking error

As a matter of fact, the control design shown in this chapter is a tiny part of the huge amount of work contained in Jean's Ph.D. work which builds on this simple design to construct a whole impressive framework of micro-grid power/frequency management system which showed totally operational and is by now operated in different regions in the world.

This chapter was simply a kind of *teaser* that should incite the reader to consult Jean's work [1] for a deeper examination of the way a thoroughly elaborated uncertainties management can be a game changer in real-life.

Beyond the specific application to micro-grids that is mentioned in the present chapter, the *Take-Home-Message* here is that the role of the so-called *conventional* control-related approaches should never be under-estimated when tackling real-life challenging problems.

Moreover, those works which tackle seemingly simple problems such as the one consisting in controlling *small uncertain systems* with destabilizing internal phenomena via the simple control structures (Sect. 4.4), should never be looked at as marginal works that are far from the real-life complexity. This is because the solutions to the latter can be generally built using a clever assembly of simple solutions to several sub-problems.

Acknowledgements The material of this chapter is based on a joint work with my former Ph.D. student Jean Dobrowolski and colleagues and friend Seddik Bacha from the Grenoble's power systems' lab (G2Elab) and David Gualino from `Schneider-Electric Industrie` company. While the ideas and concepts were jointly elaborated, all the tedious work of implementing these ideas on real-life systems is due exclusively to Jean Dobrowolski. He successfully transformed abstract theorems into real-life successes!

Reference

1. Dobrowolski, J.: Modlisation, contrle/commande et certification d'un micro-rseau lectrique dcentralis avec entres exognes alatoires et informations contraintes. Ph.D. thesis (2017). http://www.theses.fr/2017GREAT111

Glossary

Bellman Principle of Optimality The principle stating that the remaining part of an optimal trajectory is optimal with regard to the reached state. This principle holds for perfect model and infinite prediction horizon.

Closed-Loop Simulation A simulation of a controlled dynamical system in which the input takes the value provided by some feedback control law.

Clustering A Machine Learning (ML) operation consisting of creating groups of subsets of an original set based on the inter-distance between the members of the different subset. Clustering is one of the four major ML paradigms, namely: regression, classification, clustering, and Reinforcement Learning (RL).

Control Parameterization A map used in Model Predictive Control to reduce the number of decision variables by representing the future sequence of inputs through a reduced number of degrees of freedom.

Control Updating Period The laps of time separating two updating of the best sequence of inputs in Model Predictive Control. In some frameworks, this is a tunable parameter towards the derivation of real-time implementable MPC.

Cost Function Function used in optimization to associate to each value of the decision variable, a scalar called the *associated cost*. Optimization consists in finding the decision variable that minimizes the cost function.

Cross Validation (CV) A ML technique used to choose the hyper-parameters of a model. It consists in subdividing the training data into a number of subsets, say n; then to fit the model using $n-1$ of these subsets and evaluate its performance on the remaining one; repeating this operation n times enables to evaluate the quality of a model for a given setting of its hyper-parameters. Optimizing CV performance in the hyper parameters provides a tuning method for the hyper-parameters.

Dynamic Output Feedback A feedback law that incorporates an internal state that evolves according to a dynamic relationship (generally an Ordinary Differential

M. Alamir, *Nonlinear Control of Uncertain Systems*,
https://doi.org/10.1007/978-3-031-93287-8

Equation) in which only the available measurements are used. The output of the feedback is a function of the measurement and the internal state.

Extended State Estimator A State Estimator (see below) that operates on an extended dynamical model in which the original state vector is extended by adding a new state (generally with zero dynamic) to each parameter to be estimated.

Feedback Law A function that delivers the value that should be assigned to the control input. This value is based on the measurements and is generally computed so that the controlled system behaves in a desired way.

Fixed-Point Iteration Iteration in which a quantity satisfying a circular relationship of the form $y = F(y)$ is searched for by computing the next estimation based on the value of the previous estimation.

Full State Estimation The process of reconstructing the whole state vector by a state estimator. This is opposed to the partial state estimation formalized via the so-called estimation target where only a function of the state (and control) is to be reconstructed.

High Gain Approach A control technique in which one relies on the amplitude of a term, generally obtained by multiplying (gain) some error by a large (high) coefficient, in order to be sure that it always dominates some other terms that are generally badly known but associated to known bounds.

Integral Action A term that is present in the output of a feedback law that accounts for the integral of the error between the desired value of some regulated variable and its actual measured value. This action avoids steady error that might be due to specific classes of uncertainties.

Learning Data Used in ML to refer to the set of pairs of features and labels that are used to fit a specific model. The learning data is generally split into training data and test data in order to check the generalization power of the resulting model.

Machine Learning (ML) A scientific paradigm consisting of looking for a map that represents a relationship between a set of input argument called features and a targeted output called label.

Model Mismatch Difference between the nominal model used in a design and the true system. The mismatch can be due to errors on the parameters of the model (parametric uncertainties) or it can be due to non (correctly) modeled phenomena.

Model Predictive Control (MPC) A control design methodologies that consists in repetitively solve Optimal Control Problems (OCP) to compute the optimal sequence of actions and applying the first action in the optimal sequence to the controlled system.

Moving-Horizon Estimator (MHE) A particular implementation of a state estimator in which the state reconstruction is based on an on-line repetitive solutions of OCP aiming at reconstructing the unknown state by minimizing the output prediction error.

MPC Formulation The choice of the MPC ingredients such as the cost function, the terminal penalty, the prediction horizon, the weighting on the soft constraints, the control updating period, the under-sampling used in the internal prediction, etc. The choice of the formulation has strong impact on the performance and the real-time compatibility of the resulting MPC feedback.

Observation Target A quantity that is expressed as a function of the state and control input that one tries to reconstruct using the available measurement. This leads to the concept of partial estimation with opposition to that of full state estimation where the whole state vector is to be reconstructed.

Optimal Control Problem (OCP) A problem where one looks for the best choice of a set of degrees of freedom that minimizes some cost function while meeting a set of constraints.

Over-Fitting A phenomenon consisting, when fitting a ML model with too many degrees of freedom, to excessively account for every specific detail in the learning data in a way that results in a bad extrapolation capabilities when unseen data is used as arguments to the model.

Parametric Uncertainties A term that refers to differences, between a nominal model and the true model, that are only due to different values of the parameters being used. This differs from model mismatches that cannot be captured by tuning the parameters.

Parameterized Feedback A feedback law that is defined up to the choice of a set of parameters. The most famous example is the Proportional–Integral–Derivative (PID) feedback which is defined up to the choice of the parameters k_p, k_i and k_d coefficients.

Probabilistic Certification A framework that enables to certify that a property is satisfied up to some precision with a specified level of confidence. The framework generally associated to each pair of precision and confidence the number of trials to be successful in order to get the certificate.

Quadratic Programming (QP) A class of optimization problem in which the cost function is quadratic and the constraints are affine in the decision variables. These are considered to day as *easy-to-solve* problem unless their dimension (number of decision variables or number of constraints) is extremely large.

Randomized Optimization A framework that uses the probabilistic certification in order to optimize a set of design parameters so that some cost function is minimized while some property is satisfied within some precision and with some confidence level.

Random Sampling A process that produces randomly values of some random variable according to some presumable statistical distribution (uniform, Gaussian, etc.) This process is intensively used to generate random samples for the simulation, the measurement noise representation or the probabilistic certification.

Real-Time Compatibility The property of an algorithm to be able to terminate within the allowable computation time when implemented on a computational target.

Relative Degree of a System w.r.t an Output Map The number of derivations of the output that makes the control input appears explicitly in the resulting expression. In other words, it is the first derivative of the output that is directly impacted by the input.

Robust Constraint Satisfaction The property of a constraint, that depends on some random factors, to be satisfied for *all* possible realization of these factors. This kind of concept is generally almost impossible to guarantee, except in very specific situations. It is now commonly replaced by the concept of probabilistic certification of satisfaction.

Rule-Based Strategy A strategy that is human friendly which implements generally a sequence of `if-then-else` rules in order to compute its decision. This is generally considered as a baseline strategy that is generally non scalable and which the *advanced* solutions try to beat.

Stage Cost The penalty function that is defined at each instant along the prediction horizon in MPC and which contributes to the cost function by the sum of its values along the prediction horizon. This differs from the terminal cost that applies only to the final state.

State Estimator An algorithm that tries to reconstruct the values of the state vector using the available measurements delivered by the sensors.

Stochastic Optimal Control Problem (SOCP) An optimal control problem in which the definition of the cost function and/or the constraints involves statistical properties (expectation, standard deviation, etc.) of some quantities.

Terminal Cost The penalty in an MPC cost function that is defined depending on the value of the terminal state on the prediction horizon. It differs from the stage cost which is integrated over all the instants of the prediction horizon.

Index

M. Alamir, *Nonlinear Control of Uncertain Systems*,
https://doi.org/10.1007/978-3-031-93287-8

MIX
Papier aus verantwortungsvollen Quellen
Paper from responsible sources
FSC® C105338

If you have any concerns about our products,
you can contact us on
ProductSafety@springernature.com

In case Publisher is established outside the EU,
the EU authorized representative is:
Springer Nature Customer Service Center GmbH
Europaplatz 3, 69115 Heidelberg, Germany

Printed by Libri Plureos GmbH
in Hamburg, Germany